HISTORICAL ENVIRONMENTAL VARIATION IN CONSERVATION AND NATURAL RESOURCE MANAGEMENT

HISTORICAL ENVIRONMENTAL VARIATION IN CONSERVATION AND NATURAL RESOURCE MANAGEMENT

Edited by

John A. Wiens
PRBO Conservation Science
Petaluma, CA, USA
School of Plant Biology
University of Western Australia
Crawley, WA, Australia

Gregory D. Hayward
USDA Forest Service
Alaska Region, Anchorage, AK, USA
USDA Forest Service
Rocky Mountain Region
Lakewood, CO, USA

Hugh D. Safford
USDA Forest Service
Pacific Southwest Region
Vallejo, CA, USA
Department of Environmental Science and Policy
University of California
Davis, CA, USA

Catherine M. Giffen
USDA Forest Service
National Office
Washington, DC, USA

(W)WILEY-BLACKWELL

A John Wiley & Sons, Ltd., Publication

This edition first published 2012 © 2012 by John Wiley & Sons, Ltd.

Blackwell Publishing was acquired by John Wiley & Sons in February 2007. Blackwell's publishing program has been merged with Wiley's global Scientific, Technical and Medical business to form Wiley-Blackwell.

Registered office: John Wiley & Sons, Ltd, The Atrium, Southern Gate, Chichester, West Sussex, PO19 8SQ, UK

Editorial offices: 9600 Garsington Road, Oxford, OX4 2DQ, UK
      The Atrium, Southern Gate, Chichester, West Sussex, PO19 8SQ, UK
      111 River Street, Hoboken, NJ 07030-5774, USA

For details of our global editorial offices, for customer services and for information about how to apply for permission to reuse the copyright material in this book please see our website at www.wiley.com/wiley-blackwell.

The right of the author to be identified as the author of this work has been asserted in accordance with the UK Copyright, Designs and Patents Act 1988.

Library of Congress Cataloging-in-Publication Data

Historical environmental variation in conservation and natural resource management / edited by John A. Wiens . . . [et al.].
    p. cm.
  Includes bibliographical references and index.
  ISBN 978-1-4443-3792-1 (cloth) – ISBN 978-1-4443-3793-8 (pbk.) 1. Landscape ecology. 2. Natural resources–
Co-management. I. Wiens, John A.
  QH541.15.L35H57 2012
  333.7–dc23
                         2012007159

A catalogue record for this book is available from the British Library.

Wiley also publishes its books in a variety of electronic formats. Some content that appears in print may not be available in electronic books.

Set in 9/11 pt PhotinaMT by Toppan Best-set Premedia Limited
Printed and bound in Malaysia by Vivar Printing Sdn Bhd

1 2012

Front Cover: Great Basin bristlecone pine trees (*Pinus longaeva*) in the Patriarch Grove of the White Mountains, eastern California. Bristlecone pines growing in the White Mountains are the oldest known trees in the world, with individuals reaching ages upwards of 5000 years. Dry conditions in the White Mountains also result in exceptional preservation of remnant wood. By cross-dating dead wood with living trees, the tree-ring chronology for the White Mountains extends back almost 12,000 years, providing an exceptional example of a historical legacy. Photograph by Peter M. Brown, Rocky Mountain Tree-Ring Research.
Cover Design By: Steve Thompson.

CONTENTS

Contributors, vii

Foreword, x

Preface, xii

Acknowledgments, xiv

SECTION 1 BACKGROUND AND HISTORY, 1
JOHN A. WIENS

1 Setting the stage: theoretical and conceptual background of historical range of variation, 3
WILLIAM H. ROMME, JOHN A. WIENS, AND HUGH D. SAFFORD

2 Development of historical ecology concepts and their application to resource management and conservation, 19
WAYNE PADGETT, BARBARA SCHRADER, MARY MANNING, AND TIMOTHY TEAR

SECTION 2 ISSUES AND CHALLENGES, 29
HUGH D. SAFFORD

3 Challenges in the application of historical range of variation to conservation and land management, 32
GREGORY D. HAYWARD, THOMAS T. VEBLEN, LOWELL H. SURING, AND BOB DAVIS

4 Historical ecology, climate change, and resource management: can the past still inform the future? 46
HUGH D. SAFFORD, GREGORY D. HAYWARD, NICOLE E. HELLER, AND JOHN A. WIENS

5 What is the scope of "history" in historical ecology? Issues of scale in management and conservation, 63
JOHN A. WIENS, HUGH D. SAFFORD, KEVIN MCGARIGAL, WILLIAM H. ROMME, AND MARY MANNING

6 Native Americans, ecosystem development, and historical range of variation, 76
GREGORY J. NOWACKI, DOUGLAS W. MACCLEERY, AND FRANK K. LAKE

7 Conservation and resource management in a changing world: extending historical range of variation beyond the baseline, 92
STEPHEN T. JACKSON

SECTION 3 MODELING HISTORIC VARIATION AND ITS APPLICATION FOR UNDERSTANDING FUTURE VARIABILITY, 111
ROBERT E. KEANE

8 Creating historical range of variation (HRV) time series using landscape modeling: overview and issues, 113
ROBERT E. KEANE

9 Modeling historical range of variability at a range of scales: an example application, 128
KEVIN MCGARIGAL AND WILLIAM H. ROMME

SECTION 4 CASE STUDIES OF APPLICATIONS, 147
GREGORY D. HAYWARD

10 Regional application of historical ecology at ecologically defined scales: forest ecosystems in the Colorado Front Range, 149
THOMAS T. VEBLEN, WILLIAM H. ROMME, AND CLAUDIA REGAN

11 Incorporating concepts of historical range of variation in ecosystem-based management of British Columbia's coastal temperate rainforest, 166
ANDY MACKINNON AND SARI C. SAUNDERS

12 Incorporating HRV in Minnesota national forest
    land and resource management plans: a
    practitioner's story, 176
    *MARY SHEDD, JIM GALLAGHER, MICHAEL
    JIMÉNEZ, AND DUANE LULA*

13 Applying historical fire-regime concepts to
    forest management in the western United
    States: three case studies, 194
    *THOMAS E. DEMEO, FREDERICK J. SWANSON,
    EDWARD B. SMITH, STEVEN C. BUTTRICK,
    JANE KERTIS, JEANNE RICE, CHRISTOPHER D.
    RINGO, AMY WALTZ, CHRIS ZANGER,
    CHERYL A. FRIESEN, AND JOHN H.
    CISSEL*

14 Using historical ecology to inform
    wildlife conservation, restoration, and
    management, 205
    BETH A. HAHN AND JOHN L. CURNUTT

15 River floodplain restoration experiments offer a
    window into the past, 218
    *RAMONA O. SWENSON,
    RICHARD J. REINER, MARK REYNOLDS,
    AND JAYMEE MARTY*

16 Streams past and future: fluvial responses to
    rapid environmental change in the context of
    historical variation, 232
    *DANIEL A. AUERBACH, N. LEROY POFF, RYAN R.
    MCSHANE, DAVID M. MERRITT, MATTHEW I.
    PYNE, AND THOMAS K. WILDING*

17 A framework for applying the historical
    range of variation concept to ecosystem
    management, 246
    *WILLIAM H. ROMME, GREGORY D. HAYWARD,
    AND CLAUDIA REGAN*

**SECTION 5   GLOBAL
PERSPECTIVES, 263**
JOHN A. WIENS

18 Ecological history guides the future of
    conservation: lessons from Africa, 265
    A.R.E. SINCLAIR

19 Ecological history has present and future ecological
    consequences – case studies from Australia, 273
    DAVID LINDENMAYER

20 A view from the past to the future, 281
    KEITH J. KIRBY

21 Is the historical range of variation relevant to
    rangeland management? 289
    BRANDON T. BESTELMEYER

22 Knowing the Fennoscandian taiga: ecohistorical
    lessons, 297
    YRJÖ HAILA

**SECTION 6   CHALLENGES FOR THE
FUTURE, 305**

23 Reflections on the relevance of history in a
    nonstationary world, 307
    JULIO L. BETANCOURT

24 The growing importance of the past in
    managing ecosystems of the future, 319
    *HUGH D. SAFFORD, JOHN A. WIENS, AND
    GREGORY D. HAYWARD*

Index, 329

Colour plate pages fall between pp. 162 and 163

Companion website

This book has a companion website:

www.wiley.com/go/wiens/historicalenvironmentalvariation

with Figures and Tables from the book

CONTRIBUTORS

DANIEL A. AUERBACH, *Department of Biology, Colorado State University, Fort Collins, CO 80523, USA; Graduate Degree Program in Ecology, Colorado State University, Fort Collins, CO 80523, USA. [daa@colostate.edu]

BRANDON T. BESTELMEYER, USDA-ARS Jornada Experimental Range, New Mexico State University, Las Cruces, NM, 88003, USA. [bbestelm@nmsu.edu]

JULIO L. BETANCOURT, US Geological Survey National Research Program Water Resources Division, 1955 E. 6th St. Tucson, AZ 85719, USA. [jlbetanc@usgs.gov]

STEVEN C. BUTTRICK, The Nature Conservancy, 821 SE 14th Avenue, Portland, OR 97214, USA. [sbuttrick@tnc.org]

JOHN H. CISSEL, Joint Fire Science Program, 3833 South Development Avenue, Boise, ID 83705, USA. [John_Cissel@blm.gov]

JOHN L. CURNUTT, USDA Forest Service, Eastern Region, 626 East Wisconsin Avenue, Milwaukee, WI 53203, USA. [jcurnutt@fs.fed.us]

BOB DAVIS, USDA Forest Service, Southwestern Region, 333 Broadway Blvd. SE, Albuquerque, NM 87102, USA. [bdavis03@fs.fed.us]

THOMAS E. DEMEO, USDA Forest Service, Pacific Northwest Regional Office, 333 SW First Avenue, Portland, OR 97204, USA. [tdemeo@fs.fed.us]

CHERYL A. FRIESEN, USDA Forest Service, Central Cascades Adaptive Management Partnership, 3106 Pierce Parkway, Suite D, Springfield, OR 97477, USA. [cfriesen@fs.fed.us]

JIM GALLAGHER, USDA Forest Service, Chippewa National Forest (retired), 200 Ash Avenue NW, Cass Lake, MN 56633, USA. [kanjim@paulbunyan.net]

BETH A. HAHN, USDA Forest Service, Northern Region, 200 East Broadway, Missoula, MT 59802, USA. [bhahn@fs.fed.us]

YRJÖ HAILA, Department of Regional Studies, University of Tampere 33014, Tampere, Finland. [yrjo.haila@uta.fi]

GREGORY D. HAYWARD, *USDA Forest Service, Alaska Region, 3301 C Street, Anchorage, AK 99504, USA; USDA Forest Service, Rocky Mountain Region, Lakewood, CO 80401, USA. [ghayward01@fs.fed.us]

NICOLE E. HELLER, Climate Central, 895 Emerson Street, Palo Alto, CA 94301, USA. [nheller@climate-central.org]

STEPHEN T. JACKSON, *Department of Botany 3165, Aven Nelson Building, 1000 E. University Ave.,

* Indicates current address.

Laramie, WY 82071, USA; Program in Ecology 4304, University of Wyoming, 1000 E. University Ave., Laramie, WY 82701, USA. [jackson@uwyo.edu]

MICHAEL JIMÉNEZ, USDA Forest Service, Superior National Forest, 8901 Grand Avenue Place, Duluth, MN 55808, USA. [mjimenez@fs.fed.us]

ROBERT E. KEANE, USDA Forest Service, Rocky Mountain Research Station, Missoula Fire Sciences Laboratory, 5775 Highway 10 West Missoula, MT 59808, USA. [rkeane@fs.fed.us]

JANE KERTIS, USDA Forest Service, Siuslaw National Forest, 4077 SW Research Way, Corvallis, OR 97331, USA. [jkertis@fs.fed.us]

KEITH J. KIRBY, Natural England, 3rd Floor, Touthill Close, City Road, Peterborough, PE1 1XN, UK. [keith.kirby@naturalengland.org.uk]

FRANK K. LAKE, USDA Forest Service, Pacific Southwest Research Station, Redding Silviculture Laboratory, 3644 Avtech Parkway, Redding, CA 96002, USA. [franklake@fs.fed.us]

DAVID LINDENMAYER, Fenner School of Environment and Society, The Australian National University, Canberra, ACT, 0200, Australia. [David.Lindenmayer@anu.edu.au]

DUANE LULA, USDA Forest Service, Superior National Forest (retired), 8901 Grand Avenue Place, Duluth, MN 55808, USA. [phoebemrc@msn.com]

DOUGLAS W. MACCLEERY, USDA Forest Service, Washington Office, Forest Management (retired), P.O. Box 96090, Washington, DC 20090-6090, USA; *1010 Danton Lane, Alexandria, VA 22308, USA. [ram668@aol.com]

ANDY MACKINNON, BC Ministry of Natural Resource Operations, Coast Area Operations, 4300 North Road, Victoria, BC, Canada V8Z 5J3. [andy.mackinnon@gov.bc.ca]

MARY MANNING, USDA Forest Service, Northern Region, 200 East Broadway, Missoula, Montana 59807-7669, USA. [mmanning@fs.fed.us]

JAYMEE MARTY, The Nature Conservancy, 555 Capitol Ave., Suite 1290, Sacramento, CA 95814, USA. [jmarty@tnc.org]

KEVIN MCGARIGAL, Department of Natural Resources Conservation, University of Massachusetts, Amherst, MA 01003, USA. [mcgarigalk@nrc.umass.edu]

RYAN R. MCSHANE, *Department of Biology, Colorado State University, Fort Collins, CO 80523, USA; Graduate Degree Program in Ecology, Colorado State University, Fort Collins, CO 80523, USA. [ryan.mcshane@colostate.edu]

DAVID M. MERRITT, Graduate Degree Program in Ecology, Colorado State University and USDA Forest Service, Watershed, Fish, and Wildlife Staff, Natural Resource Research Center, 2150 Centre Ave., Bldg A, Suite 368, Fort Collins, CO 80526, USA. [dmerritt@fs.fed.us]

GREGORY J. NOWACKI, USDA Forest Service, Eastern Regional Office, 626 E. Wisconsin Avenue, Milwaukee, WI 53202, USA. [gnowacki@fs.fed.us]

WAYNE PADGETT, Bureau of Land Management, Utah State Office, 440 W. 200 S., Salt Lake City, UT 84101, USA. [wayne_padgett@blm.gov]

N. LEROY POFF, *Department of Biology, Colorado State University, Fort Collins, CO 80523, USA; Graduate Degree Program in Ecology, Colorado State University, Fort Collins, CO 80523, USA. [poff@lamar.colostate.edu]

MATTHEW I. PYNE, *Department of Biology, Colorado State University, Fort Collins, CO 80523, USA; Graduate Degree Program in Ecology, Colorado State University, Fort Collins, CO 80523, USA. [mpyne@rams.colostate.edu]

CLAUDIA REGAN, USDA Forest Service, Rocky Mountain Region, 740 Simms Street, Lakewood, CO 80225, USA.

RICHARD J. REINER, The Nature Conservancy, 500 Main Street, Chico, CA 95928, USA. [rreiner@tnc.org]

MARK REYNOLDS, *The Nature Conservancy, 201 Mission Street, 4th Floor, San Francisco, CA 94105, USA.* [mreynolds@tnc.org]

JEANNE RICE, *USDA Forest Service, Mt. Hood National Forest, 16400 Champion Way, Sandy, OR 97055, USA.* [jrice@fs.fed.us]

CHRISTOPHER D. RINGO, *TetraTech CES, Mt. Vernon, WA 98273, USA.* [cdringo@fs.fed.us]

WILLIAM H. ROMME, *Warner College of Natural Resources, Colorado State University, Fort Collins, CO 80523, USA.* [romme@warnercnr.colostate.edu]

HUGH D. SAFFORD, **USDA Forest Service, Pacific Southwest Region, Vallejo, CA 94592, USA; Department of Environmental Science and Policy, University of California, Davis, CA 95616, USA.* [hdsafford@ucdavis.edu]

SARI C. SAUNDERS, *BC Ministry of Natural Resource Operations, Coast Area Operations, 2100 Labieux Rd., Nanaimo, BC, Canada V9T 6E9.* [sari.saunders@gov.bc.ca]

BARBARA SCHRADER, *USDA Forest Service, Alaska Region, PO Box 21628, Juneau, AK 99802-1628, USA.* [bschrader@fs.fed.us]

MARY SHEDD, *USDA Forest Service, Superior National Forest, 8901 Grand Avenue Place, Duluth, MN 55808, USA.* [mshedd@fs.fed.us]

A.R.E. SINCLAIR, *Beatty Biodiversity Centre, 6270 University Boulevard, University of British Columbia, Vancouver, BC, Canada V6T 1Z4.* [Sinclair@zoology.ubc.ca]

EDWARD B. SMITH, *The Nature Conservancy, 114 North San Francisco Street, Suite 205, Flagstaff, AZ 86001, USA.* [esmith@tnc.org]

LOWELL H. SURING, *Northern Ecologic LLC, 10685 County Road A, Suring, WI 54174, USA.* [lowell@northern-ecologic.com]

FREDERICK J. SWANSON, *USDA Forest Service, Pacific Northwest Research Station, 3200 SW Jefferson Way, Corvallis, OR 97331, USA.* [fswanson@fs.fed.us]

RAMONA O. SWENSON, **Cardno ENTRIX, 701 University Ave., Suite 200, Sacramento, CA 95825, USA; The Nature Conservancy of California, Cosumnes River Preserve, 13501 Franklin Blvd., Galt, CA 95632, USA.* [ramona.swenson@cardno.com]

TIMOTHY TEAR, *The Nature Conservancy, 1210 Hempstead Road, Niskayuma, NY 12309, USA.* [ttear@tnc.org]

THOMAS T. VEBLEN, *Department of Geography, University of Colorado, Boulder, CO 80309, USA.* [thomas.veblen@colorado.edu]

AMY WALTZ, *The Nature Conservancy, 115 NW Oregon, Suite 30, Bend, OR 97701, USA.* [awaltz@tnc.org]

JOHN A. WIENS, *PRBO Conservation Science, 3820 Cypress Dr #11, Petaluma, CA 94954, USA; School of Plant Biology, University of Western Australia, 35 Stirling Highway, Crawley, WA 2006, Australia.* [jwiens@prbo.org]

THOMAS K. WILDING, *Department of Biology, Colorado State University, Fort Collins, CO 80523, USA.* [twilding@rams.colostate.edu]

CHRIS ZANGER, *The Nature Conservancy, 115 NW Oregon, Suite 30, Bend, OR 97701, USA.* [czanger@tnc.org]

FOREWORD

Resource managers have always relied on history to inform their decision making – either intuitively or with formal assessment of historical conditions. In land and natural resource planning, the concept of historical range of variation (HRV) has often been used to delimit conditions that are natural, desirable, and toward which management might seek to return.

Today, however, land managers operate in an era of global change driven by development, globalization, and anthropogenic climate change. The realization that the future is likely to be quite unlike the past leads many managers and scientists to question the value of historical knowledge and, in particular, the value of HRV assessments. "Stationarity is dead," they say, and posit that reference conditions from the past may no longer be sustainable and may not be appropriate guides for decisions about the future.

What does the HRV mean in a time of rapid change? Is the concept of HRV still useful? Does HRV capture the right historical insights about ecological systems that are relevant to current decisions and future consequences? If so, then how can land and resource managers use assessments of HRV to inform decision making? These are the key questions this book addresses.

This book, written by a unique mixture of scientists and managers, uses a combination of essays, case studies, synthesis, and analysis to address specific challenges in the effective application of historical ecology to land management. It examines how history has informed resource management under the assumption that variability is not driven by or overwhelmed by directional change, and it clarifies and underlines how history is still relevant to conservation and resource management in a rapidly changing world.

In fact, historical ecology and its derivative tool HRV are necessary in understanding temporal dynamics that shape the response of a system to drivers of change, both natural and anthropogenic. The authors affirm HRV as a legitimate planning tool, but they acknowledge the difficult challenges that must be addressed to enable rigorous application of HRV assessments to land management. Foremost is the need to understand past climatic variation at the appropriate spatial and temporal scales.

While not aiming to be a "guide" or "cookbook," this volume nonetheless provides the insights necessary for using historical ecology in land and resource management. The authors embrace the idea that resource management is strongly context specific. Therefore, local ecology, geography, and social conditions critically dictate important characteristics of resource management.

Adapting to drivers of change and their associated stressors requires increasingly sophisticated and disciplined management of risks and uncertainties. The authors argue effectively that understanding the nature of variability and employing robust measures of variation is part of the bedrock of effectively managing risks. As we build new tools for risk-based management, the insights we gain from better understanding the histories of ecosystems and their relationships to environmental change will help us develop improved management options that are "built to last" and will continue to meet protection and restoration objectives across a range of stressor profiles and extreme events.

Martin Luther King, Jr. wrote of the "long arc of history." In natural systems, that arc might look more like a wave, where the oscillations reflect inherent variability. Future forces of change may reshape that variability, and our understanding of the historical conditions that anchor that curve will help us make sense of how severely it can be altered in the future. Even in a changing world, our ecological theories and models are based on what we learn from past and current observations. Although we cannot return to the past, we can, and must, continue to benefit from understanding it.

David A. Cleaves
US Forest Service Climate Change Advisor
Washington, DC

PREFACE

It is said that those who ignore history are doomed to repeat it. That seems increasingly unlikely, as burgeoning human populations and global economies are transforming the face of the Earth, changing climate, and altering the distributions of plants and animals at an alarming rate. This prospect has led both scientists and philosophers to wonder whether "history," in the traditional sense, is no longer relevant. What good are the lessons of history if the future is so different that they do not apply? This is the question that the contributors to this book attempt to answer.

But history, and its relevance to the future, is only one side of the coin. The other is environmental variation. It is news to no one that natural environments vary over time. Changes in the seasons, dry years and wet years, yearly variations in wildfire frequency and severity, El Niño and La Niña climatic oscillations, and extreme events such as cyclones or floods testify to the variability of nature. Often, this variability occurs over timescales that exceed our individual experiences. To cope with this variability, concepts such as "historical range of variation," "range of natural variability," or "acceptable range of variation" have been used to guide ecological restoration, conservation, and land management.

Human expectations for the future have always been based in large part on past experience. More often than not, the window on this past experience is defined by individual lifetimes – generation length or something close to it. At the same time, current land management in many parts of the world is predicated on comparisons with historical conditions that are assumed to represent "natural" or "properly functioning" conditions, which occurred many generations in the past.

Regardless of how they are defined, such reference conditions can provide a vision of the conservation or management goal and a means to measure progress toward that vision. Historical ecology has emerged as a coherent discipline.

Concepts of land management using historical ecology are often difficult to express in operational terms, however, and our knowledge of history is often sketchy and incomplete. Moreover, models of future climate change project the emergence of novel, "no-analog" conditions, for which past conditions may not provide a reasonable guide. As a result, the concepts have sometimes been awkwardly employed and misunderstood. Some scientists, managers, and conservationists have questioned whether history is even relevant in a rapidly changing world. In view of such skepticism, and the varied ways historical ecology has been applied in land management and conservation, it is important to determine the constraints on how historical reference information can be used in future management. We must also remember, however, that the future is always contingent on the past. We should aim to develop models and perspectives that build from a historical context, using ecological responses to past environmental variability to shed light on the range of potential responses to future conditions.

The chapters in this book explore the utility of historical ecology and, more specifically, assessments of the historical range of variation (HRV, in agency parlance) to inform land management and conservation and to develop concepts related to understanding future ranges of variability. The treatments are built around the following observations or premises:

• Ecological systems are dynamic, varying in ways that change over multiple time scales. As a result, what is "normal" is difficult to define.

• Understanding the dynamics of ecosystems can benefit from an examination of ecological patterns over multiple temporal scales, from years to centuries to tens of millennia.

• The notion of stationarity – that systems vary over time with a constant mean and variance – is inconsistent with the historical record and with current ecological thinking. Things have never been "the same," and rates of change are likely to increase in a future dominated by massive changes in climate, land use, disturbance regimes, and global economics.

• The window of past ecological variation generally does not provide a fixed target or goal for conservation, management, or restoration. Attempting to return systems to some former state is not sustainable over the long term.

• Nonetheless, history provides lessons that are essential to informing conservation and management policies and actions. Understanding the past is a key to managing for the future.

• History tells us that different species are likely to respond to environmental changes in different ways. As conditions change more rapidly in the future, shifts in geographic distributions of species, reproduction, phenology, and migratory movements may contribute to the destabilization of biological communities, producing new forms of species interactions that lack direct analogs in the historical record.

• Historical records tell us that nature is full of thresholds, "tipping points" that can precipitate sudden changes in system properties, produce surprises, and thwart the most well-intentioned management. Thresholds or extreme environmental events may also act to synchronize the temporal clock among many elements of a system. When this happens, the synchronicity may act to amplify responses to environmental changes over broad areas.

• "History" is a continuous stream through time rather than a set of past reference points. Organisms and ecological systems respond to this stream at distinctive scales. The scales of the system of interest, the management or conservation objectives, and the scales of historical variation must align. Mismatches can easily lead to false conclusions about what led to what.

• Our knowledge of history diminishes as we look farther into the past. But the "long past" can provide important insights into some ecological processes that will not be revealed by a shorter time reference. Because uncertainty increases and relationships become more difficult to understand over longer time lines, adopting a historical perspective means that we must accept greater uncertainty in our documentations of patterns and inferences about causes.

• Case studies of the application of historical ecology to resource management demonstrate the powerful insights that come through examining historical context. Similarly, case studies provide insights into pitfalls and issues in the application of historical ecology to management and conservation.

• Because people have a tendency to regard as "normal" that which they experience while growing up, it is important to communicate the reality that personal experience represents only a snapshot of a dynamic world. This understanding is essential to mobilizing public and political support for confronting broadscale and long-term environmental change. It is important to convey to the public and decision makers the nature of historical variation and what it does and does not tell us.

The genesis of this book was a 3-day workshop held in April 2008 at the National Conference Center in Lansdowne, Virginia, under the sponsorship of the United States Forest Service and The Nature Conservancy. The primary focus of the workshop was to develop a background on historical ecology and its use in resource (primarily forest) management and to develop guidelines that would enable managers and conservationists to use historical information in planning, monitoring, and evaluation activities. Here we build upon this foundation to broaden the scope, while retaining the emphasis on the use of historical ecology in conservation and management. The work is firmly rooted in the cumulative experience of scientists and land managers in both governmental and private sectors – those who must deal with critical natural-resource issues in a dynamic social, political, and environmental context.

John A. Wiens
Gregory D. Hayward
Hugh D. Safford
Catherine M. Giffen

ACKNOWLEDGMENTS

The 2008 workshop was made possible by generous support from The Nature Conservancy and the US Forest Service, National Forest System. Without their sponsorship and the energetic participant of over 40 scientists and managers, the synthesis of ideas represented in this book would not have been possible. We particularly thank the invited speakers from the April workshop who are not represented in the pages of this volume, including Greg Aplet, Virginia Dale, Peter Landres, Ron Neilson, Wayne Owen, and Anne Zimmermann. Their insights and new ways of looking at old ideas were refreshing. Participants in the workshop included representatives from the US Forest.

Service from throughout the nation (Alaska to Georgia), representatives from The Nature Conservancy from several states, The Wilderness Society, NatureServe, US Geological Survey, Oak Ridge National Laboratory, and faculty from Colorado State University. We thank the participants and their employers for their support.

Work leading to publication of this volume has been supported by the US Forest Service Pacific Southwest Region, Rocky Mountain Region, Alaska Region, and National Headquarters. Several individuals beyond the many authors contributed substantially to the volume. In particular, we thank Claudia Regan for her inspiration and dedication to the workshop and this subsequent book. Chris Iverson championed the workshop and was critical to securing funding for that effort and for securing resources to move forward from the workshop and complete this book. Richard Holthausen and Wayne Owen played critical roles in conceiving and designing the workshop. Within the US Forest Service, Publishing Arts helped with graphics. Drew Paryzer prepared the index.

As with most efforts of this sort, the difficult task of reviewing chapters and providing critical feedback fell to our friends and colleagues. We thank the many reviewers who took time from their families and work to improve the quality of this effort.

Finally, we express our sincere thanks for the many resource-management practitioners and scientists who looked to historical ecology for insights into management problems and helped pioneer the practice of using historical ecology in resource conservation and management.

Section 1

Background and History

John A. Wiens

PRBO Conservation Science, Petaluma, CA, USA
University of Western Australia, Crawley, WA, Australia

History is part of all human endeavors. Even fresh ideas, innovative technologies, or revolutions in scientific thinking build on a history of experiences and thinking. Historical ecology and, more specifically, the development of historical range of variation (HRV) as a tool in resource management did not spring unheralded from the mists of time. Yet, for the better part of the nineteenth century and well into the middle of the twentieth century, both variation and past history were largely viewed as bothersome details. Variation was equated with statistical variance, something that clouded the average or normal conditions and that could be reduced through appropriate sampling design and clever statistics. History was often thought of in the context of ecological succession, an orderly progression from disturbance to a stable, sustaining ecological system.

Underlying all of this was the age-old belief in a balance of nature. Natural systems, according to this idealist philosophy, had an inherent tendency toward balance – stable communities that would vary little about a fixed long-term mean. The rise of equilibrium-based mathematical theory in ecology during the 1950s and 1960s helped to formalize this view of nature.

But variation and history were always there, lurking in the background, confounding study designs and statistical analyses and, rather often, sowing the seeds of controversy. Ecologists studying what they thought were stable systems at different times or places saw different things. Attempts to reconcile these disparate

Historical Environmental Variation in Conservation and Natural Resource Management, First Edition. Edited by John A. Wiens, Gregory D. Hayward, Hugh D. Safford, and Catherine M. Giffen.
© 2012 John Wiley & Sons, Ltd. Published 2012 by John Wiley & Sons, Ltd.

observations with theory, or to define unitary targets for management actions, began to falter. Long-term studies revealed the dynamism of ecosystems, and probing into historical records showed that past events left their imprint on contemporary patterns. Ecologists, and then resource managers and conservationists, began to worry about variation and history, and to consider how to incorporate them into their thinking, theories, and practices.

The two chapters in this section provide background on the development of historical ecology and HRV, first from a conceptual perspective, and then from a more pragmatic management perspective. History tells us that it is important to know where we have been to understand how we got to where we are now. These chapters show us how a knowledge of how historical ecology and HRV came to be and have been applied can provide a foundation for thinking about how best to use historical insights in managing and conserving natural resources in a rapidly changing world.

Chapter 1

SETTING THE STAGE: THEORETICAL AND CONCEPTUAL BACKGROUND OF HISTORICAL RANGE OF VARIATION

William H. Romme,[1] John A. Wiens,[2,3] and Hugh D. Safford[4,5]

[1]Colorado State University, Fort Collins, CO, USA
[2]PRBO Conservation Science, Petaluma, CA, USA
[3]University of Western Australia, Crawley, WA, Australia
[4]USDA Forest Service, Pacific Southwest Region, Vallejo, CA, USA
[5]University of California, Davis, CA, USA

Historical Environmental Variation in Conservation and Natural Resource Management, First Edition. Edited by John A. Wiens, Gregory D. Hayward, Hugh D. Safford, and Catherine M. Giffen.
© 2012 John Wiley & Sons, Ltd. Published 2012 by John Wiley & Sons, Ltd.

History is nothing but assisted and recorded memory.

George Santayana

The times they are a-changin'.

Bob Dylan

1.1   INTRODUCTION

The above quotations, from two philosophers of different times and modes of expression, capture the essential tension that underlies this book. "History" is the template on which the present is founded, but what we know of history is determined by human knowledge and the interpretation (or filtering) of past events. As we probe more deeply into the past, our knowledge becomes more fragmentary, uncertainty increases. Yet history holds precious clues to why things are the way they are today, and therefore how we might act to keep them that way or alter them, and how we might avoid mistakes of the past. History provides essential lessons for the conservation and management of natural resources.

At the same time, times are changing, and changing rapidly. The climate is warming. The rapid expansion of human populations and economies has dramatically depleted the extent of the earth's natural and seminatural habitats. Land-use change and habitat fragmentation have reduced the availability and connectivity of suitable habitat for native plants and wildlife. Although the future is by definition uncertain (Fig. 1.1), it seems increasingly likely that the future will be quite different from the present or the past. Even if history has been a useful (albeit less than certain) guide for conservation and management up to now, will it be hopelessly compromised, and thus irrelevant, in a radically different future?

This is the question that carries through this book. In this chapter, we set the stage for what follows by providing a perspective on historical ecology and environmental variation, briefly reviewing some of the issues in the application of history to resource management, and introducing some of the key themes developed more comprehensively in the following chapters. A general thesis of this book is that the future is built on foundations laid in the past, no matter the rapidity, profundity, or direction of potential change. The mechanisms by which organisms and ecosystems respond to global change in the future will be those by which they

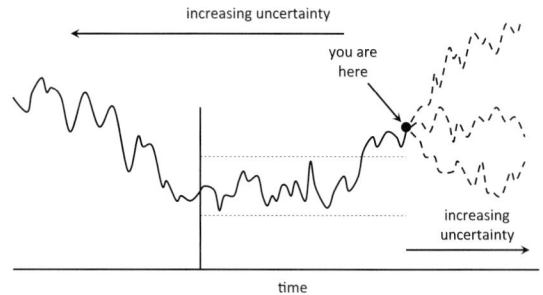

Fig. 1.1 Hypothetical example of historical variation. During most of the recent past (solid line to the right of the vertical bar), the system has remained within a defined range of variation (horizontal dashed lines), which might be considered the historical range of variation. Recently, the system has shifted beyond that historical range; this is what prompts the perception that the past range of variation represents a "natural" state and a suitable target for management. In the more distant past, however, the system may have undergone quite different variations (here shown as a long-term trend). There may be multiple future trajectories of the system, only some of which are concordant with historical variation (either recent or long past). Uncertainty about the state of the system (and thus, by some accounts, the value of historical information) increases as one goes more deeply into the past, or very far at all into the future.

have responded in the past. Although the future is an unknown place, we do well to remember that the value of historical knowledge and past experience is highest where the lay of the land is least familiar.

1.2   HISTORICAL ECOLOGY AND HISTORICAL RANGE OF VARIATION (HRV)

Resource management across the United States and other parts of the world is often predicated on restoring ecosystem patterns and processes understood from analyses of historical conditions. In practice, desired conditions for ecological restoration, conservation, and resource management are commonly derived from historical "reference states" (Meffe & Carroll 1994; Egan & Howell 2001). Because most current ecosystems have been highly altered by human use, ecological management tends to focus on the past. Managers and scientists use information about places and processes revealed through historical ecology to generate assessments of ecological trends, serve as the basis for benchmarks of ecological integrity, develop and parameterize models of ecological processes, and provide targets for preservation, restoration, resource management, and policy (Landres et al. 1999; Swetnam et al. 1999; Egan & Howell 2001). Historical ecology is particularly useful for documenting mechanisms of temporal change and providing context to interpret the meaning of this change for current and future management. For example, reconstructions of long-term vegetation change, based on pollen preserved in lake-bottom sediments or macrofossils within packrat middens, reveal that Pleistocene and early Holocene community assemblages were generally dissimilar to widespread plant communities today (Jackson 2006). This historical perspective tells us that current communities probably will not simply migrate intact to higher latitudes and elevations with global warming, but that future species assemblages will likely differ compositionally from contemporary assemblages, just as they did after the major climatic changes at the end of the Pleistocene. Similarly, knowledge that piñon pine (*Pinus monophylla* and *Pinus edulis*) migrated northward from Pleistocene refugia in southern Arizona and northern Mexico, and that this migration continues today (Miller & Wigand 1994; Swetnam et al. 1999), tells us that recent piñon expansion into grasslands and shrublands at the northern limit of its current range may represent a process of natural biogeographic spread rather than an unnatural response to land-use practices (Romme et al. 2009).

Restoration ecology and resource management incorporate historical ecology based on the premise that the ecological conditions most likely to preserve native species or conserve natural resources are those that sustained them in the past, when ecosystems were presumed to be less affected by people. Environments varied then, but the variations were more "natural," so species were likely to have adapted evolutionarily to those variations. The expectation, then, is that by managing an ecosystem within the bounds of historical (i.e. "natural") variation for key ecosystem patterns and processes, sustainability and persistence should be fostered (Manley et al. 1995; Egan & Howell 2001; Wiens et al. 2002). Less explicitly, the use of historical ecology to inform contemporary practices assumes that current environmental envelopes are much as they were in the past, absent the rather massive effects of humans in transforming the landscape (Turner et al. 1990). This manifests the persistent belief, expressed in disciplines as varied as natural philosophy, systems dynamics, and organic gardening, in an overall balance of nature (Pimm 1991; Kricher 2009).

Historical ecology involves the general application of historical information to a wide array of questions and issues in both basic and applied ecology. The use of historical information in natural resource conservation, restoration, and management has become formalized in the concept of HRV.[1] In its simplest form, HRV is characterized as the range of some condition or process (e.g. disturbance regime, stand structure, patch size, diversity) that has occurred over some specified period in the past (Box 1.1). In this book, we borrow from Landres et al. (1999), Keane et al. (2009), and others to define HRV as

> The variation of ecological characteristics and processes over scales of time and space that are appropriate for a given management application.

In contrast to the generality of historical ecology, the HRV concept focuses on a distilled subset of historical ecological knowledge developed for use by resource managers; it represents an explicit effort to incorporate a historical perspective into management and conservation. Of necessity, the concept of HRV involves specification of the historical period over which to characterize natural variability. In the United States, the reference period for HRV assessments has often been restricted to the few centuries preceding European settlement (Aplet & Keeton 1999). Accordingly, many proposed land-management actions have treated the period immediately prior to settlement as an explicit target and have sought to re-create conditions thought to be characteristic of this period.

Multiple terms and subtly different concepts have evolved to describe and characterize HRV, but all get at a very similar basic idea. Box 1.1 summarizes many of the terms and definitions that have received widespread use and presents some new terms and concepts recently developed to help integrate historical ecology with social perspectives on ecosystem management; these ideas are developed later in this chapter.

---

[1] The HRV acronym has been used to refer to "historical range of variability" and to "historical range of variation" (Box 1.1). We use the latter terminology throughout this book.

Box 1.1 Some terms and definitions that have been used to describe the HRV concept.

Term	Definition	Reference
Historical range of variation (HRV)	The variation of ecological characteristics and processes over scales of space and time that are appropriate for a given management application	This book
Historical range of variability	"the estimated range of some ecological condition or process that occurred in the past . . . often expressed as a probability distribution of likely states . . . denotes a dynamic set of boundaries between which most native biodiversity variables have persisted – with fluctuations – through time and across space"	Binkley and Duncan (2010)
Historical range of variability	"HRV . . . represents how vegetation is structured (e.g. tree density), how it varies spatially and temporally, and how fire functions (e.g. fire size, intervals) with little effect of people, except where people have been a significant structuring force"	Baker (2009, p. 3)
Historical range of variability	"the variability of regional or landscape composition, structure, and disturbances, during a period of time of several cycles of the common disturbance intervals, and similar environmental gradients, referring, for the United States, to a period prior to extensive agricultural or industrial development"	Hann and Bunnell (2001, p. 394)
Historical range of variation	"the range of variation over a period of record, ideally encompassing multiple generations of dominant plants, a time span that would also encompass much of the relevant variation in animal populations and physical factors"	White and Walker (1997, p. 343)
Historical range of variability	"characterizes fluctuations in ecosystem conditions or processes over time . . . define[s] the bounds of system behavior that remain relatively consistent over time"	Morgan et al. (1994, p 88)
Historic range of variability	"the spatial and temporal variation in composition, structure and function experienced in an ecosystem from about 1600 to 1850, when the influences of European-Americans were minimal [within the land area being evaluated]"	Dillon et al. (2005, p. 1)
Historical range and varlablllty	"the variation of historical ecosystem characteristics and processes over time and space scales that are appropriate for the management application"	Keane et al. (2009, p. 1026)
natural range of variability (NRV)	The ecological conditions and processes within a specified area, period of time, and climate, and the variation in these conditions, that would occur without substantial influence from human mechanisms"	Hann and Bunnell (2001, p. 394)
Reference conditions	"the spectrum of ecosystem conditions (i.e. structure, composition, and function) found within a defined area over a specified time period preceding Euro-American settlement"	Stephenson (1999, p. 1253)

Term	Definition	Reference
Reference conditions	"Reference conditions characterize the variability associated with biotic communities and native species diversity. They provide insights to important questions such as the natural frequency, intensity, and scale of forest disturbances; the age-class distribution of forest trees; and the abundance or rareness of plant or animal species within an ecosystem."	Kaufmann et al. (1994, p. 7)
Natural range of variation	"the range of variation in the absence of human influence"	White and Walker (1997, p. 343)
Natural variability . . . essentially synonymous with the terms range of natural variation/variability, natural range of variability, historical range of variability, and reference variability	"the ecological conditions, and the spatial and temporal variation in these conditions, that are relatively unaffected by people, within a period of time and geographical area appropriate to an expressed goal"	Landres et al. (1999, p. 1180)
Future range of variability (FRV)	"the estimated range of some ecological condition or process that may occur in the future – a dynamic set of boundaries on some condition or process that may occur in the future . . . [may be] expressed as a probability distribution of likely states"	Binkley and Duncan (2010)
Social range of variability (SRV)	"the range of an ecological condition that society finds acceptable at a given time . . . [reflecting] the suite of resource-management options that most people consider acceptable"	Duncan et al. (2010)
Social range of variability (SRV)	"the range of an ecological condition that society finds acceptable at a given time . . . [may be] expressed as a distribution of public acceptability"	Binkley and Duncan (2010)
Ecological range of variability (ERV)	"the estimated range of some ecological condition as a function of the biophysical forces, such as fires and hurricanes, and the social forces, e.g. burning, harvest, and development, that affect the area"	Duncan et al. (2010)

The concept of HRV and assessments of HRV for specific places have been developed and applied most often in forested ecosystems of the western United States, where the impacts of Euro-American settlement occurred relatively recently (generally mid- to late 1800s) and where long-lived trees are available for relatively precise reconstructions of past forest structure and disturbance regimes. Insights from historical ecology also have figured prominently in land-management and policy decisions in places where a formal HRV assessment has not been developed, such as in the forested landscapes of New England where European settlement occurred very early (beginning in the 1600s) and almost all of the original forest has been cut at least once or even removed for agriculture (Foster et al. 1996). Thus, a formal HRV assessment is only one of many ways in which ecological history can inform present-day resource management.

Where a local HRV assessment has been developed, with comparisons of historical versus current conditions, it is tempting to assume that the HRV condition therefore must be the desired management target. This follows immediately from the assumption that the ecological systems are in a dynamic equilibrium, varying about an unchanging long-term mean (Fig. 1.1). Conditions that existed during the HRV reference period may indeed be appropriate targets for some ecosystems or for some elements of an ecosystem (e.g. as a coarse-filter strategy for maintaining overall ecological integrity and resilience; Hunter 1990; Haufler et al. 1996; Holling & Meffe 1996). It is increasingly recognized, however, that returning some or even many characteristics of specific landscapes to the HRV condition may be neither ecologically feasible nor socially desirable, especially in the face of climate change, escalating land-use impacts, and spread of invasive species. Millar and Woolfenden (1999) and Millar et al. (2007) discuss issues of technical feasibility, pointing out that restoration and maintenance of some of the ecological conditions that prevailed during the "Little Ice Age" (which is the reference period for many HRV studies) would become increasingly difficult, expensive, and uncertain of success as we move into a progressively warmer world. A case in point would be restoration of salmon (*Oncorhynchus* spp.) populations in southern California streams where the species was present historically but where future water temperatures are expected to be too warm for salmon (Millar, pers. comm. 2008). Duncan et al. (2010) illustrate the social fallacy of uncritically assuming HRV to be the management target by pointing out that a high-severity fire regime might be the "natural" or historical norm in a wildland–urban interface area, but that society would never accept large destructive fires as a management objective. Although such issues have led some to dismiss the HRV concept as irrelevant, most would argue that HRV is irrelevant only if used naively. We briefly mention several well-conceived and useful applications of HRV later in this chapter, and others are discussed elsewhere in this book.

The concept of HRV has been described and illustrated in several excellent review articles, including a special section in *Ecological Applications* in 1999 (Parsons et al. 1999) and a recent summary by Keane et al. (2009). The relevance of historical ecology and HRV to the broader framework of ecosystem management was also reviewed in earlier articles that are still very pertinent, notably Kaufmann et al. (1994), Morgan et al. (1994), and Christensen et al. (1996). Rather than duplicate those detailed reviews, we instead highlight a few key concepts and issues that remain at the forefront of thinking about HRV, especially in the context of coping with future environmental change.

1.3 SOME CRITICAL ISSUES AND LIMITATIONS OF THE HRV CONCEPT

In the last decade or two, several serious conceptual and data-related issues regarding the use of HRV in resource management have emerged. It is these concerns that have led some to believe that HRV is not a useful tool in real-life management. Three issues are particularly important: the role of humans in reference ecosystems, the amount and quality of data available, and the dramatic and often irreversible alterations of ecosystems being wrought by climate change, invasive species, and intensifying human land use. These concerns are addressed briefly below, and a variety of examples of dealing with them effectively are presented in greater detail in the other chapters in this book.

The role of humans in reference ecosystems

An early criticism of the HRV concept was that it seemed to overlook the significant environmental impacts of indigenous people during the historical reference period. Developed as it was by people mainly of European background and affluent socioeconomic status, the idea of HRV was suggested to have misanthropic or even racist overtones, especially when it was expressed as the "range of *natural* conditions" (Box 1.1). This is a major reason why the term "*historical range of variation*" is used most often today: it avoids the knotty question of what is "natural" and allows for a potentially important history of human influence within reference ecosystems. The issue now is one of determining just what that influence was. On one side of the question are scholars and popular writers who argue that pre-Columbian Native American people[2]

---

[2] Throughout this book, we use "Native American" to refer to North American indigenous peoples.

substantially altered North American ecosystems almost everywhere. On the other side are authors describing America as a pristine wilderness essentially untouched by humans (see Vale 2002). Neither extreme position is likely to be correct. Just as climate and topography vary across the continent, so too the activities and impacts of Native Americans varied across space and through time (Vale 2002). There were places and times where the ecological impacts of indigenous people were likely quite intense, as near a Flathead village along the Yellowstone River, and other places where impacts probably were negligible, as in the unproductive and game-poor lodgepole pine (*Pinus contorta*) forests on the Yellowstone Plateau (Vale 2002). The challenge for ecologists and managers doing HRV work today is to determine objectively the kinds and magnitudes of human influences on ecological structure and function in a specific place during a specific time in the past.

Although the specific ecological impacts of indigenous people in North America prior to the arrival of Europeans will always be somewhat uncertain (Duncan et al. 2010), there is no question that the trappers, loggers, miners, grazers, farmers, industrialists, and settlers who swept across the continent beginning in the sixteenth century ushered in a new era of enormous ecological change. It is these changes against which HRV provides an especially meaningful benchmark. This is why the reference period in HRV assessments is often the several centuries just prior to Euro-American settlement; this usually is the time for which we have the most complete information about ecological conditions as they existed before the storm. Impacts of indigenous people during that reference period may or may not need to be included as important components of HRV, depending on the particular locale and specific ecological questions being addressed (see Nowacki et al., Chapter 6, this book).

The amount and quality of data available for HRV summaries and assessments

The availability of accurate and relevant information is a limiting factor in many facets of resource management, and historical assessments are no exception. This topic has been discussed both in the reviews cited above and in articles and reports dealing with historical assessments conducted in specific places. Box 1.2 summarizes some of the major issues related to sources,

strengths, and limitations of historical ecological data and HRV information in particular.

There are perhaps three key issues to consider when evaluating historical conditions that existed in any particular place. First is the "fading record," the idea that we generally have more information and can more readily reconstruct missing information for more recent time periods than for long-distant times (Swetnam et al. 1999). The record fades with time because the necessary materials for historical reconstruction gradually decay over time: things like old fire-scarred trees, ancient dead wood, and packrat middens are continually lost to wood harvest, wildfires, natural biological and physical decomposition, and weather. A major reason why HRV assessments are particularly useful in forests of the western United States is because the generally dry climate and relatively recent Euro-American settlement mean that these paleo-materials have been lost at a slower rate than in other places having a moister climate and/or a longer period of Euro-American land use.

The second key issue is what we might call the "selective record" of what is available – and unavailable – for any time in the past (Santayana's "assisted and recorded memory"). We have good information in many western forests about pre-1850 stand structure, composition, and fire frequency because trees that were alive at that time are still extant, either as very old living individuals or as well-preserved dead wood. But even where we have the best record of canopy structure and dynamics, we usually have little or no direct information about the herbaceous understory, invertebrate communities, soil microbial communities, or biogeochemical processes. These latter elements of an ecosystem, arguably as important as the trees, simply do not leave any tangible evidence for us to interpret today. Instead, we must make logical inferences about these components, which is not necessarily a bad thing in itself. It is important to remember, however, that this information is inferential rather than direct.

Finally, we cannot escape the fact that our historical information usually comes from localized places (e.g. a particular forest stand containing old trees and fire scars, or a particular lake with datable sediments), but the landscape of management interest typically is much larger. Therefore, we face issues of extrapolation and interpolation as we scale up from the sources of data to the area of application (White & Walker 1997; see Wiens et al., Chapter 5, this book). Moreover, the

Box 1.2 Some major sources and limitations of HRV information. The examples and references listed are illustrative, but are by no means comprehensive.

Source of information	Major limitations	Reference or example
Written records and maps	• Usually available only for a relatively recent period • Variation in accuracy, detail, and scope of records	Foster et al. (1996), Kulakowski and Veblen (2006)
Historic photographs	• Usually available only for a relatively recent period • Photos were often taken to depict cultural features (e.g. bridges) rather than ecological conditions	Swetnam et al. (1999), Zier and Baker (2006)
Ring chronologies and fire scars in very old trees or dead wood	• Often unavailable or represent only a relatively recent period, especially where harvest has been intense or decomposition is rapid	Moore et al. (1999), Stephenson (1999)
Charcoal, pollen, and macrofossils preserved, for example, in lake sediments or packrat middens	• Limited spatial distribution • May depict only very local conditions that are not representative of the larger landscape	Betancourt et al. (1990), Jackson (2006)
Landscape simulation models	• Sensitive to the empirical data used to build the model • May be difficult to validate	Keane et al. 2002, 2009), Nonaka and Spies (2005)
Inference from modern conditions of ecosystem parameters	• Lack of pre-twentieth-century databases for comparison and validation • Issues of scale, extrapolation, and interpolation	Veblen and Donnegan (2005), White and Walker (1997)

available reference sites may not be representative of the landscape as a whole; fire-scarred trees may be concentrated on drier sites where fires historically burned at lower intensity rather than on more productive sites that supported higher-intensity fires, or packrat middens may represent the localized flora of the rocky area itself instead of the open expanses that surround the rocks (Swetnam et al. 1999). To a greater or lesser degree, every individual place in the world is different from every other place because of differences in local environment and history, many of which are subtle. The challenge to ecologists is to combine insights from data sets that are incomplete to synthesize a broader picture of ecosystem structure and dynamics (White & Walker 1997). This is no small task, although several chapters in this book illustrate innovative approaches that have been applied successfully to the problem of characterizing the whole from the available parts.

Facing a future of climate change, invasive species, and accelerating human land-use impacts

Given the scope and magnitude of current and impending ecological changes taking place throughout the planet, it is no surprise that critics question whether historical ecology, and HRV in particular, still has any relevance to resource management. Indeed, if the only

---

Box 1.3 Examples of how the HRV concept can be utilized effectively in resource management, planning, and policymaking, despite anticipated changes in climate, land use, and invasive species (see text for explanations). The list is meant to be illustrative rather than comprehensive.

Application	Example
Identifying ecological conditions that will enhance ecological resilience in the face of changing climate and fire regimes	Restoring pre-1900 structure to southwestern ponderosa pine forests (e.g. Allen et al. 2002; Fulé 2008)
Evaluating the ecological feasibility of management goals that society may find desirable	Using prescribed burning to reduce fuels and prevent severe fires in wildland-interface zones (e.g. Romme et al. 2004)
Designing management plans that maintain ecological integrity while producing economic commodities	Mimicking historical spatial and temporal patterns of forest disturbance, in programs designed to simultaneously harvest timber and maintain old-growth forest habitats (e.g. Cissel et al. 1999)
Educating the public and facilitating collaborative discourse and decision making about resource-management goals and methods	Integrating historical ecological insights with social perceptions and desires to explore a "future range of variability" (e.g. Duncan et al. 2010)
Others . . . see chapters in this book	

---

application of HRV were to try to return ecosystems to replicas of pre-European conditions, then this criticism would be unassailable. The fallacy is in thinking of HRV as a specific target for a desired state of ecological systems, rather than using the historical record to assess how (and why or whether) such systems responded to environmental variations in the past as a way of informing current and future conservation management. Chapter 4 addresses this issue more fully. In the following section, however, we briefly develop some examples to illustrate that applications of the HRV concept are far broader and richer than a simplistic effort to reconstruct some idealized past state.

1.4 SOME EXAMPLES OF THE APPLICATION OF HRV IN A CHANGING WORLD

To set the stage for what will follow in subsequent chapters, we briefly examine four examples of useful

and well-conceived applications of the HRV concept (summarized in Box 1.3). These examples and this discussion are meant to be illustrative rather than comprehensive.

HRV helps in identifying ecological conditions that will restore ecological resilience in the face of changing climate and fire regimes

The frequency and extent of severe, stand-replacing fires in Arizona ponderosa pine (*Pinus ponderosa*) forests have increased dramatically in the past two decades, and many of the burned forests appear to have been converted, perhaps permanently, to shrublands and grasslands (Savage & Mast 2005; Strom & Fulé 2007). The failure of the forests to regenerate after fire is ironic, because HRV data reveal that ponderosa pine was one of the most frequently burned vegetation types prior to the twentieth century, with typical fire-return intervals of a decade or less (Moore et al. 1999). The key distinction is that historical fires usually were low-intensity surface burns. Ponderosa pine is well adapted to a fire regime of this kind: thick bark and a high crown prevent serious injury in mature trees, while frequent burning reduces fuel loads and maintains a low-density stand structure by selectively killing small trees and shrubs. Ponderosa pine should be able to survive – even thrive – in a future world of even more frequent fire if only the forests can be restored to the low-density, low-fuel condition that prevailed before heavy grazing and fire-suppression programs removed fire from the ecosystem after ca. 1880 and allowed unusually dense and vulnerable fuel structures to develop (Ecological Restoration Institute – http://www.eri.nau.edu/). Importantly, we need not necessarily return this ecosystem to an exact replica of its 1880 condition: it may be sufficient only to set the forests on a trajectory toward restoration of the key structural and functional elements that created historical resilience (Allen et al. 2002; Fulé 2008; Diggins et al. 2010).

HRV helps in evaluating the ecological feasibility of management goals that society may find desirable

Residents of rural or suburban communities surrounded by coniferous forests are increasingly cognizant of their vulnerability to destructive wildfires (Theobald & Romme 2007). Successful programs of fire mitigation via thinning and prescribed low-intensity burning have been developed and implemented for ponderosa pine forests (as just described), and it would seem logical that similar techniques could be applied to other coniferous forest types, such as lodgepole pine forests. An examination of historical conditions and disturbance regimes, however, raises questions about the likely effectiveness of a ponderosa pine-type fuel reduction program in lodgepole pine forests. Lodgepole pine forests were typically dense and fires were usually high intensity during the pre-1900 reference period (Sibold et al. 2006); fire size and severity were controlled far more by weather and climate than by fuel characteristics (Schoennagel et al. 2004). Drawing on this information about the historical fire regime as well as recent fire experience, Romme et al. (2004) identified both operational and

ecological problems that would be faced in attempting to mitigate fire hazards in lodgepole pine forests using mechanical thinning and prescribed low-intensity burning. Operational problems include a surface-fuel structure that is not conducive to spreading surface fires, plus a narrow window of weather conditions within which sustained surface fire is even possible – between being too wet to burn at all and so dry that the fire spreads into the canopy. Ecological problems include the tendency of residual lodgepole pine trees to fall down after thinning of dense stands, plus the fact that the native flora is well adapted to high-severity burns recurring at long intervals (e.g. serotinous cones in lodgepole pine and dormant seeds of shrubs and herbs that are stimulated by fire to germinate), but may not be able to tolerate frequent re-burning given the kind of fire environment in which the flora developed. Romme et al. (2004) recommend instead a fire-mitigation program that partly emulates the historical fire regime, creating fuel breaks with small but high-severity burns (ignited either by managers or by lightning) in strategically located areas at times when weather conditions allow fire spread to be controlled, and regulating spatial patterns of exurban development to keep vulnerable structures out of areas at highest fire risk. Similar understanding of the ecology of fire in chaparral stands in southern California, which also burn naturally at high severity but have been heavily invaded by suburban homes, has led to similar recommendations regarding development zoning and strategic placement of fuel treatments (Safford 2007).

HRV helps in designing management plans that maintain ecological integrity while producing economic commodities

Management of Douglas-fir (*Pseudotsuga menziesii*) forests in the Pacific Northwest during the latter half of the twentieth century emphasized timber production but included old-growth forest reserves intended to maintain overall biological diversity and to preserve habitat for the spotted owl (*Strix occidentalis*). In response to concerns about the effectiveness of the existing reserve system, Cissel et al. (1999) used spatial modeling techniques to simulate the future landscape structure that would result from continuation of the current system of static reserves (the "interim plan") and compared this outcome with the results of a more dynamic management plan in which spatial and temporal harvest patterns would more closely resemble the patterns produced by the historical fire regime (the "landscape plan"). Notably, the landscape plan did not attempt to recreate the precise landscape structure that existed at any specific time in the past; instead, it generally mimicked the historical disturbance frequency, patch size, and spatial distribution of disturbance patches. The simulations revealed several ecological advantages of the historically informed landscape plan: average size of old-growth reserves was greater, and reserves were more equally distributed across different topographic settings when compared with projected results for the static interim plan. A disadvantage of the landscape plan was higher costs for planning and implementing timber harvests. However, this disadvantage might be partially offset by the value of timber that would be harvested with the landscape plan.

HRV helps in educating the public and facilitating collaborative discourse and decision making about resource-management goals and methods

In the not-too-distant past, public resource management was very much a top-down affair: the experts in government agencies identified goals and options, and the public had a very limited role in responding to what was presented to them. One of the major trends of the past few decades is the development of more effective public engagement in resource-management decisions. Three tasks are central to success in this process: (1) providing an ecological context that informs people of what would be ecologically possible or impossible, (2) informing land managers of people's desires and expectations for public lands, and (3) creatively integrating these two kinds of information into an ecologically realistic and socially acceptable plan of action. HRV assessments are ideally suited for the first task if the assessment is presented in understandable terms and not as an uncritically assumed target for the future. For the second task, the idea of a "social range of variability (SRV)" or "the suite of resource-management options that most people consider acceptable" has been recently introduced (Box 1.1). For the

third task, Duncan et al. (2010) propose collaborative exploration of a "future range of variability (FRV)" (Box 1.1). By depicting both HRV and SRV as probability distributions, a region of overlap is identified, representing a combination of conditions that are potentially both ecologically feasible and socially acceptable (Fig. 1.2). This range of ecologically and socially compatible conditions can be further refined by considering likely future developments, such as climate change or economic incentives or disincentives for timber harvest, to produce an "ecological range of variability (ERV)" (Box 1.1). A key aspect of the exploration of an FRV is that ERV is never static; consequently, FRV will also be dynamic as social and ecological conditions change over time. This is illustrated for the case of old-growth forest cover in Fig. 1.3. Nevertheless, neither ERV nor FRV can change without limits. This is where HRV is especially useful: it helps identify key constraints on what is ecologically possible and informs and thereby broadens the perspective through which people decide what is ecologically desirable.

1.5 PRÉCIS

These four examples illustrate some ways in which the HRV concept can be a valuable tool in resource management, even in the face of a dramatically changing future. Although we have emphasized examples from terrestrial ecosystems, historical ecology and the HRV concept apply equally in aquatic ecosystems, as developed in Chapters 15 and 16 (Box 1.4). Despite the inherent limitations of all historical data and perspectives, HRV and, more broadly, an understanding of history nevertheless provide an ecological context for resource-management and conservation decisions that cannot be obtained from any other source. Rather than being a naïve concept focused only on an unattainable past, history, properly used, can enable resource management and conservation to move into an uncertain future. In the following 22 chapters, 54 authors from a wide variety of resource-management agencies, nongovernmental organizations, universities, and research institutes provide testament to the multifaceted and highly relevant stanchion that historical ecology and HRV provide to current and future conservation and resource management.

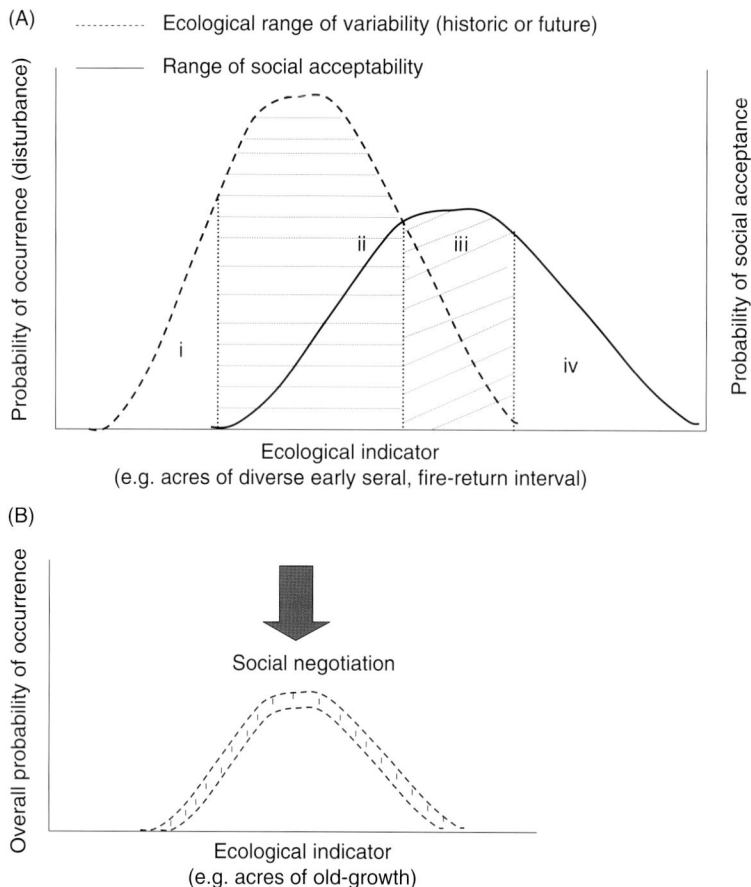

Fig. 1.2 Conceptual diagram illustrating the interaction of the range of variation for an ecological condition that reflects the disturbance of ecosystems by biophysical and human forces and their rate of recovery with the range of conditions that are considered socially acceptable. Figure 1.2A shows a hypothetical set of these relationships for some condition. Zone i represents ecological conditions that would occur without investment/intervention to prevent them but that do not have social acceptance; in Zone ii, the likelihood of occurrence is greater than the likelihood of acceptance; in Zone iii, the likelihood of occurrence is less than the likelihood of acceptance; and Zone iv represents conditions that would not occur without investment/intervention to enable them even though a segment of society wants them. To the degree that the two ranges do not overlap, social pressure/negotiation may be expected to change the shape of the ecological probabilities curve. This negotiation leads to the range of variation actually experienced (Fig. 1.2B). The exact probability distribution can only be estimated, but there is always uncertainty; hence, this distribution is represented by a fuzzy area between the two curves. Looking back in time, Fig. 1.2B would be called the historical range of variation. Looking forward in time, Fig. 1.2B would be called the future range of variation. The curve produced in Fig. 1.2B is potentially highly dynamic. Reprinted, with permission, from Duncan et al. (2010).

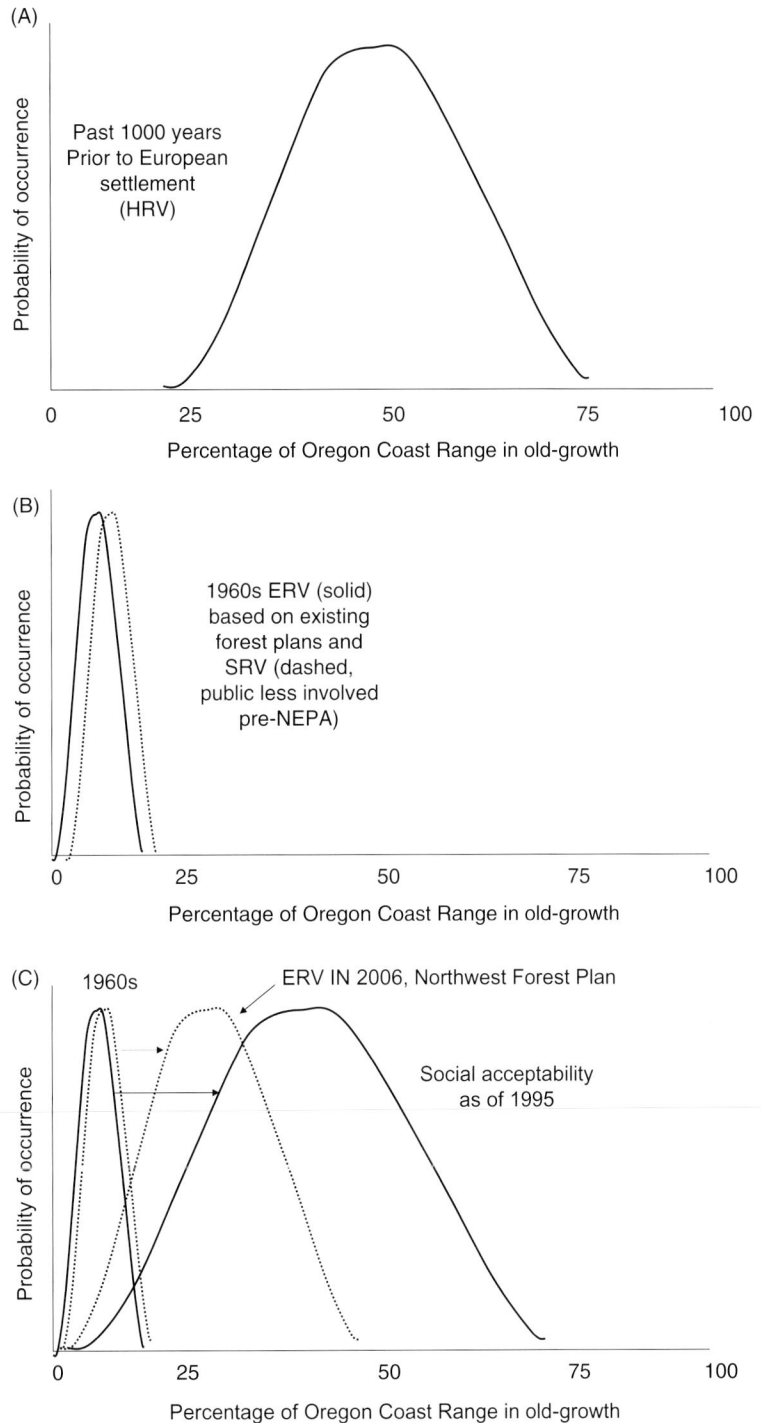

Fig. 1.3 (A) Probability distribution associated with percentage of old-growth forests in the Oregon Coast Range and the expression of the social range of acceptability of old-growth representation during the 1960s prior to passage of the National Environmental Policy Act (NEPA), which ensured public participation in planning processes. (B) Changes in local/regional social values during the mid-1990s (solid arrow and curves) led to a shift in the ecological range of variation (ERV) over the region (dashed arrow and curves) as federal land managers developed new forest plans for the Pacific Northwest region following NEPA policies and other environmental laws and political processes. (C) Because the social range of variation (SRV) continues to diverge from the ecological range of variation as the Northwest Forest Plan is implemented, policies can be expected to continue to shift the range of variation of old-growth forests farther to the right. Reprinted, with permission, from Duncan et al. (2010).

Box 1.4 Application of historical ecology in aquatic and terrestrial systems.

The science and application of historical ecology share many common elements across wet and dry environments. Resource managers concerned with conservation of both aquatic and terrestrial resources independently began using historical ecology to influence management decisions during the 1960s (Landres et al. 1999). In both systems, some practitioners employed "reference areas" to establish targets for ecological restoration (Karr 1981; Holling & Meffe 1996; Shields et al. 2003; Humphries & Winemiller 2009), while others focused on the ecological understanding provided by HRV assessments to provide context for management decisions. Clearly, issues of scale, system dynamics, and the importance of understanding processes that occur over extended periods motivated interest in historical ecology. Similarities in application of historical ecology within aquatic and terrestrial systems extend from motivations to methods, issues, and benefits realized. Aquatic ecologists use sediment cores in lakes, soil cores in valley bottoms, and backhoe trenches in areas of deposition to trace the history of biotic and geomorphological processes (e.g. Walters & Merritts 2008). In terrestrial systems, dendrochronology provides analogous time series (Grissino-Mayer et al. 2004). Historical survey records, aerial photo-graph series, chronosequence approaches, and historical narratives are used by both aquatic and terrestrial managers to examine landscape patterns (e.g. Harding et al. 1998; Egan & Howell 2001). The challenge of locating historical legacies that reliably indicate the history of disturbance events or changes in species composition plagues scientists working in both aquatic and terrestrial systems. Likewise, separating patterns resulting from aboriginal or recent human activity from other ecological processes clouds understanding; examples include milldams in the eastern United States or European piedmont (Walters & Merritts 2008) and aboriginal fire in Northeastern forests (Lorimer 2001). Locating evidence for the spatial extent of disturbance events and making inferences from incomplete historical data also limit scientists in both types of environments. The use of historical ecology has revised several long-standing paradigms and changed management practices in both aquatic (Poff et al. 1997) and terrestrial (Foster et al. 1998) resource management. The apparent lack of dialogue among terrestrial and aquatic practitioners on this topic is therefore puzzling. We anticipate that the benefits gained through expanding this dialogue will lead to advances in management of aquatic and terrestrial resources.

REFERENCES

Allen, C.D., Savage, M., Falk, D.A., et al. (2002). Ecological restoration of southwestern ponderosa pine ecosystems: a broad perspective. *Ecological Applications*, **12**, 1418–1433.

Aplet, G.H. & Keeton, W.S. (1999). Application of historical range of variability concepts to biodiversity conservation. In *Practical Approaches to the Conservation of Biological Diversity* (ed. R. Baydack, H. Campa, and J. Haufler), pp. 71–86. Island Press, Washington, DC, USA.

Baker, W.L. (2009). *Fire Ecology in Rocky Mountain Landscapes*. Island Press, Washington, DC, USA.

Betancourt, J.L., Van Devender, T.R., & Martin, P.S. (1990). *Packrat Middens: The Last 40 000 Years of Biotic Change*. University of Arizona Press, Tucson, AZ, USA.

Binkley, D. & Duncan, S.L. (2010). The past and future of Colorado's forests: connecting people and ecology. *Ecology and Society*, **14**(2), article 9. http://www.ecologyandsociety.org/vol14/iss2/art9/.

Christensen, N.L., Bartuska, A., Brown, J.H., et al. (1996). The scientific basis for ecosystem management. *Ecological Applications*, **6**, 665–691.

Cissel, J.H., Swanson, F.J., & Weisberg, P.J. (1999). Landscape management using historical fire regimes: Blue River, Oregon. *Ecological Applications*, **9**, 1217–1231.

Diggins, C., Fulé, P.Z., Kaye, J., & Covington, W.W. (2010). Future climate affects management strategies for maintaining forest restoration treatments. *International Journal of Wildland Fire*, **19**, 903–913.

Dillon, G.K., Knight, D.H., & Meyer, C.B. (2005). Historic range of variability for upland vegetation in the Medicine Bow National Forest, Wyoming. General Technical Report RMRS-GTR-139. USDA Forest Service, Fort Collins, CO, USA.

Duncan, S.L., McComb, B.C., & Johnson, K.N. (2010). Integrating ecological and social ranges of variability in conservation of biodiversity: past, present, and future. *Ecology and Society*, **15**(1), article 5. http://www.ecologyandsociety.org/vol15/iss1/art5/ES-2009-3025.pdf.

Egan, D. & Howell, E.A. (eds.). (2001). *The Historical Ecology Handbook: A Restorationist's Guide to Reference Ecosystems*. Island Press, Washington, DC, USA.

Foster, D.R., Knight, D.H., & Franklin, J.F. (1998). Landscape patterns and legacies resulting from large, infrequent forest disturbances. *Ecosystems*, **1**, 497–510.

Foster, D.R., Orwig, D.A., & McLachlan, J.S. (1996). Ecological and conservation insights from reconstructive studies of temperate old-growth forests. *Trends in Ecology and Evolution*, **11**, 419–424.

Fulé, P.Z. (2008). Does it make sense to restore wildland fire in changing climate? *Restoration Ecology*, **16**, 526–531.

Grissino-Mayer, H.D., Romme, W.H., Floyd, M.L., & Hanna, D.D. (2004). Climatic and human influences on fire regimes of the southern San Juan Mountains, CO, USA. *Ecology*, **85**, 1708–1724.

Hann, W.J. & Bunnell, D.L. (2001). Fire and resource management planning and implementation across multiple scales. *International Journal of Wildland Fire*, **10**, 389–403.

Harding, J.S., Benfield, E.F., Bolstad, P.V., Helfman, G.S., & Jones, E.B.D. III (1998). Stream biodiversity: the ghost of land use past. *Proceedings of the National Academy of Sciences USA*, **95**, 14843–14847.

Haufler, J.B., Mehl, C.A., & Roloff, G.J. (1996). Using a coarse-filter approach with species assessment for ecosystem management. *Wildlife Society Bulletin*, **24**(2), 200–208.

Holling, C.S. & Meffe, G.K. (1996). Command and control and the pathology of natural resource management. *Conservation Biology*, **10**, 328–337.

Humphries, P. & Winemiller, K.O. (2009). Historical impacts on river fauna, shifting baselines, and challenges for restoration. *BioScience*, **59**, 673–684.

Hunter, M.L. (1990). *Wildlife, Forests, and Forestry*. Prentice-Hall, Englewood Cliffs, NJ.

Jackson, S.T. (2006). Vegetation, environment, and time: the origination and termination of ecosystems. *Journal of Vegetation Science*, **17**, 549–557.

Karr, J.R. (1981). Assessment of biotic integrity using fish communities. *Fisheries*, **6**, 21–27.

Kaufmann, M.R., Graham, R.T., Boyce, D.A., et al. (1994). An ecological basis for ecosystem management. General Technical Report RM-GTR-246. USDA Forest Service, Fort Collins, CO, USA.

Keane, R.E., Hessburg, P.F., Landres, P.B., & Swanson, F.J. (2009). The use of historical range and variability in landscape management. *Forest Ecology and Management*, **258**, 1025–1037.

Keane, R.E., Parsons, R.A., & Hessburg, P.F. (2002). Estimating historical range and variation of landscape patch dynamics: limitations of the simulation approach. *Ecological Modelling*, **151**, 29–49.

Kricher, J. (2009). *The Balance of Nature: Ecology's Enduring Myth*. Princeton University Press, Princeton, NJ.

Kulakowski, D. & Veblen, T.T. (2006). Historical range of variability for forest vegetation of the Grand Mesa National Forest, Colorado. USDA Forest Service, Golden, CO, USA. http://warnercnr.colostate.edu/images/docs/cfri/cfri_grand_mesa.pdf.

Landres, P.B., Morgan, P., & Swanson, F.J. (1999). Overview of the use of natural variability concepts in managing ecological systems. *Ecological Applications*, **9**, 1179–1188.

Lorimer, C.C. (2001). Historical and ecological roles of disturbance in eastern North American forests: 9000 years of change. *Wildlife Society Bulletin*, **29**, 425–439.

Manley, P.N., Brogan, G.E., Cook, C., et al. (1995). Sustaining ecosystems: a conceptual framework. Publication R5-EM-TP-001. USDA Forest Service, San Francisco, CA, USA.

Meffe, G.K. & Carroll, C.R. (1994). *Principles of Conservation Biology*. Sinauer Associates, Inc., Sunderland, MA.

Millar, C.I., Stephenson, N.L., & Stephens, S.L. (2007). Climate change and forests of the future: managing in the face of uncertainty. *Ecological Applications*, **17**, 2145–2151.

Millar, C.I. & Woolfenden, W.B. (1999). The role of climate change in interpreting historical variability. *Ecological Applications*, **9**, 1207–1216.

Miller, R.F. & Wigand, P.E. (1994). Holocene changes in semi-arid pinyon-juniper woodlands. *BioScience*, **44**, 465–474.

Moore, M.M., Covington, W.W., & Fulé, P.Z. (1999). Reference conditions and ecological restoration: a southwestern ponderosa pine perspective. *Ecological Applications*, **9**, 1266–1277.

Morgan, P., Aplet, G.H., Haufler, J.B., Humphries, H.C., Moore, M.M., & Wilson, W.D. (1994). Historical range of variability: a useful tool for evaluating ecosystem change. *Journal of Sustainable Forestry*, **2**, 87–111.

Nonaka, E. & Spies, T.A. (2005). Historical range of variability in landscape structure: a simulation study in Oregon, USA. *Ecological Applications*, **15**, 1727–1746.

Parsons, D.J., Swetnam, T.W., & Christensen, N.L. (1999). Invited feature: uses and limitations of historical variability concepts in managing ecosystems. *Ecological Applications*, **9**, 1177–1178.

Pimm, S.L. (1991). *The Balance of Nature? Ecological Issues in the Conservation of Species and Communities*. University of Chicago Press, Chicago, IL, USA.

Poff, N.L., Allan, J.D., Bain, M.B., et al. (1997). The natural flow regime. *BioScience*, **47**, 769–784.

Romme, W., Allen, C., Bailey, J., et al. (2009). Historical and modern disturbance regimes of piñon-juniper vegetation in the western United States. *Rangeland Ecology and Management*, **62**, 203–222.

Romme, W.H., Turner, M.G., Tinker, D.B., & Knight, D.H. (2004). Emulating natural forest disturbance in the wildland-urban interface of the Greater Yellowstone Ecosystem of the United States. In *Emulating Natural Forest Landscape Disturbances: Concepts and Applications* (ed. A.H. Perera, L.J. Buse, and M.G. Weber), pp. 243–250. Columbia University Press, New York, USA.

Safford, H.D. (2007). Man and fire in Southern California: doing the math. *Fremontia*, **35**, 25–29.

Savage, M. & Mast, J.N. (2005). How resilient are southwestern ponderosa pine forests after crown fires? *Canadian Journal of Forest Research*, **35**, 967–977.

Schoennagel, T., Veblen, T.T., & Romme, W.H. (2004). The interaction of fire, fuels, and climate across Rocky Mountain forests. *BioScience*, **54**, 661–676.

Shields, F.D., Copeland, R.R., Klingeman, P.C., Doyle, M.W., & Simon, D. (2003). Design for stream restoration. *Journal of Hydraulic Engineering*, **129**, 575–584.

Sibold, J.S., Veblen, T.T., & Gonzales, M.E. (2006). Spatial and temporal variation in historic fire regimes in subalpine forests across the Colorado Front Range in Rocky Mountain National Park, Colorado, USA. *Journal of Biogeography*, **32**, 631–647.

Stephenson, N.L. (1999). Reference conditions for giant sequoia forest restoration: structure, process, and precision. *Ecological Applications*, **9**, 1253–1265.

Strom, B.A. & Fulé, P.Z. (2007). Pre-wildfire fuel treatments affect long-tem ponderosa pine forest dynamics. *International Journal of Wildland Fire*, **16**, 128–138.

Swetnam, T.W., Allen, C.D., & Betancourt, J.L. (1999). Applied historical ecology: using the past to manage for the future. *Ecological Applications*, **9**, 1189–1206.

Theobald, D.M., & Romme, W.H. (2007). Expansion of the US wildland-urban interface. *Landscape and Urban Planning*, **88**, 340–354.

Turner, B.L. II., Clark, C., Kats, R.W., Richards, J.F., Matthews, J.T., & Meyer, W.B. (eds.). (1990). *The Earth as Transformed by Human Action*. Cambridge University Press, Cambridge, UK.

Vale, T.R. (2002). The pre-European landscape of the United States: pristine or humanized? *Fire, Native Peoples, and the Natural Landscape* (ed. T.R. Vale), pp. 1–39. Island Press, Washington, DC, USA.

Veblen, T.T. & Donnegan, J.A. (2005). Historical range of variability for forest vegetation of the national forests of the Colorado Front Range. USDA Forest Service, Golden, CO, USA. http://warnercnr.colostate.edu/publications-and-reports/.

Walters, R.C. & Merritts, D.J. (2008). Natural streams and the legacy of water-powered mills. *Science*, **139**, 299–304.

White, P.S. & Walker, J.L. (1997). Approximating nature's variation: selecting and using reference information in restoration ecology. *Restoration Ecology*, **5**, 338–349.

Wiens, J.A., van Horne, B., & Noon, B.R. (2002). Integrating landscape structure and scale into natural resource management. In *Integrating Landscape Ecology into Natural Resource Management* (ed. J. Liu and W.W. Taylor), pp. 23–67. Cambridge University Press, Cambridge, UK.

Zier, J.L. & Baker, W.L. (2006). A century of vegetation change in the San Juan Mountains, Colorado: an analysis using repeat photography. *Forest Ecology and Management*, **228**, 251–262.

Chapter 2

DEVELOPMENT OF HISTORICAL ECOLOGY CONCEPTS AND THEIR APPLICATION TO RESOURCE MANAGEMENT AND CONSERVATION

Wayne Padgett,[1] *Barbara Schrader,*[2] *Mary Manning,*[3] *and Timothy Tear*[4]

[1]Bureau of Land Management, Utah State Office, Salt Lake City, UT, USA
[2]USDA Forest Service, Alaska Region, Juneau, AK, USA
[3]USDA Forest Service, Northern Region, Missoula, Montana, USA
[4]The Nature Conservancy, Niskayuma, NY, USA

Historical Environmental Variation in Conservation and Natural Resource Management, First Edition. Edited by John A. Wiens,
Gregory D. Hayward, Hugh D. Safford, and Catherine M. Giffen.
© 2012 John Wiley & Sons, Ltd. Published 2012 by John Wiley & Sons, Ltd.

2.1 INTRODUCTION

Since early in the evolution of natural resource management, competing utilitarian and preservationist philosophies have been at play in shaping policies for managing public lands (Keiter 1994). Gifford Pinchot and John Muir, two environmental leaders in the late nineteenth and early twentieth centuries (Fig. 2.1), held markedly different views of humans and their relationships to the environment. Pinchot, first chief of the US Forest Service (USFS), was considered a conservationist because he focused on the sustainable use of natural resources for human benefits. Muir, founder of the Sierra Club, was a preservationist: someone who believed in protecting the environment

from human impacts for its own sake so future generations could enjoy and experience nature as had their predecessors.

Historically, these views of conservation and preservation have led to divergent and sometimes conflicting policies affecting public lands. Today, land management includes elements of both preservation and conservation; each view influences how we approach and apply historical range of variation (HRV) in management activities. In this chapter, we outline the milestone laws and decisions that ultimately led to the use of concepts associated with HRV in the management of public and private lands. We focus on how the role of HRV evolved along with land-use planning in the public and private sector and supports goals and

Fig. 2.1 Left: President Theodore Roosevelt with conservationist Gifford Pinchot (October 1907) standing on the deck of the steamer *Mississippi*, during a tour of the Inland Waterways Commission (US Forest Service). Right: President Roosevelt with preservationist John Muir (ca. 1903) on Glacier Point, Yosemite Valley, California (Underwood and Underwood). Images are in public domain courtesy of Library of Congress, Rare Book and Special Collections Division.

aspirations for healthy ecosystems, regardless of ownership. We propose that, through building stronger collaborations and social contracts among government agencies, nongovernment organizations, and private entities that support adaptive management and the use of the best available science, it will be possible to reduce legal barriers and provide more flexible management practices to improve the stewardship of lands and natural resources in increasingly uncertain times.

2.2 LEGAL AND REGULATORY ACTIONS THAT LED TO THE USE OF HRV: A TIMELINE

Conservation of managed and natural landscapes has a rich history in the United States and has evolved into a complex of activities now dictated by national and state laws. It was often the divergent views of the conservationist and the preservationist that forged the way for this changing legal landscape, although in some cases it was the common view that lands were neither being preserved nor conserved that determined the legal and regulatory directions for how lands are managed today.

Federal land management and conservation in the United States can be traced to the late nineteenth century, amid concerns that broadscale clear-cutting of forests was resulting in potential shortages of timber, and that water supply and flood-control issues were linked to these forest-management practices (Loomis 2002). Under the Forest Reserve Act of 1891, transfer of public-domain lands to federal control was authorized in response to these growing concerns and, at the same time, the concept of scientific management of forests was introduced to the dialogue. The Organic Act of 1897 provided an institutional framework for the establishment of federal forest reserves and directed that they be managed concurrently for timber, watershed, and forest protective purposes. This law provided the statutory foundation for the idea that management of public lands needed to be for more than resource extraction alone. Indeed, the National Forests, which came into being in 1905, were managed more for their watershed values than timber until the economic boom following the Second World War. Prior to this war, timber was primarily harvested from private lands and national-forest timber was used only as a reserve. It was only after the depletion of those private resources that timber harvests on national-forest lands began to increase significantly (Fedkiw 1998).

Initially, all national-forest lands were located in the American West, where most of the public-domain lands occurred. In 1911, the Weeks Act was passed, appropriating funds to purchase approximately 6 million acres of private lands in the eastern United States. This law, which could be applied to any state, resulted in the protection and improvement of watersheds with the establishment of 52 National Forests in 26 eastern states. In addition, nearly 20 million acres of land were purchased to establish National Forests and Grasslands across 41 states and Puerto Rico (USDA Forest Service 2011).

It was becoming clear that expanding populations and associated resource uses impacted lands other than those covered by forests. The growing livestock industry in the West was resulting in ever-more despoiled conditions of the western rangelands (Sayre & Fernandez-Gimenez 2003). President Theodore Roosevelt signed the Taylor Grazing Act in 1934 to address issues of overgrazing and the degradation of rangelands in the West. This act established the Grazing Division within the Department of Interior (renamed Grazing Service in 1939). Then, in 1946, the Grazing Service and the General Land Office merged to form the Bureau of Land Management (BLM) whose mission included the restoration of rangeland conditions in the West (Ross 2006).

By the middle of the twentieth century, Americans increasingly visited the National Forests for leisure and recreation, prompting the need for legislation that would broaden the scope of appropriate uses on public lands. In 1960, Congress passed the Multiple-Use Sustained-Yield Act (MUSYA) that declared management of the National Forests was to focus equally on recreation, range, timber, watershed protection, and wildlife. Although MUSYA merged Pinchot's conservationist views of managing our natural resources with Muir's protectionist ideas, in practice, the former was heavily weighted over the latter.

Further federal legislation expanded the focus on the environment and on the ecological conditions of the nation's natural resources. The National Environmental Policy Act of 1969 (NEPA), signed into law by President Richard Nixon, was both simple and comprehensive (Aim 1988). Among other things, NEPA required that environmental-impact statements be prepared for federal actions having significant effects on the environment. It was through these environmental laws that not only the description of existing conditions, but also the concept of desired future conditions of natural resources, was introduced. This concept was fundamental to the development of HRV. NEPA required federal agencies to enter into a dialogue with the public about proposed actions, a dialogue that often ended up in the courts. Environmental organizations and citizens' groups began to use NEPA as a tool not only to force federal compliance with environmental laws and regulations, but also to demand the incorporation of transparent science in land- and resource-management decision making.

In 1975, citizens in West Virginia protested the USFS's practice of clear-cutting forest lands on the Monongahela National Forest because of their concerns over the impacts on hunting, fishing, and other recreational uses of the forest (Ruth 1997). These citizens sought recourse through the courts when they found the USFS unresponsive to their concerns (Steelman 2007). In *Izaak Walton League v. Butz*, the courts ruled that the USFS violated the Organic Act of 1897 in their use of clear-cutting as a means of timber harvest. The National Forest Management Act of 1976 (NFMA) was a direct response to the Monongahela decision and required that the National Forests develop 10- to 15-year management plans in concert with the public.

2.3 BEYOND THE LEGAL DEBATE – HRV INFORMS RESTORATION PLANNING

The USFS released its first planning rule in 1982 in response to NFMA. This rule guided the development, amendment, and revision of the land-management plans on all National Forests and Grasslands in the United States (Federal Register 1982). Although the USFS planning direction in the 1980s continued to focus on maximizing "net present value" of commodity resources (W. Connelly, pers. comm., 2009), NFMA provided a critical link to the ultimate use of HRV in management of public lands. Alabama's Conecuh National Forest Land Management Plan was one of the first to be released and focused on the restoration of long-leaf pine (*Pinus palustris*) ecosystems. This forest was one of the first to incorporate concepts of HRV into their planning efforts because, before restoration actions could take place, this effort required an understanding of how these ecosystems had worked historically.

Soon after, in the Pacific Northwest, Franklin and Forman (1987) described an evolution of thought regarding landscape ecology and suggested a means for addressing fragmentation of habitats for interior-dependent species. The authors suggested that the emphasis on dispersing small clearcuts throughout a landscape should be reduced for economic reasons (the amount of roads necessary to access the harvest areas) as well as because of the resulting fragmentation of habitats. They felt it was important to identify and reserve large patches of old-growth forest in the landscape to maintain habitat for key species, habitat that was quickly disappearing because of the current timber harvest practices.

On the heels of this movement, in the early 1990s, "New Perspectives" were identified in the management of forest ecosystems (Hemstrom 1991; Swanson & Franklin 1992), challenging long-standing forest-management practices in the Pacific Northwest. These new perspectives were based on the results of long-term ecosystem research in managed and unmanaged forests, moving beyond a stand to a landscape scale. The approach incorporated a convergence of the conservationist views of Pinchot and the preservationist views of Muir. When the USFS adopted the New Perspectives program, Kessler et al. (1992) noted that this new direction for managing public lands reflected an increased awareness, concern, and expectations from society for the management of ecosystems for a variety of goods and services, not only for forest products, livestock grazing, and recreation, but also for things such as biological diversity, ecological functions, and the general integrity of natural environments. There was "… a shift in management focus from sustaining yields of competing resource outputs to sustaining ecosystems."

New Perspectives soon evolved into what the USFS termed "ecosystem management." Swanson et al. (1993) examined the use of natural variability of ecosystems as a guide for ecosystem management, noting that "the use of natural variability as a reference point in ecosystem management is not an attempt to turn managed landscapes into wilderness areas or return them to any single pre-existing condition." Rather, they felt that by bringing landscapes within their natural range of variability, managers could better balance ecological objectives with commodity production. They also believed that natural variability could provide a basis for designing management prescriptions as well as reference points for assessing the effects of ecosystem management. HRV provided a "tangible model of sustainable ecosystems" (Morgan et al. 1994). It was an important tool for managers to understand how ecosystems were sustained before the impacts of humans reached a point exceeding a threshold of naturalness.

Outside the USFS, similar trends occurred. Groves (2003) noted that during the 1990s, at least four critical developments shaped conservation planning and management. First was the concept of nonequilibrium theory, which emphasized the role of natural disturbances and ecological processes in shaping nature. Second was the growing list of endangered species that highlighted the need for more than single-species management. Third was the recognition that biodiversity operates over multiple spatial and temporal scales across different levels of biological organization. And finally, the rise of landscape ecology brought with it new tools and methods for addressing fragmentation and connectivity.

Over time, even larger multiforest efforts involving the use of HRV were undertaken. For example, the Northwest Forest Plan (Shindler & Mallon 2009) and Sierra Nevada Framework (USDA Forest Service 2001), both of which involved many millions of hectares of forestland, brought HRV into focus in the 1990s by using historical conditions to rate current conditions and trends. Two more recent examples of HRV informing planning are provided by the Mark Twain National Forest in Missouri and the Chequamegon-Nicolet National Forest (CNNF) in Wisconsin. Both forests are in portions of the United States that have seen tremendous alteration since European settlement, and both face challenges in using HRV in the management of their lands. Both used HRV to inform their analysis of existing conditions and develop realistic and attainable desired conditions. Through their use of HRV, they were able to understand how and why current ecosystem components had changed over time. Both forests used historical information to help develop management goals, defining and describing desired conditions based on the best available information. Neither forest attempted to restore conditions to presettlement conditions, but both used the concepts of HRV to move toward "sustainable" ecosystems.

The Chequamegon-Nicolet National Forest stated specifically that:

> Development of the description of RNV [Range of Natural Variation] does not

imply that the CNNF intends to return the area to historical conditions; indeed, it is impossible to do so and may be undesirable within the context of achieving multiple-use objectives. The purpose of examining historical conditions is to identify ecosystem factors that formerly sustained species and communities that are now reduced in number, size, or extent, or which have been changed functionally. Maintaining or restoring some lands to closely resemble historic systems, and including some structural or compositional components of the historic landscape within actively managed lands, can help conserve important elements of biological diversity. (USDA Forest Service 2004)

Thus, HRV was not intended as a rigid template for conservation and management goals, but rather to understand how the range of conditions (e.g. processes, composition, and pattern) affected habitat for flora and fauna of associated ecosystems over space and time.

Principles and Practices on Private Lands

Shaped by the early writings of Muir and Pinchot, several nongovernmental organizations (NGO) in the United States also developed ways of incorporating HRV into planning and management. While some in the NGO community filed lawsuits against federal land-management agencies, others designed programs and implemented projects using concepts of HRV. Although the legal battles and court decisions were common drivers for government agencies to shift management practices, NGOs generally had greater freedom to manage their lands without the threat of litigation.

The Nature Conservancy (TNC) began to adopt HRV principles into its fire-management planning in the late 1980s (R. Meyers, pers. comm., 2010). This influence was based on the Proceedings of the Conference of Fire Regimes and Ecosystem Properties held in 1978 in Hawaii (Mooney et al. 1981). The use of HRV as a concept evolved as fire ecologists grappled with issues of scale and its relation to fire management. Morgan et al. (2001) pointed out that:

quantifying departures of current fire regimes from historical fire regimes is a way to evaluate ecosystem change and prioritize fuels management action, based upon the concept of historical or natural range of variability.... Management options may differ on those sites where wildland fire can be used readily now (those sites within the historical range of variability, HRV), from those where the only possibility is mechanical fuels management first (far outside HRV) and those where some combination of wildland fire use and pre-scribed fire will be most effective (those near but outside of HRV).

2.4 COLLABORATIVE CONSERVATION AND HRV APPLICATION

It was because of the focus on the historical role that fire had played in ecosystems that NGOs and federal agencies began to work more closely together. Through a collaborative partnership in the early 2000s, the Department of Interior, USFS, US Geological Survey, and TNC developed a map of ecosystems for the continental United States and their degree of departure from historic fire regimes. Mapping of fire-regime condition class (FRCC) for all lands within the US boundaries was an important step in working across political and social boundaries (see Chapter 13, this book). This illustrated an important public and private effort that was the culmination of many years of close collaboration.

While advances in fire management were occurring, freshwater research and management evaluated the role of ecosystem management and the importance of disturbance regimes in the context of large river conservation (Sparks 1995) as well as in smaller anadromous fish habitats (Reeves et al. 1995). The 1993 flood of the Mississippi River (Fig. 2.2) was pivotal in focusing attention on the importance of rivers and their flood cycles. Advances in freshwater ecology showed that stream flow was a master variable that shaped species' abundances and distributions and influenced the ecological integrity of river and stream systems. Recently, more attention has been paid to natural-flow regimes and to the distributions of habitats and water quality over both space and time (Poole et al. 2004).

This realization launched a productive period of collaborative research and management. As the ecological concepts of natural flows matured (e.g. Sparks et al. 1998), methods and tools to support management decisions were also being developed. These tools relied on historical data to define ranges of variation in freshwater flows to assess the degree of alteration and associated impairment (e.g. Richter et al. 1997a,b). Perhaps the most clear and dramatic use of HRV concepts was the illustration that the impact of dam construction and water management on altering freshwater flows could be understood by a consideration of historical variation – referred to as the "range of variation approach" (Fig. 2.3).

These tools resulted in a transition from describing the historic or natural flows of a river to prescribing a flood regime for managed rivers that would improve freshwater biodiversity health and ecological integrity (e.g. Richter & Richter 2000). These concepts were later refined to incorporate the principles of adaptive management into the process (Richter et al. 2003; 2006).

The importance of establishing social contracts – social range of variability (SRV)

As previously noted, public-land management is laden with litigation, frequently resulting in gridlock (Reynolds 2002). There is growing evidence, however, that federal agencies are capable of gaining support for disturbance-based management (Shindler & Mallon 2009). As scientific knowledge of landscapes and ecosystems develops and land managers incorporate this information in approaches to managing complex and dynamic systems, a lack of trust from the public remains a significant barrier to implementing proposed management pathways. This trust is critical in allowing land management to occur. The best-designed framework, based on the best available science, will fail unless the public understands and accepts the concepts, rationale, and approaches to management being proposed (Shindler & Mallon 2009).

An emerging companion concept to HRV is SRV: "the range of an ecological condition that society finds acceptable at any given time" (Duncan et al. 2010). Incorporating SRV into land-management planning and developing a natural resource management and decision framework composed of both biophysical and social perspectives will allow flexibility and depth in examining uncertainties, trade-offs, and costs.

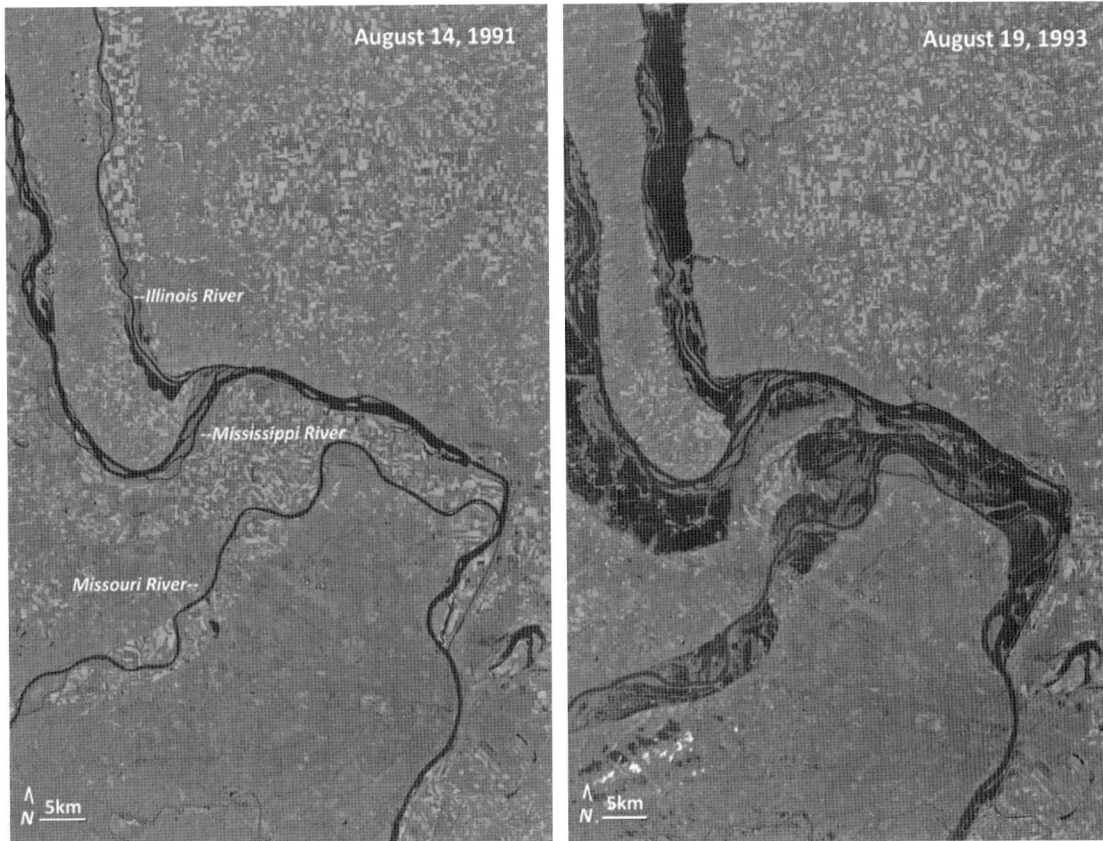

Fig. 2.2 The Mississippi River near St. Louis, MO in August 14, 1991 (left) and the flooding of the river on August 19, 1993 (right). Source: http://earthobservatory.nasa.gov.

Management options can be approached through a conceptual model overlaying ecological variability with measures of social acceptability. Social implications and cautions in employing historical ecology concepts in land management include the recognition that HRV is a tool, not a target; spatial scale and ownership objectives constrain the use of HRV; and centralized, single-focus management is a poor fit for dynamic systems (Thompson et al. 2009). SRV can be a critical component of successful collaborative conservation and land management.

Building stronger collaborations between private and public sectors, academia, and management will allow land managers to use HRV as part of best available science, reducing legal barriers in decision making. Although government agencies are generally more constrained and shaped by litigation than NGOs, together these parallel tracks, derived from the initial thinking of Pinchot, Muir, and others, have built a compelling body of knowledge and literature surrounding the use of HRV in conservation. As Tear et al. (2005) highlighted, the use of best available science to set objectives in conservation is a critical component for success at state and national levels. Critical reviews of HRV models (e.g. Keane et al. 2002) further support this idea. As we move to a period in which management decisions of potentially greater magnitude and consequence for society will need to be made, we must move away from a history in which government agencies and NGOs advanced similar concepts but with minimal overlap, to a place where broader collaboration is the norm.

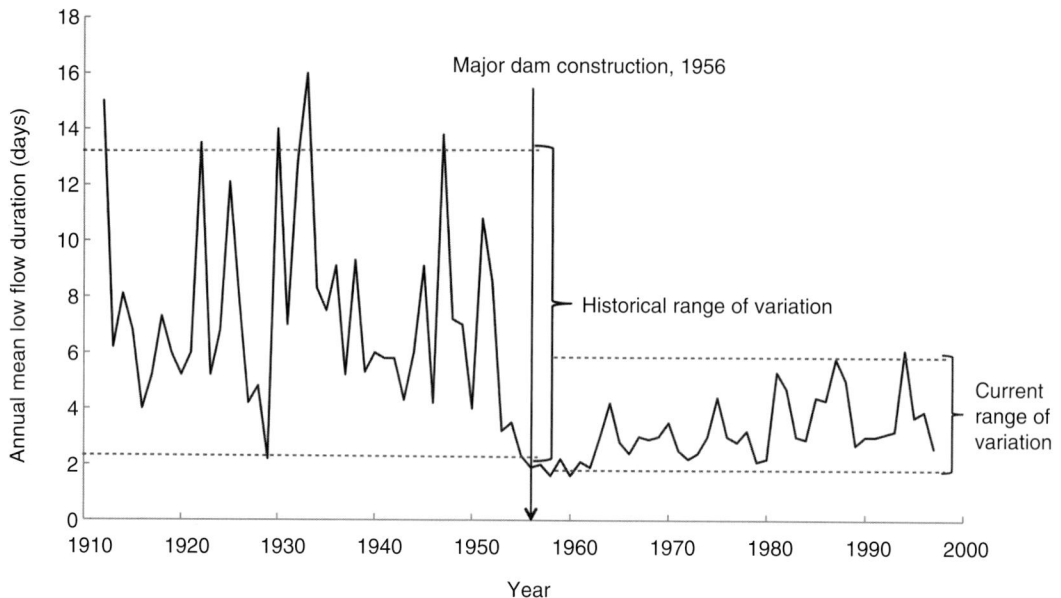

Fig. 2.3 Based on river-flow measurements from the Roanoke River in North Carolina, Richter et al. (1997b) used their "range of variability approach" to quantify the degree of change in freshwater flows before and after dam construction in 1956 (adapted from Richter et al. 2003).

There is a growing sense of urgency surrounding ecosystem conservation and management in the face of climate change. Science and management must respond to the increasing availability of information regarding climate change and keep pace with appropriate and timely decision making. Being able to rely on HRV as an accepted approach that stands up in a court of law as an important part of best available science is a critical step.

2.5 CHALLENGES AND THE FUTURE OF LAND MANAGEMENT

Many of the challenges facing management in the future are addressed in more detail in later chapters of this book. In the face of climate change, it is important to learn from past experiences and iteratively incorporate the lessons learned into future management actions. Such adaptive management is "the necessary lens through which natural resource management must be conducted" (Millar et al. 2007). It is critical that we not simply strive to reach potentially unachievable conditions, such as those that occurred histori-

cally, but that we learn from the past with our eyes open to what the future might bring. Acknowledging that global climatic patterns are shifting and affecting ecosystem processes in different ways demands an adaptation in thinking about ecosystem function. Incorporating ranges of historic and future variation into land-management planning requires addressing levels of uncertainty associated with shifting climatic patterns and associated disturbances that may be far outside the historic range of variation previously experienced. Scientists and land managers alike must overcome the common misconception that HRV and desired future conditions should be one and the same.

Critics of the value of HRV in resource management also argue that conditions have changed significantly since European settlement in North America, that it is difficult to interpret information from the past, and that past patterns and processes are irrelevant in today's world because of changing climates (Landres et al. 1999). It is this perception that has sometimes resulted in the dismissal of HRV as a concept irrelevant to land managers in a world of landscapes changing in response to climate change. Nonetheless, the concepts of natural variability provide a framework for improving our understanding of ecological systems, of how they function, and how they respond to management actions. Although the environment is clearly changing at a rapid rate, our understanding of the composition, structure, patterns, and response to disturbances of historical ecosystems remains our best means for evaluating the effects of management actions on ecological systems today. Historical ranges of variation are intended to inform, but not describe, desired conditions.

REFERENCES

Aim, A.L. (1988). NEPA: past, present, and future. *EPA Journal*, **January/February**, 32. http://www.epa.gov/history/topics/nepa/01.htm.

Duncan, S.L., McComb, B.C., & Johnson, K.N. (2010). Integrating ecological and social ranges of variability in conservation of biodiversity: past, present, and future. *Ecology and Society*, **15**(1), 5.

Federal Register. (1982). National Forest System Land Management Planning. Vol. 47 FR 43026, September 30, 1982. 36 CFR Part 219.

Fedkiw, J. (1998). Managing multiple uses on national forests, 1905–1995: a 90-year learning experience and it isn't finished yet. USDA Forest Service. Washington, DC, USA. http://www.foresthistory.org/ASPNET/Publications/multiple_use/index.htm.

Franklin, J.F. & Forman, R.T.T. (1987). Creating landscape patterns by forest cutting: ecological consequences and principles. *Landscape Ecology*, **1**, 5–18.

Groves, C. (2003). *Drafting a Conservation Blueprint: A Practitioner's Guide to Planning for Biodiversity*. Island Press, Washington, DC, USA.

Hemstrom, M.A. (1991). New perspectives in forestry: an objectives-based approach. In *Proceedings of the New Perspectives Workshop*, Petersburg, AK (ed. M.J. Copenhagen). USDA Forest Service, Juneau, AK, USA.

Keane, R.E., Parsons, R.A., & Hessburg, P.F. (2002). Estimating historical range and variation of landscape patch dynamics: limitations of the simulation approach. *Ecological Modeling*, **151**, 29–49.

Keiter, R.B. (1994). Beyond the boundary line: constructing a law of ecosystem management. *University of Colorado Law Review*, **65**, 293–333.

Kessler, W.B., Salwasser, H., Cartwright, C.W., & Caplan, J.A. (1992). New perspectives for sustainable natural resources management. *Ecological Applications*, **2**, 221–225.

Landres, P.B., Morgan, P., & Swanson, F.J. (1999). Overview of the use of natural variability concepts in managing ecological systems. *Ecological Applications*, **9**, 1179–1188.

Loomis, J. (2002). *Integrated Public Lands Management: Principles and Applications to National Forests, Parks and Wildlife Refuges, and BLM Lands*. 2nd ed. Columbia University Press, New York, NY, USA.

Millar, C.I., Stephenson, N.L., & Stephens, S.L. (2007). Climate change and forests of the future: managing in the face of uncertainty. *Ecological Applications*, **17**, 2145–2151.

Mooney, H.A., Bonnicksen, T.M., Christensen, N.L., Lotan, J.E., & Reiners, W.A. (technical coordinators). 1981. Fire regimes and ecosystem properties: proceedings of the conference, Honolulu, HI. General Technical Report WO-26. USDA Forest Service, Washington, DC, USA.

Morgan, P., Aplet, G., Haufler, J., Humphries, H., Moore, M., & Wilson, W. (1994). Historical range of variability: a useful tool for evaluating ecosystem change. *Journal of Sustainable Forestry*, **2**(1), 87–111.

Morgan, P., Hardy, C.C., Swetnam, T.W., Rollins, M.G., & Long, D.G. (2001). Mapping fire regimes across time and space: understanding coarse and fine-scale fire patterns. *International Journal of Wildland Fire*, **10**, 329–342.

Poole, G., Dunham, J., Keenan, D., et al. (2004). The case for regime-based water quality standards. *BioScience*, **54**, 155–161.

Reeves, G.H., Benda, L.E., Burnett, K.M., Bisson, P.A., & Sedell, J.R. (1995). A disturbance-based ecosystem approach to maintaining and restoring freshwater habitat of evolutionarily significant units of anadromous salmonids in the Pacific Northwest. *American Fisheries Society Symposium*, **17**, 334–349.

Reynolds, K.M. (2002). Social acceptability of natural resource decision-making processes. In *Proceedings of the Wood Compatibility Workshop* (ed. D. Johnson and R. Haynes), pp. 245–252. General Technical Report PNW-GTR-563. USDA Forest Service, Portland, OR, USA.

Richter, B.D., Baumgartner, J.V., Powell, J., & Braun, D. (1997a). A method for assessing hydrologic alteration within ecosystems. *Conservation Biology*, **10**, 1163–1174.

Richter, B.D., Baumgartner, J.V., Wigington, R., & Braun, D. (1997b). How much water does a river need? *Freshwater Biology*, **37**, 231–249.

Richter, B.D., Mathews, R., Harrison, D.L., & Wigington, R. (2003). Ecologically sustainable water management: managing river flows for ecological integrity. *Ecological Applications*, **13**, 206–224.

Richter, B.D. & Richter, H.E. (2000). Prescribing flood regimes to sustain riparian ecosystems along meandering rivers. *Conservation Biology*, **14**, 1467–1478.

Richter, B.D., Warner, A.T., Meyer, J.L., & Lutz, K. (2006). A collaborative and adaptive process for developing environmental flow regulations. *River Research and Applications*, **22**, 297–318.

Ross, J. (2006). FLPMA turns 30: the Bureau of Land Management also celebrates its 60th birthday. *Rangelands*, **28**, 16–23.

Ruth, L. (1997). Conservation and Controversy: National Forest Management, 1960–1995. Chapter 7 in D.C. Erman (general editor) and the SNEP Team. Status of the Sierra

Nevada: the Sierra Nevada ecosystem project. US Geological Survey Digital Data Series DDS-43 Version 1.0.

Sayre, N.F. & Fernandez-Gimenez, M. (2003). The genesis of range science, with implications for current development policies. In *Proceedings of the 7th International Rangelands Congress, Durban, SA* (ed. N. Allsopp, A.R. Palmer, S.J. Milton, K.P. Kirkman, G.I.H. Kerley, C.R. Hurt, and C.J. Brown).

Shindler, B. & Mallon, A.L. (2009). Public acceptance of disturbance-based forest management: a study of the Blue River landscape strategy in the Central Cascades adaptive management area. Research Paper PNW-RP-581. USDA Forest Service, Portland, OR, USA.

Sparks, R.E. (1995). Need for ecosystem management of large rivers and their floodplains. *BioScience*, **45**, 168–182.

Sparks, R.E., Nelson, J.C., & Yin, Y. (1998). Naturalization of the flood regime in regulated rivers. *Bioscience*, **48**, 706–720.

Steelman, T.A. (2007). The Monongahela controversy and decision. In *Forests and Forestry in America: An Encyclopedia* (ed. F. Cubbage). Society of American Foresters, Bethesda, MD, USA. http://encyclopediaofforestry.org/index.php/The_Monongahela_Controversy_and_Decision#The_Monongahela_Controversy_and_Decision.

Swanson, F.J. & Franklin, J.F. (1992). New forestry principles from ecosystem analysis of Pacific Northwest Forests. *Ecological Applications*, **2**, 262–274.

Swanson, F.J., Jones, J.A., Wallin, D.O., & Cissel, J.H. (1993). Natural variability: implications for ecosystem management. In Ecosystem management: principles and applications, Vol. II (M.E. Jensen and P.S. Bourgeron, technical editors), pp. 80–94. General Technical Report PNW-GTR-318. USDA Forest Service, Wenatchee, WA, USA.

Tear, T.H., Kareiva, P., Angermeier, P., et al. (2005). How much is enough? The recurrent problem of setting measurable objectives in conservation. *BioScience*, **55**, 835–849.

Thompson, I., Mackey, B., McNulty, S., & Mosseler, A. (2009). Forest resilience, biodiversity, and climate change: a synthesis of the biodiversity/resilience/stability relationship in forest ecosystems. Technical Series No. 43. Secretariat of the Convention on Biological Diversity, Montreal, Canada.

USDA Forest Service. (2001). Record of decision: Sierra Nevada forest plan amendment environmental impact statement. USDA Forest Service, Vallejo, CA, USA.

USDA Forest Service. (2004). Appendix D, General assessment of historical range of variability, Chequamegon-Nicolet National Forest final environmental impact statement, 2004 land and resource management plan. R9-CN-FEIS. USDA Forest Service, Rhinelander, WI, USA.

USDA Forest Service. (2011). Weeks Act Centennial 2011. USDA Forest Service, Washington, DC, USA. (http://www.fs.fed.us/land/staff/weeks-act.html).

Section 2

Issues and Challenges

Hugh D. Safford

USDA Forest Service, Pacific Southwest Region, Vallejo, CA, USA
University of California, Davis, CA, USA

In 1999, a special issue of *Ecological Applications* was dedicated to historical ecology. In this issue, Landres et al. (1999) outlined the concept of HRV (what they called "natural variability") and its use in resource management. Landres et al. listed three broad barriers to the use of historical ecology in the management of ecological systems: (1) the relevance of history and HRV concepts to environments that are different today from what they once were; (2) insufficient amount or quality of historical information; and (3) difficulties in managing dynamism in ecological systems. Over time, the second barrier is becoming progressively less challenging, as we develop better ways of looking into, interpreting, and modeling the past. The third barrier includes a grab bag of issues, including the scales of observation and management,

social and political will, and intrinsic difficulties related to managing for disturbance, especially in ecosystems characterized by high disturbance intensities or frequencies (Morgan et al. 1994; Landres et al. 1999). These latter barriers can arise from complications inherent to human attempts to understand and manipulate ecosystems and, as such, are likely to be (semi-) permanent features of the resource-management landscape. The first barrier, on the other hand, is a growing problem. Accelerating rates of climate and landscape change pose a major threat to the interpretation and application of historical ecological information simply because they progressively decrease the similarity between contemporary and historical reference ecosystems. The growing challenge of the first barrier is a major focus of this book.

Historical Environmental Variation in Conservation and Natural Resource Management, First Edition. Edited by John A. Wiens, Gregory D. Hayward, Hugh D. Safford, and Catherine M. Giffen.

The chapters of this section contain some overlap, which is testament to a general understanding that is developing vis-à-vis the proper use of historical information in resource management and conservation in a rapidly changing world. Early HRV exposés (e.g. Morgan et al. 1994; Landres et al. 1999) described the limitations of historical ecology, but the rapidity of recent increases in human populations, greenhouse gas emissions, air temperatures, and landscape degradation has put these limitations in sharper focus. The chapters in this section deal with these challenges head-on, from a variety of viewpoints. Although their topic areas are disparate, the chapters arrive at the common conclusion that historical information, properly used, will be as important to resource management and conservation in the future as it is today.

Chapter 3 (Hayward et al.) provides a synthetic introduction to most of the major challenges facing users of historical ecology and HRV in the twenty-first century. These include conceptual challenges, challenges to defining HRV, and challenges to using HRV. Hayward et al. introduce all of the key themes of this and following sections of this book, including the problem of stationarity, changing climates, spatial and temporal scales, the effects of humans on natural systems, and the relative roles of modeling and empirical data (covered in more detail in the following modeling section). Chapter 3 underlines the importance of moving away from HRV as a management target and focusing instead on the "contextual" information it provides for understanding mechanisms, directions, and rates of change. Hayward et al. make the case that historical information is our only empirical source of insight into the temporal dynamics of ecosystems over periods of time longer than a few decades. As such, historical ecology and HRV are necessary foundation stones for resource-management and conservation strategies that take rapid temporal change into account.

Chapter 4 (Safford et al.) centers on the question of climate change, which has made Landres et al.'s "relevance" barrier increasingly germane. Some authors have decried the use of historical reference conditions as management targets, and some resource managers are no longer confident that historical ecology has much to offer them in a changing world. Safford et al. note that the current debate surrounding historical ecology parallels an age-old debate within human history and, like other chapters in this section, demonstrates that the principal issue is not the fundamental value of historical ecology, but rather the ways in which historical information is used. The assumption of stationarity or equilibrium is a major stumbling block in many applications of HRV, and Chapter 4 attempts to clarify where such an assumption is reasonable and where it is not (Chapter 23 provides additional perspective on the stationarity issue). Safford et al. discuss how a historical frame of reference remains important to resource management, even in view of – and, indeed, because of – rapid global change. The authors focus on three broad areas of intersection between science and management where historical ecology plays a foundational role: understanding the patterns and mechanisms of temporal dynamics in ecosystems, recognizing and accounting for the roles of humans in ecosystem dynamics, and appreciating the overriding importance of maintaining ecosystem functions and services even as focal species wax and wane with global change.

Levin (1992) argued that issues related to spatial, temporal, and organizational scales form the central problem area in ecology. Levin further argued that applied challenges in ecology, including the causes and outcomes of climate change, are unapproachable without lines of reasoning that integrate phenomena operating at different scales. In Chapter 5, Wiens et al. explore the meaning and importance of scale for the understanding and application of historical ecology. They note that there are different dimensions of scale "filtering" in ecology: relating to the environment, the ecosystem and its response to the environment, and the scales at which humans observe and manipulate both of the former. These different filters must be aligned to properly understand and use historical information in resource management and conservation. Chapter 5 discusses the possibility of ecological threshold responses to global change and includes recommendations for managers and conservationists, including the suggestion that global change will force historical ecologists to look much more deeply into the past than has been the custom; both of these topics are further explored in Chapter 7. Wiens et al. conclude that history may not be sufficient on its own to sustainably manage future ecosystems, but it is certainly not "dead."

The theme of human history and its linkage to ecological history and current and future resource management is taken up in Chapter 6 (Nowacki et al.). Chapter 6 focuses on North America, documenting the relative importance of human impacts on historical ecosystems across the continent and exploring the relevance of North American human history to the HRV concept. Compared to other inhabited continents, human history is at its shortest in the Americas, and many researchers and managers have assumed that human effects on past ecosystems were slight enough to be ignored (the so-called pristine myth; Denevan 1992). Nowacki et al. show that this assumption is ill-founded for much of the continent. Indeed, profound human impacts on some North American ecosystems can be traced back to the beginning of the Holocene (the extinction of the Pleistocene megafauna and the introduction of anthropogenic fire being two salient examples). Chapter 6 provides evidence that the effects of past humans on North American ecosystems were correlated with ecosystem productivity, with the greatest impacts occurring along the warmer coasts and the interior of the southern United States. Nowacki et al. finish with an assessment of the importance of Native American history to application of the HRV concept in a changing world. As they put it, "developing an appreciation of the past roles of humans in modifying 'natural' ecosystems in North America is also a first step toward better understanding and managing the roles humans will play in future ecosystem dynamics."

The final chapter in this section, Chapter 7 (Jackson), comes from the paleoecological perspective ("deep history"). Jackson outlines five major areas of active investigation in paleoecology (historical human influences; the impacts of past faunal extinctions; threshold dynamics and "novel ecosystems"; climate variability in the distant past, before typical HRV baselines; and the origins of ecosystems). He concludes that HRV as

typically applied has some key inadequacies in the face of rapid global change. As in the other chapters in this section, Jackson warns against the implicit assumption of stationarity that plagues some applications of HRV, and he joins other chapters in underlining the key role of scale in tempering our perceptions of ecosystems and their dynamics. Like Chapter 4, Jackson foresees a future where managing for the preservation of specific ecosystems becomes progressively less tenable, to be replaced by goals driven by the maintenance of ecological goods and services. Chapter 7 introduces the concept of "extended" HRV (eHRV), which recognizes the irony that traditional HRV, although developed to integrate ecological dynamism into what had been a very "static" resource-management ethos, has itself become "static" as the limited time periods considered in most HRV assessments do not incorporate sufficient ecological variability to inform management in a future that may be radically different from what we see today.

REFERENCES

Denevan, W.M. (1992). The pristine myth: the landscape of the Americas in 1492. *Annals of the Association of American Geographers*, **82**, 369–385.

Landres, P.B., Morgan, P., & Swanson, F.J. (1999). Overview of the use of natural variability concepts in managing ecological systems. *Ecological Applications*, **9**, 1179–1188.

Levin, S.A. (1992). The problem of pattern and scale in ecology. *Ecology*, **73**, 1943–1967.

Morgan, P., Aplet, G.H., Haufler, J.B., Humphries, H.C., Moore, M.M., & Wilson, W.D. (1994). Historical range of variability: a useful tool for evaluating ecosystem change. *Journal of Sustainable Forestry*, **2**, 87–111.

Chapter 3

CHALLENGES IN THE APPLICATION OF HISTORICAL RANGE OF VARIATION TO CONSERVATION AND LAND MANAGEMENT

Gregory D. Hayward,[1,2] *Thomas T. Veblen,*[3] *Lowell H. Suring,*[4] *and Bob Davis*[5]

[1]USDA Forest Service, Alaska Region, Anchorage, AK, USA
[2]USDA Forest Service, Rocky Mountain Region, Lakewood, CO, USA
[3]University of Colorado, Boulder, CO, USA
[4]Northern Ecologic LLC, Suring, WI, USA
[5]USDA Forest Service, Southwestern Region, Albuquerque, NM, USA

3.1   INTRODUCTION

Historical ecology has influenced land-management decisions for decades. An enlightened paper by Aldo Leopold in 1924 used key observations from historical ecology to develop insights regarding management of grazing and fire in Arizona (Leopold 1924). More recently, the US Forest Service (Forest Service Handbook 1909.12 _40_43) and National Park Service (Unnasch et al. 2009), who manage a significant portion of the public land in North America, formalized the use of historical ecology in planning. The scientific literature includes significant praise for historical range of variation (HRV) and its application to conservation and land management (e.g. Parsons et al. 1999; Wong & Iverson 2004; Jackson & Hobbs 2009). More importantly, historical ecology has influenced the day-to-day work of land managers practicing silviculture, prescribed burning, habitat management, and grazing management through improved understanding of disturbance processes. Explicit application of HRV in land management is still evolving.

HRV assessments have been used to establish targets[1] for desired conditions (e.g. Cissel et al. 1994) and as context to understand current conditions (Seastedt et al. 2008), but the use of HRV in conservation and land management has also been questioned (L.C. Brett unpublished report; Klenk et al. 2009). If agencies and organizations seek to employ HRV effectively in wildland management, resource managers must develop a better appreciation for HRV and its application. The use of HRV, and historical ecology more broadly, is hampered by several issues – some conceptual, some methodological. Recognition of these issues can help managers incorporate history into their work more effectively. In this chapter, we review the issues that act as barriers to using knowledge from historical ecology to enrich land management. Our goal is to identify and clarify issues and barriers, providing an overview of issues addressed more comprehensively later in this book. We begin by examining the "premises of HRV," discussing the history and background from which managers apply historical ecology to conservation and land management. The overview is followed by a section addressing issues associated with development of HRV assessments and their application to land management.

[1] Target in the context of this chapter refers to established goals, desired conditions, or objectives.

3.2   PREMISES OF HRV

Although the importance of using historical ecology to inform land management has long been recognized by individual managers (e.g. Leopold 1924), efforts to formally integrate historical ecology into land-management decisions date primarily from the early 1990s (Grumbine 1994; Christensen et al. 1996). The idea that ecosystems should be managed in a manner consistent with their historical structure and functioning was proposed as an alternative to management paradigms guided primarily by resource extraction. In that context, and for the purposes of this chapter, HRV of ecological conditions can be defined as outlined in Chapter 1 as *the variation of ecological characteristics and processes over scales of time and space that are appropriate for a given management application*. This definition emphasizes that HRV describes a body of knowledge about historical conditions without any explicit prescription for how that body of knowledge should be applied; the application of HRV to land-management decisions must be flexible and context-specific.

Despite uncertainties about how to use HRV assessments (Klenk et al. 2009), there is agreement among scientists and managers that historical ecology provides essential insights for decision making. This consensus is based on the premise, supported by ecological theory and practice, that comprehending the causes and consequences of temporal variability in ecosystems is crucial to successful resource management (Christensen et al. 1996; Landres et al. 1999). Ecological science has demonstrated the importance of natural disturbances in determining ecosystem conditions and trajectories: equilibrium conditions are rare or short-lived in most ecosystems (Glenn-Lewin et al. 1992; White & Jentsch 2001). The shift away from a previous emphasis on equilibrium toward acceptance of variation resonated in the 1990s with the practical experiences of many land managers. In 1995, two influential papers written by managers explicitly recognized the importance of disturbance ecology in resource management (Averill et al. 1994; Manley et al. 1995). As Averill et al. (1994) put it:

> Awareness and understanding of disturbance ecology . . . is essential in understanding the consequences of management choices. The more we attempt to maintain an ecosystem in a static condition, the less likely we are to achieve what we intended.

Ecosystem management explicitly recognizes the dynamic character of ecosystems and stresses embracing natural variability for decision making (Christensen et al. 1996; Landres et al. 1999). Ecosystem management places a priority on the sustainability of ecosystem services as a precondition for determining management goals. This implies an intergenerational responsibility to pass on ecosystems that provide valuable resources for humans (Lubchenco et al. 1991). Goals consistent with the sustainability imperative must be explicitly stated in terms of "desired future trajectories" for particular ecosystem components and processes. Application of HRV provides a necessary foundation for establishing these goals.

Careful application of historical ecology to resource management requires a systematic assessment of HRV relevant to the land-management unit under consideration. An HRV assessment is a synthetic document, written specifically for practitioners and interested publics, that provides insight into fluctuating ecological conditions in relation to variability in major drivers of ecological change, which include climate variability, natural disturbances, and human activities (Veblen & Donnegan 2005; Romme et al. 2009). Importantly, the application of historical ecology and HRV is not dependent on the assumption that past conditions provide appropriate management targets. Rather, HRV concepts provide a *context* for management goals. This conclusion stems from three premises that are well supported by theory and practice:

1 *Knowledge of past natural variation provides an essential reference for evaluating impacts of modern land-use practices such as grazing, fire suppression, and logging on current ecosystem conditions and processes* (Swanson et al. 1994; Landres et al. 1999). Evaluation of the ecological effects of land-use practices depends on perceiving how natural disturbances and climate variability have affected ecosystem dynamics in a particular landscape. For example, contrasting effects of land-use practices on contemporary patterns of fuel types in ponderosa pine (*Pinus ponderosa*) ecosystems demonstrate the importance of developing HRV assessments for particular geographic areas and of being cautious in extrapolating conclusions from studies executed in other regions (Sherriff & Veblen 2006; Klenner et al. 2008).
2 *Past natural disturbances have played key roles in structuring contemporary ecosystems, and will continue to do so in the future* (White & Jentsch 2001). Broadscale legacies from past disturbances account for much of the spatial heterogeneity of landscape structure, which

in turn strongly influences the spread and severity of disturbances (Turner et al. 1993). For example, retrospective studies have revealed the importance of natural disturbance in the nineteenth century that created landscape templates that influenced late twentieth-century bark beetle and wildfire activity (Veblen et al. 1994; Bigler et al. 2005).
3 *Hypotheses about the drivers and mechanisms of contemporary and future ecological change can be developed and tested with historical ecological data* (Swetnam et al. 1999; Keane et al. 2009). Historical ecological studies document how changes in land use, such as grazing or elimination of fires set by aboriginal populations, have affected fire regimes of particular ecosystem types in the past (Gruell 1985; Savage & Swetnam 1990). Likewise, retrospective studies have revealed differential wildfire responses to major ocean–atmosphere oscillations and their influence on climate across different forest ecosystem types from low-elevation dry ponderosa pine woodlands to cool, mesic subalpine forests (Swetnam & Betancourt 1990; Sherriff & Veblen 2008; Taylor et al. 2008). Such retrospective studies are essential for developing a mechanistic understanding of ecological changes, which in turn supports the development of simulation models of future landscape dynamics driven by climate variability and changes in land-use practices (Flannigan et al. 2009).

These three premises strongly support the application of HRV to conservation and land-management decisions even in the context of climate change and ecological effects of invasive species (Harris et al. 2006, Chapter 4, this book). The insights into historical ecology synthesized in an HRV assessment are intended to inform discussions of potential management goals that incorporate social values for decision making. Knowledge of historical ecology improves the value judgments involved in a deliberative decision-making process, but adoption of management goals should not be dictated by environmental history.

3.3 CHALLENGES TO DEVELOPING HRV ASSESSMENTS

HRV assessments represent the tool most often used by managers to bring insights from historical ecology to land management. Deciding what topics to cover in an HRV assessment and how to frame the complex knowledge of historical ecology are difficult challenges. In this section, we explore several conceptual and operational decisions that must be made in defining HRV for particular management settings.

Conceptual challenges to developing HRV assessments

Does the application of HRV concepts to conservation and land management imply assumptions of ecological equilibrium or stasis?

The short answer to this question is "no." The longer answer is that untenable equilibrium assumptions often promote inappropriate uses of HRV information to set static goals.

The notion that equilibrium or stasis was the norm in the natural landscape prior to major impacts from European settlement in North America is no longer tenable in the light of several decades of research on environmental history (White & Jentsch 2001; Jackson et al. 2009b). Despite consensus among ecologists and managers that equilibrium assumptions are inappropriate, in practice, there is a tendency to adopt the assessment conclusions as targets for restoration purposes, especially where snapshot reconstructions of landscapes are available (Hann et al. 2004). Some of this tendency is explained by a deeply rooted cultural belief that a "balance of nature" should be the norm – the human psyche may find permanency more intellectually tractable than unending change (Botkin 1990). Indeed, the "naturalness" that is assumed to be represented by pre-European settlement landscapes resonates with large sectors of the American public as a model for environmental management (Higgs 1997; Cole & Yung 2010). Thus, value-laden perceptions are often mixed with science-based criteria in the potential application of HRV to land management.

Although an assumption of static equilibrium is clearly inappropriate, more nuanced equilibrium concepts that incorporate disturbance are relevant to the application of HRV (DeAngelis & Waterhouse 1987; Turner et al. 1993). Recognition of the scale-dependent nature of landscape equilibrium, particularly the idea of a minimum dynamic area, can provide useful guidance for land-management decision making in disturbance-prone environments (Fig. 3.1). As illustrated by Turner et al. (1993) and Shugart (2005), an equilibrium in vegetation cover can be observed in disturbance-prone landscapes, but the spatial extent of observation necessary to observe an equilibrium will

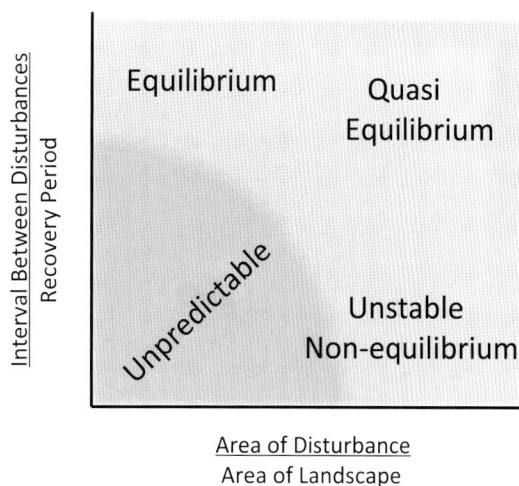

Fig. 3.1 Disturbance extent and frequency along with patch recovery rate all interact with the spatial extent of a landscape to determine whether a stable distribution of vegetation types (patches) will be observed over time. The ideas of "minimum dynamic area" (Turner et al. 1993) and "shifting mosaic steady state" of ecosystem dynamics (Bormann and Likens 1979) both result from this complex relationship. As illustrated, the extremes of the relationship form easily classified states of equilibrium and instability. Outside these extreme poles of the relationship, however, the specific characteristics of disturbance size, disturbance intensity, disturbance frequency, and system recovery along with the size of landscape determine whether relative stability versus continuous change is likely to be observed. As suggested by Turner et al. (1993; fig. 5), the state space near the origin is unpredictable with the relationship becoming more clear at the extremes.

depend on disturbance frequency and extent. Therefore, if managers seek to establish a system of natural areas sufficiently large to include a mosaic of all the structural units related to natural disturbances, knowledge of frequency and size variation in disturbances and recovery processes is essential (Pickett & Thompson 1978). Historical ecology provides insights about the size, frequency, spatial distribution, and ecological effects of infrequent large events.

Under what circumstances should HRV be defined to represent a management target?

The use of historical reference conditions to guide management has become a widespread practice, and there has been much research aimed at reconstructing static stand or landscape conditions that potentially could serve as management targets (Egan & Howell 2001, Hann et al. 2004). However, the inclusion of temporal *variability* of ecosystem elements in applications of HRV is a relatively recent and challenging practice (Swanson et al. 1994; Landres et al. 1999). As noted by Millar et al. (2007), many managers "use the range of historical ecosystem conditions as a management target, assuming that by restoring and maintaining historical conditions they are maximizing chances of maintaining ecosystems." This raises the general question: When will a range of reference conditions described in an HRV assessment provide appropriate management targets? Or, alternatively, when are HRV assessments more appropriately used as ecological context to guide development of desired conditions?

When the expressed goal is to manage landscapes in ways that sustain ecosystems and ecosystem services, HRV assessments are needed to comprehend temporal dynamics and establish an ecological context for desired conditions. This perspective – using HRV to provide context – might be thought of as providing a mechanistic understanding of how ecosystems respond to environmental changes. Examples of this approach can be found in the central Rocky Mountains where HRV assessments have been developed to support land-management planning and conservation by the US Forest Service (Regan 2005). This approach is based on the recognition that an appreciation for causes and consequences of spatiotemporal variability is essential in adopting management goals aimed at sustainability and promoting ecosystem resilience (Christensen et al. 1996). Confronting ecological variation encourages land managers to consider a range of possible management outcomes as opposed to pursuing goals associated with a single resource, as was common under previous paradigms (Thompson et al. 2008).

As demonstrated in the Rockies and elsewhere (see Chapters 12 and 14, this book), HRV does not need to be a target to be relevant to land-management planning but rather it can bring ecological knowledge to the development of desired conditions. In fact, under many circumstances, it may not be appropriate, for both biophysical and social reasons, to use reference conditions defined by HRV to establish specific resource-management targets. For instance, climate change, land development, and invasive species each influence ecosystems in ways that present ecological barriers to management actions designed to "restore" reference conditions. Furthermore, social values such as fire protection or resource extraction motivate desired conditions that may require environmental characteristics different from the past (see Chapter 1, this book). Attaining those social goals is constrained by ecological capability. Land-management decision making involves negotiating a complex mix of competing cultural, economic, social, and ecological considerations (Klenk et al. 2009). Consequently, if the focus of the desired condition emphasizes production of a specific commodity or sustaining a specific population, the HRV of other characteristics in the system may not be appropriate targets for management (Landres et al. 1999; Morgan 2004). In light of these challenges, the application of HRV to provide ecological context (rather than set targets) encourages managers to evaluate current conditions as the starting place for multiple possible futures and craft land-management goals that are ecologically achievable (Unnasch et al. 2009).

Alternatively, if the expressed goal is species conservation through a reserve design, short-term ecological restoration, or historical preservation, management to maintain an ecosystem within particular bounds defined by an HRV assessment may be the appropriate target. We can envision HRV defining a target to conserve microrefugia as a short-term strategy in the face of climate change. Similarly, when managing habitat for rare species, quality habitat may appear to result primarily from historical disturbance processes and managing for alternative states may be risky. Under these circumstances, the cost and effort to achieve management goals and restore reference conditions may be extremely high in light of climate change and other stressors. An assessment of the disparity between historical conditions and current conditions may provide insights into the cost and probability of success. Finally, HRV may define a legitimate target when establishing or maintaining historic vegetation. Management of battlefields and other cultural treasures (e.g. Park Service goals at Gettysburg National Military Park) represent noteworthy examples.

Attributes and processes to address in HRV assessments

There is a wide range of ecological attributes and processes for which historical information may provide essential background for land-management decision making (Egan & Howell 2001). For example, evidence of historical fire effects can be found in (1) tree-ring fire scars (Swetnam et al. 1999), (2) tree and stand ages delimiting boundaries of fires (Sibold et al. 2006), and (3) sedimentary charcoal deposits (Gavin et al. 2007). Historical vegetation composition and structure can be reconstructed from (1) pollen and other micro- and macrofossils in sediments (Whitlock et al. 2008), (2) tree-ring stand reconstructions (Fulé et al. 2002), (3) land survey records (Habeck 1994), (4) historical vegetation maps (Kulakowski et al. 2004), and (5) repeat photography (Gruell 1985; Veblen & Lorenz 1991). Less frequently, the demography or genetic structure of populations might be characterized. Reconstructions of past ecological conditions from all of these sources have significant limitations when applied to HRV for managing broad landscapes (Swetnam et al. 1999; Williams & Baker 2010). These data sources often refer to uncertain spatial domains (sedimentary records), are censored by destruction of evidence by recent disturbances (tree rings), are temporally limited and may be biased toward cultural sites (historical photographs), are unavailable for certain ecosystem types (tree rings), or are generally not available for most management units (historical vegetation maps). Hence, effectively crafted HRV assessments rely on the integration of diverse evidence collected for a range of ecological attributes to illustrate patterns of ecological change at the spatial and temporal scales necessary to inform conservation or land-management questions.

Stand structure and stand age-class distribution

The historical information perhaps most frequently requested by land managers relates to past stand structures (densities, basal areas, tree spatial patterns) and frequency distributions of stand age-classes (Harrod et al. 1999; Kaufmann et al. 2000). Although this information can provide useful context for guiding management decisions, several key limitations must be recognized. Stand- and age-class reconstructions derived from projecting current stand characteristics backwards in time require numerous assumptions about the representativeness of surviving old stands, decay and disappearance of evidence, and spatial extent of poorly documented disturbances. Reconstructions must be interpreted recognizing the time dependency of such data. For example, in landscapes characterized by a high variance in fire occurrence,

stand-age data reconstructed for any relatively short period (e.g. from a decade to a century) may reflect only a small range of conditions typical of landscape structure over a multicentury period (Sibold et al. 2006). Reasons for departures between modern and historical age-class distributions are often not well understood. For example, the relative roles of historical logging and grazing versus modern fire suppression in shaping current stand structures in western North America are often conflated and are the subjects of substantial debate (Sherriff & Veblen 2006; Baker et al. 2007).

Choosing attributes and processes to characterize HRV

Selection of attributes and processes to include in an HRV assessment must consider management goals and ecological relationships for the landscape under study. For example, if timber production is a major land use, it is appropriate to focus on attributes that affect timber management. Hence, it would be desirable to compare past variation to current variation for attributes such as tree density by size class, proportion of the landscape in different cover types, extent of tree diseases, forest-floor depth, and patch sizes (e.g. Dillon et al. 2005). Historical data on these attributes permit comparisons of the effects of timber harvesting with those of natural disturbances such as blowdown or fires, and provides a science foundation for assessing consequences of management.

Although managers may be interested in historical descriptions of as many ecosystem attributes as feasible, priority should be given to attributes and processes that most clearly describe ecological change – what might be called keystone attributes and processes. The use of pattern to infer process has long been a fundamental research strategy in ecology (Watt 1947), which directly applies to HRV assessments. To be useful, reconstructed patterns should be mechanistically related to processes through observations and experiments conducted in real time (Davies et al. 2009). Ultimately, understanding ecological processes will be far more useful to managers than simply reconstructing past patterns with uncertain causes.

HRV assessments are most useful when they distinguish the effects of past management from the effects of climate variation on current ecological conditions. To achieve this goal requires that HRV assessments explicitly examine climate trends within the study unit (e.g. Veblen & Donnegan 2005). A lack of climate data often impedes characterization of local climate. Consequently, it is useful to interpret the local (usually short) climate records in the context of longer records from regional syntheses (Cayan 1996).

Both climate and land use mediate ecological change through influences on disturbance regimes. Evaluating the ecological effects of disturbances, however, is challenging. For example, dates of past fires are readily determined by tree-ring fire-scar dating. However, reconstruction of the effects of past fires on tree population dynamics (mortality, survival, and new establishment) is often limited to the most recent fires due to the destruction of evidence from older events by more recent fires or land management (Sherriff & Veblen 2006; Baker et al. 2007).

Quantification of indicators of past disturbance (e.g. tree-ring fire scars, sedimentary charcoal) is essential in an HRV assessment, but consideration of contingencies that determine ecological responses to disturbances is equally important, albeit methodologically more challenging. Multiple disturbances in rapid sequence can have more severe effects than the same disturbance events separated in time (Paine et al. 1998). Disturbances such as severe bark beetle outbreaks or blowdown that alter seed availability of fire-adapted tree species are likely to alter the ecosystem response to subsequent fire. Thus, studies of recent disturbance sequences and their interactions, conducted in real time, need to be integrated with retrospective studies (e.g. Bigler et al. 2005; Kulakowski & Veblen 2007).

The actual metrics used to describe ranges of variations in ecological conditions and processes will necessarily vary with the type of ecological component. Appreciation of the circumstances that produced past episodes of unusually rapid change is typically more useful than a knowledge of the overall average values. Although averages and ranges need to be described in an HRV assessment, careful descriptions of temporal sequences and extreme events are particularly helpful. Extreme events can have long-lasting impacts in the landscape that create the template for subsequent ecological changes. For example, large fires in the mid-1800s in the central Rocky Mountains created age-class distributions that strongly influenced the current bark beetle pandemic, which killed most large lodgepole pine (*Pinus contorta*) across more than a million hectares (Bigler et al. 2005). Knowledge of these extreme fires explains current landscape patterns more effectively than summary statistics (such as mean fire-return interval).

Unfortunately, many of the ecological attributes and processes that provide keen insight into system dynamics are the most difficult to quantify. Therefore, before initiating an HRV assessment, it is critical to define the purpose of the assessment and to evaluate whether the objectives can be met through characterization of the attributes most likely to be quantified.

Bounding the scale of an assessment

Spatial scale

The focus of HRV assessments is usually bounded by jurisdictional limits of a land-management unit, but data collection must include larger areas. Characterization of a broader spatial context is essential for evaluating influences from outside the primary landscape of interest. The spatial extent of the evaluation area needs to be defined by management goals and concerns and the spatial dynamics of dominant ecological processes. For example, if the focus is on wide-ranging wildlife, the evaluation must be conducted at a broader scale than the range of the most mobile species (Cissel et al. 1994). If the focus is on how management decisions may affect values in surrounding areas, such as vulnerability of surrounding exurban settlement to fire, then cross-boundary analyses are required. Hence, bounding the geographic extent of an HRV assessment requires consideration of management application and the environment being analyzed.

A hierarchical, multiscale approach is useful to establish the appropriate spatial extent for HRV assessments (Turner et al. 1993; Wu & Loucks 1995). Considerations of hierarchy theory may be helpful in making these decisions (King 1997). Regardless of the management question, awareness of the context set by higher level features and the rate-determining characteristics of the lower level features are critical for understanding the focal level. For example, most HRV assessments of terrestrial landscapes should consider both a stand scale (i.e. an area of relatively homogeneous vegetation and site factors) and broader scales.

Our previous discussion of equilibrium assumptions raises a critical issue for some HRV assessments. Evaluating the dynamics of systems requires explicit recognition of the scale-dependent nature of landscape equilibrium and the idea of a minimum dynamic area (Fig. 3.1; see Chapter 5, this book).

Temporal scale

There is no standard, optimal time period for the varied purposes of HRV assessments across the wide range of potential management applications (Landres et al. 1999). It is useful to adopt a sliding timescale for different processes and attributes. The choice of temporal scales must consider the extent and grain of the ecological elements of interest (see Chapter 5, this book). In order to provide critical context, the range of temporal scales should extend beyond the interval necessary to capture important dynamics in the elements of interest. Again, considerations of hierarchy theory are helpful. Although the focus of many HRV assessments may be on the past two centuries of modern human land use, understanding dynamics across longer periods may change one's perception of current dynamics. Longer-term climate reconstructions hint at disturbance dynamics over extended periods. For example, the widespread megadrought of the mid- to late-sixteenth century strongly suggests the occurrence of widespread fires in the Rocky Mountain region at that time (Gray et al. 2003). Similarly, although tree-ring reconstructions of mountain pine beetle outbreaks are not available for the sixteenth century, temperature and moisture conditions that would have been conducive to bark beetle outbreaks are known to have occurred throughout the southern Rockies during that time (Gray et al. 2003; Raffa et al. 2008).

Incorporating the consequences of Native American cultures

There is a long history of debate about the abundance of indigenous peoples and their ecological impacts in North America before Euro-American settlement (e.g. Denevan 1992; Vale 2002, see Chapter 6, this book). Much of the debate is motivated by the aesthetic and spiritual interpretation of "naturalness" and how indigenous peoples fit with those philosophical conceptualizations (Anderson 1991). This has contributed to emotionally charged controversies about the severity and extent of ecological impacts of indigenous cultures (Kay 1995) and overgeneralizations about their ecological impacts (Denevan 1992). The ecological impacts of Native Americans undoubtedly varied spatially and tem-

porally and require site-specific evaluations of the evidence of impacts (Vale 2002). No default generalization about impacts is appropriate, other than the expectation of variability in the location and timing of impacts.

Archeological and ethnographic evidence supports the view that, at least in some places, ecological conditions were significantly modified by indigenous peoples through biotic collection and dissemination, use of fire, agriculture, and hunting (Stewart 1942, 2002; Vale 2002). Evidence can be used to assess the impact of past cultures in particular areas following three criteria: the amount of cultural energy required to maintain the system in its present state, the extent to which the system would change if humans were removed from the scene, and the proportion of the fauna and flora composed of native versus nonnative species (Anderson 1991). Thus, if humans frequently ignited fires that had extensive and severe impacts in a region where lightning-ignited fires were scarce, the human impact would be judged to be high. Places where agriculture was practiced or where species were extirpated by biotic collection by hunter-gatherer communities would also rank high, whereas sites rarely visited by people and not burned by anthropogenic fires would rank low in impact. Recognizing these differential levels of human impacts acknowledges the role of humans in ecosystem dynamics. Ultimately, insight into historical human influences on the environment becomes part of an HRV assessment by expanding the understanding of variation in ecological composition, structure, and process over time.

Role of ecological modeling and ecological inference in developing HRV assessments

Many applications of historical ecology to land management focus on broadscale characteristics, particularly disturbance processes of systems. Making sense of the interactions of these disturbance agents and the resulting temporal and spatial dynamics is challenging. Empirically derived syntheses of HRV often suffer from temporal and spatial limitations of the available data. In many ecological systems, natural archives of environmental history, such as tree rings or sedimentary records, are not available to provide historical data. Spatially explicit assessment of HRV through simulation modeling and ecological inferences based on knowledge of current systems represent powerful solutions to this challenge (McGarigal & Romme 2005).

The choice between an assessment built on direct evaluation of empirical data versus use of simulation modeling to support land-management planning rests on the depth and rigor of existing data and the ecological questions of interest (Keane et al. 2009). Simulation models parameterized from existing empirical data offer a way of learning that indirectly addresses some critical limitations of empirical assessments (White & Walker 1997; Keane et al. 2007). By transcending the physical constraints of field studies, simulation modeling expands the temporal and spatial extent of analysis. Using simulation modeling to quantify HRV can assist in identifying thresholds and can be used to explore possible future scenarios (Keane et al. 2007; Kennedy & Wimberly 2009).

Many spatially explicit simulation models are available for quantifying patch dynamics (e.g. Mladenoff & Baker 1999; Keane et al. 2004). However, spatial models require significant parameterization, and estimates for many important parameters may not be available. An alternative approach employs nonspatial modeling (e.g. Hemstrom et al. 2007), although this approach risks mischaracterizing disturbance processes that result from spatially explicit mechanisms. The choice of modeling approaches is not trivial and should be accompanied by a clear examination of the consequences of using a model that accounts for the interactions among patches.

Although simulation modeling can help expand spatial and temporal inferences to meaningful ecological extents, when information to parameterize models is not available, ecological inference may represent the most effective approach to examine ecosystem dynamics (Dillon et al. 2005; Regan 2005). Several approaches may be used to infer historical conditions from ecological data. Most often, judgments about the conditions that existed in the past are based on knowledge of plant adaptations and short-term observations of disturbance dynamics. Observations of disturbance dynamics in conjunction with information on past climates, current vegetation, and land-use history can provide clues about environmental conditions that existed in the past through careful use of inference (Yuan 2007).

Using historical ecology in land management in the context of climate change

Although land managers applying HRV struggle with issues such as identification of the appropriate refer-ence period or matching the spatial extent of analysis with the process of interest, the use of HRV in land management enjoys general support in both applied and scientific communities (Wong & Iverson 2004; Regan 2005). Recently, however, concerns have arisen about the application of historical ecology to land-management planning in the context of climate change, leading some to question whether the insights provided by HRV assessments are relevant (Millar et al. 2007; Klenk et al. 2009). There are two aspects for these concerns. The first, described for water resource managers, was a provocative call for a new management paradigm by Milly et al. (2008) (see Chapters 4 and 23, this book). Many applications of historical ecology, including water management, were predicated on an assumption of climatic stationarity. Environmental variables were assumed to fluctuate about a stable mean within an unchanging envelope of variability: the trajectory of each variable, if measured at an appropriate temporal scale, was flat. In a world of rapidly changing climate, however, assumptions based on stationarity fail. Some would suggest the application of HRV is severely compromised by this failure.

The second case for abandoning HRV assessments in the face of climate change becomes apparent when future climate models and species distribution or "bioclimatic envelop" models link to predict community composition in the future (Rehfeldt et al. 2006). Placed in the context of current and paleo-distributions of species, such analyses suggest that many landscapes will support plant associations that do not have historical analogs (Williams & Jackson 2007). If future ecological systems will likely represent novel ecological conditions, what use are HRV assessments?

Does awareness that "stationarity is dead" or that many future ecological systems will not have historical analogs render HRV assessments impotent? As demonstrated throughout this book, the answer depends on how land management applies historical ecology and the resulting HRV assessments. As applied by some practitioners in certain situations, stationarity *is* a critical assumption (e.g. Hann et al. 2004). When HRV assessments represent a target for future ecological conditions, stationarity is assumed and a careful evaluation of trends in local climate necessarily becomes part of a thorough evaluation of likely outcomes. Depending on the system and time horizon for management, the application of HRV under this circumstance may be misguided. Nonetheless, managers can use HRV assessments without defining restoration targets. An understanding of nonstationarity and its consequences can lead to application of HRV to provide context for setting conservation or land-management goals by describing the dominant disturbance processes and the temporal dynamics of systems. The first issue, the problem of stationarity, may therefore be somewhat of a red herring. It is important to be aware of the assumption of stationarity, but applications of HRV to land-management planning need not assume stationarity.

The second argument for abandoning HRV – that climate change will result in novel systems and backward-looking HRV assessments will provide few useful insights – similarly suffers from a narrow perspective about the application of historical ecology to land management. The insight that climate change may result in novel ecological systems stems directly from historical ecology (Williams & Jackson 2007). As described in Chapters 4 and 7, HRV assessments can be an appropriate tool for evaluating the consequences of moving toward novel systems and to predict the potential rates and direction of those changes. Just as community ecology provides insight into predator–prey relationships, and population ecology identifies drivers of population growth, historical ecology defines the temporal dynamics of systems.

During the past several decades, land managers embraced the application of landscape ecology. Discussions of spatial scale are commonplace in land-management offices. As paleoecologists and other historical ecologists provide a richer appreciation for temporal dynamics, similar discussions of temporal scale must become common. HRV assessments can bring the discussion of temporal dynamics to land managers in a form that is accessible and influences decisions. Human-induced climate change provides the motivation for looking backward at rates of change, geographic synchrony in those changes, and the ecological patterns and processes associated with climate drivers (Jackson et al. 2009a, p. 64). Casting aside HRV while developing resource plans in a world of changing climate is akin to developing plans for migratory corridors without insights from landscape ecology.

3.4 CONCLUSION

We suggest that using historical ecology in land management and conservation is fundamentally no different from the application of other forms of ecological knowledge. HRV assessments characterize long-term ecological processes and patterns – the dynamics that drive ecosystem change. When applied critically, HRV assessments help practitioners examine assumptions and provide insight into ecological mechanisms that inform the articulation of desired conditions. Although desired conditions should define ecological circumstances capable of sustaining ecological services and biological diversity, these may be difficult to conceive clearly in a world of rapid change. An understanding of both rates of change and mechanisms for change in the past provides the ecological basis for models necessary to craft realistic, concrete scenarios for planning. For instance, historical ecology demonstrates that widespread geographic synchrony in drought and other stressors result in synchronous disturbances at continental scales (see Chapter 23, this book). These disturbances result in rapid changes in vegetation communities and are often associated with both the expansion and contraction of species ranges. These insights have direct implication for the timing, intensity, and targets of management associated with invasive species. As another example, historical ecology demonstrates that many ecosystems in the arid Rocky Mountains are young and changing rapidly (Jackson et al. 2009a). Compared to more mature ecosystems, such as the temperate rain forests of the Pacific coast, these systems are likely to be on a trajectory of directional change, regardless of climate change. The outcomes of restoration and active management to produce commodities (or the ecological services derived from these systems) will change because of the trajectory of old and young systems. Insights from historical dynamics can aid in predicting those differences.

The goals and objectives of land management have changed over time and will continue to change as social and political environments change. Conservation and land-management planning now consider outcomes over longer time frames. The careful use of historical ecology will ensure that management is grounded in the ecological patterns and processes that occur at a variety of temporal scales, including those beyond the scope of direct human observation.

Rather than relying on experience gained through a single professional career, HRV assessments can provide managers with information across broad temporal scales in a form accessible to the public and relevant to management. For instance, scenario planning (Peterson et al. 2003), an approach widely used in business and economic settings, can be used in conjunction with historical ecology to produce novel insights to inform planning. As illustrated by Jackson et al. (2009a), scenario planning in conjunction with historical ecology can alert managers of "inevitable surprises." Prior to the 1988 fires in Yellowstone National Park, for example, such a large disturbance event seemed unlikely to most managers and the public. Understood through a perspective of historical ecology, however, disturbance dynamics of the lodgepole-pine system would suggest that, like a 100-year flood, such a disturbance was inevitable. Scenario planning, enlightened by a long-term perspective, could yield contingency plans for events such as the current bark beetle outbreak in the Rocky Mountains, that are rare, but inevitable.

In the end, appropriate use of HRV assessments in land management depends on careful, respectful communication among practitioners, stakeholders, and scientists. As illustrated in Chapter 12, this dialogue is often difficult, as it challenges beliefs and employs unfamiliar vocabulary. In this chapter we have outlined several considerations for building HRV assessments that relate directly to critical management issues and meet high standards of science. Successful application, as with most human endeavors, will depend most on open, patient communication and an interest in understanding the world from multiple perspectives.

ACKNOWLEDGMENTS

We gratefully acknowledge an array of people who helped motivate and improve this chapter, in particular the organizers and participants of the 2008 workshop that defined this book. Patricia Maloney, Fred Swanson, Peter Landres, John Wiens, an anonymous reviewer, and Hugh Safford provided critical input on the manuscript.

REFERENCES

Anderson, J. (1991). A conceptual framework for quantifying naturalness. *Conservation Biology*, **5**, 348.

Averill, R.D., Larson, L., Saveland, J., Wargo, P., Williams, J., & Bellinger, M. (1994). Disturbance processes and ecosystem management. USDA Forest Service, Lakewood, CO, USA. http://www.treesearch.fs.fed.us/pubs/5139.

Baker, W.L., Veblen, T.T., & Sherriff, R.L. (2007). Fire, fuels, and restoration of ponderosa pine-Douglas-fir forests in the Rocky Mountains, USA. *Journal of Biogeography*, **34**, 251–269.

Bigler, C., Kulakowski, D., & Veblen, T.T. (2005). Multiple disturbance interactions and drought influence fire severity in Rocky Mountain subalpine forests. *Ecology*, **86**, 3018–3029.

Bormann, F.H. & Likens, G.E. (1979). *Pattern and Process in a Forested Ecosystem*. Springer-Verlag, New York.

Botkin, D.B. (1990). *Discordant Harmonies: A New Ecology for the Twenty-First Century*. Oxford University Press, New York.

Cayan, D.R. (1996). Interannual climate variability and snowpack in the western United States. *Journal of Climate*, **9**, 928–948.

Christensen, N.L., Bartuska, A.M., Brown, J.H., et al. (1996). The report of the Ecological Society of America committee on the scientific basis for ecosystem management. *Ecological Applications*, **6**, 665–691.

Cissel, J.H., Swanson, F.J., McKee, W.A., & Burditt, A.L. (1994). Using the past to plan the future in the Pacific Northwest. *Journal of Forestry*, **92**, 30–31.

Cole, D.N. & Yung, L. (eds.). (2010). *Beyond Naturalness: Rethinking Park and Wilderness Stewardship in an Era of Rapid Change*. Island Press, Washington, DC, USA.

Davies, K.W., Svejcar, T.J., & Bates, J.D. (2009). Interaction of historical and nonhistorical disturbances maintains native plant communities. *Ecological Applications*, **19**, 1536–1545.

DeAngelis, D.L. & Waterhouse, J.C. (1987). Equilibrium and nonequilibrium concepts in ecological models. *Ecological Monographs*, **57**, 1–21.

Denevan, W.M. (1992). The pristine myth: the landscape of the Americas in 1492. *Annals of the Association of American Geographers*, **82**, 369–385.

Dillon, G.K., Knight, D.H., & Meyer, C.B. (2005). Historic range of variability for upland vegetation in the Medicine Bow National Forest, Wyoming. General Technical Report RMRS-GTR-139. USDA Forest Service, Fort Collins, CO, USA.

Egan, D. & Howell, E.A. (eds.). (2001). *The Historical Ecology Handbook*. Island Press, Washington, DC, USA.

Flannigan, M.D., Krawchuk, M.A., de Groot, W.J., Wotton, M., & Gowman, L.M. (2009). Implications of changing climate for global wildland fire. *International Journal of Wildland Fire*, **18**, 483–507.

Fulé, P.Z., Covington, W.W., Moore, M.M., Heinlein, T.A., & Waltz, A.E.M. (2002). Natural variability in forests of the Grand Canyon, USA. *Journal of Biogeography*, **29**, 31–47.

Gavin, D.G., Hallett, D.J., Hu, F.S., et al. (2007). Forest fire and climate change in western North America: insights from sediment charcoal records. *Frontiers in Ecology and the Environment*, **5**, 499–506.

Glenn-Lewin, D.C., Peet, R.K., & Veblen, T.T. (eds.). (1992). *Plant Succession: Theory and Prediction*. Chapman and Hall, London, UK.

Gray, S.T., Betancourt, J.L., Fastie, C.L., & Jackson, S.T. (2003). Patterns and sources of multidecadal oscillations in drought-sensitive tree-ring records from the central and southern Rocky Mountains. *Geophysical Research Letters*, **30**, 491–494.

Gruell, G.E. (1985). Indian fires in the interior West: a widespread influence. In *Proceedings, Symposium and Workshop on Wilderness fire, Missoula, MT* (J. E. Lotan, B. M. Kilgore, W. C. Fischer, and R. W. Mutch, technical coordinators), pp. 68–74. General Technical Report INT-182. USDA Forest Service, Ogden, UT, USA.

Grumbine, R.E. (1994). What is ecosystem management? *Conservation Biology*, **8**, 27–38.

Habeck, J.R. (1994). Using General Land Office records to assess forest succession in ponderosa pine-Douglas-fir forests in western Montana. *Northwest Science*, **68**, 69–78.

Hann, W., Shlisky, A., Havlina, D., et al. (2004). Interagency fire regime condition class guidebook. Version 1.3.0. http://www.frcc.gov.

Harris, J.A., Hobbs, R.J., Higgs, E., & Aronson, J. (2006). Ecological restoration and global climate change. *Restoration Ecology*, **14**(2), 170–176.

Harrod, R.J., McRae, B.H., & Hartl, W.E. (1999). Historical stand reconstruction in ponderosa pine forests to guide silvicultural prescriptions. *Forest Ecology and Management*, **114**, 433–446.

Hemstrom, M.A., Merzenich, J., Reger, A., & Wales, B. (2007). Integrated analysis of landscape management scenarios using state and transition models in the upper Grande Ronde River subbasin, Oregon, USA. *Landscape and Urban Planning*, **80**, 198–211.

Higgs, E.S. (1997). What is good restoration ecology? *Conservation Biology*, **11**, 338–348.

Jackson, S.T. & Hobbs, R.J. (2009). Ecological restoration in the light of ecological history. *Science*, **325**, 567–569.

Jackson, S.T., Gray, S.T., & Shuman, B. (2009a). Paleoecology and resource management in a dynamic landscape: case studies from the Rocky Mountain headwaters. *Paleontological Society Papers*, **15**, 61–80.

Jackson, S.T., Betancourt, J.L., Booth, R.K., & Gray, S.T. (2009b). Ecology and the ratchet of events: climate variability, niche dimensions, and species distributions. *Proceedings of the National Academy of Sciences*, **106**(Suppl. 2), 19685–19692.

Kaufmann, M.R., Regan, C.M., & Brown, P.M. (2000). Heterogeneity in ponderosa pine/Douglas-fir forests: age and size structure in unlogged and logged landscapes of central Colorado. *Canadian Journal of Forest Research*, **30**, 698–711.

Kay, C.E. (1995). Aboriginal overkill and native burning: implications for modern ecosystem management. *Western Journal of Applied Forestry*, **10**, 121–126.

Keane, R.E., Parsons, R., & Rollins, M.G. (2004). Predicting fire regimes across multiple scales. In *Emulating Natural Disturbances: Concepts and Techniques* (ed. A. Perera and L. Buse), pp. 55–68. Columbia University Press, New York, USA.

Keane, R.E., Rollins, M., & Zhu, Z.L. (2007). Using simulated historical time series to prioritize fuel treatments on landscapes across the United States: the LANDFIRE prototype project. *Ecological Modelling*, **204**, 485–502.

Keane, R.E., Hessburg, P.F., Landres, P.B., & Swanson, F.J. (2009). The use of historical range and variability (HRV) in landscape management. *Forest Ecology and Management*, **258**, 1025–1037.

Kennedy, R.S.H. & Wimberly, M.C. (2009). Historical fire and vegetation dynamics in dry forests of the interior Pacific Northwest, USA, and relationships to Northern Spotted Owl (*Strix occidentalis caurina*) habitat conservation. *Forest Ecology and Management*, **258**, 554–566.

King, A.W. (1997). Hierarchy theory: a guide to system structure for wildlife biologists. In *Wildlife and Landscape Ecology: Effects of Pattern and Scale* (ed. J.A. Bissonette), pp. 185–212. Springer-Verlag, New York, USA.

Klenk, N.L., Bull, G.Q., & MacLellan, J.I. (2009). The "emulation of natural disturbance" (END) management approach in Canadian forestry: a critical evaluation. *Forestry Chronicle*, **85**, 440–445.

Klenner, W., Walton, R., Arsenault, A., & Kremsater, L. (2008). Dry forests in the southern interior of British Columbia: historic disturbances and implications for restoration and management. *Forest Ecology Management*, **10**, 1711–1722.

Kulakowski, D. & Veblen, T.T. (2007). Effect of prior disturbances on the extent and severity of a 2002 wildfire in Colorado subalpine forests. *Ecology*, **88**, 759–769.

Kulakowski, D., Veblen, T.T., & Drinkwater, S. (2004). The persistence of quaking aspen (*Populus tremuloides*) in the Grand Mesa area, Colorado. *Ecological Applications*, **14**, 1603–1614.

Landres, P., Morgan, P., & Swanson, F. (1999). Overview of the use of natural variability concepts in managing ecological systems. *Ecological Applications*, **9**, 1179–1188.

Leopold, A. (1924). Grass, brush, timber, and fire in southern Arizona. *Journal of Forestry*, **22**, 1–10.

Lubchenco, J., Olson, A.M., Brubaker, L.B., et al. (1991). The sustainable biosphere initiative: an ecological research agenda. *Ecology*, **72**, 371–412.

Manley, P.N., Brogan, G.E., Cook, C., et al. (1995). Sustaining ecosystems: a conceptual framework. Publication R5-EM-TP-001. USDA Forest Service, San Francisco, CA, USA.

McGarigal, K. & Romme, W.H. (2005). *Historic Range of Variability in Landscape Structure and Wildlife Habitat: Uncompahgre Plateau Landscape*. University of Massachusetts, Amherst, MA, USA.

Millar, C.I., Stephenson, N.L., & Stephens, S.L. (2007). Climate change and forests of the future: managing in the face of uncertainty. *Ecological Applications*, **17**, 2145–2151.

Milly, P.C.D., Betancourt, J., Falkenmark, M., Hirsch, R.M., Kundzewicz, Z.W., Lettenmaier, D.P., & Stouffer, R.J. (2008).

Stationarity is dead: whither water management? *Science*, **319**, 573–574.

Mladenoff, D.J. & Baker, W.L. (1999). *Spatial Modeling of Forest Landscape Change: Approaches and Applications*. Cambridge University Press, Cambridge, UK.

Morgan, P. (2004). Back to the future: the value of history in understanding and managing dynamic landscapes. In Views from the ridge: considerations for planning at the landscape scale (ed. H. Gucinski, C. Miner, and B. Bittner), pp. 78–84. General Technical Report PNW-GTR-596. USDA Forest Service, Portland, OR, USA.

Paine, R.T., Tegner, M.J., & Johnson, E.A. (1998). Compounded perturbations yield ecological surprises. *Ecosystems*, **1**, 535–545.

Parsons, D.J., Swetnam, T.W., & Christensen, N.L. (1999). Uses and limitations of historical variability concepts in managing ecosystems. *Ecological Applications*, **9**, 1177–1178.

Peterson, G.D., Cumming, G.S., & Carpenter, S.R. (2003). Scenario planning: a tool for conservation in an uncertain world. *Conservation Biology*, **17**, 358–366.

Pickett, S.T.A. & Thompson, J.N. (1978). Patch dynamics and the design of nature reserves. *Biological Conservation*, **13**, 27–37.

Raffa, K.F., Aukema, B.H., Bentz, B.J., Carroll, A.L., Hicke, J.A., Turner, M.G., & Romme, W.H. (2008). Cross-scale drivers of natural disturbances prone to anthropogenic amplification: the dynamics of bark beetle eruptions. *BioScience*, **6**, 501–517.

Regan, C. (technical coordinator). (2005). Protocol for developing terrestrial ecosystem current landscape condition assessments. USDA Forest Service, Golden, CO, USA.

Rehfeldt, G.E., Crookston, N.L., Warwell, M.V., & Evans, J.S. (2006). Empirical analysis of plant–climate relationships for the western United States. *International Journal Plan Science*, **167**, 1123–1150.

Romme, W.H., Floyd, L., & Hanna, D. (2009). *Historical Range of Variability and Current Landscape Condition Analysis: South Central Highlands Section, Southwestern Colorado & Northwestern New Mexico*. Colorado State University, Fort Collins, CO, USA.

Savage, M. & Swetnam, T.W. (1990). Early 19th century fire decline following sheep pasturing in a Navajo ponderosa pine forest. *Ecology*, **71**, 2374–2378.

Seastedt, T.R., Hobbs, R.J., & Suding, K.N. (2008). Management of novel ecosystems: are novel approaches required? *Frontiers in Ecology and Environment*, **6**, 547–553.

Sherriff, R.L. & Veblen, T.T. (2006). Ecological effects of changes in fire regimes in *Pinus ponderosa* ecosystems in the Colorado Front Range. *Journal of Vegetation Science*, **17**, 705–718.

Sherriff, R.L. & Veblen, T.T. (2008). Variability in fire-climate relationships in ponderosa pine forests in the Colorado Front Range. *International Journal of Wildland Fire*, **17**, 50–59.

Shugart, H.H. (2005). Equilibrium versus non-equilibrium landscapes. In *Issues and Perspectives in Landscape Ecology* (ed. J.A. Wiens and M.R. Moss), pp. 36–41. Cambridge University Press, Cambridge, UK.

Sibold, J.S., Veblen, T.T., & Gonzalez, M.E. (2006). Spatial and temporal variation in historic fire regimes in subalpine forests across the Colorado Front Range in Rocky Mountain National Park. *Journal of Biogeography*, **32**, 631–647.

Stewart, O.C. (1942). Culture element distributions: XVIII Ute-Southern Paiute. *University of California Anthropological Records*, **6**, 231–356.

Stewart, O.C. (2002). *Forgotten Fires: Native Americans and the Transient Wilderness*. University of Oklahoma Press, Norman, OK, USA.

Swanson, F.J., Jones, J.A., Wallin, D.O., & Cissel, J.H. (1994). Natural variability: implications for ecosystem management. Eastside forest health assessment, Vol. II: ecosystem management: principles and applications (M.E. Jensen and P.S. Bourgeron, technical editors), pp. 80–94. General Technical Report PNW-318. USDA Forest Service, Portland, OR, USA.

Swetnam, T.W. & Betancourt, J.L. (1990). Fire-southern oscillation relations in the Southwestern United States. *Science*, **249**, 1017–1020.

Swetnam, T.W., Allen, C.D., & Betancourt, J.L. (1999). Applied historical ecology: using the past to manage for the future. *Ecological Applications*, **9**, 1189–1206.

Taylor, A.H., Trouet, V., & Skinner, C.N. (2008). Climatic influences on fire regimes in montane forests of the southern Cascades, California, USA. *International Journal of Wildland Fire*, **17**, 60–71.

Thompson, J.R., Duncan, S.L., & Johnson, K.N. (2008). Is there potential for the historical range of variability to guide conservation given the social range of variability? *Ecology and Society*, **14**(1), 18. http://www.ecologyandsociety.org/vol14/iss1/art18/.

Turner, M.G., Romme, W.H., Gardner, R.H., O'Neil, R.V., & Kratz, T.K. (1993). A revised concept of landscape equilibrium: disturbance and stability on scaled landscapes. *Landscape Ecology*, **8**, 213–227.

Unnasch, R.S., Braun, D.P., Comer, P.J., & Eckert, G.E. (2009). The ecological integrity assessment framework: a framework for assessing the ecological integrity of biological and ecological resources of the National Park System. USDI National Park Service, Washington, DC, USA.

Vale, T.R. (2002). The pre-European landscape of the United States: pristine or humanized? In *Fire, Native Peoples, and the Natural Landscape* (ed. T.R. Vale), pp. 1–39. Island Press, Washington, DC, USA.

Veblen, T.T. & Donnegan, J.A. (2005). Historical range of variability for forest vegetation of the National Forest of the Colorado Front Range. USDA Forest Service, Fort Collins, CO, USA. http://www.fs.fed.us/r2/projects/scp/tea/HRVFrontRange.pdf.

Veblen, T.T. & Lorenz, D.C. (1991). *The Colorado Front Range: A Century of Ecological Change*. University of Utah Press, Salt Lake City, UT, USA.

Veblen, T.T., Hadley, K.S., Nel, E.M., Kitzberger, T., Reid, M.S., & Villalba, R. (1994). Disturbance regime and disturbance interactions in a Rocky Mountain subalpine forest. *Journal of Ecology*, **82**, 125–135.

Watt, A.S. (1947). Pattern and process in the plant community. *Journal of Ecology*, **35**, 1–22.

White, P.S. & Jentsch, A. (2001). The search for generality in studies of disturbance and ecosystem dynamics. *Progress in Botany*, **62**, 399–450.

White, P.S. & Walker, J.L. (1997). Approximating nature's variation: selecting and using reference information in Restoration Ecology. *Restoration Ecology*, **5**, 338–349.

Whitlock, C., Marlon, J., Briles, C., Brunelle, A., Long, C., & Bartlein, P. (2008). Long-term relations among fire, fuel, and climate in the northwestern US based on lake-sediment studies. *International Journal of Wildland Fire*, **17**, 72–83.

Williams, J.W. & Jackson, S.T. (2007). Novel climates, no-analog communities, and ecological surprises: past and future. *Frontiers in Ecology and the Environment*, **5**, 475–482.

Williams, M.A. & Baker, W.L. (2010). Bias and error in using survey records for ponderosa pine landscape restoration. *Journal of Biogeography*, **37**, 707–721.

Wong, C. & Iverson, K. (2004). Range of natural variability: applying the concept to forest management in central British Columbia. *BC Journal of Ecosystems and Management*, **4**, 1–14.

Wu, J. & Loucks, O. (1995). From balance of nature to hierarchical patch dynamics: a paradigm shift in ecology. *Quarterly Review of Biology*, **70**, 439–466.

Yuan, L.L. (2007). Using biological assemblage composition to infer the values of covarying environmental factors. *Freshwater Biology*, **52**, 1159–1175.

Chapter 4

HISTORICAL ECOLOGY, CLIMATE CHANGE, AND RESOURCE MANAGEMENT: CAN THE PAST STILL INFORM THE FUTURE?

Hugh D. Safford,[1,2] *Gregory D. Hayward,*[3,4]
Nicole E. Heller,[5] *and John A. Wiens*[6,7]

[1]USDA Forest Service, Pacific Southwest Region, Vallejo, CA, USA
[2]University of California, Davis, CA, USA
[3]USDA Forest Service, Alaska Region, Anchorage, AK, USA
[4]USDA Forest Service, Rocky Mountain Region, Lakewood, CO, USA
[5]Climate Central, Palo Alto, CA, USA
[6]PRBO Conservation Science, Petaluma, CA, USA
[7]University of Western Australia, Crawley, WA, Australia

Historical Environmental Variation in Conservation and Natural Resource Management, First Edition. Edited by John A. Wiens, Gregory D. Hayward, Hugh D. Safford, and Catherine M. Giffen.
© 2012 John Wiley & Sons, Ltd. Published 2012 by John Wiley & Sons, Ltd.

The future ain't what it used to be.
                                    Yogi Berra

4.1 INTRODUCTION

Global mean annual air temperatures at the earth's surface are predicted to rise by as much as 6.4°C in the next century, creating climatic conditions unprecedented in at least the last 2 million years (IPCC 2007a; Fig. 4.1). Levels of CO_2 in the atmosphere are at their highest in at least 650 000 years, and the long residence time of CO_2 in the atmosphere means that the earth is locked into global warming and its results for decades or centuries to come even if international efforts to drastically reduce greenhouse gas (GHG) emissions succeed (IPCC 2007a). At the same time as climates warm, rapid expansion of human populations and economies has dramatically reduced the extent

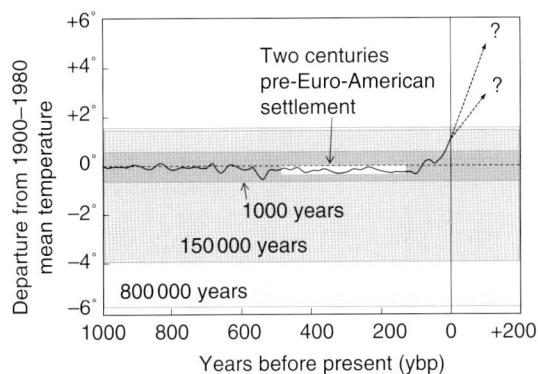

Fig. 4.1 Historical ranges of variation for global mean annual temperature, compared with the temperature record for the last 1000 years (Bradley 2000) and moderate and extreme warming projections over the next century as modeled by IPCC (2007a). Light gray area is reconstructed range of variation for last 800 000 years; medium gray area is reconstructed range of variation for last 150 000 years (both from Tausch et al. 1993). Dark gray area is reconstructed range of variation for the 1000 years before 1980; the dashed horizontal line is the mean temperature for the twentieth century before 1980 (both from Bradley 2000). The thin white area is the range of variation for the two centuries preceding Euro-American settlement in North America, the most typical reference period for HRV assessment in the United States. The "Little Ice Age" stretched from c. 600 to 100 ybp; the "Medieval Warm Period" stretched from c. 1100 to 650 ybp.

of the earth's natural and seminatural habitats. Land-use change has reduced the availability of suitable habitat for native plants and wildlife, and in many places, fragmentation of habitat has led to landscapes characterized by terrestrial archipelagos of seminatural ecosystems that are only weakly connected via dispersal and migration. These changes decrease the ability of biota to respond to increasing levels of ecological stress that are resulting from a combination of anthropogenic trends, including altered disturbance regimes, air and water pollution, atmospheric deposition, and biotic invasion (Noss 2001; Sanderson et al. 2002).

Worldwide, traditional practices of ecosystem management are largely dependent on the characterization of "reference states," which may constitute targets or desired conditions for management activities. Because human alterations to ecosystems have been so pervasive, fully functional contemporary reference ecosystems are rare, and reference states must often be defined based on historical conditions. There is much knowledge to be won by understanding the past histories of natural systems. For example, information gained through historical ecological studies (from pollen sedimentation, packrat middens, tree rings, photographs, historical survey maps, etc.) has been used to, among other things, document and analyze ecological trends, build and validate models of ecological processes, and provide targets for preservation, restoration, and resource extraction activities (Landres et al. 1999; Swetnam et al. 1999; Egan & Howell 2001). Understanding a location's history also provides a philosophical grounding, a "sense of place," that some argue is fundamental to the establishment of a sustainable relationship between people and the earth (Stegner 1992).

One of the time-honored fundaments of restoration ecology and resource management has been the implicit assumption that the historical range of variation (HRV [Box 4.1]; also referred to as the range of natural variation) represents a reasonable set of bounds within which contemporary ecosystems should be managed. The underlying premise is that the ecological conditions most likely to preserve native species or conserve natural resources are those that sustained them in the past, when ecosystems were presumably less affected by people (Manley et al. 1995; Egan & Howell 2001; Wiens et al. 2002). However, rapid and profound changes in climates and land use, as well as other anthropogenic stressors, may threaten the validity of some of the ways in which humans use historical information in resource management (Box 4.1). In the

Box 4.1 Definitions and background for restoration and climate change related terms used in the text.

Term	Definition (sources)	Examples of use in resource management	Implications of global change
Historical ecology	The study of the ecological past. Includes paleoecology, archival research (natural and documentary), long-term experiments, instrumental records, and so on. Historical ecology is often focused on describing past processes or patterns before the onset of some form of degradation, human or otherwise (Swetnam et al. 1999; Egan & Howell 2001).	Innumerable. See chapters in this book, and references therein. Historical ecology forms the basis for our understanding of ecosystem change through time, and biological and physical components of this change.	Dynamic and uncertain future requires better understanding of mechanisms driving ecological processes and patterns and their direct and indirect links to climate. Historical ecology is the principal source of this knowledge.
Historical range of variation (HRV)	The variation of ecological characteristics and processes over scales of space and time that are appropriate for a given management application. HRV analysis is a segment of the field of historical ecology focused on statistical description of the variability in composition, structure, and function of an ecosystem during some (ecologically relevant) time period in the past (Morgan et al. 1994; Manley et al. 1995; Landres et al. 1999).	There are hundreds. See chapters in this book and references therein. Use of HRV by land and resource managers has tended to focus on HRV-derived reference conditions under assumptions of stationarity (see below), but HRV analyses do not genetically depend on stationarity assumptions, nor do they need to generate static reference condition outputs. HRV is a tool, and limitations to its use are largely products of decisions made by its users (related to, for example, the spatiotemporal scale studied, the variables considered, the management questions or biases involved).	Uses of HRV data assuming stationarity are becoming less tenable. Global change requires assessment of HRV on broader scales of space and time, in order to better identify past conditions that resemble likely future conditions in terms of climate or other drivers. HRV assessments and their users must clearly identify the limitations of HRV outputs (as designed in a given case) under global change scenarios. Use of HRV to study mechanisms of ecological change will become more important; use of HRV to identify specific historic reference conditions will become less important.

Term	Definition (sources)	Examples of use in resource management	Implications of global change
Stationarity	The concept that ecosystems fluctuate within an unchanging envelope of variability. A stationary time series is one whose statistical properties (mean, variance, patterns of autocorrelation, etc.) are constant over time, that is, it is free of trends, periodicity, or thresholds (see below). True stationarity rarely, if ever, occurs in nature, although temporal segments of a time series may approximate stationarity (Koutsoyiannis 2006; Milly et al. 2008).	Stationarity assumptions often underlie the use of reference conditions in restoration and management (see below). For example, stationarity assumptions form the basis for hydrologic management (e.g. reservoir management, seasonal streamflow assumptions, 100-year floodplain mapping) and related things like flood insurance rates and property values.	Ecological time series which may have usefully approximated stationarity in the past may not do so in the future, due to changes in climate or other environmental drivers. This may occur because of, for example, a temporal trend in the mean (or variance, etc.), and/or a change in the bounds of statistical variance which characterizes the time series.
Historical reference conditions	Derived from the stationarity assumption, where desired future conditions for an ecosystem are simply those which characterized the target ecosystem in the past, before degradation. In US federal land management, historical reference conditions (HRCs) are usually derived from the centuries immediately before Euro-American arrival, or some other time period that represents quasi-stationary behavior in the variable(s) of interest (Morgan et al. 1994; Landres et al. 1999; Egan & Howell 2001).	HRCs are a fundament of traditional ecological restoration. HRCs may be based on high and low bounds in the variability of the ecological variable of interest, or, less advisably, on the mean. Assuming relative continuity of ecological conditions (climate, human impacts, etc.) between the reference period and the current period, HRCs are a powerful management tool, especially where they explicitly recognize the range of statistical variability in ecosystem patterns and processes.	Using HRCs as an endpoint target for management falls victim to the problems identified above for stationarity. If future environmental conditions are not similar to past environmental conditions, then returning an ecosystem to the reference condition may be impossible, unsustainable, or even detrimental. HRCs based on mean values will be less tenable under global change than those incorporating variability. In some cases, HRCs may serve as useful "waypoint" targets for restoration of ecosystems (e.g. "preparing" degraded ecosystems for the effects of global change).

Continued

Term	Definition (sources)	Examples of use in resource management	Implications of global change
Ecological integrity	Generally defined as the degree to which all ecosystem components and their interactions are represented and functioning. Determined based on "properly functioning" reference sites, which are often derived from historical data and HRV. Functional integrity is usually given more weight than compositional integrity, as individual (non-keystone) species can enter or leave most ecosystems without significantly affecting overall ecosystem function. Characterization and measurement of ecological integrity is dependent on subjective human judgments (Woodley et al. 1993; Karr 1996).	As applied in management, ecological integrity is very dependent on notions of HRV or range of natural variation. The US Forest Service Watershed Condition Classification program ranks river watersheds nationally with respect to their "geomorphic, hydrologic and biotic integrity" relative to their "potential natural condition." The US National Park Service Ecological Integrity Assessment Framework and The Nature Conservancy's Measures of Success framework are other examples.	"Integrity" is difficult to define without allusion to reference conditions; as such, it has same issues as HRV (above). Changes in environmental conditions may make attainment of "integrity" difficult or impossible, since the model for integrity is derived from past conditions.
Ecological resilience	The amount of disturbance an ecosystem can absorb without shifting to a different "stable state" (a domain in which a specific set of mutually reinforcing ecological processes and structures maintain the ecosystem within a definable range of variation). Exceeding the resilience of an ecosystem will theoretically lead over a "threshold" to a new stable state, where a different set of ecological processes and structures will be engaged (resulting in a new range of variation) (Peterson et al. 1998; Gunderson 2000).	Maintenance of the capacity of an ecosystem to "snap back" to a desired, or at least well-known, state. Many forest thinning projects in the western United States are designed to keep severe fires from leading to threshold conditions. Invasive species control work in both terrestrial and aquatic ecosystems is focused on resilience as well.	"Resilience" has become a major buzzword in the new managing-for-climate-change parlance. Management for climate change adaptation (see below) is focused to a great extent on maintaining or creating "resilience." Resilience management focused on ecosystem function and structure will be more successful than management focused on species composition.

Term	Definition (sources)	Examples of use in resource management	Implications of global change
Ecosystem services	The conditions and processes through which ecosystems (and their components) sustain and fulfill human life. Ecosystem services include provisioning services (food, water, fuel), regulating services (water purification, carbon sequestration, etc.), supporting services (seed dispersal, nutrient cycling, etc.), and cultural services (spiritual inspiration, recreation, etc.) (Daily et al. 1997, MEA 2005).	New York City purchase and eco-management of Catskills watershed area to preserve clean drinking water. Protection or planting of riparian forest buffers to filter agricultural runoff from urban water sources around the Chesapeake Bay. In Australia, national science agency (CSIRO) is leading nationwide Ecosystem Services Project to valuate and sustainably manage ecosystem services underpinning key agricultural industries and other benefits like recreation and tourism.	In management, some shift in focus from single species conservation to ecosystem processes and products may prove necessary. Certain ecosystem services or their levels of supply may change significantly under global change.
Climate change adaptation	Beneficial adjustments in natural or human systems in response to actual or expected climatic stimuli or their effects. Actions that reduce the vulnerability of an ecosystem to climate change and increase ecosystem resistance or resilience to climate-driven changes in ecological conditions (IPCC 2007b).	In reforestation efforts, use of seeds from climate zones better aligned with future predicted climates. In road management, increasing culvert size to allow passage of higher streamflows. Designation of habitat corridors to permit species migration in response to climate change.	Because climate change mitigation is unlikely to stop global warming soon, climate change adaptation is becoming a central focus area for resource management.

last decade, as the scale and pace of climate change have become more apparent, a rising chorus of authors has called into question the value and legitimacy of uncritical applications of historical reference conditions to contemporary and, especially, future land and resource management (e.g. White & Walker 1997; Harris et al. 2006; Millar et al. 2007; Craig 2010; Stephenson et al. 2010). At the same time, many of these authors explicitly recognize the fundamental, and even growing, value of historical ecological data to ecosystem management in the twenty-first century and beyond. Many land and resource managers are therefore confused: should we use historical data or not? In what ways? To what ends? In this chapter, we attempt (1) to clarify the nature of the problem, and (2) to provide clearer guidance to managers about promising uses of historical ecology and HRV analyses in land and resource management in the face of climate change.

4.2 CLIMATE CHANGE AND THE APPLICATION OF HISTORICAL ECOLOGY: WHAT IS THE PROBLEM?

For as long as the concept of HRV has been a central management tenet, there has been debate about its application. One focus of criticism is the relevance of applying HRV targets in environments that are different from what they were in the past. Landres et al. (1999) and others (e.g. White & Walker 1997; Swetnam et al. 1999; Harris et al. 2006; Millar et al. 2007) have subdivided this criticism into three basic problems:

1 Every point in space and time is unique, and climates are continually changing. Hence, descriptions of past patterns and processes provide insufficient or even erroneous reference for future management.

2 Management goals based on HRV often focus on recreating past environments and maintaining them in a static condition. This is a recipe for failure in a rapidly changing world.

3 Humans have changed natural ecosystems to such an extent that there are no truly pristine areas left on the planet. As a result, information derived from the past is difficult to interpret and apply.

These criticisms rest on the common notion that the past is an imperfect guide to the future, especially when we have good information about a future characterized by strong directional change (e.g. warming temperatures, drying trends, increased CO_2). These concerns relate to a more fundamental question in ecology about whether it is ever reasonable to assume equilibrium states for ecological communities (Connell & Sousa 1983; Levin 1992). Although the idea that ecosystems function under permanent stable equilibria was refuted decades ago, many widespread management applications are nonetheless based on assumptions of environmental stationarity – the idea that the long-term mean is more or less invariant and the range of past conditions encompasses current and future conditions as well (Box 4.1; "A" in Fig. 4.2). The reasoning is that, although true environmental stationarity

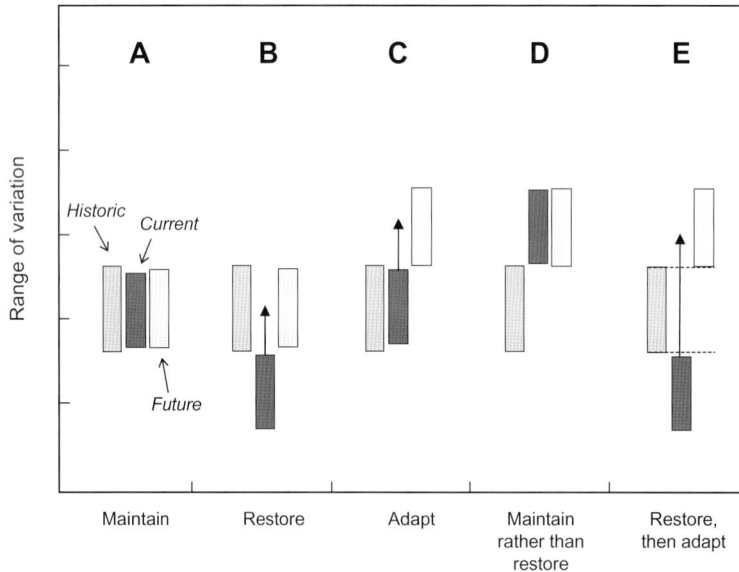

Fig. 4.2 Comparison of historic (medium gray), current (dark gray), and projected future (light gray) ranges of variation for a given ecological variable. The vertical length of the bars represents the range of variation. Possible management responses for each case are given along the x-axis. The assumption is made that the presumed future range of variation is the desired future management condition, but this may not always be the case. For example, under "C," rather than adapting, society could decide to (try to) maintain conditions within HRV; under "D," society could decide to return the system to HRV. In all cases, society could decide to leave the current condition as is. "E" is a common situation in many ecosystems, where human management has diminished some ecological component, pattern, or process, and the current departure from HRV is expected to increase in the future (forest fire frequency in dry-summer parts of the western United States is a good example). In our view, the use of reference condition-based targets as restoration "waypoints" is warranted in these conditions, but the long-term ("endpoint") focus should be on further adaptation to changing conditions. These are extreme cases and most real-world examples will include partial overlap among past, current, and future ranges of variation. Inspiration for this figure comes from Fig. 1 in Landres et al. (1999).

may not exist over the long term, the periodicity or rate of change may be slow enough compared to human experience to permit the useful assumption of stationarity. The best-known example of this assumption being put to use is probably "hydrologic stationarity", which posits that hydrologic variables like streamflow and maximum flood vary about a time-invariant mean and can be reasonably estimated from the instrumental record. In essence, hydrologic stationarity assumes that the future will be statistically indistinguishable from the past (Box 4.1; "A" in Fig. 4.2). Relatively stable hydrologic dynamics over much of the twentieth century facilitated the successful application of this statistical assumption to managing water systems. However, strong directional changes in surface and groundwater hydrology over the last two to three decades have led to a general realization that "stationarity is dead" and that hydrologic prediction must adapt to rapidly changing baselines (Milly et al. 2008; see "C" and "D" in Fig. 4.2). The challenge now is to redesign management strategies for a future that is both dynamic and uncertain.

The same concerns are raised with respect to stationarity in applications of historical ecology to terrestrial resource management (Harris et al. 2006; Millar et al. 2007; Stephenson et al. 2010). For example, in its simplest form, HRV is characterized as the range of some condition or process (e.g. fire-return interval, stand structure, patch size, diversity) that has occurred in the past. Although HRV does not inherently require an assumption of stationarity, many applications of HRV do (Box 4.1). In doing an HRV analysis, one must explicitly define the temporal extent of the historical period over which to characterize natural variability (Wiens et al., Chapter 5, this book). In the United States, the reference period for HRV assessment has often been restricted to the one or two centuries immediately prior to European settlement (Fig. 4.1). Accordingly, many proposed land-management actions have treated the period immediately prior to Euro-American settlement as an explicit target, and have sought to recreate ecological patterns – and sometimes, but much more rarely, processes – thought to be characteristic of this period.

Given the tendency to assume stationarity, a fundamental tension occurs as resource managers look to apply historical ecology in a world of rapid climate change. In many places, site conditions by the end of the twenty-first century may occupy climatic niches that do not exist today (so-called "no-analogue" conditions; "C" and "E" in Fig. 4.2). For example, Saxon et al. (2005) and Williams et al. (2007) project that by 2100, between 25% and 50% of the United States will support climatic domains with no contemporary analogue. Hobbs et al. (2006), Williams and Jackson (2007), and others argue convincingly that widespread novel ecosystems are inevitable outcomes of future climatic shifts. In the face of such dramatic ecosystem change, the pervasive idea that sustainable land and resource management is best undertaken by maintaining the environment within the range of pre-European conditions is difficult to defend. As Stephenson et al. (2010) put it:

> Our world has entered an era in which keystone environmental drivers – those that define the possible range of characteristics of a protected area – simply have no analog in the past, no matter how distantly we look.... Although the range of past ecosystem conditions remains a valuable source of information about the forces that shape ecosystems, it no longer automatically serves as a sensible target for restoration and maintenance of ecosystems.

4.3 LESSONS FROM HISTORY

At its foundation, the current dispute about the value and proper use of historical reference conditions is a reprise of the 2500-year-old debate about the value and proper use of human history. As far back as the fifth and fourth centuries BC, Greek historians had fundamentally different approaches to the study of history and different perceptions of history's value to society. For example, Herodotus's focus was on the preservation of great events of individual heroism, while Thucydides and Aristotle stressed the importance of the study of history to understanding the development of both contemporary and future events (Kelley 1991). Questions regarding the relevance of the past – whether history really ever repeats itself or whether each contemporary (and future) event is fundamentally unique – have featured in debates among historians (and practitioners of most professions, for that matter) for millennia. As in almost all such debates, the only reasonable answer is that strict adherence to either position is unjustifiable. Every event is unique, but every incident is also born from the roots of antecedent events (Sprugel 1991; Tausch et al. 1993). Questions related to topics as disparate as low voter turnout in the United States, rates of anorexia among teenagers, the Jewish–Muslim conflict in the Middle East, persistent accelerator failure in Toyota cars, the spread of an invasive aquatic plant, and the causes of spousal conflict are all fundamentally unapproachable without a thorough appreciation for the role of past events and trends. As P.N. Stearns (1998) put it, "How can we evaluate war if the nation is at peace, unless we use historical materials?" In an essay written for the American Historical Society in 1985, W.H. McNeill noted that,

> . . . the study of history does not lead to exact prediction of future events. Though it fosters practical wisdom, knowledge of the past does not permit anyone to know exactly what is going to happen. Looking at some selected segment from the past in order to find out what will occur "next time" can mislead the unwary, simply because the complex setting within which human beings act is never twice the same. Consequently, the lessons of history, though supremely valuable when wisely formulated, become grossly misleading when oversimplifiers try to transfer them mechanically from one age to another, or from one place to another. Anyone who claims to perform such a feat is sadly self-deceived. Practical wisdom requires us instead to expect differences as well as similarities, changes as well as continuities – always and everywhere.

The lesson for resource management is obvious: resource managers should no longer uncritically treat ecological patterns documented by historical studies as default management targets. In a rapidly changing world, management policies and practices that rely heavily on assumptions of ecological stationarity must be suspect. Management plans that seek to recreate past conditions and to artificially maintain a management area in that state over the long term will require increasing investments in energy, time, and money, and may well fail in the end. Historical ecology and HRV analyses will always be keystones of our understanding of environmental change, but under rapid global change, HRV can no longer be uncritically

treated as equivalent to "desired condition." There are situations and ecosystems where historical reference conditions continue to provide useful management guidance and even interim targets (Fig. 4.2), but in most places, the use of historical information to substantiate *stationary* land and resource-management practices is truly a thing of the past.

4.4   APPLYING HISTORICAL ECOLOGY IN A CHANGING WORLD: BACK TO THE FUTURE

Although specific historical reference conditions are becoming less useful as management targets, we agree with many recent authors (e.g. Harris et al. 2006; Jackson & Hobbs 2009; Keane et al. 2009; Stephenson et al. 2010; Cole et al. 2010a) that the fundamental value of historical ecological information to resource management is unchanged, and indeed its value increases as we sail farther into uncharted climatic and environmental waters. In this section, we highlight three areas and examples where historical data provide fundamental insights unavailable to managers "mired in the present." In this era of rapid global change, we believe that a historical frame of reference remains elemental for (1) comprehending the temporal dynamics of ecosystem processes and patterns (as opposed to specifying static targets for management goals); (2) recognizing and understanding human interactions with ecosystems (as opposed to defining the essential "natural" state of ecosystems); and (3) refocusing our management efforts on critical ecosystem processes (rather than conserving specific biological players).

Comprehending the temporal dynamics of ecosystems

As species distributions change in response to shifting climate, resulting ecosystems will ultimately consist of novel species combinations and possibly fundamentally altered ecosystem processes (Williams & Jackson 2007). Shifts in ecological processes are likely to render efforts to restore ecosystems to prior conditions difficult or impossible. In this era of rapid environmental change, historical ecology is less likely to provide a target for management conditions but more likely to provide understanding of ecological dynamics through time, and thus some level of predictability of ecological response to changing climates in the future (Swetnam et al. 1999; Keane et al. 2009; Stephenson et al. 2010).

Climate–wildfire relations in western North America provide a classic example of how historical ecology can supply a template for understanding temporal changes in ecological processes. Over the last few decades, wildland fires in western North America have become progressively more frequent, larger, and more destructive of human property, and federal and state governments now spend billions of dollars annually in wildfire control and prevention (Westerling et al. 2006; Miller et al. 2009). Many researchers are engaged in the search for mechanisms that drive fire occurrence and behavior in order to provide some predictability under changing climates. Historical ecological data show that climate strongly controlled fire regimes before Euro-American settlement of the West (Swetnam & Betancourt 1998; Taylor & Skinner 2003). In general, presettlement fire histories show that years of widespread fire tended to follow years of low precipitation, which would result in lower soil and fuel moistures and higher probabilities of successful ignition in the subsequent fire season. More recent data from the mid- to late-twentieth century have shown similar patterns, with different climate–fire correlations characterizing different geographic regions (Westerling et al. 2003; Littell et al. 2009). Now researchers have closed the temporal loop and linked the historical record with twentieth- and twenty-first-century data, allowing the development of region-specific predictive models for fire area based on three centuries or more of climate–fire relations (e.g. Westerling & Swetnam 2003). In this case, the historical data played a major role in developing and calibrating the fire–climate models. More importantly, the historical data extended the data record to the period preceding Euro-American settlement, thus allowing analysis of "natural" fire–climate relations before the introduction of widespread human alterations of the landscape (e.g. agriculture, grazing, and logging), fire regime (e.g. fire suppression and human ignition), and climate (e.g. warmer temperatures, early spring melting). This research reveals that fire area and frequency are still substantially controlled by interannual climate variability, even within the highly modified human landscape. At the same time, the long data record illustrates the regional- and ecosystem-based differences in fire response to climate, and helps managers and scientists to better understand where direct human influences on fire and forest fuels

are most likely to influence fire occurrence and behavior (Schoennagel et al. 2004; Littell et al. 2009).

In another example, Swetnam et al. (1999) studied the temporal dynamics of *Pinus edulis* and *Pinus remota* (piñon pine) populations in the southwestern United States. Swetnam et al. (1999) overlaid demographic trends in piñon pine populations on regional reconstructions of climate from tree rings. Their historical analysis suggested that tree mortality during extreme drought periods opened recruitment niches and released younger age-classes from resource competition; strong recruitment pulses tended to occur in the first wet period after a drought and mortality event. The profound drought of the 1950s, thought to be the worst in the last millennium, killed thousands of square kilometers of piñon, leading to high seedling survivorship and a massive pulse in piñon recruitment in the 1970s and 1980s. Since that time, establishment of piñon has been sustained by a long spate of warm wet springs associated with tropical Pacific warming (Swetnam & Betancourt 1998). By the middle of the twenty-first century, this surge in recruitment will greatly change contemporary forest structure. Swetnam et al. (1999) predict that without the comprehensive knowledge of demographics and historical climate in the region, the tendency would be to ascribe patterns of forest change to direct anthropogenic causes (such as grazing or fire suppression), leading possibly to misguided management responses.

Recognizing and understanding human interactions with ecosystems

Some applications of the HRV concept focus on historical periods in which the impacts of humans were presumed to be weak. Indeed, historical or natural variation is sometimes described as the condition of being apart from humans (Kaufmann et al. 1994; Manley et al. 1995; Romme et al., Chapter 1, this book; Box 1.1). This is ironic, as the use of the term "historic" in HRV was originally meant to differentiate it from "natural" variability, so as to permit the inclusion of past human influences on the environment (Morgan et al. 1994; Egan & Howell 2001). Any emphasis on the pristine as the target condition is problematic for the management of modern ecosystems, as few, if any, contemporary ecosystems are pristine or have been for centuries or millennia (Sprugel 1991; Tausch et al. 1993; Sanderson et al. 2002; Hobbs et al. 2006; Willis

& Birks 2006). Ecosystems can be placed along a gradient from intensively managed to passively managed to "natural," but their location on this gradient is spatially and temporally dynamic. Human activities have altered ecosystems for millennia, for example by simplification, degradation, disturbance, or species introduction. The pace and scope of anthropogenic change have increased dramatically over the last 50 years (Vitousek et al. 1997; Sanderson et al. 2002), and GHG-driven changes in the global climate clearly signal the death knell for the notion that any ecosystems on Earth exist outside the sphere of human influence. Humans and nonhuman species simply cannot be disentangled from each other, so most managed ecosystems are best understood as complex socio-ecological systems (Zavaleta & Chapin 2010). As Hobbs et al. (2010) write, "if nature evolved with human input, the absence of all human influence may not be desirable in some cases." For instance, current resource managers in the United States set fires to create habitat mosaics, similar to the methods of historic populations of Native Americans. However, recognition of this shared tradition to manage the land is relatively new, as the notion of a nature without human influence has guided management practices in the New World for much of the last century, reflecting the inaccurate belief that indigenous people had very little "meaningful" impact on the landscape (Nowacki et al., Chapter 6; Jackson, Chapter 7, this book). On the contrary, in Europe, Asia, and Africa, where human history stretches back tens to hundreds of thousands of years, management for ecological restoration has never really been about restoring a prehuman past (Bakker et al. 2000; Jackson & Hobbs 2009).

The management of modern ecosystems requires different references for sustainability than the historical conditions of a past that did not include pervasive human global change, including rapid GHG-driven changes in climate and atmospheric and oceanic chemistry. Harris et al. (2006) warns that "valuing the past when the past is not an accurate indicator for the future may fulfill a nostalgic need but may ultimately be counterproductive in achieving realistic and lasting restoration outcomes." Historical ecology can help to discriminate between "natural" and "cultural" causes of environmental change (Swetnam et al. 1999), and it can help identify locations and situations where human impacts have deeply and possibly irreversibly altered ecosystems. No-analogue species assemblages by themselves will not necessarily involve an upheaval

of ecosystem processes and biotic relationships, but where they do, systems may cross a threshold into a domain of qualitatively different dynamics (Box 4.1; Romme et al., Chapter 1; Wiens et al., Chapter 5, this book). Historical evidence can potentially help to determine whether a sustainable return to some previous state may be possible and well-advised (e.g. "B" in Fig. 4.2), possible but ill-advised ("D" in Fig. 4.2), or simply impossible.

As an example, monitoring data on nitrogen deposition over the last century and a half show that human activities have greatly increased nitrogen availability in terrestrial and aquatic ecosystems in many parts of the industrialized world (Holland et al. 2005). These inputs profoundly change the outcomes of many ecological processes and interactions and constrain our ability to restore and manage ecosystems. Aber et al. (1989) described the various ecosystem effects of "nitrogen saturation," ranging from reduced water quality, to disruptions of physiological function in forests, to increased acidity and cation leaching from soils, to altered competitive relationships. Nitrogen deposition is a major impediment to persistence and restoration of herbaceous vegetation in many places, as fast-growing weedy annual (and often exotic) species invade previously nutrient-poor sites and outcompete native species that had previously dominated (Choi et al. 2008). Atmospheric N inputs have also contributed to eutrophication of lakes and estuaries, reduced fishery productivity, abetted invasive species, and provoked other fundamental changes in the ecology of aquatic ecosystems (Rabalais 2002). In both terrestrial and aquatic ecosystems, apparently irrevocable changes in the physical and biotic environments can make the setting and attainment of historically based management targets an exercise in futility.

Other illustrations of the value of using historical data to recognize human impacts abound. For example, dendrochronology, fire scars, historical photographs, early land surveys, and other historical records were used to ascertain that mid-nineteenth-century grazing practices in the southwestern United States greatly reduced forest-surface fuel loads in *Pinus ponderosa* forests and contributed to a dramatic decrease in the frequency of fire, a change that had previously been ascribed purely to fire-exclusion policies (Savage 1991; Moore et al. 1999; Swetnam et al. 1999). The detection of this important grazing–fire interaction led to a restoration framework that includes recommendations for grazing practices that allow the recuperation of herbaceous fuels at the forest surface (Moore et al. 1999). In another example, Jackson et al. (2001) created a well-dated time series of ecosystem structure based on biological, biogeochemical, physical, and historical proxies to help clarify the underlying causes and rates of ecological change in a variety of coastal aquatic communities, ranging from kelp forests and coral reefs to offshore benthic communities. The historical records reveal long-term disturbance from the overfishing of large vertebrates and shellfish. The major changes provoked by overfishing occurred before the advent of ecological science and thus could only be documented through historical analysis. Jackson et al. (2001) concluded that the historical evidence suggested that large-scale restoration of ecologically important but currently rare ocean species was possible in some ecosystems, but not without a major reduction in other human perturbations.

Given the pervasiveness of human influences on ecosystems, it is wise to use contemporary ecosystems as reference systems whenever possible. We use historical data principally to understand ecological events and processes that we cannot observe firsthand, but directional changes in the baseline state (climate, air, water, soil, etc.) mean that historical conditions may make poor templates for the future. To compensate, contemporary reference ecosystems that are functioning as we desire should form part of the package of information that underlies restoration and resource management. Given the scale of habitat transformation in many regions, however, managers may need to exercise creativity in finding appropriate reference ecosystems.

Refocusing management efforts on critical ecosystem processes

Biologists are beginning to recognize that resource-management objectives are better defined as "motion pictures" rather than "snapshots" (Dunwiddie 1992) and that "saving all of the parts" (Leopold 1949) is not necessarily more important than understanding how the parts interact and interrelate (Falk et al. 2006). The HRV concept was developed to ensure that ecosystem functions, especially disturbance processes, were incorporated into management (Morgan et al. 1994; Landres et al. 1999). However, as currently practiced, conservation management often focuses on preservation of specific species, species assemblages, or some

relatively static notion of the habitat required to maintain populations. In light of rapid climate change, a different perspective is developing, one that is more focused on management of ecosystem structure and function rather than specific species or their habitat (Hunter et al. 1988; Harris et al. 2006; Hobbs et al. 2006; Stephenson et al. 2010). This perspective emphasizes the ecological function or ecological integrity of a site (Box 4.1), and is less concerned with the identities, numbers, or arrangements of biota.

One version of this perspective focuses exclusively on the physical habitat and the long-term maintenance of landscape features and physical processes as "ecological arenas." This approach concentrates on hydrological, pedological, geological, and pyrological processes, rather than on the temporary biological occupants (species) that are affected by these processes, because history shows that (1) "during periods when climate changes are large, communities are too ephemeral to be considered important biological entities in their own right" (Hunter et al. 1988), and (2) biotic responses to climate change usually occur on a more rapid timeline than those of the physical habitat. In the "ecological arena" approach, the focus of management is maintenance of an ecosystem's "dynamic capacity to respond adaptively" to exogenous influences (Costanza et al. 1993), that is, on its resilience to change (Box 4.1). An important feature of the ecosystem-process perspective is an enhanced emphasis on the provision and maintenance of basic ecosystem services, including those that directly benefit mankind. These include not only essential goods such as food, fuel, and water, but other less tangible benefits such as climate regulation, air and water purification, and soil creation (Box 4.1; Daily et al. 1997). Historical ecology forms a basis for this perspective (e.g. expected levels of ecosystem services are based on historical levels, and "properly functioning" ecosystem processes are only definable with an eye to the past), and a foundation of understanding that will sustain long-term management for ecosystem sustainability as climates and environments continue to change (Hunter et al. 1988; Millar & Woolfenden 1999).

Although the focus of resource management and restoration is per force shifting from the past to the future (Egan & Howell 2001; Harris et al. 2006; Choi et al. 2008), the value of historical information is not reduced; rather, the nature of its value is changed. In North America, management has often sought to recreate patterns characteristic of historic (predegradation) reference conditions, and historical information helped provide a blueprint for the desired outcome. In the future, the management emphasis in many ecosystems will shift from one of historical fidelity to one of ecological integrity, resilience, and delivery of services (Box 4.1; Millar et al. 2007; Stephenson et al. 2010; Cole et al. 2010b). In this changed management environment, the role of historical ecology becomes one of informing a management response to global change rather than resisting global change. Historical ecology can, among other things, identify important broadscale and long-term processes that influence local ecological outcomes under different climate conditions or disturbance regimes; provide clues to mechanisms underlying ecosystem dynamics and resilience (i.e. why have some systems persisted through climatic changes in the past?); guide the development and validation of predictive models; suggest appropriate future trajectories; and inform us if current conditions are anomalous and worthy of management intervention (Tausch et al. 1993; Landres et al. 1999; Millar & Woolfenden 1999; Swetnam et al. 1999; Cole et al. 2010a). In the end, historical ecology represents our clearest window into ecological patterns and processes that occur at temporal scales beyond the scope of human observation.

4.5 IS TRADITIONAL REFERENCE CONDITION-BASED MANAGEMENT DEAD?

Although the focus of conservation, restoration, and resource management is understandably shifting from the past to the future, from composition to function, and from equilibrium to dynamism, we believe that management targets based on past, predegradation conditions continue to be worthy goals in some places and contexts. In all cases, however, management targets must be thoughtfully articulated to be robust to uncertainty and compatible with trajectories of global change.

Any projection of future conditions is by definition uncertain. Although calibrated models of climate change agree on a range of plausible future mean temperatures, temperature extremes are impossible to predict with any certainty, and the extremes are often the primary drivers of ecological responses to climate (Easterling et al. 2000). Precipitation predictions vary widely, and for some places, models projecting increased precipitation are balanced by other models predicting decreased precipitation. As many authors have noted,

uncertainties in future projections of climate change and its impacts are almost certainly higher than uncertainties in the spatial and temporal accuracy of historical ecological data (Willis & Birks 2006; IPCC 2007a; Keane et al. 2009; Lawler et al. 2010). Thus, the argument has been made that because our understanding of historical ecosystems is usually much greater than that for most novel or emerging ecosystems, setting short-term targets based on known, historic ecosystems may minimize the risk of making things worse (Jackson & Hobbs 2009; Romme et al., Chapter 1, this book). Nevertheless, the key is to understand that projected trajectories for many places will lead beyond the HRV in climate – measured over thousands of years – in the next century (Fig. 4.1; "C" through "E" in Fig. 4.2). Reference condition-based management targets may be better seen as "waypoints" rather than "endpoints" ("E" in Fig. 4.2), where the short-term goal is simply pointing the ecosystem in the right direction via restoration work, but the long-term management goal is ecosystem adaptation to changing conditions.

One of the most frequent climate-adaptation recommendations in the literature (rank 3 of 113 in Heller & Zavaleta 2009) is to focus on the mitigation of ecosystem threats *other* than climate. As Noss (2001) notes, "climate change is not currently the greatest threat to (ecosystems) but adds another layer of stress to species and ecosystems already suffering from poor land-use practices." HRV analysis can help to identify salient departures from past variations in ecosystem composition, and structures and/or functions that are not sustainable under contemporary or projected future climates. Mitigating more proximate ecosystem stressors (disturbance frequencies and severities, fragmentation, invasive species, pollution, habitat degradation) identified in such an analysis may be the preferred course of "least regret" for many ecosystem managers beset by the uncertainty of future trajectories of climate, ecosystems, human populations, budgets, staffing, and politics (Noss 2001; Hannah et al. 2002; Lawler et al. 2010).

Under a changing global climate, a major issue is determining when and where historic ecosystems continue to provide reasonable management targets. A case in point might be relatively stable, "pseudo-equilibrial" ecosystems (e.g. ecosystems driven largely by local, biological interactions) where climate change and land-use change are not projected to greatly alter conditions. Such places do exist, although they may be rare and spatially discontinuous. Examples include "refugial" areas in tropical and temperate rainforest, which are hypothesized to have served as colonization sources for surrounding areas affected by cold or drought during Pleistocene climate changes (Prance 1982; Eeley et al. 1999), or wetland ecosystems in temperate mountains in locations that persisted through earlier dry and/or warm climate periods. High-resolution climate modeling in mountain watersheds suggests that features such as cold-air drainage may act to greatly reduce warming in deep valleys (Dobrowski et al. 2009), and such places may serve as biotic refugia in the future as well.

Another way to extend the value of traditional HRV applications may be to adjust the temporal or spatial theaters of their application to reflect changing conditions. Various authors argue that clinging to static goals based on historical conditions narrowly defined may actually undermine ecosystem resilience and adaptability (Harris et al. 2006; Aplet & Cole 2010). The new challenge is to find historical (or contemporary) information that can provide references for conditions that are relevant to the directional changes underway. For example, Millar and Woolfenden (1999) discuss how the period preceding major Euro-American settlement, a common reference period for restoration, coincided with the so-called Little Ice Age, a relatively cold period from 1400 to 1900 AD (Fig. 4.1). Using historical information on conditions during an anomalously cold period as the reference for restoration and management as we go into a warm period certainly does not make sense. Millar and Woolfenden (1999) suggest the Medieval Warm Period, 900–1350 AD, may provide more relevant information for appropriate management goals in the twenty-first century (see Fig. 4.1). Other authors suggest simply using much longer time spans (e.g. millennia instead of centuries) as reference periods for estimated historical natural variation than are used in current practice (Fig. 4.1; Jackson, Chapter 7, this book).

In some cases, we may also be able to extend the life of certain HRV assessments by redefining the spatial domain to which they apply. For example, Fulé (2008), working in semiarid fire-prone conifer forests in the southwestern United States, suggested that restoration treatments based on historical reference conditions for drier, lower elevation sites could be applied to wetter, higher elevation sites that are predicted to become drier and warmer in the future. This would help facilitate projected changes in vegetation composition and structure and reduce the probability of severe ecosystem disturbance. This "assisted migration" of the restoration target is an application of the "waypoint" rather than "endpoint" point of view ("E" in Fig. 4.2), where the object is to realign the target ecosystem with projected future changes by focusing on a historic reference condition for a different place or different time period that qualitatively coincides with the projected future condition of the target.

4.6 CONCLUSION

As the old adage states, "change is constant." In many places, however, land and resource management over the last century has ignored this adage and proceeded as if ecosystems were permanent and historic patterns and processes could simply be recreated or preserved ad infinitum. Although current climate trends are largely a human artifice, and although they portend many negative ecological, social, economic, and political consequences, they also serve to remind us of the dynamism of ecosystems. The assumption of stationarity in most ecological systems is unlikely to get us where we want to be. As climates and ecosystems change, we are likely to have more management success focusing on function rather than form, process rather than pattern, and ecological resilience and integrity rather than on historical fidelity. The use of historical ecology will not and should not diminish, but the use of past reference conditions as static targets, explicit or implicit, will become less and less justifiable as the rapidity of ecological change increases.

To conclude, one thing seems clear: even with a thorough understanding of historical cause and effect, we cannot predict the future with sufficient certainty to avoid mistakes in our management of land and resources. Most human learning is experiential (Kolb 1984) and progresses through trial and error. This militates for the adoption of more experimental management procedures, where different and credible hypotheses of change are translated into different management responses carried out on different parts of the landscape, and responses to these management experiments are thoroughly measured and evaluated. Humans are also tuned more to learn from unexpected than expected outcomes (Zaghloul et al. 2009). One of the lessons of history is that unexpected change and unpredictable outcomes have been driving ecological and evolutionary variability since the beginning of time. Species with the broadest capacity for adaptive response to unexpected change are most often the winners on the evolutionary roulette wheel. As climates and other conditions change, the sustainability of ecosystem patterns and processes will depend on a similar human capacity for adaptive response.

ACKNOWLEDGMENTS

We gratefully acknowledge the constructive criticism of Curt Flather, Greg Aplet, and two anonymous reviewers.

REFERENCES

Aber, J.D., Nadelhoffer, K.J., Steudler, P., & Melillo, J.M. (1989). Nitrogen saturation in northern forest ecosystems. *Bioscience*, **39**, 378–386.

Aplet, G.H. & Cole, D.N. (2010). The trouble with naturalness: rethinking park and wilderness goals. In *Beyond Naturalness: Rethinking Park and Wilderness Stewardship in an Era of Rapid Change* (ed. D.N. Cole and L. Yung), pp. 12–33. Island Press, Washington, DC, USA.

Bakker, J.P., Grootjans, A.P., Hermy, M., & Poschlod, P. (eds.). (2000). How to define targets for ecological restoration? *Journal of Applied Science*, **3**, 1–72.

Bradley, R.S. (2000). Past global changes and their significance for the future. *Quaternary Science Reviews*, **19**, 391–402.

Choi, Y.D., Temperton, V.M., Allen, E.B., et al. (2008). Ecological restoration for future sustainability in a changing environment. *Ecoscience*, **15**, 53–64.

Cole, D.N., Higgs, E.S., & White, P.S. (2010a). Historical fidelity: maintaining legacy and connection to heritage. In *Beyond Naturalness: Rethinking Park and Wilderness Stewardship in an Era of Rapid Change* (ed. D.N. Cole and L. Yung), pp. 125–141. Island Press, Washington, DC, USA.

Cole, D.N., Millar, C.I., & Stephenson, N.L. (2010b). Responding to climate change: a toolbox of management strategies. In *Beyond Naturalness: Rethinking Park and Wilderness Stewardship in an Era of Rapid Change* (ed. D.N. Cole and L. Yung), pp. 179–198. Island Press, Washington, DC, USA.

Connell, J.H. & Sousa, W. (1983). On the evidence needed to judge ecological stability or persistence. *American Naturalist*, **121**, 789–824.

Costanza, R., Wainger, L., Folke, C., & Maler, K. (1993). Modeling complex ecological economic systems. *BioScience*, **43**, 545–555.

Craig, R.K. (2010). Stationarity is dead–long live transformation: five principles for climate change adaptation law. *Harvard Environmental Law Review*, **34**, 9–75.

Daily, G.C., Alexander, S., Ehrlich, P.R., et al. (1997). Ecosystem services: benefits supplied to human societies by natural ecosystems. *Issues in Ecology*, **2**, 1–18.

Dobrowski, S.Z., Abatzoglou, J., Greenberg, J., & Schladow, G. (2009). How much influence does landscape-scale physiography have on air temperature in a mountain environment? *Agricultural and Forest Meteorology*, **149**, 1751–1758.

Dunwiddie, P.W. (1992). On setting goals: from snapshots to movies and beyond. *Restoration Management Notes*, **10**(2), 116–119.

Easterling, D.R., Meehl, G.A., Parmesan, C., Changnon, S.A., Karl, T.R., & Mearns, L.O. (2000). Climate extremes: observations, modeling and impacts. *Science*, **289**, 2068–2074.

Eeley, H.A.C., Lawes, M.J., & Piper, S.E. (1999). The influence of climate change on the distribution of indigenous forest in Kwazulu-Natal, South Africa. *Journal of Biogeography*, **26**, 595–617.

Egan, D. & Howell, E.A. (eds.). (2001). *The Historical Ecology Handbook: A Restorationist's Guide to Reference Ecosystems*. Island Press, Washington, DC, USA.

Falk, M.A., Palmer, D.A., & Zedler, J.B. (2006). Integrating restoration ecology and ecological theory: a synthesis. In *Foundations of Restoration Ecology* (ed. M.A. Falk, D.A. Palmer, and J.B. Zedler), pp. 341–345. Island Press, Washington, DC, USA.

Fulé, P.Z. (2008). Does it make sense to restore wildland fire in changing climate? *Restoration Ecology*, **16**, 526–531.

Gunderson, L.H. (2000). Ecological resilience – in theory and application. *Annual Review of Ecology and Systematics*, **31**, 425–439.

Hannah, L., Midgely, G.F., & Millar, D. (2002). Climate change-integrated conservation strategies. *Global Ecology and Biogeography*, **11**, 485–495.

Harris, J.A., Hobbs, R.J., Higgs, E., & Aronson, J. (2006). Ecological restoration and global climate change. *Restoration Ecology*, **14**, 170–176.

Heller, N.E. & Zavaleta, E.S. (2009). Biodiversity management in the face of climate change: a review of 22 years of recommendations. *Biological Conservation*, **142**, 14–32.

Hobbs, R.J., Arico, S., Aronson, J., et al. (2006). Novel ecosystems: theoretical and management aspects of the new ecological world order. *Global Ecology and Biogeography*, **15**, 1–7.

Hobbs, R.J., Zavaleta, E.S., Cole, D.N., & White, P.S. (2010). Evolving ecological understandings: the implications of ecosystem dynamics. In *Beyond Naturalness: Rethinking Park and Wilderness Stewardship in an Era of Rapid Change* (ed. D.N. Cole and L. Yung), pp. 34–49. Island Press, Washington, DC, USA.

Holland, E.A., Braswell, B.H., Sulzman, J., & Lamarque, J.-F. (2005). Nitrogen deposition onto the United States and Western Europe: synthesis of observations and models. *Ecological Applications*, **15**, 38–57.

Hunter, M.L. Jr., Jacobson, G.L. Jr., & Webb, T. III. (1988). Paleoecology and the coarse-filter approach to maintaining biological diversity. *Conservation Biology*, **2**, 375–385.

Intergovernmental Panel on Climate Change (IPCC). (2007a). *Climate Change 2007: The Physical Science Basis*. Cambridge University Press, Cambridge, UK.

Intergovernmental Panel on Climate Change (IPCC). (2007b). *Climate Change 2007: Impacts, Adaptation and Vulnerability*. Cambridge University Press, Cambridge, UK.

Jackson, J.B.C., Kirby, M.X., Berger, W.H., et al. (2001). Historical overfishing and the recent collapse of coastal ecosystems. *Science*, **293**, 629–638.

Jackson, S.T. & Hobbs, R.J. (2009). Ecological restoration in the light of ecological history. *Science*, **325**, 567–568.

Karr, J.R. (1996). Ecological integrity and ecological health are not the same. In *Engineering within Ecological Constraints* (ed. P.C. Schulze), pp. 97–109. National Academy of Engineering, Washington, DC, USA.

Kaufmann, M.R., Graham, R.T., Boyce, D.A., et al. (1994). An ecological basis for ecosystem management. General Technical Report RM-GTR-246. USDA Forest Service, Fort Collins, CO, USA.

Keane, R.E., Hessburg, P.F., Landres, P.B., & Swanson, F.J. (2009). The use of historical range and variability (HRV) in landscape management. *Forest Ecology and Management*, **258**, 1025–1037.

Kelley, D.R. (ed.). (1991). *Versions of History from Antiquity to the Enlightenment*. Yale University Press, New Haven, CT, USA.

Kolb, D.A. (1984). *Experiential Learning: Experience as the Source of Learning and Development*. Prentice Hall, Englewood Cliffs, NJ, USA.

Koutsoyiannis, D. (2006). Nonstationarity versus scaling in hydrology. *Journal of Hydrology*, **324**, 239–254.

Landres, P.B., Morgan, P., & Swanson, F.J. (1999). Overview of the use of natural variability concepts in managing ecological systems. *Ecological Applications*, **9**, 1179–1188.

Lawler, J.J., Tear, T.H., Pyke, C., et al. (2010). Resource management in a changing and uncertain climate. *Frontiers in Ecology and the Environment*, **8**, 35–43.

Leopold, A. (1949). *A Sand County Almanac – with Essays on Conservation from Round River*. Ballantine Books, New York, USA.

Levin, S.A. (1992). The problem of pattern and scale in ecology. *Ecology*, **73**, 1943–1967.

Littell, J.S., McKenzie, D., Peterson, D.L., & Westerling, A.L. (2009). Climate and wildfire area burned in western U.S. ecoprovinces, 1916–2003. *Ecological Applications*, **19**, 1003–1021.

Manley, P.N., Brogan, G.E., Cook, C., et al. (1995). Sustaining ecosystems: a conceptual framework. Publication R5-EM-TP-001. USDA Forest Service, Vallejo, CA, USA.

McNeil, W.H. (1985). Why study history? American Historical Society webpage. http://www.historians.org/pubs/archives/whmcneillwhystudyhistory.htm.

Millar, C.I. & Woolfenden, W.B. (1999). The role of climate change in interpreting historical variability. *Ecological Applications*, **9**, 1207–1216.

Millar, C.I., Stephenson, N.L., & Stephens, S.L. (2007). Climate change and forests of the future: managing in the face of uncertainty. *Ecological Applications*, **17**, 2145–2151.

Millennium Ecosystem Assessment (MEA). (2005). *Ecosystems and Human Well-Being: Synthesis*. Island Press, Washington, DC, USA.

Miller, J.D., Safford, H.D., Crimmins, M., & Thode, A.E. (2009). Quantitative evidence for increasing forest fire severity in the Sierra Nevada and southern Cascade Mountains, California and Nevada, USA. *Ecosystems*, **12**, 16–32.

Milly, P.C.D., Betancourt, J., Falkenmark, M., Hirsch, R.M., Kundzewicz, Z.W., Lettenmaier, D., & Stouffer, R.J. (2008). Stationarity is dead: whither water management? *Science*, **319**, 573–574.

Moore, M.M., Covington, W.W., & Fulé, P.Z. (1999). Reference conditions and ecological restoration: a southwestern ponderosa pine perspective. *Ecological Applications*, **9**, 1266–1277.

Morgan, P., Aplet, G.H., Haufler, J.B., Humphries, H.C., Moore, M.M., & Wilson, W.D. (1994). Historical range of variability: a useful tool for evaluating ecosystem change. *Journal of Sustainable Forestry*, **2**, 87–111.

Noss, R.F. (2001). Beyond Kyoto: forest management in a time of rapid climate change. *Conservation Biology*, **15**, 578–590.

Peterson, G., Allen, C., & Holling, C.S. (1998). Ecological resilience, biodiversity, and scale. *Ecosystems*, **1**, 6–18.

Prance, G.T. (1982). *The Biological Model of Diversification in the Tropics*. Columbia University Press, New York, USA.

Rabalais, N.N. (2002). Nitrogen in aquatic ecosystems. *Ambio*, **31**, 102–112.

Sanderson, E.W., Jaiteh, M., Levy, M.A., Redford, K.H., Wannebo, A.V., & Woolmer, G. (2002). The human footprint and the last of the wild. *BioScience*, **52**, 891–904.

Savage, M. (1991). Structural dynamics of a southwestern pine forest under chronic human influence. *Annals of the Association of American Geographers*, **81**, 271–289.

Saxon, E., Baker, B., Hargrove, W., Hoffman, F., & Zganjar, C. (2005). Mapping environments at risk under different global climate change scenarios. *Ecology Letters*, **8**, 53–60.

Schoennagel, T., Veblen, T.T., & Romme, W.H. (2004). The interaction of fire, fuels, and climate across Rocky Mountain forests. *BioScience*, **54**, 661–676.

Sprugel, D.G. (1991). Disturbance, equilibria, and environmental variability: what is "natural" vegetation in a changing environment? *Biological Conservation*, **58**, 1–18.

Stearns, P.N. (1998). Why study history? American Historical Society webpage. http://www.historians.org/pubs/free/whystudyhistory.htm.

Stegner, W. (1992). *A Sense of Place*. Random House, New York, USA.

Stephenson, N.L., Millar, C.I., & Cole, D.N. (2010). Shifting environmental foundations: the unprecedented and unpredictable future. In *Beyond Naturalness: Rethinking Park and Wilderness Stewardship in an Era of Rapid Change* (ed. D.N. Cole and L. Yung), pp. 50–66. Island Press, Washington, DC, USA.

Swetnam, T.W., Allen, C.D., & Betancourt, J.L. (1999). Applied historical ecology: using the past to manage for the future. *Ecological Applications*, **9**, 1189–1206.

Swetnam, T.W. & Betancourt, J.L. (1998). Mesoscale disturbance and ecological response to decadal climatic variability in the American Southwest. *Journal of Climate*, **11**, 3128–3147.

Tausch, R.J., Wigand, P.E., & Burkhardt, J.W. (1993). Plant community thresholds, multiple steady states, and multiple successional pathways: legacy of the quaternary? *Journal of Range Management*, **46**, 439–447.

Taylor, A.H. & Skinner, C.N. (2003). Spatial patterns and controls on historical fire regimes and forest structure in the Klamath Mountains. *Ecological Applications*, **13**, 704–719.

Vitousek, P.M., D'Antonio, C.M., Loope, L.L., Rejmánek, M., & Westbrooks, R. (1997). Introduced species: a significant component of human-caused global change. *New Zealand Journal of Ecology*, **21**, 1–16.

Westerling, A.L. & Swetnam, T.W. (2003). Interannual to decadal drought and wildfire in the western United States. *EOS*, **84**, 545–560.

Westerling, A.L., Brown, T.J., Gershunov, A., Cayan, D.R., & Dettinger, M.D. (2003). Climate and wildfire in the western United States. *Bulletin of the American Meteorological Society*, **84**, 595–604.

Westerling, A.L., Hidalgo, H.G., Cayan, D.R., & Swetnam, T.W. (2006). Warming and earlier spring increase western U.S. forest wildfire activity. *Science*, **313**, 940–943.

White, P.S. & Walker, J.L. (1997). Approximating nature's variation: selecting and using reference information in restoration ecology. *Restoration Ecology*, **5**, 338–349.

Wiens, J.A., van Horne, B., & Noon, B.R. (2002). Integrating landscape structure and scale into natural resource management. In *Integrating Landscape Ecology into Natural Resource Management* (ed. J. Liu and W.W. Taylor), pp. 23–67. Cambridge University Press, Cambridge, UK.

Williams, J.W. & Jackson, S.T. (2007). Novel climates, no-analog communities, and ecological surprises. *Frontiers in Ecology and the Environment*, **5**, 475–482.

Williams, J.W., Jackson, S.T., & Kutzbacht, J.E. (2007). Projected distributions of novel and disappearing climates by 2100 AD. *Proceedings of the National Academy of Sciences*, **104**, 5738–5742.

Willis, K.J. & Birks, H.J.B. (2006). What is natural? The need for a long-term perspective in biodiversity conservation. *Science*, **314**, 1261–1265.

Woodley, S.J., Key, J., & Francis, G. (eds.). (1993). *Ecological Integrity and the Management of Ecosystems*. St. Lucie Press, Delray Beach, FL, USA.

Zaghloul, K.A., Blanco, J.A., Weidemann, C.T., McGill, K., Jaggi, J.L., Baltuch, G.H., & Kahana, M.J. (2009). Human substantia nigra neurons encode unexpected financial rewards. *Science*, **323**, 1496–1499.

Zavaleta, E.S. & Chapin, F.S. III. (2010). Resilience frameworks: enhancing the capacity to adapt to change. In *Beyond Naturalness: Rethinking Park and Wilderness Stewardship in an Era of Rapid Change* (ed. D.N. Cole and L. Yung), pp. 142–158. Island Press, Washington, DC, USA.

Chapter 5

WHAT IS THE SCOPE OF "HISTORY" IN HISTORICAL ECOLOGY? ISSUES OF SCALE IN MANAGEMENT AND CONSERVATION

John A. Wiens,[1,2] *Hugh D. Safford,*[3,4]
Kevin McGarigal,[5] *William H. Romme,*[6] *and*
Mary Manning[7]

[1]PRBO Conservation Science, Petaluma, CA, USA
[2]University of Western Australia, Crawley, WA, Australia
[3]USDA Forest Service, Pacific Southwest Region, Vallejo, CA, USA
[4]University of California, Davis, CA, USA
[5]University of Massachusetts, Amherst, MA, USA
[6]Colorado State University, Fort Collins, CO, USA
[7]USDA Forest Service, Northern Region, Missoula, Montana, USA

5.1 INTRODUCTION

Understanding and applying concepts such as historical range of variation (HRV) in management and conservation requires a specification of what is "history" and what is "variation." Both are matters of scale. Depending on one's perspective, history is what happened yesterday, last week, a month ago, or years, decades, centuries, millennia, or even longer in the past. Variation in any factor of interest depends on the timescale as well. Because variation is often expressed as a departure from some average condition, whether that condition remains stable or itself varies depends on the time frame adopted. Spatial scale is also important: both history and variation are likely to differ between a small, local area and a larger, regional area. Whether historical variation is used to guide management (e.g. Landres et al. 1999), assess recovery from environmental disturbances (e.g. Wiens 1995), or set targets for restoration actions (e.g. White & Walker 1997), determining appropriate scales should be a central concern.

More often than not, however, the scales of investigation or management are determined arbitrarily, on the basis of factors related more to logistics, the time frame of human perception, or the availability of data than to the dynamics of the system of interest. Consider just temporal scale. Much of the information used to test the ecological theories on which management and conservation are based, for example, comes from 1- to 3-year studies constrained by academic degree requirements or grant-funding cycles. Such studies can provide only a snapshot of variation, so in order to generalize the results, one must assume that system dynamics are in equilibrium. Long-term studies (10–40 years or more) can document variation within the specified time window, but such studies have been conducted for only a handful of species or ecosystems and are likewise limited by the duration of funding support or the interest or persistence of the investigator (Likens 1989; Irland et al. 2006). In the management arena, US National Forest Land Management Plans generally consider a 10- to 15-year planning horizon (as established by the National Forest Management Act), and strategic plans of agencies such as the US Geological Survey or US Fish and Wildlife Service are cast in terms of 5-year periods covering a decade or two.

More to the point of this book, the timescales used to define HRV differ depending on the attributes used to characterize variation, the availability of data, the particular management directives or policies, and an often-unstated perception of what past time period is

relevant to the present (Millar & Woolfenden 1999a). In North America, the period of relevant history is often considered to be the two to three centuries before the arrival of European colonists, when disturbance dynamics are assumed to have been "natural," and the beginnings of the transformation of the continent (Whitney 1994) (although this excludes the previous effects of indigenous Native Americans on the landscape; Denevan 1992; but see Barrett et al. 2005). The comparable demarcation points for history are quite different in Europe or, for that matter, Australia or parts of Africa or South America. Paleoecologists think of timescales in yet different ways (e.g. Schoonmaker 1998; Jackson, Chapter 7, this book).

There are two points to be made. First, the scales used to define time periods over which historical variation is considered cover a broad range, in part reflecting different purposes, objectives, constraints, perceptions, and attributes of interest. Second, no matter which scale is used, the patterns and magnitude of variation observed will be dictated and circumscribed by that scale, and any management or conservation practices that are guided by historical variation will be similarly constrained.

Our objective in this chapter is to delve more deeply into some of the issues related to scale and historical variation and the implications for incorporating historical ecology into resource management and conservation. These issues complicate applications of historical ecology, and the specter of rapid changes in ecological systems wrought by climate change and land-use change only adds to the difficulty. Our perspective is that, despite these difficulties, historical ecology is fundamental to understanding the current and likely future status of the ecosystems we manage and value, because it provides insight into the temporal and spatial mechanisms by which ecosystems respond to global change. At the very least, historical information can inform resource managers about the level of departure of current from past conditions, which can assist in articulating goals and desired conditions and identifying appropriate management options. Regardless of how the future plays out, planning for it requires knowing where we are today, and knowing where we are today requires knowing how we got here.

5.2 SCALE IN ECOLOGY

First, some background on what scale means and how it has been used in ecology. Although ecologists use "scale" or "scaling" in a variety of ways (Peterson & Parker 1998; Wu 2007), scale generally refers to the dimensions in space or time in which a study is conducted or a concept or theory is intended to apply. Saying that a study area is 10 ha or was conducted over 2 years specifies the spatial or temporal scale of the study.

There are two components of scale: *grain* and *extent* (Wiens et al. 1993; Fortin & Dale 2005). In a spatial context, grain refers to the resolution of a study or analysis or the finest unit of observation (e.g. the smallest grid-cell size), while extent defines the overall area in which a study or analysis is conducted. For example, an investigator might study the distribution and survival of oak (*Quercus* spp.) seedlings in a 100-ha woodlot (the extent) by sampling several dozen quadrats of $1 m^2$ (the grain). The temporal analog might be conducting the study over 3 years (the extent) by sampling every month (the grain). Considerations of grain and extent, as components of scale, relate closely to the definition of both history and variation in determination of the HRV. Extent defines the span of history considered, and grain determines what one sees of the variation that actually occurs.

Grain and extent are selected by an investigator or manager as part of a study design or management application. Beyond this, however, natural phenomena often vary nonlinearly in relation to scale. Scale domains (Wiens 1989) define ranges of scale within which the linkages between ecological processes (causes) and patterns (effects) remain relatively stable before shifting to some different process–pattern relationship as the scale is changed and a "scaling threshold" is passed (Fig. 5.1). Such scale-dependent domains and thresholds are well-known in population dynamics (May 1974) and chaos theory (Gleick 1987), where they represent transitions to a different state space (Gunderson & Holling 2002).

In addition to drawing attention to the importance of grain and extent and threshold dynamics, the last two decades of research and thinking about scaling in ecology have led to several insights about the factors that influence the choice of appropriate scales. These are not new or novel insights, but they bear on how history can inform management and conservation practice.
• The inherent dynamics of the ecological system determine in part the appropriate scale of investigation or management. These dynamics (e.g. population fluctuations, frequency of disturbances) are what produce the ecologically relevant variations. The scale(s) on

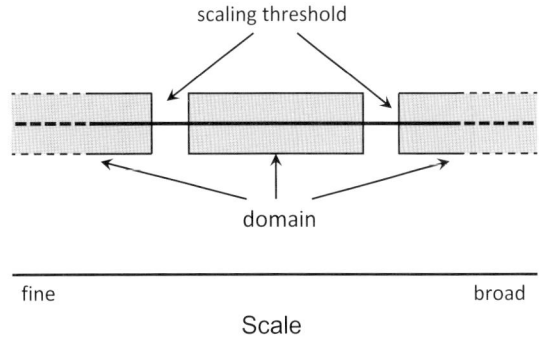

Fig. 5.1 Domains of scale define regions of the scale spectrum within which ecological attributes (e.g. organisms, processes, spatial patterns) and pattern–process relationships remain relatively consistent. Adjacent domains are separated by transitional zones or thresholds in which system dynamics may be unpredictable. If the focus is on phenomena within a given scale domain, studies conducted at finer scales may fail to include critical components, whereas studies conducted at broader scales may fail to reveal the pattern or causal relationships because such linkages are averaged out or are characteristic only of the given domain.

which a system is viewed will determine how or whether these dynamics and variations are revealed. A single-year study with monthly observations will not capture variations associated with multiyear droughts or El Niño Southern Oscillation (ENSO) oceanographic cycles.
• Different system attributes, such as weather, climate, fire, or the life-history strategies of different species, vary in their characteristic scale domains in space and time (Fig. 5.2). The appropriate scale(s) for assessing variation in these attributes will therefore differ. There is no "one size fits all" scale.
• The appropriate scale will also differ depending on the question asked and the focus of the question. A study of the physiology of an individual tree, for example, will require a different scale in space and time to capture critical cause–effect relationships than one focused on the controls of nitrogen cycling in forest ecosystems (Fig. 5.2).
• Policies that govern the protection or management of natural resources are frequently implemented opportunistically, often at the scales of administrative units (e.g. individual Wildlife Refuges, US Forest Service Regions) or budget cycles. The implicit nature of scale in these instances often makes it difficult to learn why some policies succeed and others fail. Permeating all of

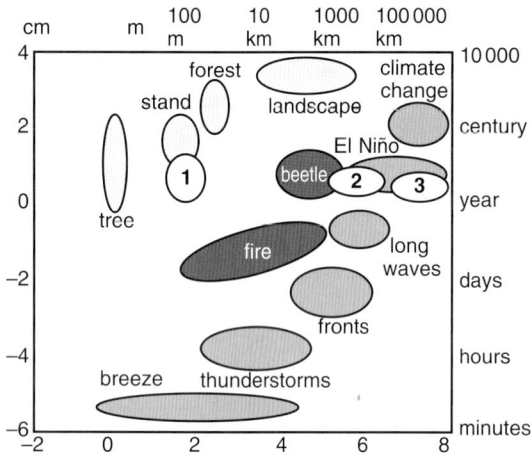

Fig. 5.2 Generalized space-timescale domains of features of forests, disturbances associated with fire or beetle outbreaks, and atmospheric factors (after Holling et al. 2002). The scale domains for different forest-management approaches are also shown (1 = forest thinning or prescribed burning; 2 = regional forest plan; 3 = National Forest policy). (From *Panarchy* edited by Lance H. Gunderson and C.S. Holling. Copyright © 2002 Island Press. Reproduced by permission of Island Press, Washington, DC, USA.)

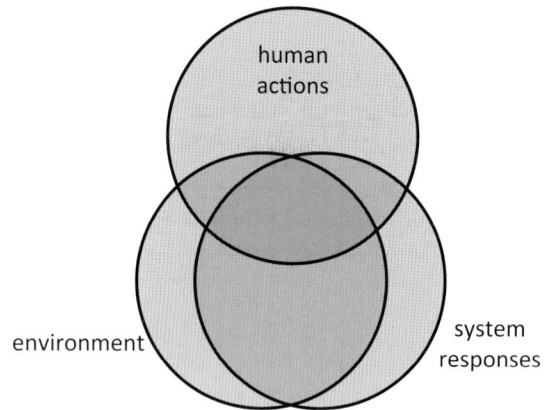

Fig. 5.3 Venn diagram of the relationships between the scales (areas of the circles) of variation in the environment, in the responses of ecological systems to the environment, and in the application of human actions (e.g. management, conservation, resource extraction, policy). The scales only partially overlap because some variations in the environment occur at scales to which ecological systems do not respond or some attributes of ecological systems occur at scales too fine or too broad for management or conservation actions. The areas of overlap indicate matching of scales. The large overlap between the scales of environmental variation and responses of ecological systems reflect the general tendency of organisms or biological processes to be adjusted to environmental variations, either through short-term responses or evolutionary adaptation. The lesser overlap between these scales and that of human actions reflects the general practice of determining the scale of these actions on the basis of the size of management units, the prevue of policy, duration of funding, and so on, rather than the inherent scaling of environmental variation and ecological responses.

this are the scales of human perception. Most people view the things that affect their lives (and over which they have some control) at local spatial scales or short-term temporal scales. They tend to be more detached from international (or even national) affairs or things that may affect them a decade or two hence. There is an important asymmetry here, in that most people have a short-term view of the future but a much longer, even nostalgic, view of the past. This is one reason why it is so difficult to generate public concern over the potential impacts that climate change may have in 2050. Like it or not, our perceptual scaling is often imposed on the natural systems we study or attempt to manage or conserve.

The upshot of these multiple layers of scaling is that what one perceives about natural systems – their dynamics, history, and variations – is filtered by scale. Roughly speaking, there are three dimensions of scale filtering that must be considered: the scales at which the environment varies, the scales at which the system operates and responds to environmental variations, and the scales at which we observe and manipulate these dynamics (Fig. 5.3). The first two scaling filters

are usually functionally intermeshed: through either short-term or evolutionary adaptation, ecological systems tend to adjust to the scales of relevant variation in environmental factors. The third scaling filter, however, involves human scalings that are often detached from those of the environment and the natural systems. As a result, the scales of both academic studies of these systems and the actions taken to manage or conserve them may be mismatched to the natural scaling of the systems. Management may go astray when there is a lack of congruence between the scale of management actions and the scales of variation in the environment and in the dynamics of the natural system. For example, treatments to protect western conifer forests from losses to bark beetles (*Scotylus, Dendroctonus,* and *Ips* spp., among others) typically are conducted at the scale of individual trees (applying insecticides) or stands (thinning to reduce tree-to-tree competition and stress), and important research has been conducted to support management at this spatial scale (e.g. Amman & Logan 1998; Schmid & Mata 2005). However, the bark beetle outbreaks now occurring across much of western North America are being driven primarily by changing climatic conditions (warmer temperatures) occurring at a continental or global scale (Raffa et al. 2008), which are overwhelming the effects of stand-level forest treatments.

5.3   SOME IMPLICATIONS OF SCALE IN HISTORICAL ECOLOGY

The perception of system dynamics

What one sees of the dynamics of an ecological system is a function of the scale on which the system is viewed. Consider, for example, the three hypothetical scenarios of system dynamics depicted in Fig. 5.4. These scenarios represent distinctly different forms of system dynamics. In "A" the system is in a dynamic equilibrium in which variation occurs within limits about an average condition that does not change over a long time period. This is what is envisioned in the concept of stationarity (Milly et al. 2008). In "B" the system is undergoing a long-term, continuous change; variation occurs about an average condition that shows a trend over time. In "C" the system exhibits sudden, discontinuous changes as the dynamics pass thresholds or tipping points. Threshold dynamics are well known in many ecological systems (Beisner et al. 2003; Bestelmeyer 2006; Groffman et al. 2006), and they have spawned an array of state-and-transition models (Westoby et al. 1989; Bestelmeyer et al. 2009) that deal with the nonlinear dynamics of complex systems. If these systems were to be viewed only over a short timescale (a in Fig. 5.4), their dynamics and variations would appear to be much the same. Expanding the temporal scale (b in Fig. 5.4) might enable one to distinguish between the long-term stability of scenario A and the trend of scenario B, but the dynamics of scenario C would be indistinguishable from those of scenario A. Only over a long temporal scale (relative to the dynamics of the system; c in Fig. 5.4) would one detect the abrupt shifts in system state that accompany thresholds.

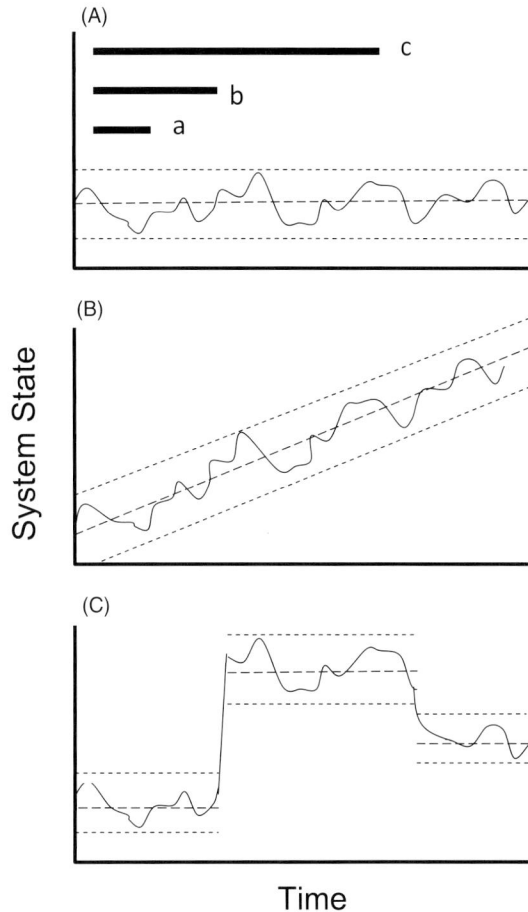

Fig. 5.4 Three hypothetical examples of the temporal variation in a system property. A = dynamic equilibrium; B = continuous trend over time; C = system dynamics characterized by thresholds and shifts in system state. The dashed line indicates the long-term mean; the dotted lines envelope the range of variation. Bars a, b, and c illustrate different scales on which the system dynamics might be observed. Although the magnitude of variation is shown as unchanging over time in these examples, it is more likely to change, perhaps especially as a system approaches a threshold in system dynamics (see text).

As the recognition of nonequilibrium dynamics has become more widespread in ecology (Wiens 1984; Rohde 2005), it has become evident that the equilibrium envisioned in scenario A of Fig. 5.4 may be unlikely except over short time spans in ecosystems that recover rapidly from disturbances or that are intensively managed to maintain the dynamics within a restricted envelope of variation. Scenario B may typify ecosystems undergoing recovery, restoration, or succession after disturbances such as forest fires (Turner et al. 2003; Hobbs & Suding 2009), but it is often coupled with the presumption that, absent recurrent disturbance, system dynamics will eventually stabilize and thereafter resemble scenario A. If the disturbances are large and frequent (relative to recovery times), however, the system may resemble scenario C, changing suddenly to a different configuration and different dynamics when a threshold is passed. Such changes have been documented where severe grazing and/or changes in fire frequency fostered the establishment of invasive plants, resulting in the replacement of a shrub-dominated community by grassland or vice versa (Bestelmeyer et al. 2004; Chambers et al. 2007). Systems with dynamics such as in scenarios A, B, or C are likely to require different management, even if the management objectives are the same. Use of an inappropriate scale can lead to a misinterpretation of system dynamics, resulting in management that is ineffective or worse (Cumming et al. 2006).

Scales a, b, and c in Fig. 5.4 indicate the temporal scale extent. This is the span of "history" included in assessments of HRV. But the grain of temporal observation may also be important, since this is what determines the variation component. By definition, variations that occur at a scale finer than the observation grain cannot be detected. As the grain becomes coarser, more of the inherent variation in a system may be filtered out. Use of a 10-year running mean to portray annual precipitation, for example, smoothes out the peaks and troughs of wet and dry years. Such extreme events can have disproportionate effects on ecological systems, particularly if the physiological tolerances of organisms are exceeded. For example, bark beetle outbreaks may be terminated by a short period (a week or so) of extreme cold (below ca. $-20°C$), which kills the beetle larvae within the trees, especially if the cold spell occurs in early or late winter (Logan & Powell 2001; Logan et al. 2003). Daily or weekly temperature data are required to detect this key ecological event; average monthly or annual temperature values would be unlikely to capture the critical short-duration cold period that terminated an outbreak. These points emphasize the importance of assessing what scales are relevant to the ecological system of interest and the conservation or management objectives. We are not interested in documenting variation simply because it

is there, but rather because it affects properties of interest. The art of dealing with scale involves "getting it just right." If the grain is too fine, we may be overwhelmed by a deluge of extraneous detail and fail to see the important factors, but if the scale is too coarse, we may average away critical dynamics and likewise fail to see the important factors. If the extent is too brief or small, we may fail to capture the true historical or spatial dynamics of the system, but if it is too broad, we may confound our understanding of the system (and thus management effectiveness) by including the effects of factors that no longer operate or that are not amenable to management. It is perhaps trite, but also necessary, to emphasize the importance of a deep understanding of the system – Barbara McClintock's "feeling for the organism" (Keller 1983) – if one is to use a scale (grain and extent) that will capture the relevant variations over a relevant span of history.

Nonlinearities in history

Some years ago, Harold Blum (1951) published a book entitled *Time's Arrow and Evolution* in which, drawing from arguments in physics and cosmology, he addressed the irreversibility of temporally chronological processes such as evolution. Although history may sometimes appear to repeat itself, in reality, there is no going back – time's arrow points only forward. But this does not mean that history follows linear pathways. Father Time does not shoot straight. History follows a convoluted pathway, in which particular events at particular time may have decisive impacts on the future trajectories of history. This is the "butterfly effect" of chaos theory and quantum physics, in which the temporal trajectory of complex systems is sensitive to initial conditions (Gleick 1987). In fact, any historical trajectory represents a series of possible branch points stemming from critical events (or even seemingly benign events) at particular points in time (Fig. 5.5). The characteristics of an ecological system are affected by what has come before. Knowing what has come before is perhaps the most compelling reason for paying attention to historical variation when we attempt to manage or conserve a system.

Sudden shifts or tipping points in the wandering trajectory of history may define the boundaries of domains of scale (Wiens 1989). Patterns and variations within scale domains may be governed by a common set of processes. This consistency in process–pattern rela-

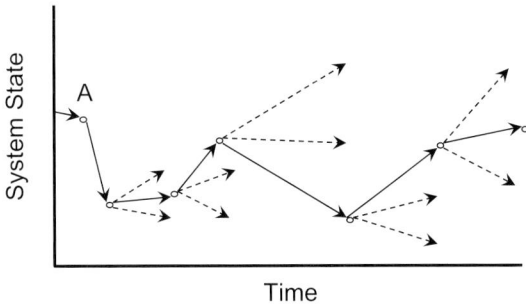

Fig. 5.5 Alternative trajectories of change in system state over time. Following a disturbance event (A), the system at subsequent points might follow any of several trajectories; which trajectory is followed determines the subsequent system state. Over time, a series of potential branching points may create the potential for the system to display a wandering trajectory. History is a sequence of such branch points and alternative trajectories of change.

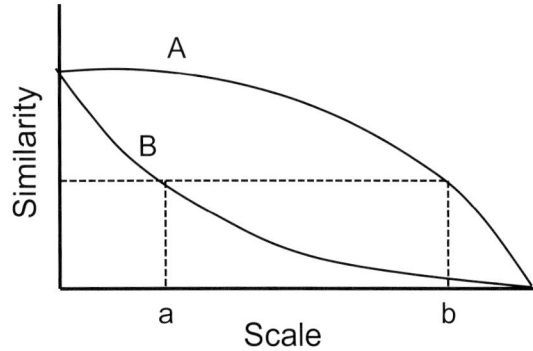

Fig. 5.6 The similarity of a system to its condition at some initial point tends to decrease as the space or timescale over which similarity is calculated increases. Different systems (or different components of the same system) may follow different similarity-decay functions (A and B). In this hypothetical example, the two systems exhibit the same level of similarity to the initial point at quite different scales (a and b).

tionships engenders similarity, which means that there is a reasonable likelihood that causes and effects that have operated in the past will continue to operate now and in the future. This is the premise of using HRV to guide management and conservation actions. Changing scale, however, may increase the likelihood of encountering and crossing thresholds between different sets of factors that control ecological dynamics (e.g. Fig. 5.2). Thus, as we go farther back in time (or a greater distance in space), similarity diminishes – there is a distance-decay function to scale (Nekola & White 1999). Depending on the decay rate, the scale domains of history that may encompass sufficiently similar conditions to inform present practices will differ (Fig. 5.6). In some instances, these scale domains may be lengthy. For example, Millar and Woolfenden (1999b) argue that the major ecosystem components and processes in the Sierra Nevada of California have remained broadly similar over the last ca. 4000 years.

Within these domains, system dynamics may be inherently resilient to disturbance: the systems are able to absorb the potentially disruptive effects of disturbances without being forced out of the "comfort zone" of variation (Beisner et al. 2003). Resilience acts through feedback mechanisms to maintain the composition or functioning of ecological systems within limits despite environmental stresses, disturbances, and disruptions (Peterson et al. 1998). Consequently, resilience is a much talked-about trait of ecological systems, and some approaches to conservation and

management are explicitly focusing on enhancing the resilience of ecological systems, particularly in the context of climate change (Millar et al. 2007). If ecological systems are more resilient to change within scaling domains, then aligning the scale used to define historical variation for management with the scale of such domains may be particularly effective.

If only we knew how to define scaling domains *a priori*! Ah, there's the rub. Although it has been suggested that the magnitude and frequency of variations in system properties may increase as an ecological system approaches a threshold (Wiens 1992; Carpenter & Brock 2006; Scheffer et al. 2009), it is in the nature of thresholds that there is often no warning that they are about to be crossed. Once a system is pushed over a threshold or beyond a tipping point, however, the composition and functioning of the system may be so altered that it is fundamentally changed – it is a different system. Returning the system to its former condition may not be possible, even with massive management efforts. For example, cheatgrass (*Bromus tectorum*), a nonnative annual, has invaded shrubsteppe vegetation in much of the Great Basin of western North America, where it has profoundly altered the historical disturbance regime and caused widespread ecological degradation (Mack 1981; D'Antonio & Vitousek 1992). Widely spreading fires formerly were infrequent in the shrubsteppe, largely because bare ground between individual shrubs and bunchgrasses created a discontinuous fuel bed. However, cheatgrass is a winter annual grass that senesces in early spring, creating a continuous fuel bed of easily ignited fuel. Fires within cheatgrass-dominated areas can spread rapidly over large areas. A positive feedback becomes established, in which more frequent fire promotes more abundant cheatgrass, which in turn promotes more frequent fire. As a result, the historical fire-return interval (30–110 years) has been drastically shortened to 1–5 years. The fire-intolerant native flora eventually may be extirpated, at which point the system is converted to annual grassland. Once established, cheatgrass is very difficult to eradicate; as long as it is abundant, attempts to reestablish the native shrubs and bunchgrasses will not likely succeed because of recurrent cheatgrass-fueled fires (Whisenant 1990; Chambers et al. 2007). Peterson (in Holling & Gunderson 2002) gives several examples of other systems displaying such alternative stable states.

5.4   SCALES IN TIME AND SPACE INTERACT

So far we have emphasized temporal scaling because of its close relationship with historical variation. But a glance at Fig. 5.2 reveals that there are parallels in the characteristic temporal and spatial scale domains for many factors. Some, such as weather or individual reproduction, operate at "small-fast" scales where biophysical processes that control physiology or behavior dominate, while others, such as landscape or biome composition and structure, may be dictated largely by climatic, geomorphological, or biogeographical processes that operate at "large-slow" scales (Holling et al. 2002). Obviously, holding the temporal scale (i.e. history) constant while changing the spatial scale (or vice versa) will alter the factors controlling the dynamics of a system, confounding attempts to understand what produces the historical variation and compromising the applicability of historical information to current or future management.

The magnitude and pattern of temporal variation in a system may also be influenced by the form of spatial variation, and this interaction may determine whether one sees high or low variance in system attributes (and manages accordingly). For example, many populations of plants and animals have a patchy population structure. Local subpopulations may be loosely connected by dispersal and immigration, with the result that their dynamics are partially but not entirely independent of one another, as envisioned in metapopulation or source-sink theory (Hanski & Gaggiotti 2004; Liu et al. 2011). The similar concept of shifting-mosaic steady-state dynamics (Bormann & Likens 1995) has been developed to explain the high variation in successional state among patches in a forested landscape while the landscape as a whole appears to be much more stable. If the spatial scale on which these dynamics are viewed is fine, one will see the variation among the patches or subpopulations, whereas a broader scale perspective will reveal the relative stability of the landscape or metapopulation (Turner et al. 1993). Whether one sees the forest or the trees is a matter of scale, and it affects the variation component of historical variation.

When one defines a time period in which to assess historical variation, there is also an implicit presumption that the spatial scaling of important factors or processes has remained much the same over that period. We know that this is not often the case. Over the past two centuries, for example, forest clearing and subsequent afforestation have reduced and then increased forest cover in the northeastern United States (Irland 1999), whereas in much of the midwestern and western United States, land-use conversion has led to a progressive loss and isolation of both forests and native grasslands (Curtis 1956; Sharpe et al. 1987; George & Dobkin 2002). These changes would have altered the size and frequency of forest and grassland disturbance due to fire even without the additional effects of a century of fire suppression. As a result of these spatial changes, the temporal dynamics of these systems have also changed. As the spatial extent of forested areas has shrunken over time, the scale of the temporal domain over which variation is gauged for management application should also become smaller, at least if our interest is in historical variations that are similar enough to contemporary conditions to be relevant. Of course, there continue to be valid reasons to understand ecosystem dynamics from before fragmentation, but the contemporary applicability of such knowledge may be limited. Either way, history is a function of spatial as well as temporal scale.

The lingering effects of past history

History, of course, is not something that is nicely partitioned off wherever we decide to place our scale boundaries. History also leaves its legacies, elements that reflect conditions at some past time but which no longer hold. The oldest trees in an old-growth sequoia (*Sequoiadendron giganteum*) forest, for example, germinated as seedlings thousands of years ago, when the climate was different and the imprint of humans on the landscape was still light, and they grew and aged through periods in which fire frequencies varied substantially from one century to another. Their dominance in the overstory of a contemporary forest may have little to do with current environmental conditions or disturbance regimes. Another example: Lost Forest is an isolated stand of ponderosa pine (*Pinus ponderosa*) in central Oregon that is a relict of a more extensive forest that contracted during a hot dry period thousands of years ago (Chadwick & Eglitis 2007). Although annual precipitation is now less than what is usually required to maintain a ponderosa pine forest, Lost Forest persists because of unusual soil properties that retain sufficient water (Moir et al. 1973). The biota of every region on earth is the result of the complex intermingling of travelers who came at different times under different conditions.

Such legacies complicate attempts to understand why a particular set of species occurs in a particular area. They also make it difficult to define an appropriate scale for assessing historical variation. No matter how time is sliced, the period selected to represent variation that is relevant to current conditions will be tainted by interlopers from previous periods. Extending the timescale back four centuries would include the effects of the "Little Ice Age"; a millennium would include the Medieval Warm Period. Some of the trees still present in Wytham Woods in England would have germinated and grown under those conditions (see Kirby, Chapter 20, this book). The shorter the timescale considered, the more likely that legacies will be present. What is "short," however, must be measured relative to the longevity of the organisms: a short time to an oak may be an infinity to a caterpillar feeding upon its leaves.

5.5 CONCLUSIONS: DEALING WITH SCALE

Everything about ecology is scale dependent at some level. Individuals, populations, communities, and ecosystems of different types operate on characteristic domains of scale. These domains differ for different ecological processes such as feeding, reproduction,

habitat selection, trophic flows, nitrogen cycling, and so on. What one attempts to manage or conserve at a particular place and time is a compilation of the multiple scales on which the components of the system function and respond to the environment, with some broader scale elements (e.g. landscape effects, historical legacies) thrown in. And overriding all of it are filters of human perceptions and societal mandates, which are also scale dependent. There is great potential for mismatching the scales of management with the scales most relevant to the biology and ecology of the systems of interest.

So what messages should a manager or conservationist take away from our musings about scale, particularly as they relate to the use of historical variation to guide their actions? Here are some:

• Avoid arbitrary definitions of the scale of history. The appropriate scales depend on the targets of management and the biology and ecology of those targets, the management objectives or questions asked, and the policy and societal imperatives and constraints.

• Be open to management at multiple scales. Because the above factors all operate within different scale domains, there is no single best scale for management; managers and conservationists will need to adopt a much broader, multiscale perspective than they have in the past.

• Consider the scales of the dominant disturbance processes as a starting point for determining appropriate scales for management. Ecological insight, based on an understanding of natural history, can help to identify those ecological processes that most clearly determine the status of the focal ecosystem features. These disturbance processes can be used to establish provisional scales for management or conservation.

• Recognize that the size of a management unit profoundly influences our perception of the inherent dynamics of the system. A land unit may appear as a dynamic, steady-state system or as an unstable, non-equilibrium system, depending on the spatial and temporal scales at which the natural disturbance regime operates relative to the size of the unit and the time frame over which the dynamics of the system are observed. Thus, an extensive forested landscape in which the natural-disturbance regime is dominated by frequent small tree-fall gaps that fill in quickly will appear to be in a state of equilibrium, whereas a small protected area within a landscape where the natural fire regime is one of infrequent but large stand-replacing fires may be perceived as destroyed by the first large fire that occurs.

• Be diligent in consulting the historical record – as far back as information is available – before developing resource-management plans. There are many examples of misguided management actions and restoration plans that did not properly account for the full range of ecosystem variability or its connection to exogenous drivers like climate, raising questions about the current or future effectiveness of the actions or plans.

5.6 CODA: SCALING CONSERVATION AND MANAGEMENT IN A CHANGING WORLD

Although the future is by definition uncertain, the one thing that does seem certain is that the future will be different from the past. The accelerating pace of climate change and land-use change will see to that. There is more to it than changes occurring more rapidly than in the past, however. As species shift distributions, disappear, or evolve in response to the changing environment, existing assemblages will be torn apart and reassembled into new, novel configurations that have not been seen before (Hobbs et al. 2006; Williams et al. 2007; Stralberg et al. 2009).

We have discussed above the multiple factors that make it difficult to define appropriate scales of time or space to inform contemporary conservation and management. How can we then hope to avoid compounding mismatches of scale in the future, when the dynamics of populations, communities, and ecosystems shift to different scales or pass thresholds into new, unexplored domains of scale?

We can begin with some speculations about how future environmental changes might affect the scaling of the elements of Fig. 5.2. Although the scales of forest units from conifer needles to landscapes may remain much the same, the scales on which the "slow-broad" weather and climate factors (e.g. long waves, El Niño Southern Oscillation, climate change) operate are likely to shift toward faster but even broader scales. For example, model projections suggest that episodes of extreme temperatures, rainfall, or droughts may become more frequent, longer lasting, and more widespread (Climate Change Science Program 2008). In response to changes in regional climates, outbreaks of insects such as spruce budworm (*Choristoneura* spp.) may become more widespread and more persistent and forest fires may become more frequent. To be effective, management and conservation must anticipate how

the biological elements or disturbance processes of ecosystems will shift in their time-space scaling dimensions.

The challenge is exacerbated because future environmental change will be driven by a combination of land-use change and climate change (Pielke et al. 2002; Jetz et al. 2007). These forces operate across multiple scales. Changes in land use are responsive to broadscale policies, such as farm policies that encourage uniformly intensive agricultural practices over large regions or the fire-exclusion policies that were applied to almost all federal lands. Such policies are usually implemented, however, at a local scale of tens of hectares to a few hundred square kilometers. On the other hand, changes in climate are driven by broadscale continental or global atmospheric and oceanographic dynamics, which percolate down to affect regional climate and local weather (or, in marine systems, sea level, salinity, and alkalinity). As land use continues to fragment landscapes, the spatial scale of management may shrink (to focus on the smaller remnant patches) or expand (to focus on entire landscape mosaics). The global drivers of climate change may also necessitate a broadening of scale, especially if one is to consider the causes as well as the consequences of change. The fragmentation of landscapes due to land use and the reshuffling of biological communities resulting from climate change may reduce the distance in time or space over which systems display similar composition or dynamics; the distance-decay functions in Fig. 5.6 will shift from A toward B.

As a result of these environmental changes, there may be shrinkage in the size of the scale domains over which one can assume some degree of uniformity in ecological process–pattern relationships, and thus expect historical variation to be relevant. Moreover, the likelihood that more and more "no-analog" assemblages and ecosystems will emerge may suggest that past history is no longer sufficient to adequately inform future conservation and management practices. Yet history is not dead. We may no longer be able to expect historical variation to prescribe the domain of future variation, but there is still much to be learned from an examination of how environments have varied in the past and how ecological systems have responded to these variations through adaptation, movements and reassembly of communities, or, in some cases, extinction (see Chapter 7, this book). These lessons can be learned from the many scales of history discussed elsewhere in this book. Paleoecology can indicate how large-magnitude environmental changes in the distant past affect biotas, although the temporal grain of this perspective is limited by the available data. Historical ecology can use fine-grain information from the past decades or centuries to assess responses to environmental variations in greater detail, although the range of variation encompassed by these records tends to be inversely related to the time span of the data. Despite such limitations, information on historical variation at multiple scales, combined with the projections of models of climate change, fire dynamics, species distributions, and the like, may provide a foundation for scenario analyses of possible future trajectories of ecosystem change. Just as landscape ecology provided new insights into ecological processes at a range of spatial scales, historical ecology provides a means for understanding the temporal dynamics of systems more clearly and, through that understanding, obtaining sharper insights into the potential characteristics of ecosystems in the future.

REFERENCES

Amman, G.D. & Logan, J.A. (1998). Silvicultural control of mountain pine beetle: prescriptions and the influence of microclimate. *American Entomologist*, **44**, 166–177.

Barrett, S.W., Swetnam, T.W., & Baker, W.L. (2005). Indian fire use: deflating the legend. *Fire Management Today*, **65**(3), 31–34.

Beisner, B.E., Haydon, D., & Cuddington, K.L. (2003). Alternative stable states in ecology. *Frontiers in Ecology and the Environment*, **1**, 376–382.

Bestelmeyer, B.T. (2006). Threshold concepts and their use in rangeland management and restoration: the good, the bad, and the insidious. *Restoration Ecology*, **14**, 325–329.

Bestelmeyer, B.T., Hedrick, J.E., Brown, J.R., Trujillo, D.A., & Havstad, K.M. (2004). Land management in the American Southwest: a state-and-transition approach to ecosystem complexity. *Environmental Management*, **34**, 38–51.

Bestelmeyer, B.T., Tugel, A.J., Peacock, G.L., et al. (2009). State-and-transition models for heterogeneous landscapes: a strategy for development and application. *Rangeland Ecology and Management*, **62**, 1–15.

Blum, H.F. (1951). *Time's Arrow and Evolution*. Princeton University Press, Princeton, NJ, USA.

Bormann, F.H. & Likens, G.E. (1995). *Pattern and Process in a Forested Ecosystem*. 2nd ed. Springer-Verlag, New York, USA.

Carpenter, S.R. & Brock, W.A. (2006). Rising variance: a leading indicator of ecological transition. *Ecology Letters*, **9**, 308–315.

Chadwick, K.L. & Eglitis, A. (2007). Health assessment of the Lost Forest Research Natural Area. USDA Forest Service, Bend, OR, USA. http://www.fs.fed.us/r6/nr/fid/pubsweb/lost-forest.pdf.

Chambers, J.C., Roundy, B.A., Blank, R.R., Meyer, S.E., & Whittaker, A. (2007). What makes Great Basin sagebrush ecosystems invasible by *Bromus tectorum? Ecological Monographs*, **77**, 117–145.

Climate Change Science Program. (2008). Abrupt climate change. A report by the U.S. Climate Change Science Program and the Subcommittee on Global Change Research (P.U. Clark and A.J. Weaver, coordinating lead authors; E. Brook, E.R. Cook, T.L. Delworth, and K. Steffen, chapter lead authors), 244 pp. U.S. Geological Survey, Reston, VA, USA. http://web.mac.com/dannysatterfield/climatechange/Resources_files/sap3-4-final-report-all.pdf.

Cumming, G.S., Cumming, D.H.M., & Redman, C.L. (2006). Scale mismatches in social-ecological systems: causes, consequences, and solutions. *Ecology and Society*, **11**(1), 14. http://www/ecologyandsociety.org/vol11/iss1/art14/.

Curtis, J.T. (1956). The modification of mid-latitude grasslands and forests by man. In *Man's Role in Changing the Face of the Earth* (ed. L. Thom Jr.), pp. 721–736. University of Chicago Press, Chicago, IL, USA.

D'Antonio, C.M. & Vitousek, P.M. (1992). Biological invasions by exotic grasses, the grass/fire cycle, and global change. *Annual Review of Ecology and Systematics*, **23**, 63–87.

Denevan, W.M. (1992). The pristine myth: the landscape of the Americas in 1492. *Annals of the Association of American Geographers*, **82**, 369–385.

Fortin, M.-J. & Dale, M.R.T. (2005). *Spatial Analysis: A Guide for Ecologists*. Cambridge University Press, Cambridge, UK.

George, T.L. & Dobkin, D. (eds.). (2002). The effects of habitat fragmentation on western bird populations. *Studies in Avian Biology*, **25**, 4–7.

Gleick, J. (1987). *Chaos: making a new science*. Vintage, New York, USA.

Groffman, P.M., Baron, J.S., Blett, T., et al. (2006). Ecological thresholds: the key to successful environmental management or an important concept with no practical application? *Ecosystems*, **9**, 1–13.

Gunderson, L.H. & Holling, C.S. (2002). *Panarchy: Understanding Transformations in Human and Natural Systems*. Island Press, Washington, DC, USA.

Hanski, I. & Gaggiotti, O.E. (eds.). (2004). *Ecology, Genetics, and Evolution of Metapopulations*. Elsevier, Amsterdam, The Netherlands.

Hobbs, R.J., & Suding, K.N. (2009). *New Models for Ecosystem Dynamics and Restoration*. Island Press, Washington, DC, USA.

Hobbs, R.J., Arico, S., Aronson, J., et al. (2006). Novel ecosystems: theoretical and management aspects of the new ecological world order. *Global Ecology and Biogeography*, **15**, 1–7.

Holling, C.S. & Gunderson, L.H. (2002). Resilience and adaptive cycles. In *Panarchy: Understanding Transformations in Human and Natural Systems* (ed. L.H. Gunderson and C.S. Holling), pp. 25–62. Island Press, Washington, DC, USA.

Holling, C.S., Gunderson, L.H., & Peterson, G.D. (2002). Sustainability and panarchies. In *Panarchy: Understanding Transformations in Human and Natural Systems*. In (ed. L.H. Gunderson and C.S. Holling), pp. 63–102. Island Press, Washington, DC, USA.

Irland, L.C. (1999). *The Northeast's Changing Forest*. Harvard University Press, Cambridge, MA, USA.

Irland, L.C., Camp, A.E., Brissette, J.C., & Donohew, Z.R. (2006). Long-term silvicultural and ecological studies: results for science and management. GISF Research Paper 005. Yale University, Global Institute of Sustainable Forestry, New Haven, CT, USA.

Jetz, W., Wilcove, D.S., & Dobson, A.P. (2007). Projected impacts of climate and land-use change on the global diversity of birds. *PLoS Biology*, **5**(6), e157. doi: 10.1371/journal.pbio.0050157.

Keller, E.F. (1983). *A Feeling for the Organism*. W.H. Freeman and Company, New York, USA.

Landres, P., Morgan, P., & Swanson, F.J. (1999). Overview and use of natural variability concepts in managing ecological systems. *Ecological Applications*, **9**, 1179–1188.

Likens, G.E. (ed.) (1989). *Long-Term Studies in Ecology: Approaches and Alternatives*. Springer-Verlag, New York, USA.

Liu, J., Hull, V., Morzillo, A., & Wiens, J. (eds.). (2011). *Sources, Sinks, and Sustainability*. Cambridge University Press, Cambridge, UK.

Logan, J.A. & Powell, J.A. (2001). Ghost forests, global warming, and the mountain pine beetle (Coleoptera: Scolytidae). *American Entomologist*, **47**, 160–173.

Logan, J.A., Régnière, J., & Powell, J.A. (2003). Assessing the impacts of global climate change on forest pests. *Frontiers in Ecology and the Environment*, **1**, 130–137.

Mack, R.N. (1981). Invasion of Bromus tectorum L. into Western North America: an ecological chronicle. *Agro-Ecosystems*, **7**, 145–165.

May, R.M. (1974). Biological populations with nonoverlapping generations: stable points, stable cycles, and chaos. *Science*, **186**, 645–647.

Millar, C.I. & Woolfenden, W.B. (1999a). The role of climate change in interpreting historical variability. *Ecological Applications*, **9**, 1207–1216.

Millar, C.I. & Woolfenden, W.B. (1999b). Sierra Nevada forests: where did they come from? Where are they going? What does it mean? In *Transactions of the 64th North American Wildlife and Natural Resources Conference*. pp. 206–236.

Millar, C.I., Stephenson, N.L., & Stephens, S.L. (2007). Climate change and forests of the future: managing in the face of uncertainty. *Ecological Applications*, **17**, 2145–2151.

Milly, P.C.D., Betancourt, J., Falkenmark, M., Hirsch, R.M., Kundzewicz, Z.W., Lettenmaier, D.P., & Stouffer, R.J. (2008). Stationarity is dead: whither water management? *Science*, **319**, 573–574.

Moir, W.H., Franklin, J.F., & Maser, C. (1973). Lost forest research natural area: supplement 3 to Federal research natural areas in Oregon and Washington, a guidebook for scientists and educators. Miscellaneous publication 271. USDA Forest Service, Portland, OR, USA.

Nekola, J.C. & White, P.S. (1999). The distance decay of similarity in biogeography and ecology. *Journal of Biogeography*, **26**, 867–878.

Peterson, D.L. & Parker, V.T. (eds.). (1998). *Ecological Scale: Theory and Applications*. Columbia University Press, New York.

Peterson, G., Allen, C., & Holling, C.S. (1998). Ecological resilience, biodiversity, and scale. *Ecosystems*, **1**, 6–18.

Pielke, R.A., Sr., Marland, G., Betts, R.A., et al. (2002). The influence of land-use change and landscape dynamics on the climate system: relevance to climate-change policy beyond the radiative effect of greenhouse gases. *Philosophical Transactions of the Royal Society of London A*, **15**(360), 1705–1719.

Raffa, K.F., Aukema, B.H., Bentz, B.J., Carroll, A.L., Hicke, J.A., Turner, M.G., & Romme, W.H. (2008). Cross-scale drivers of natural disturbances prone to anthropogenic amplification: the dynamics of bark beetle eruptions. *BioScience*, **58**, 501–517.

Rohde, K. (2005). *Nonequilibrium Ecology*. Cambridge University Press, Cambridge, UK.

Scheffer, M., Bascompte, J., Brock, W.A., et al. (2009). Early-warning signals for critical transitions. *Nature*, **461**, 53–59.

Schmid, J.M., & Mata, S.A. (2005). Mountain pine beetle-caused tree mortality in partially cut plots surrounded by unmanaged stands. Research Paper RMRS-RP-54. USDA Forest Service, Fort Collins, CO, USA.

Schoonmaker, P.K. (1998). Paleoecological perspectives on ecological scale. In *Ecological Scale: Theory and Applications* (ed. D.L. Peterson and V.T. Parker), pp. 79–103. Columbia University Press, New York.

Sharpe, D.M., Guntenspergen, G.R., Dunn, C., Leitner, L.A., & Stearns, F. (1987). Vegetation dynamics in a southern Wisconsin agricultural landscape. In *Landscape Heterogeneity and Disturbance. Ecological Studies*, Vol. 64 (ed. M.G. Turner), pp. 137–155. Springer-Verlag, New York.

Stralberg, D., Jongsomjit, D., Howell, C.A., Snyder, M.A., Alexander, J.D., & Wiens, J.A. (2009). Re-shuffling of species with climate disruption: a no-analog future for California birds? *PLoS ONE*, **4**(9), e6825. doi: 10.1371/journal.pone.0006825.

Turner, M.G., Romme, W.H., Gardner, R.H., O'Neill, R.V., & Kratz, T.K. (1993). A revised concept of landscape equilibrium: disturbance and stability in scaled landscapes. *Landscape Ecology*, **8**, 213–227.

Turner, M.G., Romme, W.H., & Tinker, D.B. (2003). Surprises and lessons from the 1988 Yellowstone fires. *Frontiers in Ecology and the Environment*, **1**, 351–358.

Westoby, M., Walker, B., & Noy-Meir, I. (1989). Opportunistic management for rangelands not at equilibrium. *Journal of Range Management*, **42**, 266–274.

Whisenant, S.G. (1990). Changing fire frequencies on Idaho's Snake River Plains: ecological and management implications. In *Proceedings: Symposium on Cheatgrass Invasion, Shrub Die-off and Other Aspects of Shrub Biology and Management*. (E.D. McArthur, E.M. Romney, S.D. Smith, and P.T. Tueller, compilers), pp. 4–10. General Technical Report GTR-INT-276. USDA Forest Service, Ogden, UT, USA.

White, P.S., & Walker, J.L. (1997). Approximating nature's variation: Selecting and using reference information in restoration ecology. *Restoration Ecology*, **5**, 338–349.

Whitney, G.G. (1994). *From Coastal Wilderness to Fruited Plain: A History of Environmental Change in Temperate North America 1500 to the Present*. Cambridge University Press, New York, USA.

Wiens, J.A. (1984). On understanding a non-equilibrium world: myth and reality in community patterns and processes. In *Ecological Communities: Conceptual Issues and the Evidence* (ed. D.R. Strong Jr., D. Simberloff, L.G. Abele, and A.B. Thistle), pp. 439–457. Princeton University Press, Princeton, NJ, USA.

Wiens, J.A. (1989). Spatial scaling in ecology. *Functional Ecology*, **3**, 385–397.

Wiens, J.A. (1992). Ecological flows across landscape boundaries: a conceptual overview. In *Landscape Boundaries: Consequences for Biotic Diversity and Ecological Flows* (ed. A.J. Hansen and F. di Castri), pp. 217–235. Springer-Verlag, New York, USA.

Wiens, J.A. (1995). Recovery of seabirds following the Exxon Valdez oil spill: an overview. In *Exxon Valdez Oil Spill: Fate and Effects in Alaskan Waters*. ASTM Special Technical Publication 1219 (ed. P.G. Wells, J.N. Butler, and J.S. Hughes), pp. 854–893. American Society for Testing and Materials, Philadelphia, PA, USA.

Wiens, J.A., Stenseth, N.C., Van Horne, B., & Ims, R.A. (1993). Ecological mechanisms and landscape ecology. *Oikos*, **66**, 369–380.

Williams, J.W., Jackson, S.T., & Kutzbacht, J.E. (2007). Projected distributions of novel and disappearing climates by 2100 AD. *Proceedings of the National Academy of Sciences USA*, **104**, 5738–5742.

Wu, J. (2007). Scale and scaling: a cross-disciplinary perspective. In *Key Topics in Landscape Ecology* (ed. J. Wu and R.J. Hobbs), pp. 115–142. Cambridge University Press, Cambridge, UK.

Chapter 6

NATIVE AMERICANS, ECOSYSTEM DEVELOPMENT, AND HISTORICAL RANGE OF VARIATION

Gregory J. Nowacki,[1] *Douglas W. MacCleery,*[2,3] *and Frank K. Lake*[4]

[1]USDA Forest Service, Eastern Regional Office, Milwaukee, WI, USA
[2]USDA Forest Service, Washington Office, Forest Management (retired), Washington, DC, USA
[3]Alexandria, VA, USA
[4]USDA Forest Service, Pacific Southwest Research Station, Redding, CA, USA

Historical Environmental Variation in Conservation and Natural Resource Management, First Edition. Edited by John A. Wiens, Gregory D. Hayward, Hugh D. Safford, and Catherine M. Giffen.
© 2012 John Wiley & Sons, Ltd. Published 2012 by John Wiley & Sons, Ltd.

6.1 INTRODUCTION

In North America, human beings have not always been considered "legitimate players" in natural ecosystems, often being dismissed as irrelevant, unnatural, or simply "noise" in the system. This is in contrast to the continents outside the Americas, where human history is very long and disentangling human history from ecosystem history is essentially impossible, not to mention nonsensical (Bakker et al. 2000; Jackson & Hobbs 2009). The debate over whether humans are natural parts of North American ecosystems lingers today (Comer 1997; Haila 1997; Hunter 1997), but in our view, it is imperative to include humans in discussions of the ecological past, recognizing that human influences on ecological history vary from significant to slight depending on the location and time frame (Vale 2002). In this chapter, we describe the relative importance of human effects on past ecosystems in North America and, in light of the evidence, discuss the relevance of human history to the historical range of variation (HRV) concept. A spectrum of human–environmental relations is discussed, including human predation on wildlife populations, and the effects of deliberate burning and agriculture/horticulture on vegetation. Human–environmental relations vary among ecosystems and cultures and are illustrated for a representative set of ecoregions across the United States.

6.2 HUMAN ENTRY INTO THE "NEW WORLD"

Throughout the Pleistocene, North American ecosystems reorganized rhythmically with glacial cycles roughly every 100 000 years, driven by an array of processes absent of humans (Pielou 1991; Delcourt 2002). The mix of drivers changed at the end of the last glacial period when human arrival ushered in the Holocene Epoch. Most Native Americans are the direct descendants of northeast Asians (Meltzer 2003) who entered the New World via the Bering Land Bridge approximately 17–22 000 years ago (Mann & Hamilton 1995). Upon arrival, humans brought with them three key things that would profoundly change the American landscape: (1) tools (including weapons), (2) dogs (Fiedel 2005), and (3) fire (Stewart 2002).

Populations pooled in Eastern Beringia (Alaska and Yukon Territory), a cul-de-sac of sorts, until ice sheet retreat sometime between 13 000–14 000 years ago (Mann & Hamilton 1995; Meltzer 2003). Once south of the glacial barriers, migrants dispersed swiftly across America, most likely in waves of genetically and linguistically divergent tribal groups (Rogers et al. 1990). For instance, one of the early cultures, the Clovis, may have expanded across North America in as little as 200 years (Waters and Stafford 2007). The Clovis peoples had the traits of big-game hunters, coupling a diverse toolkit, including fire and large-fluted projectiles, with high mobility and broad habitat use. Megafauna (large animals like mammoths [*Mammuthus*], mastodons [*Mammut*], and ground slouths [*Megatherium* and *Eremotherium*]) were preferentially sought as food. As conspicuous targets on the landscape that left abundant signs of their presence (trails, dung, and broken branches), megafauna could be pursued with relative ease, even as their numbers dwindled (Haynes 2002). Aided with dogs, the ability to track, engage, and slay megafauna and transport meat from kill sites would have made Clovis groups incredibly efficient and lethal hunters (Fiedel 2005). Because of all these factors, the Clovis cultures have been implicated in Late Pleistocene megafaunal extinctions (Haynes 2002).

Megafaunal extinction at the hands of humans

The extinction of megafauna was an exceptional ecological event with consequences still reverberating through ecosystems today. A total of 34 megafaunal genera went extinct in North America in the Late Pleistocene, including all mammals larger than 1000 kg (Koch & Barnosky 2006). Numerous hypotheses have been advanced to explain megafaunal extinction in North America (Koch & Barnosky 2006). One of the more durable explanations was offered by Martin (1967), entitled the "Pleistocene overkill" theory. Since initially proposed, this human-based extinction hypothesis has endured scientific debate with ongoing research and changing philosophies (Grayson & Meltzer 2003; Fiedel & Haynes 2004). Skeptics claim that severe climatic reversals and associated vegetation shifts during the Pleistocene-Holocene transition were to blame. But then, why were megafauna able to survive numerous prior glacial-to-interglacial transitions only to succumb in the last one? The only apparent difference was human presence. Human-based extinctions are not unprecedented, having occurred across the globe over millennia (Martin & Steadman 1999; Koch & Barnosky 2006).

Extinction across a range of animals, from grazers to browsers and generalists to specialists, strongly argues against climate-based vegetation shifts as the probable cause, but is wholly consistent with human-hunting models (Owen-Smith 1987; Koch & Barnosky 2006). Humans affected megafauna directly (hunting) and indirectly (through competition and habitat alteration) in the midst of postglacial climate change. Changing climate and habitat conditions may have added stress to megafauna populations, but it was humans who ultimately did them in (Haynes 2002). Without human predation, it is unlikely that a Late Pleistocene mass extinction would have occurred (Koch & Barnosky 2006).

6.3 ECOSYSTEMS AFTER MEGAFAUNA – HUMANS BEGIN TO TAKE CONTROL

Megaherbivores were the keystone species of their time, disproportionally affecting their environs and embedded food webs. Their ability to radically transform landscapes, open forest and shrub canopies, and promote habitat diversity influenced a multitude of animals at all trophic levels (Owen-Smith 1987). Large proboscideans (mammoths and mastodons) were particularly adept at habitat modification through grazing, browsing, trampling, and wallowing. Megaherbivore extinctions rippled through the animal kingdom, causing further population reductions, food web collapses, and extinctions among animals dependent on megafaunal activities or presence. Openland-dependent, nonmigrating animals and large carnivores and scavengers were most vulnerable (Haynes 2002).

As Johnson (2009, p. 2509) aptly states, "big herbivores have big effects on plants" and thus on community composition, structure, and patterns. Some North American trees such as honey locust (*Gleditsia triacanthos*), hawthorns (*Crataegus*), mesquite (*Prosopis*), and Osage-orange (*Maclura pomifera*) still carry obsolete megaherbivore defenses in the form of stout thorns (Johnson 2009; Bronaugh 2010). Upon extinction, many megaherbivore-limited food sources suddenly became plentiful (Janzen & Martin 1982); the liberation of preferentially browsed trees, especially hardwoods, is clearly seen in the pollen record (Gill

et al. 2009). Conversely, anachronistic species reliant on megaherbivores for seed scarification and dispersal declined, including Kentucky coffee tree (*Gymnocladus dioicus*), honey locust, Osage-orange, pawpaw (*Asimina triloba*), and persimmon (*Diospyros virginiana*) (Janzen & Martin 1982; Barlow 2001). In the post-megafauna world, the trajectories of ecosystem development were markedly different than any preceding interglacial period (Johnson 2009; Bronaugh 2010), leading to a complete reorganization of vegetation with new species assemblages and food webs. From this standpoint alone, humans have profoundly affected ecosystem development across the Americas since the onset of the Holocene.

Humans and fire

What characteristic best differentiates humans from other species: language, intelligence, use of tools? Some would argue it is the use of fire, as no other species has been able to master it (Goudsblom 1987). Fire is an important evolutionary force – a force that humans have used to shape environments across the globe (Sauer 1975; Bond & Keeley 2005). The domestication of fire was intrinsic to the development and spread of human culture, dating back at least 100 000 years and possibly over 1 million years (Bowman et al. 2009; Pausas & Keeley 2009). Invariably during this period, where there was man, there was fire (Sauer 1975; Pausas & Keeley 2009). The people traversing the cold Arctic environments surely brought fire technologies to North America.

The spread of fire- and weapon-toting humans across North America left an indelible mark on the paleoecological record. Disturbance regimes changed from megaherbivory to human-based fire. The rise of fire (as measured in charcoal) commensurate with megafauna decrease (measured in dung fungal spores) is reported throughout North America (Davis & Shafer 2006; Gill et al. 2009). An increase in fire may have compensated for ecological changes expected from declines in herbivory, as fire can be considered an "herbivore" of sorts (see Bond & Keeley 2005). The switch from megaherbivory to a human-mediated fire regime led to a new set of plant communities, resulting from the rebound of plants formerly suppressed by herbivory (relaxation), reduction of anachronistic plants, and the promotion of fire-adapted plants (Gill et al. 2009; Johnson 2009).

The rise of agriculture

An extensive switch in human diet ensued after megafaunal extinction. Human population in North America likely declined responding to the combined loss of a significant food source (megafauna) and deterioration of climate during the Younger Dryas. After megafaunal extinction, people switched to smaller Folsom points designed for smaller prey. Subsistence technologies diversified, allowing for greater utilization of resources, especially plants.

With the advent of agriculture about 5000 BP, eastern North America became one the five major centers of independent agricultural development in the Americas (Smith & Yarnell 2009). During this time, people transitioned from highly mobile hunter-gatherers to more sedentary, agriculture-based societies (at least in those areas that supported agriculture). By 3800 BP, agriculture in the eastern United States had become more intense, sophisticated, and responsible for an increasing share of total human nutrition. Human societies responded by growing and becoming more complex. Agrarian lifestyles flourished in warmer, more productive southern portions of the continent. Especially after 800 AD, Three Sister agriculture (composed of maize [*Zea*], beans [*Phaseolus*], and squash [*Cucurbita*]) arose in southeastern North America, eventually spreading to many parts of what is now the United States by the sixteenth century.

Across all cultural phases of the Holocene, fire continued to be a key tool in food acquisition, especially in maintaining open habitats, fostering berry- and nut-producing shrubs and trees, game hunting, and land clearance for gardening and habitation (Doolittle 2000; Williams 2000; Abrams & Nowacki 2008). Upon their arrival along the coasts, Europeans did not "discover" an untrammeled wilderness but rather a kaleidoscope of ecosystems reflecting the many native peoples who lived there (Heizer 1978; Patterson & Sassaman 1988; Suttles 1990). Europeans found the preconditioned landscapes (old Indian villages, gardens, and fire-maintained clearings) most suitable for settlement as they afforded the best chances for survival (Mann 2005). Such locations were often the most productive and geographically strategic on the landscape. Butzer (1990, p. 27) wrote that in North America, there existed "a pre-European cultural landscape, one that represented the trial and error as well as the achievement of countless human generations. It is upon this imprint that the more familiar Euro-American landscape was grafted, rather than created anew."

6.4 HUMAN-MODIFIED LANDSCAPES AT THE EVE OF EUROPEAN CONTACT

By the time Europeans arrived in America, indigenous peoples had occupied the continent for many millennia. Human effects on North American landscapes were pervasive in the most productive regions (Southeast, West Coast) and limited in the least productive regions such as the boreal North and interior West (Pyne 1982; Patterson & Sassaman 1988; Vale 2002). Pre-European population density would appear to be a good proxy for human ecological impact. Tribal cultures were an integral part of productive landscapes, serving as active change agents who intentionally managed resources (Stewart 1963; Nicholas 1988). Native Americans affected ecosystem development in enumerable ways, including hunting, gathering, fishing, agriculture, arboriculture (active planting and dissemination of desired woody species), wood gathering, village and trail construction, and habitat manipulation (Williams 2000; Abrams & Nowacki 2008). Fire was used for many of the aforementioned activities, ultimately being the tool of choice to manage landscapes for many socioeconomic benefits (Pyne 1983; Williams 2000).

Through examples spanning North America from east to west, the remainder of this chapter illustrates the diverse and often profound influence Native Americans had on ecosystems across the continent. We geographically organize our review of historical human effects on ecosystems, using standard ecological divisions of North America as a template (Cleland et al. 2007; Fig. 6.1).

Eastern North America

A large percentage of global vegetation does not reflect climax conditions set by climate, but rather represents subclimax conditions largely driven by fire (Bond et al. 2005). This was certainly the case in the eastern United States where fire-dependent communities, such as tallgrass prairies and oak (*Quercus*), pine (*Pinus*), and oak-pine woodlands, dominated at the time of European contact (Frost 1993; Hamel & Buckner 1998; Nowacki & Abrams 2008). These principally

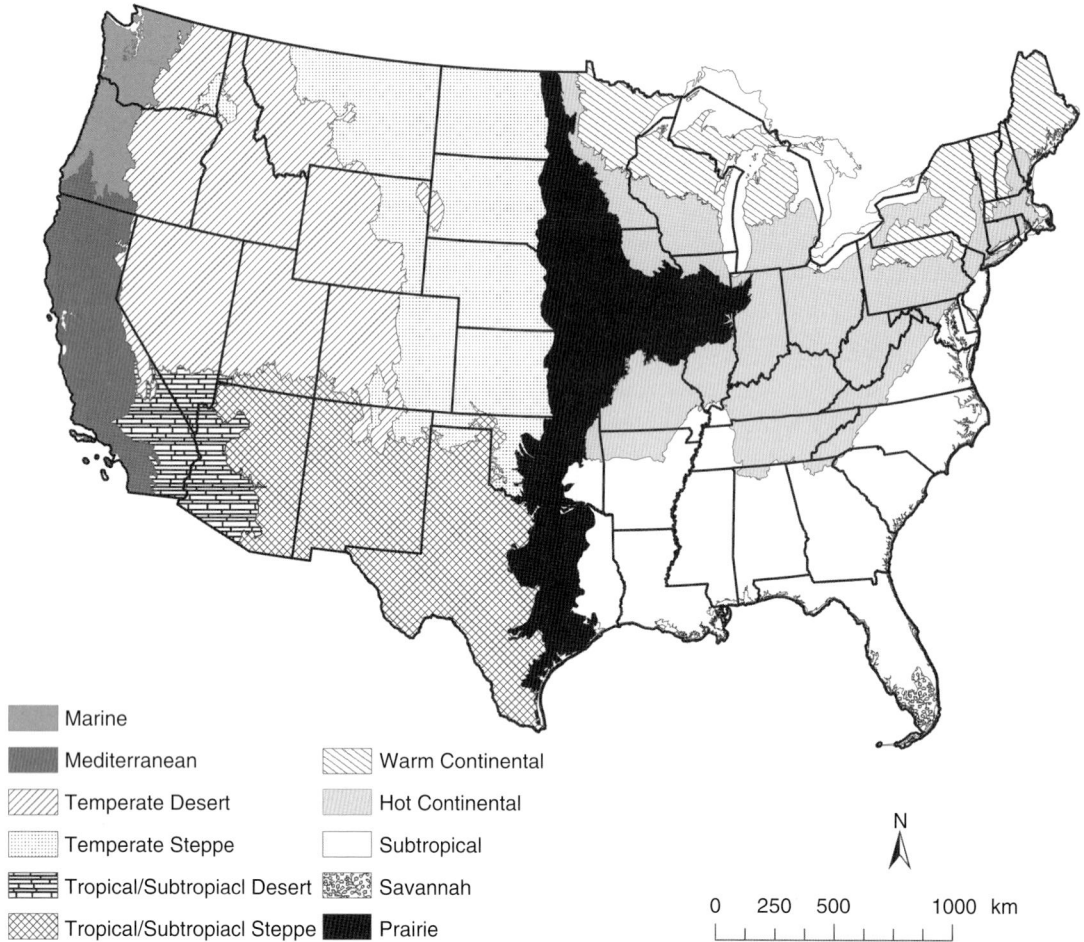

Fig. 6.1 Ecological Divisions of the continental United States (source: Cleland et al. 2007).

open vegetation types sharply contrast with the closed-canopy broadleaf forests that would have otherwise prevailed (Bond et al. 2005). Fire was the principal mechanism controlling vegetation expression in the East, as without it most forested systems would have been composed of late-successional species such as hemlock (*Tsuga*), fir (*Abies*), maple (*Acer*), basswood (*Tilia*), beech (*Fagus*), and magnolia (*Magnolia*). Fires were largely attributed to Native Americans as most were dormant season burns when lightning was at a minimum (Barden 1997; Lorimer 2001; Abrams & Nowacki 2008). Moreover, once started, Native Americans had little incentive or means to put fires out,

hence explaining the large spatial extent of fire-dependent ecosystems across the East.

At earliest European contact, human modifications of landscapes were not uniform along the East Coast, but followed a north-south productivity and human population gradient (Driver & Massey 1957; Patterson & Sassaman 1988). In the far north, Native Americans were more oriented to hunting, fishing, and gathering since the cool, short growing season was less conducive to agriculture. The low carrying capacity of the land restricted human populations, and in turn their environmental effects. One exception was wild rice (*Zizania*) husbandry, which thrived within this lake-studded region. There is some evidence that maize consumption and possible cultivation took place in the southern portions of the boreal forest, but again environmental effects would have been minimal (Boyd & Surette 2010). Human ignitions only supplemented the principally lightning-based fire regime.

Warm Continental Division: conifer-northern hardwood systems

Land impacts were more conspicuous where boreal conifers intermingled with temperate broadleaf trees (Warm Continental Division, Fig. 6.1), although effects were primarily confined to areas around villages, encampments, and trails (Campbell & Campbell 1994; Clark & Royall 1995). Here, the prevailing cool and moist climate coupled with the pyrophobic hardwoods (wet, flaccid leaf litter) greatly suppressed fire, thwarting human's most potent land-altering tool. Consequently, Native Americans focused on smaller, more manageable burns immediately surrounding their river- and lake-side villages, creating disturbance patches or cultural "islands" within a sea of climax northern hardwoods (Table 6.1; Mann et al. 1994). Western New York and northwestern Pennsylvania are representative, where small clearings and oak-dominated forests occurred in areas previously occupied by the Iroquois, especially along rivers and pathways (Ruffner & Abrams 2002; Black et al. 2006). Here, the small and localized effects of Native Americans were reflected in land surveys, with activities recorded on <1% of the bounds surveyed in the 1790s; since this land survey followed Iroquois depopulation, this estimate is biased low but remains an indication of the relatively small human footprint (Marks & Gardescu 1992). The amount of land disturbance undoubtedly oscillated over time contingent with Native American movements and population levels (Ruffner & Abrams 2002; Stambaugh & Guyette 2006).

The intensity of human impacts radiated from cultural centers in concentric rings, with fields within 1 km, fuelwood collection and burning for berry production within 4–5 km, and foraging and small game hunting within 6–8 km (Williams 1989; Black et al. 2006). One string of villages along the shores of Green Bay (Wisconsin) is illustrative of the human footprint on the northern hardwood matrix (Dorney & Dorney 1989). Here, Potawatomi and Winnebago peoples established an agricultural-horticultural landscape of fields, open lands, and oak savannas and forests. Burning apparently created these vegetation conditions as soils, topography, or climate did not appreciably differ between the human-modified locale and the distal northern hardwoods.

Similar impacts were found in southern Ontario (Clark & Royall 1995; Munoz & Gajewski 2010), where the Native American footprint probably did not exceed 3.2% of the total land base (Campbell & Campbell 1994). At Crawford Lake, sediment analysis showed that increases in charcoal corresponded with increases of fire-dependent pine (*Pinus*) and oak (*Quercus*), the appearance of maize, weedy grasses, and purslane (*Portulaca*) (distinct cultural markers), and decreases of maple and beech from ca. 1350 to 1650 AD. Burning was cited as the primary factor in converting northern hardwoods to open systems of pine and oak. Furthermore, human ignitions have been implicated in the extension of oak-hickory forests up major river valleys along the southern margin of northern hardwood system (Cogbill et al. 2002; Black et al. 2006). Acorns were a highly prized, storable carbohydrate source that supplemented Native American diets, motivating land disturbances that promoted oak and associating species (e.g. blueberries) within the northern hardwood complex (Abrams & Nowacki 2008). Dunham (2009) found archeological sites disproportionally associated with oak habitats in the eastern Upper Peninsula of Michigan. Overall, human land alterations promoted biological and landscape diversity and certainly increased the abundance of shade-intolerant, fire-adapted, and culturally important plants in these Northwoods.

Hot Continental Division: central and Appalachian hardwoods

Human land modifications flourished south of the cool northern hardwood system where much of the area was either directly or indirectly under Native American control. The favorable climate of the Hot Continental Division (Fig. 6.1) supported agriculture and acorn- and nut-based cultures. Catchment analysis revealed that local cutting and burning practices (for agriculture, fuelwood, and construction materials) increased the representation of hickory (*Carya*), walnut (*Juglans*), and black locust (*Robinia pseudoacacia*) (Black & Abrams 2001), whereas broadcast burns led to the dominance of oak (*Quercus*), hickory, and American

Table 6.1 Probable climax vegetation versus actual dominant vegetation at the time of European settlement by Ecological Division.

Division	Climax vegetation[1]	Dominant presettlement vegetation[2]	Native American Influence[3]
Warm Continental	Conifer-northern hardwoods	Conifer-northern hardwoods	Local; limited to settlements and travel corridors
Hot Continental	Mixed mesophytic forests	Oak and oak-pine woodlands	Widespread; landscape burns through Native American ignitions
Subtropical	Beech-magnolia forests	Southern pine woodlands	Widespread; landscape burns through Native American ignitions
Prairie	Mixed mesophytic and oak forests	Tallgrass prairie	Ubiquitous; prairies wholly dependent on Native American burning
Temperate Steppe	Shrublands	Short and mid-grass prairie	Widespread; Native American ignitions increased fire frequency to help support grasses
Temperate Desert	Sagebrush-grass, ponderosa pine, and shade-tolerant conifers	Sagebrush-grass and ponderosa pine	Local to widespread (depending on terrain)
Tropical/ Subtropical Desert/Steppe	Shrublands and pinyon-juniper	Grasslands and pinyon-juniper	Local; effects concentrated on the most productive areas
Marine	Western hemlock-Douglas-fir forests	Douglas-fir forests	Local; prairies and shrublands maintained by Native American burning.
Mediterranean	Mixed conifer-oak forests	Oak woodlands and chaparral	Widespread; open communities maintained by Native American burning

[1] Postulated climax vegetation (uplands) in accordance with the disturbance regime without humans.
[2] Actual dominant vegetation (uplands) in accordance with the disturbance regime with humans.
[3] Stewart (2002), Abrams and Nowacki (2008), Williams (2000), and references therein.

chestnut (*Castanea*) over much of the surrounding landscape (Delcourt & Delcourt 1997; Nowacki & Abrams 2008). Without human-based fire, much of the region would have been dominated by shade-tolerant mesophytic trees, such as maple, beech, and basswood (Lorimer 1992; Table 6.1). Interestingly, the interface between oak-dominated ecosystems of this division and the northern hardwood ecosystems to the north was probably more of a function of anthropogenic fire than climate (Cogbill et al. 2002).

Descriptions of the open character of this region were very common in the notes of early observers. John Smith commented that around Jamestown, Virginia "a man may gallop a horse amongst these woods any waie, but where the creekes and Rivers shall

hinder" (Williams 1989, p. 44). Andrew White, on an expedition along the Potomac in 1633, observed that the forest was "not choked with an undergrowth of brambles and bushes, but as if laid out in by hand in a manner so open, that you might freely drive a four horse chariot in the midst of the trees" (Williams 1989, p. 44). Such observations of the open nature of oak forests from the East Coast through the Ozarks are typical of those of most early observers, who commonly spoke of the ease of riding a horse or driving a wagon under a park-like canopy.

A mix of grasslands and open woodlands was maintained specifically for game, an important protein source for indigenous peoples (Abrams & Nowacki 2008). There is evidence that Native Americans were encouraging the eastward migration of bison (*Bison bison*) through habitat manipulation at the time of European contact (Rostlund 1960; Pyne 1982). Sizable grasslands surrounded by oak-dominated woodlands once existed here. In Kentucky, a vast grassland on the Pennyroyal Plateau measured approximately 249 km (155 m) long and 19 km (12 m) wide (Lorimer 2001). In Virginia, the Shenandoah Valley was one vast grass prairie covering more than 2590 sq km (1000 sq mi) where Native Americans burned annually (Van Lear & Waldrop 1989). After burning by indigenous peoples ceased, much of this area reverted to forest and the early white settlers had to clear land that had only recently been prairie (Rostlund 1957). R.C. Anderson (1990, p. 14) writes that the eastern prairies and grasslands "would mostly have disappeared if it had not been for the nearly annual burning of these grasslands by the North American Indians." Indeed, the existence of the Prairie Peninsula (Prairie Division; Fig. 6.1) in the humid East is thought to be largely an artifact of Native American burning (Anderson 2006). Where not plowed or pastured by Europeans, the open character of the land quickly reverted to dense, closed-canopy oak forests due to fire suppression. Under today's subdued fire regime, these oak forests are undergoing mesophication and converting to climax forests of shade-tolerant maple, beech, and elm (*Ulmus*) with little understory diversity (Table 6.1) (Nowacki & Abrams 2008).

Subtropical Division: southern pines and associated ecosystems

At the time of European contact, the ecological effects of indigenous peoples were likely more significant in the Southeast than anywhere else in North America. This was due to relatively high population densities (Williams 1989), the widespread application of agriculture, as well as plant communities and climatic conditions that facilitated human use of fire.

It is difficult to accurately determine ecological conditions at European contact (1500 AD) as there were few European observers in this region at that time. One of the most important was the Hernando de Soto expedition. From 1539 to 1542, de Soto (with his 600 men, 200 horses and 300 swine) pillaged, plundered, and inadvertently spread European diseases from Florida to North Carolina, west across the Mississippi River, then down to the Gulf of Mexico (Thomas et al. 1993; Mann

2005). Even with its large numbers of men and animals, the de Soto expedition moved with relative ease throughout the Southeastern landscape, and chroniclers wrote of expansive agricultural fields, open park-like forests, numerous villages, and large numbers of people (Swanton 1939; Rostlund 1957).

In what is now Alabama, a de Soto chronicler reported that the land was "so fertile and thickly populated that on some days the Spaniards passed 10 or 12 towns, not counting those that lay on one side or the other of the road" (Rostlund 1957, p. 385). Upon arriving at the Mississippi River, de Soto found a landscape teeming with humans. The river itself was lined with villages (Mann 2005). Eerily, by the time the next explorer passed through this area more than a century later (La Salle in 1682), the entire valley had been radically transformed into a relatively silent place (Mann 2005). Where de Soto had observed scores of villages, expansive agricultural fields, and high human populations, La Salle found mostly forest with very few people or villages. The country had been depopulated (80–95%) by European diseases, and the ecology of the area substantially altered (Young & Hoffman 1993; Mann 2005; Scharf 2010).

Landscapes cleared for agriculture or routinely burned had two or more centuries to recover before the first waves of permanent Euro-American settlers arrived to find landscapes that were more "pristine" than they had been in more than a thousand years (Denevan 1992; Scharf 2010). In addition to the rapid succession of extensive old fields, the depopulation of indigenous peoples undoubtedly led to changes in wildlife populations as a major predation source was substantially reduced. For instance, bison expanded as far south as Florida and as far east as Virginia and Pennsylvania; the large numbers of bison transformed many areas as they grazed, creating large wallows and well-worn migration corridors, some of which remain visible today (Rostlund 1960; Belue 1996).

Eyewitness reports describing ecological conditions in the Southeast become more common after 1600 (Smith 1616; Lindestrom 1656, Lederer 1672, Catesby 1731; Bartram 1791). Many of these writers noted extensive "ancient" Indian plantations and abandoned fields extending for kilometers along rivers (see in particular Bartram 1791). The most common ecological conditions reported by these observers were open forests, interspersed with grasslands and meadows, with extensive cane lands along the rivers. Bartram's journals contain numerous references to "delightful groves" of open grown "stately forests" of oak, ash, hickory, walnut, and so on, as well as "vast open forests" (Bartram 1791). References to dense forests of late successional species are rare indeed. Rostlund (1957, p. 408) concludes that "the open, parklike appearance of the woodlands, undoubtedly the most common type of forest in the ancient Southeast, was mostly the work of man."

Frequent forest burning did more than create open stands of shade-intolerant trees, but created grasslands where forests would have otherwise existed. There are many references to treeless areas in the early literature, which were often referred to as "barrens," "plains," "meadows," or "savannahs." In Bartram's journal, there are numerous separate references to "vast meadows," "extensive savannas," and "large grassy plains," some of which were reported to be many kilometers in length (Rostlund 1957). Bartram reported that the Alachua Savanna in northern Florida was "a level green plain, above 15 m (24 km) over, 50 m (80 km) in circumference, with scarcely a tree to be seen" (Rostlund 1957, p. 408).

Canebrakes were also a major feature of Southern bottomlands at the time of European settlement (Platt & Brantley 1997). Indigenous people valued cane for food, shelter, baskets, and tools, especially weapons (Hamel & Chiltoskey 1975). William Bartram repeatedly remarked on canebrakes during his southeastern travels, describing "vast cane meadows," "widespread cane swamps," or "an endless wilderness of canes" (Platt & Brantley 1997, p. 10). The area of canebrakes declined rapidly in the eighteenth century from altered disturbance regimes by Euro-Americans, including cattle grazing, agricultural displacement, and changes in fire frequency. Today, cane has been virtually eliminated from the Southeastern landscape and is a contributing factor in the extinction of Bachman's warbler (*Vermivora bachmanii*) (Remsen 1986).

At the time of European contact, the Coastal Plain was dominated by open stands of large pines (Bartram 1791). Within the range of the fire-tolerant longleaf pine, which covered about 37 million hectares, fire-sensitive southern mixed broadleaved forests (beech, magnolia, semi-evergreen oaks) were restricted to moist, fire-protected locations (Frost 1993). In the southern portion of longleaf pine distribution, summer lightning fires were sufficient to maintain longleaf pine (Komarek 1964), but burning by indigenous people likely contributed to range extension of longleaf and other southern pines into topographically dissected

areas where it would not otherwise have occurred (Frost 1993). By the early 1990s, Frost (1993) estimated that about 1 million hectares remained in naturally regenerated longleaf pine, with only about 270 000 ha (<0.7% of the original range) in a condition similar to the classic open-grown, fire-maintained, longleaf pine-wiregrass community. Like canebrakes, the longleaf pine/wiregrass community is currently considered a critically endangered ecosystem (Noss et al. 1995).

The Great Plains and interior west

Native American populations became progressively more sparse west of the Mississippi River across the Great Plains (Temperate Steppe Division) through the Rocky Mountains (Temperate Desert Division; Fig. 6.1), essentially following an aridity gradient. Together with changes in topography, Native American–environmental relations varied broadly across this western expanse. The relatively flat and seasonally dry conditions in the center of the continent were naturally conducive to extensive fire. The vast grasslands that epitomized the Great Plains were promulgated by frequent surface burning (Wright & Bailey 1980), and fed large ungulates such as bison, elk (*Cervus canadensis*), deer (*Odocoileus*), and later horses (*Equus*). Ungulate populations were regulated by hunting where denser Native American populations existed, perhaps to the point of being largely relegated to the outskirts of tribal territories (Martin & Szuter 1999; Kay 2007). Well versed in prey–habitat relations, indigenous peoples promoted preferred habitat conditions through burning and were the principal igniters outside the thunderstorm season (April to September) when optimal fuel conditions (dry grass) existed. Without the addition of human ignitions, the Great Plains may have been substantially shrubbier, supporting fewer large herbivores (Arno & Gruell 1983; Stewart 2002; Table 6.1). Post-European agriculture, grazing, and fire suppression adversely affected the species composition and diversity of the Great Plains.

In the Rocky Mountains (Temperate Desert Division), resources important to indigenous peoples were geographically spread from river valley to alpine meadow. Evidence suggests that management of resource patches by indigenous people was common, although at decreasing intensity with elevation (Barrett & Arno 1982; Hessburg & Agee 2003). Many tribes practiced seasonal rounds, migrating to higher elevations during the summer, and back to winter villages in the lowlands. Along the way, both the flora and fauna of these mountainous ecosystems were affected by Native American hunting, plant gathering, and burning. Early observations of western fires compiled by Gruell (1985) revealed that many were set by indigenous peoples, especially at lower and middle elevations. Here, intentional burning helped create and maintain a mosaic of stand conditions across the landscape. Periodic burning helped promote root and berry crops, materials for basketry, forage for prey animals, and improved visibility for hunting (Stewart 2002). However, given the inherent severity of mountain and high desert environments (rugged topography, short growing seasons, temperature extremes, snow, and aridity), Native Americans probably had a much lighter touch in this region relative to other, more productive areas of North America (Vale 2002). Nevertheless, human ignitions were embedded in historical fire regimes that supported a diverse vegetation mosaic – a mosaic now being unraveled by fire suppression (Pyne 1982; Arno & Gruell 1983; Brown & Hull-Sieg 1996). Repeat photography clearly shows the dramatic recent increases in forest cover in many western landscapes (Progulske 1974; Gruell 1983; Skovlin & Thomas 1995).

West Coast

Marine Division: temperate rainforests

Lightning has been accepted as the main ignition source of fires in this topographically diverse region, with frequency of strikes increasing with distance from coast and with elevation (Agee 1993; van Wagtendonk & Cayan 2008). Fires of the mesic forests in Coast Ranges and Cascades were infrequent, stand-replacing, high-severity events (Agee 1993; Lertzman et al. 2002), although mixed severity and more frequent low-moderate severity fire regimes existed in mixed conifer-hardwood vegetation and grassland communities (Agee 1993; Weisberg & Swanson 2003; Taylor et al. 2008). Where summer lightning fires were not adequate to foster and maintain the desired condition, tribal burning was used. The extent to which Native American fires affected historical fire regimes of these mesic conifer-dominated forests is debated (Agee 1993; Boyd 1999).

The longevity and extent of burning practices by diverse tribal groups across the Pacific Northwest and California is not clear from archeological interpretations (Ames & Maschner 1999; Boyd 1999). Late Holocene tribal groups did not practice sedentary agriculture, but rather a proto-agriculture/horticulture integrated with fire regimes (Lewis 1993; Boyd 1999). Vegetation nearest villages, trail systems, and remote resource collection areas (i.e. camps) was selectively modified by altering the frequency, seasonality, and extent of fires (Boyd 1999). Cultural practices sought to increase wildlife forage, root, seed, nut, and berry crops, enhance basketry materials, and improve hunting opportunities (Boyd 1999).

Mediterranean Division: oak balanocultures

Across the major mountain ranges and valleys of California, the establishment and increased abundance of oaks (*Quercus* and *Lithocarpus*) took place in the Mid-Holocene approximately 9000–10 000 years ago (West et al. 2007; Briles et al. 2008). Oaks likely reached their present distribution during the Late Holocene (West et al. 2007).

Frequent lower intensity fires promoted oak-dominated habitats. Dendroecological studies indicate that fires occurred primarily during the summer seasonal drought, typical of the lightning ignitions (van Wagtendonk & Fites-Kaufman 2006). However, there is also evidence of spring burns, which suggests human ignitions (Skinner et al. 2009).

Estimates of precontact (circa 200 years BP) fire-return intervals range from annual to decadal events for Oregon white oak/grassland communities (McDadi & Hebda 2008), and from 5 to 20 years for mixed oak/conifer-hardwood forests of California (Skinner & Taylor 2006). Charcoal and pollen analyses suggest a reduction in fire frequency during European settlement (McDadi & Hebda 2008), leading to the conversion of oak-dominated savannas to closed-canopy forests of more mesic, fire-sensitive species. This change occurred despite broad climatic similarity between pre- and post-European eras (Byrne et al. 1991).

The rapid change in grassland/oak vegetation coincident with European settlement during a relatively stable climate provides a striking contrast to the persistence of grassland/oak vegetation over long periods of variable climate during Native American management. Subsistence economies (DeLancey & Golla 1997) associated with oak grasslands remained fairly constant over the last 2500 years in the Sacramento and

San Joaquin valleys (Elasser 1978). Those tribal groups inhabiting inland oak/grass-dominated valleys are thought to have been deliberately broadcast-burning vegetation by 3000, if not 5000 years BP (Weiser & Lepofsky 2009). Archeological evidence indicates the majority of nonmarine/littoral-adapted tribal groups developed important associations with oak-dominated habitats. Milling stones and related lithic tools (e.g. mortars and pestles) used for processing seeds, nuts, or other foods by tribal groups suggest an adaption to use acorns as early as 7000 years ago in the California North Coast Range and 4000 years BP in the San Francisco Bay Area, Sacramento Valley, and southern California (Elasser 1978; Keeley 2002; Arnold et al. 2004).

By 1300 years ago, acorns were a cultural mainstay in the Sacramento/San Joaquin Valleys and in southern California (Elasser 1978; Arnold et al. 2004). The bow and arrow was introduced by 1500–2000 years ago, and coincides with other technological changes (e.g. mortars and pestles) associated with diversified subsistence economies (acorns, roots, seeds/nuts, fish, game, etc.). The use of fire by tribal groups to promote oak dominance coincided with the need to counteract natural successional tendencies, and the need to maintain tribal economies dependent on oaks and associated species that provided valued food, materials, and medicines; various tribal groups exhibited proprietor claims to tracts of oaks and associated resources (game and bulbs/roots) within these habitats (Anderson 2007; McDadi & Hebda 2008). Inland areas codominant with oaks, and especially traversable ridge systems with southern exposure, were kept open through burning by tribal groups that depended on resources found in early seral oak grasslands (Boyd 1999; Keeley 2002).

A combination of biophysical conditions and tribal burning likely fostered oak dominance and persistence. It is unknown just how much of the oak-dominated landscape was influenced by tribal burning (Boyd 1999; Anderson 2007). Oak habitats have greatly declined since Euro-American settlement due to the cessation of tribal burning, grazing of livestock, fire suppression, industrial forestry practices (favoring conifers), and urbanization (Hosten et al. 2006; Anderson 2007).

6.5 NATIVE AMERICANS AND THE HRV CONCEPT

Precontact Native American cultures were sophisticated, dynamic, diverse social organizations with indi-

viduals who knew how to alter their surroundings for their benefit (Pyne 1983). Since their postglacial arrival, human populations and their scope of influence increased as they dispersed across the Americas, reaching a peak at the time of European contact (Mann 2005; Scharf 2010). By that time, human population density closely paralleled gradients in land productivity, ranging from low to high densities from north to south in eastern North America and from mountains and deserts (low) to fertile valleys and coastlines (high) in the West. As such, the geographical influences of Native Americans were not uniform, but reflected differences in populations, land use, cultural traditions, available resources, climate, and vegetation flammability (Kimmerer & Lake 2001; Stewart 2002; Vale 2002).

In North America, literature illustrating the various roles of early humans in environments has not been well integrated into our ecological knowledge. The reasons are many, including historic and cultural biases and political agendas, as well as the challenges associated with obtaining clear scientific evidence of environmental conditions 300 or more years ago (Stewart 1963; Pyne 1982; Williams 2000). The important role humans played was largely overlooked by most early ecologists, who described North American landscape in terms of static "climax" systems consistent with the prevailing philosophy at the time. Early paleoecologists also largely ignored the human factor, tying Holocene species movements and ecosystem reorganization exclusively to climate (Delcourt & Delcourt 2004; Scharf 2010). Shades of this oversight continue today, through our intense focus on climate change, which tends to assume that vegetation (baseline data for climate-change models) is solely reflective of climate. Indeed, a new synthesis is needed integrating ecology, paleontology, and archeology to discern the complex relations between ecosystems and embedded humans (a panarchical approach; Delcourt & Delcourt 2004). Fortunately, the acknowledgment of past humans as an ecological factor is gaining momentum, and is reflected in the use of "*historical* range of variation" rather than "*natural* range of variation." In this regard, humans are essentially acknowledged as the change agents that put the "H" in the HRV concept.

The preceding sections chronicle a rich array of Native American influences on ecosystems across time and space, beginning with their continental arrival as the new top predator. By employing weapons, dogs, and efficient hunting strategies, Native Americans proved lethal against an unwary megafaunal prey base, and played a leading role in the rapid demise of many large herbivore species. The removal of mega-herbivores had a liberating effect on plant life (especially trees) in the midst of postglacial revegetation of the North American continent. Native Americans also represented a new ignition source – an ever-constant source of flame that was less constrained (spatially, seasonally) than natural ignitions (Stewart 2002; Abrams & Nowacki 2008). Fire ultimately became the most powerful tool for landscape manipulation (Sauer 1975; Kimmerer & Lake 2001), which Native Americans used with purpose and facility. Later, as hunter-gatherer societies shifted toward more sedentary lifestyles with growing technology, direct habitat manipulation through agriculture and horticulture took place where climate and soils allowed. Through these three factors alone, Native American impacts spanned spatiotemporal scales from continental (megafaunal demise and subsequent trophic effects; Early Holocene onward) to landscape (broadcast burning; Mid-Holocene onward) to local (agriculture/horticulture; Late Holocene).

Not all human-related ecological disturbances are equally important to HRV analyses. Application of historical ecology generally relies on time-space scales relevant to current land-management decisions, focusing on shorter, more recent periods with relatively stable climates, vegetation, and disturbance regimes at a particular locale. Coupled with inherent data limitations (e.g. a fading data record with time), timelines are often restricted to the last thousand years or less. To help integrate the consequences of Native American land use into HRV analyses, data collection at a minimum should identify and record principal land-use activities associated with former tribal territories. Linking human-based disturbances (type, frequency, intensity) with plant community structure and composition allows a more complete understanding of ecosystem dynamics, further enlightening the science behind restoration.

Why is it important to understand the role of indigenous people in ecosystem dynamics described by HRV analyses? Altered disturbance regimes affecting many North American ecosystems may result in unsustainable conditions. The "pristine myth" is an enormous cultural barrier to purposeful human intervention in these systems (Hamel & Buckner 1998). Many wildernesses and protected "natural" areas exhibit vegetation conditions at odds with what would have been expected in pre-European landscapes and are on unprecedented

ecological trajectories. Without active human intervention, open woodlands of shade-intolerant species (systems that have dominated many landscapes for millennia) will continue to be replaced by closed forests of shade-tolerant species (Nowacki & Abrams 2008).

Developments in historical ecology have improved our understanding of past disturbance regimes, which are vital to predicting vegetation succession and dynamics. Indeed, many rare and endemic plant and animal species in the United States are disturbance dependent, and many of those historical disturbances were associated with Native American activities. A large proportion of the endangered ecosystems listed by Noss et al. (1995) are fire-dominated ecosystems. These include prairies, pine savannahs and barrens, tropical hardwood hammocks, and shrublands. Of 21 rare communities listed in the Ozark-Ouachita Highlands Assessment, the decline of nine was attributed to fire exclusion (USDA 1999). Nationwide, Owen and Brown (2005) found that of 186 federally listed, proposed, and candidate plant species, 25% required fire, 35% tolerated fire, 38% were not affected by fire, and only 2% were adversely affected by fire.

Resource managers are largely powerless to counter the ecological effects of the substantial land-use changes that have occurred in North America over the last three centuries. But they are not powerless to recognize and address through management activities the ecological effects of altered disturbance regimes. An understanding of natural and human influences on the development of historical landscapes is critical to effectively planning and executing projects designed to restore or conserve rare and endemic species and ecosystems (Hamel & Buckner 1998; Scharf 2010). Developing an appreciation of the past roles of humans in modifying "natural" ecosystems in North America is also a first step toward better understanding and managing the roles humans will play in future ecosystems.

REFERENCES

Abrams, M.D. & Nowacki, G.J. (2008). Native Americans as active and passive promoters of mast and fruit trees in the eastern USA. *The Holocene*, **18**, 1123–1137.
Agee, J.K. (1993). *Fire Ecology of the Pacific Northwest Forests*. Island Press, Washington, DC, USA.
Ames, K.M. & Maschner, H.D. (1999). *Peoples of the Northwest Coast: Their Archaeology and Prehistory*. Thames and Hudson, London, UK.

Anderson, M.K. (2007). Indigenous uses, management, and restoration of oaks of the far western United States. Technical Note No. 2. USDA Natural Resource Conservation Service, National Plant Data Center, Greensboro, NC, USA.

Anderson, R.C. (1990). The historic role of fire in the North American grassland. In *Fire in North American Tallgrass Prairies* (ed. S.L. Collins and L.L. Wallace), pp. 8–18. University of Oklahoma Press, Norman, OK, USA.

Anderson, R.C. (2006). Evolution and origin of the central grassland of North America: climate, fire, and mammalian grazers. *Journal of the Torrey Botanical Society*, **133**, 626–647.

Arno, S.F. & Gruell, G.E. (1983). Fire history at the forest-grassland ecotone in southwestern Montana. *Journal of Range Management*, **36**, 332–336.

Arnold, J.E., Walsh, M.R., & Hollimon, S.E. (2004). The archaeology of California. *Journal of Archaeological Research*, **12**, 1–73.

Bakker, J.P., Grootjans, A.P., Hermy, M., & Poschlod, P. (eds.). (2000). How to define targets for ecological restoration? *Journal of Applied Science*, **3**, 1–72.

Barden, L.S. (1997). Historic prairies in the Piedmont of North and South Carolina, USA. *Natural Areas Journal*, **17**(2), 149–152.

Barlow, C. (2001). Anachronistic fruits and the ghosts who haunt them. *Arnoldia*, **61**, 14–21.

Barrett, S.W. & Arno, S.F. (1982). Indian fires as an ecological influence in the northern Rockies. *Journal of Forestry*, **80**, 647–651.

Bartram, W. (1791). *Travels of William Bartram*. Dover Publications (1955 ed.), New York, USA.

Belue, T.F. (1996). *The Long Hunt: Death of the Buffalo East of the Mississippi*. Stackhole Books, Mechanicsburg, PA, USA.

Black, B.A. & Abrams, M.D. (2001). Influences of Native Americans and surveyor biases on metes and bounds witness-tree distribution. *Ecology*, **82**, 2574–2586.

Black, B.A., Ruffner, C.M., & Abrams, M.D. (2006). Native American influences on the forest composition of the Allegheny Plateau, northwest Pennsylvania. *Canadian Journal of Forest Research*, **36**, 1266–1275.

Bond, W.J. & Keeley, J.E. (2005). Fire as a global "herbivore": the ecology and evolution of flammable ecosystems. *Trends in Ecology and Evolution*, **20**, 387–394.

Bond, W.J., Woodward, F.I. & Midgley, G.F. (2005). The global distribution of ecosystems in a world without fire. *New Phytologist*, **165**, 525–537.

Bowman, D.M.J., Balch, J.K., Artaxo, P., et al. (2009). Fire in the earth system. *Science*, **324**, 481–484.

Boyd, M. & Surette, C. (2010). Northernmost precontact maize in North America. *American Antiquity*, **75**, 117–133.

Boyd, R. (ed.). (1999). *Indians, Fire and the Land in the Pacific Northwest*. Oregon State University Press, Corvallis, OR, USA.

Briles, C.E., Whitlock, C., Bartlein, P.J., & Higuera, P. (2008). Regional and local controls on postglacial vegetation and fire in the Siskiyou Mountains, northern California, USA. *Palaeogeography, Palaeoclimatology, Palaeoecology*, **265**, 159–169.

Bronaugh, W. (2010). The trees that miss the mammoths. *American Forests*, **115**, 38–43.

Brown, P.M. & Hull-Sieg, C. (1996). Fire history in interior ponderosa pine communities of the Black Hills, South Dakota, USA. *International Journal of Wildland Fire*, **6**, 97–105.

Butzer, K.W. (1990). The Indian legacy in the American Landscape. In *The Making of the American Landscape* (ed. M.P. Conzen), pp. 27–50. Unwin Hyman, Boston, MA, USA.

Byrne, R., Edlund, E., & Mensing, S. (1991). Holocene changes in the distribution and abundance of oaks in California. In *Proceedings of the Symposium on Oak Woodlands and Hardwood Rangeland Management; Davis, CA* (R.B. Standiford technical coordinator), pp. 182–188. General Technical Report PSW-126. USDA Forest Service, Berkeley, CA, USA.

Campbell, I.D. & Campbell, C. (1994). The impact of Late Woodland land use on the forest landscape of southern Ontario. *Great Lakes Geographer*, **1**, 22–29.

Catesby, M. (1731). *The Natural History of Carolina, Florida, Bahama Islands*. Beehive Press, Savannah, GA, USA.

Clark, J.S. & Royall, P.D. (1995). Transformation of a northern hardwood forest by aboriginal (Iroquios) fire: charcoal evidence from Crawford Lake, Ontario, Canada. *The Holocene*, **5**, 1–9.

Cleland, D.T., Freeouf, J.A., Keys, J.E., Nowacki, G.J., Carpenter, C.A., & McNab, W.H. (2007). Ecological subregions: sections and subsections of the conterminous United States. General Technical Report WO–76 [CD]. USDA Forest Service, Washington, DC, USA.

Cogbill, C.V., Burk, J., & Motzkin, G. (2002). The forests of presettlement New England, USA: spatial and compositional patterns based on town proprietor surveys. *Journal of Biogeography*, **29**, 1279–1304.

Comer, P.J. (1997). Letters: A "natural" benchmark for ecosystem function. *Conservation Biology*, **11**, 301–303.

Davis, O.K. & Shafer, D.S. (2006). *Sporormiella* fungal spores, a palynological means of detecting herbivore density. *Palaeogeography, Palaeoclimatology, Palaeoecology*, **237**, 40–50.

DeLancey, S. & Golla, V. (1997). The Penutian hypothesis: retrospect and prospect. *International Journal of American Linguistics*, **63**, 171–202.

Delcourt, H.R. (2002). *Forests in Peril: Tracking Deciduous Trees from Ice-Age Refuges into the Greenhouse World*. McDonald and Woodward Publishing Company, Blacksburg, VA, USA.

Delcourt, H.R. & Delcourt, P.A. (1997). Pre-Columbian Native American use of fire on southern Appalachian landscapes. *Conservation Biology*, **11**, 1010–1014.

Delcourt, P.A. & Delcourt, H.R. (2004). *Prehistoric Native Americans and Ecological Change*. Cambridge University Press, Cambridge, UK.

Denevan, W.M. (1992). The pristine myth: the landscape of the Americas in 1492. *Annals of the Association of American Geographers*, **82**, 369–385.

Doolittle, W.E. (2000). *Cultivated Landscapes of Native North America*. Oxford University Press, Oxford, UK.

Dorney, C.H. & Dorney, J.R. (1989). An unusual oak savanna in northeastern Wisconsin: the effect of Indian-caused fire. *American Midland Naturalist*, **122**, 103–113.

Driver, C.H. & Massey, W.C. (1957). Comparative studies of North American Indians. *Transactions of the American Philosophical Society*, **47**, 167–456.

Dunham, S.B. (2009). Nuts about acorns: a pilot study on acorn use in Woodland Period subsistence in the eastern Upper Peninsula of Michigan. *Wisconsin Archeologist*, **90**, 113–130.

Elasser, A.B. (1978). Development of regional prehistoric cultures. In *Handbook of North American Indians, Vol. 8: California* (ed. R.F. Heizer), pp. 37–57. Smithsonian Institution, Washington, DC, USA.

Fiedel, S.J. (2005). Man's best friend: mammoth's worst enemy? A speculative essay on the role of dogs in Paleoindian colonization and megafaunal extinction. *World Archaeology*, **37**, 11–25.

Fiedel, S. & Haynes, G. (2004). A premature burial: comments on Grayson and Meltzer's "requiem for overkill." *Journal of Archaeological Science*, **31**, 121–131.

Frost, C.C. (1993). Four centuries of changing landscape patterns in the longleaf pine ecosystem. In *Proceedings of the Tall Timbers Fire Ecology Conference No. 18* (ed. S. Hermann), pp. 17–43. Tall Timbers Research Station, Tallahassee, FL, USA.

Gill, J.L., Williams, J.W., Jackson, S.T., Lininger, K.B., & Robinson, G.S. (2009). Pleistocene megafaunal collapse, novel plant communities, and enhanced fire regimes in North America. *Science*, **326**, 1100–1103.

Goudsblom, J. (1987). The domestication of fire as a civilizing process. *Theory Culture Society*, **4**, 457–476.

Grayson, D.K. & Meltzer, D.J. (2003). A requiem for North American overkill. *Journal of Archaeological Science*, **30**, 585–593.

Gruell, G.E. (1983). Fire and vegetative trends in the Northern Rockies: interpretations from 1871–1982 photographs. General Technical Report INT-158. USDA Forest Service, Ogden, UT, USA.

Gruell, G.E. (1985). Fire on the early western landscape: an annotated record of wildland fires 1776–1900. *Northwest Science*, **59**, 97–107.

Haila, Y. (1997). Letters: a "natural" benchmark for ecosystem function. *Conservation Biology*, **11**, 300–301.

Hamel, P.B. & Buckner, E.R. (1998). How far could a squirrel travel in the treetops? A prehistory of the southern forest. In *Transactions of the 63rd North American Wildlife and Natural Resources conference, Orlando, FL*, pp. 309–315. Wildlife Management Institute, Washington, DC, USA.

Hamel, P.B. & Chiltoskey, M.U. (1975). *Cherokee Plants and Their Uses: A 400-Year History*. Herald, Sylva, NC, USA.

Haynes, G. (2002). The catastrophic extinction of North American mammoths and mastodonts. *World Archaeology*, **33**, 391–416.

Heizer, R.F. (ed.). (1978). *Handbook of North American Indians, Vol. 8: California*. Smithsonian Institution, Washington, DC, USA.

Hessburg, P.F. & Agee, J.K. (2003). An environmental narrative of Inland Northwest United States forests, 1800–2000. *Forest Ecology and Management*, **178**, 23–59.

Hosten, P.E., Hickman, O.E., Lake, F.K., Lang, F.A., & Vesely, D. (2006). Oak woodlands and savannas. In *Restoring the Pacific Northwest: The Art and Science of Ecological Restoration in Cascadia* (ed. D. Apostol and M. Sinclair), pp. 63–96. Island Press, Washington, DC, USA.

Hunter, M. Jr. (1997). Letters: a "natural" benchmark for ecosystem function. *Conservation Biology*, **11**, 303–304.

Jackson, S.T. & Hobbs, R.J. (2009). Ecological restoration in the light of ecological history. *Science*, **325**, 567–568.

Janzen, D.H. & Martin, P.S. (1982). Neotropical anachronisms: the fruits the Gomphotheres ate. *Science*, **215**, 19–27.

Johnson, C.N. (2009). Ecological consequences of Late Quaternary extinctions of megafauna. *Proceedings of the Royal Society B: Biological Sciences*, **276**, 2509–2519.

Kay, C.E. (2007). Were native people keystone predators? A continuous-time analysis of wildlife observations made by Lewis and Clark in 1804–1806. *Canadian Field-Naturalist*, **121**, 1–16.

Keeley, J.E. (2002). Native American impacts on fire regimes of the California coastal ranges. *Journal of Biogeography*, **29**, 303–320.

Kimmerer, R.W. & Lake, F.K. (2001). The role of indigenous burning in land management. *Journal of Forestry*, **99**, 36–41.

Koch, P.L. & Barnosky, A.D. (2006). Late Quaternary extinctions: state of the debate. *Annual Review of Ecology, Evolution, and Systematics*, **37**, 215–250.

Komarek, E.V. (1964). The natural history of lightning. In *Proceedings of the Third Tall Timbers Fire Ecology Conference*, pp. 139–183. Tall Timbers Research Station, Tallahassee, FL, USA.

Lederer, J. (1672). The discoveries of John Lederer in several marches from Virginia to the west of Carolina and other parts of the continent. Collected and translated by Sir William Talbot (London, 1672); reprint 1902, G.P. Humphry, Rochester NY, USA.

Lertzman, K., Gavin, D., Hallett, D., Brubaker, L., Lepofsky, D., & Mathewes, R. (2002). Long-term fire regime estimated from soil charcoal in coastal temperate rainforest. *Conservation Ecology*, **6**(2), article 5. http://www.consecol.org/vol6/iss2/art5.

Lewis, H.T. (1993). Patterns of Indian burning in California: ecology and ethnohistory. In *Before the Wilderness* (ed. T.C. Blackburn and K. Anderson), pp. 55–116. Ballena Press, Menlo Park, CA, USA.

Lindestrom, P. ([1656] 1925). *Geographia Americae*. (A. Johnson, translator). Swedish Colonial Society, Philadelphia, PA, USA..

Lorimer, C.G. (1992). Causes of the oak regeneration problem. In General Technical Report SE-84. (ed. D. Loftis and C.E. McGee). USDA Forest Service, Asheville, NC, USA.

Lorimer, C.G. (2001). Historical and ecological roles of disturbance in eastern North American forests: 9000 years of change. *Wildlife Society Bulletin*, **29**, 425–439.

Mann, C.C. (2005). *1491: New Revelations of the Americas before Columbus*. Alfred A. Knopf, New York, USA.

Mann, D.H. & Hamilton, T.D. (1995). Late Pleistocene and Holocene paleoenvironments of the North Pacific Coast. *Quaternary Science Reviews*, **14**, 449–471.

Mann, D.H., Engstrom, F.B., & Bubier, J.L. (1994). Fire history and tree recruitment in an uncut New England Forest. *Quaternary Research*, **42**, 206–215.

Marks, P.L. & Gardescu, S. (1992). Vegetation of the central Finger Lakes region of New York in the 1790s. In Late eighteenth century vegetation of central and western New York State on the basis of original land survey records (ed. C.A. Chumbley), pp. 1–35. Bulletin No. 484. New York State Museum, Albany, NY, USA.

Martin, P.S. (1967). Prehistoric overkill. In *Pleistocene Extinctions: The Search for a Cause* (ed. P.S. Martin and H.E. Wright Jr.), pp. 75–120. Yale University Press, New Haven, CT, USA.

Martin, P.S. & Steadman, D.W. (1999). Prehistoric extinctions on islands and continents. In *Extinctions in Near Time: Causes, Contexts, and Consequences* (ed. R.D.E. MacPhee), pp. 17–55. Kluwer Academic/Plenum Publishers, New York, USA.

Martin, P.S. & Szuter, C.R. (1999). War zones and game sinks in Lewis and Clark's West. *Conservation Biology*, **13**, 36–45.

McDadi, O. & Hebda, R.J. (2008). Change in historic fire disturbance in a Garry oak (*Quercus garryana*) meadow and Douglas-fir (*Pseudotsuga menziesii*) mosaic, University of Victoria, British Columbia, Canada: a possible link with First Nations and Europeans. *Forest Ecology and Management*, **256**, 1704–1710.

Meltzer, D.J. (2003). Peopling of North America. *Development in Quaternary Science*, **1**, 539–563.

Munoz, S.E. & Gajewski, K. (2010). Distinguishing prehistoric human influence on late-Holocene forests in southern Ontario, Canada. *The Holocene*, **20**, 967–981.

Nicholas, G.P. (ed.). (1988). *Holocene Human Ecology in Northeastern North America*. Plenum Press, New York, USA.

Noss, R.F., Laroe, E.T. III., & Scott, J.M. (1995). Endangered ecosystems of the United States: a preliminary assessment of loss and degradation. Biological Report 28. USDI National Biological Service, Washington, DC, USA.

Nowacki, G.J. & Abrams, M.D. (2008). The demise of fire and "mesophication" of forests in the eastern United States. *BioScience*, **58**, 123–138.

Owen, W. & Brown, H. (2005). The effects of fire on rare plants. *Fire Management Today*, **65**(4), 13–15.

Owen-Smith, N. (1987). Pleistocene extinctions: the pivotal role of megaherbivores. *Paleobiology*, **13**, 351–362.

Patterson, W.A. III. & Sassaman, K.E. (1988). Indian fires in the prehistory of New England. In *Holocene Human Ecology in Northeastern North America* (ed. G.P. Nicholas), pp. 107–135. Plenum Press, New York.

Pausas, J.G. & Keeley, J.E. (2009). A burning story: the role of fire in the history of life. *BioScience*, **59**, 593–601.

Pielou, E.C. (1991). *After the Ice Age: The Return of Life to Glaciated North America*. University of Chicago Press, Chicago, IL, USA.

Platt, S.G. & Brantley, C.G. (1997). Canebreaks: an ecological and historical perspective. *Castanea*, **62**(1), 8–21.

Progulske, D.R. (1974). Yellow ore, yellow hair, yellow pine: a photographic study of a century of forest ecology. Bulletin No. 616. South Dakota State University, Brookings, SD, USA.

Pyne, S.J. (1982). *Fire in America: A Cultural History of Wildland and Rural Fire*. Princeton University Press, Princeton, NJ, USA.

Pyne, S.J. (1983). Indian fires. *Natural History*, **2**, 6–11.

Remsen, J.V. (1986). Was Bachman's warbler a bamboo specialist? *Auk*, **103**, 216–219.

Rogers, R.A., Martin, L.D., & Nicklas, T.D. (1990). Ice-Age geography and the distribution of native North American languages. *Journal of Biogeography*, **17**, 131–143.

Rostlund, E. (1957). The myth of a natural prairie belt in Alabama: an interpretation of historical records. *Annals of the Association of American Geographers*, **47**, 392–411.

Rostlund, E. (1960). The geographic range of the historic bison in the Southeast. *Annals of the Association of American Geographers*, **50**, 395–407.

Ruffner, C.M. & Abrams, M.D. (2002). Dendrochronological investigation of disturbance history for a Native American site in northwestern Pennsylvania. *Journal of the Torrey Botanical Society*, **129**, 251–260.

Sauer, C.O. (1975). Man's dominance by use of fire. *Geoscience and Man*, **10**, 1–13.

Scharf, E.A. (2010). Archaeology, land use, pollen and restoration in the Yazoo Basin (Mississippi, USA). *Vegetation History and Archaeobotany*, **19**, 159–175.

Skinner, C.N. & Taylor, A.H. (2006). Southern Cascades bioregion. In *Fire in California's Ecosystems* (ed. N.G. Sugihara, J.W. Van Wagtendonk, K.E. Shaffer, J. Fites-Kaufman, and A.E. Thode), pp. 195–224. University of California Press, Berkeley, CA, USA.

Skinner, C.N., Abbot, C.S., Fry, D.L., Stephens, S.L., Taylor, A.H., & Trouet, V. (2009). Human and climate influences on fire occurrence in California's North Coast Range, USA. *Fire Ecology*, **5**, 76–99.

Skovlin, J.M. & Thomas, J.W. (1995). Interpreting long-term trends in Blue Mountain ecosystems from repeat photography. General Technical Report PNW-GTR-315. USDA Forest Service, Portland, OR, USA.

Smith, B.D. & Yarnell, R.A. (2009). Initial formation of an indigenous crop complex in eastern North America at 3800 B.P. *Proceedings of the National Academy of Sciences Early Edition*, **106**(16), 6561–6566. http://www.pnas.org/content/early/2009/04/03/0901846106.full.pdf.

Smith, J. ([1616] 1905). A description of New England. In *Sailors Narratives of Voyages Along the New England Coast 1526–1624* (ed. G.P. Winship), pp. 212–247. Houghton-Mifflin (1905), Boston, MA.

Stambaugh, M.C. & Guyette, R. (2006). Fire regime of an Ozark wilderness area, Arkansas. *American Midland Naturalist*, **156**, 237–251.

Stewart, O.C. (1963). Barriers to understanding the influence of use of fire by aborigines on vegetation. In *Proceedings of the Second Tall Timbers Fire Ecology Conference*, pp. 117–126. Tall Timbers Research Station, Tallahassee, FL, USA.

Stewart, O.C. (2002). The effects of burning of grasslands and forests by aborigines the world over. In *Forgotten Fires: Native American and the Transient Wilderness* (ed. H. Lewis and M.K. Anderson), pp. 67–338. University of Oklahoma Press, Norman, OK, USA.

Suttles, W. (ed.). (1990). *Handbook of North American Indians, Vol. 7: Northwest Coast*. Smithsonian Institution, Washington, DC, USA.

Swanton, J.R. (1939). Final report of the United States De Soto Expedition Commission, House Document. 71, 76th Cong., 1st session, Washington, DC, USA.

Taylor, A.H., Trouet, V., & Skinner, C.N. (2008). Climatic influences on fire regimes in montane forests of the southern Cascades, California, USA. *International Journal of Wildland Fire*, **17**, 60–71.

Thomas, D.H., Miller, J., White, R., Nabokov, P., & Deloria, P.J. (1993). *The Native Americans: An Illustrated History*. Turner Publishing, Atlanta, GA, USA.

USDA. (1999). Ozark-Ouachita highlands assessment: terrestrial vegetation and wildlife. General Technical Report SRS-35. USDA Forest Service, Asheville, NC, USA.

Vale, T. (2002). *Fire, Native People and the Natural Landscape*. Island Press, Washington, DC, USA.

Van Lear, D.H. & Waldrop, T.A. (1989). History, uses, and effects of fire in the Appalachians. General Technical Report SE-54. USDA Forest Service, Asheville, NC, USA.

van Wagtendonk, J.W. & Cayan, D.R. (2008). Temporal and spatial distribution of lightning in California in relation to large-scale weather patterns. *Fire Ecology*, **4**, 34–56.

van Wagtendonk, J.W. & Fites-Kaufman, J. (2006). Sierra Nevada bioregion. In *Fire in California's Ecosystems* (ed. N.G. Sugihara, J.W. Van Wagtendonk, K.E. Shaffer, J. Fites-Kaufman, and A.E. Thode), pp. 264–294. University of California Press, Berkeley, CA, USA.

Waters, M.R. & Stafford, T.W. Jr. (2007). Redefining the age of Clovis: implications for the peopling of the Americas. *Science*, **315**, 1122–1126.

Weisberg, P.J. & Swanson, F.J. (2003). Regional synchroneity in fire regimes of western Oregon and Washington, USA. *Forest Ecology and Management*, **172**, 17–28.

Weiser, A. & Lepofsky, D. (2009). Ancient land use and management of Ebey's Prairie, Whidbey Island, Washington. *Journal of Ethnobiology*, **29**, 184–212.

West, G.J., Woolfenden, W., Wanket, J.A., & Anderson, S. (2007). Late Pleistocene and Holocene environments. In *California Prehistory: Colonization, Culture, and Complexity* (ed. T.L. Jones and K.A. Klar), pp. 11–34. Altamira Press, New York, USA.

Williams, G.W. (2000). Introduction to aboriginal fire use in North America. *Fire Management Today*, **60**, 8–12.

Williams, M. (1989). *Americans and Their Forests: A Historical Geography*. Cambridge University Press, Cambridge, UK.

Wright, H.A. & Bailey, A.W. (1980). Fire ecology and prescribed burning in the Great Plains: a research review. General Technical Report INT-77. USDA Forest Service, Ogden, UT, USA.

Young, G.A. & Hoffman, M.P. (eds.). (1993). *The Expedition of Hernando de Soto West of the Mississippi, 1541–1543*. University of Arkansas Press, Fayetteville, AR, USA.

Chapter 7

CONSERVATION AND RESOURCE MANAGEMENT IN A CHANGING WORLD: EXTENDING HISTORICAL RANGE OF VARIATION BEYOND THE BASELINE

Stephen T. Jackson

University of Wyoming, Laramie, WY, USA

Historical Environmental Variation in Conservation and Natural Resource Management, First Edition. Edited by John A. Wiens, Gregory D. Hayward, Hugh D. Safford, and Catherine M. Giffen.
© 2012 John Wiley & Sons, Ltd. Published 2012 by John Wiley & Sons, Ltd.

7.1 INTRODUCTION

The development and implementation of the historical range of variation (HRV) concept in the 1990s represented a logical extension of two ecological insights of the 1970s and 1980s. First, ecosystems are subject to temporal variation, including periodic or quasiperiodic disturbances (Connell 1978; Pickett & White 1985). Second, individual sites may be at different successional stages following the most recent disturbance, so the landscape can be viewed as a temporally shifting spatial mosaic of successional sites (Bormann & Likens 1979; Romme & Knight 1982). Implementation of these ideas into resource management and conservation via the HRV concept was an important step forward, in two fundamental ways. First, it explicitly recognized that terrestrial ecosystems are dynamic entities at timescales of years to decades or more. Second, it recognized that Euro-American activities and management practices, particularly the disruption of variability or of disturbance regimes, have altered ecosystem properties. Management and restoration of ecosystems requires knowledge of their "natural" states, including their characteristic variability regimes.

Incorporation of the HRV approach required a shift in ecological and management thinking from the traditional "here and now" perspective to one embracing a broader range of spatial and temporal scales. Spatial scale expanded from the stand scale to incorporate a landscape perspective, and temporal scale expanded to embrace annual and decadal variability over the course of a few centuries. The temporal perspective requires a baseline, which has generally been fixed around the decades or centuries immediately preceding Euro-American forest clearance, livestock grazing, and/or fire management.

Of course, time did not begin a few centuries before Euro-American land clearance. North American ecosystems have experienced substantial changes in the 20000 years since the last glacial maximum, and those changes are relevant to the HRV concept. Climate, human activities, and herbivores are all critical determinants of the structure and composition of vegetation. All three influence disturbance regimes, both directly (drought, ignition, consumption) and indirectly (fuel loads, fire weather, vegetation structure). All three factors changed substantially in the millennia during and since deglaciation, and in the centuries preceding land clearance.

In this chapter I use a deep historical perspective to examine, critique, and extend the HRV concept. First, I address five issues arising from a current understanding of ecological history:

1 The influence of Native American activities on HRV baselines,
2 The nature of climate variability in the centuries preceding HRV baselines,
3 The origin and antiquity of ecosystems to which HRV baselines are applied,
4 The potential for environmental change to create novel ecosystems, and
5 The consequences of faunal extinctions for ecosystem structure and function.

These issues are all under active discussion in the paleoecological, paleoclimatic, and archeological communities and not all are fully resolved. I briefly review each of these issues, and argue that the HRV concept as currently applied may be inadequate for resource management and conservation in a changing environment. I propose an extension of the HRV concept to encompass greater time spans in the past and a broader array of environmental scenarios for the future. I conclude with a brief summary of the nature of environmental change and the challenges it poses for resource management and conservation in the anthropogenic greenhouse world that will emerge in the coming decades.

7.2 ECOLOGICAL HISTORY AND HRV

Native American land use in the last millennium

The role of Native American activities, including agriculture, land clearance, fire setting, and hunting, in structuring ecosystems before Euro-American occupation is under active debate. Native Americans occupied land, cultivated crops, and harvested plant and animal populations across North America; disagreement largely centers on the areal extent and magnitude of the environmental impacts. Comprehensive review of the issues and evidence is beyond the scope of this chapter (also see Nowacki et al., Chapter 6, this book); instead, I will offer some selected case studies that illustrate the issues and the dynamic nature of human activities in the centuries preceding Euro-American occupation.

The extent to which changes in Native American activities might have influenced HRV baselines depends on the nature and magnitude of the human impacts relative to the spatial scale at which the baselines are defined. North America was neither a uniform wilderness nor an intensive cultural landscape in prehistoric times (Vale 2002). Some regions – for instance, the great river valleys of the Southeast and Midwest – included landscapes cleared and managed for agriculture and habitation by large, sedentary populations. Before 1400 AD, the maize-based Mississippian culture extended from its heartland in the Mississippi and Ohio River valleys to the Piedmont and Gulf Coastal Plain (Hudson 1997; Milner 2004; Pauketat 2009). The regional extent of forest clearance is difficult to gauge owing to the dearth of suitable sites for pollen analysis in the core Mississippian regions.

Cronon (1983) argued from historical accounts that much of southern New England was a cultural landscape with extensive forest clearance, but paleoecological studies fail to verify Cronon's conjecture – they indicate closed-canopy forest at a regional scale (Foster et al. 2002; Motzkin & Foster 2002; Parshall & Foster 2002). The clearest fossil evidence for forest clearance in the Great Lakes/St. Lawrence corridor is from Crawford Lake in southern Ontario, where the data indicate a small-scale, transient clearing for agriculture and habitation (McAndrews 1988; see Munoz and Gajewski 2010 for other examples). This case is revealing: Crawford Lake is small and hence sensitive to local vegetation patterns, land clearance was very close to the lake, and annually laminated sediments were analyzed at closely spaced intervals, so the brief occupation was detectable (see discussion in Jackson 1997). The more coarsely resolved pollen records typical of the region are poor sensors of localized patches of transient activity, but they clearly rule out widespread deforestation in the Great Lakes–St. Lawrence–New England corridor.

Precontact agriculture was widely practiced in the southwestern United States, although the pueblo cultures of the region declined sharply well before 1492 AD, at least in part owing to drought. The extent to which these societies affected vegetation beyond the habitation centers remains unclear. Firewood cutting at the Chaco Canyon population center led to local extinction of piñon pine (*Pinus edulis*) (Betancourt and Van Devender 1981), and the Chaco people hauled spruce and fir timbers 75–100 km for construction (English et al. 2001). Native Americans undoubtedly set fires throughout the interior western United States, but most of the region is lightning prone, and human impact on natural fire regimes and forest vegetation appears to have been minimal (Swetnam et al. 1999; Allen 2002; Baker 2002). Similar conclusions have been reached for forests of the Oregon Coast Ranges (Whitlock & Knox 2002) and Sierra Nevada (Parker 2002). Vegetation of the California coastal ranges, however, may have been shaped by Native American burning (Keeley 2002), and prehistoric fire events and vegetation shifts in Yosemite Valley have been attributed to Native American activities (Woolfenden 1996).

Native American societies across much of North America underwent a series of transformations in the centuries before first contact. For example, maize cultivation was rapidly adopted across the eastern United States between 800 and 1100 AD, coinciding with shifts in social and political organization. Maize-centered economies rapidly replaced mixed-crop agriculture in the Ohio and Upper Mississippi River valleys and hunting/foraging economies in the Gulf and Atlantic coastal regions and the southern Great Lakes region (Smith 1989). These cultural transformations may have had profound landscape-level effects in some regions (e.g. Mississippian population centers) and more subtle effects in others (via changes in burning, hunting, and foraging practices). The environmental consequences of these changes are poorly understood, but they underscore the dynamic nature of cultural practices and land use before European occupation.

Climate change drove or influenced some of the cultural transformations. For example, a severe multidecadal drought in the thirteenth century accompanied agricultural failures and collapse of the pueblo cultures of the Four Corners area (Dean 1988; Gray et al. 2004). The decline of the agricultural Mississippian societies of the Mississippi and Ohio valleys may be linked to a series of multidecadal droughts centered in the region between 1340 and 1460 AD (Cook et al. 2007).

European impacts on North American ecosystems may have long preceded the intensive and extensive forest clearances and grazing inception traditionally used to identify HRV baselines. In particular, widespread transmission of novel diseases (e.g. smallpox, influenza, measles) among Native American populations in the sixteenth century led to population reductions, societal collapses, and interruption of land-use practices decades to centuries before European settlement (Denevan 1992; Mann 2005). DeSoto's expedition (1538–1543), for example, caused widespread societal disruption and population decline over much of the southeastern United States owing to violence and novel diseases (Milanich & Hudson 1993, Hudson 1997). Consequent reduction in human-set fires might have led to changes in forest composition and structure decades to centuries before European settlement.

Disease was not the only European influence that spread into the continental interior well before settlers. Cultural artifacts (particularly firearms), fur-trade economies, and horses altered Native American ways of life in the sixteenth, seventeenth, and eighteenth centuries. The fur and firearms trades went hand in hand starting in the late seventeenth century. Direct population reduction of beaver and other furbearers followed, and some grazing and browsing mammals may have also declined locally or regionally. Fur-trade economies often intensified intertribal conflicts, with substantial disruption and displacement of agricultural and mixed cultures (Parkman 1867; Hunt 1940). The northward spread of horses from Spanish settlements drove a major eighteenth-century transition in hunting cultures throughout the Great Plains, leading to more effective harvesting of bison (Shimkin 1986).

Undoubtedly, ecological baselines across North America shifted in the past thousand years as Native American populations and cultures responded to natural environmental changes and the direct and indirect stresses induced by European invasion of the continent. The magnitude of Native American impacts on terrestrial ecosystems has been exaggerated by some and ignored by others. Critically needed are assessments, on a local to regional basis, of the extent of specific kinds of human impacts in specific places at specific times (e.g. Vale 2002), integrating and reconciling available documentary, archeological, dendroecological, and paleoecological records.

Climate change and variability in the last millennium

Climate variation is often envisioned by ecological managers and modelers as stationary variation about a mean. However, climate variation is seldom, if ever, stationary, regardless of temporal scale (Milly et al. 2008; Jackson et al. 2009a). Tree-ring records, which have annual resolution but extend back hundreds of years, indicate that no two years within a decade are alike, no two decades within a century are alike, and no two centuries within the last millennium have been alike – each represents a unique realization of a complex climate system that is continually evolving (Fig. 7.1). Nonstationarity of annual, decadal, multidecadal, and centennial climate variability is probably characteristic of the entire postglacial period; it is well-documented for the past 500–1500 years across North America (http://www.ncdc.noaa.gov/paleo/pdsi.html).

The nonstationary behavior of climate variability at annual to centennial timescales has important consequences for HRV application, at least for ecosystem and biodiversity properties that are under climatic control. Identification of a standard baseline for variability (e.g. a fixed range, standard deviation, or mean landscape state) may prove elusive in an evolving system. A 100- or 300-year window may be inadequate to capture the full range of natural climate variability to which a system may be exposed. The twentieth-century record is increasingly recognized as an inadequate window for assessment of natural flood and drought hazard (Milly et al. 2008) and, by extension, any individual century may be inadequate as a resource management or policy baseline.

Climate ultimately controls many ecosystem disturbances. Fires are linked to surface fuel loads, fire weather, and drought. Pest and pathogen outbreaks are often driven by drought- or wet-induced susceptibility. Windthrow in forests is related to storm origin, intensity, and frequency. Characterizing disturbance regimes (or population-recruitment regimes) poses a dilemma. On the one hand, it requires a time window sufficiently broad to encompass the full range of potential variability within the system. This broadening increases the sample size or range of experience, as amply documented by tree-ring precipitation and streamflow reconstructions (Fig. 7.1) (Meko et al. 2007; Watson et al. 2009). On the other hand, however, broadening the window by extending it back in time will eventually encompass a different variability regime (different mean state and/or variance). Each time series within a regime represents a finite, and possibly incomplete, set of possible realizations.

A final concern raised by climatic nonstationarity is that ecosystems, particularly forests and woodlands, are subject to response lags and legacy effects. Ecological responses may lag climate changes or events by years to millennia (National Research Council 2005). Past climate regimes, episodes, and extremes may leave ecological signatures that can persist for years to millennia (Jackson et al. 2009a). Annual or decadal climate extremes, for instance, can induce widespread population mortality or recruitment, so age-structure

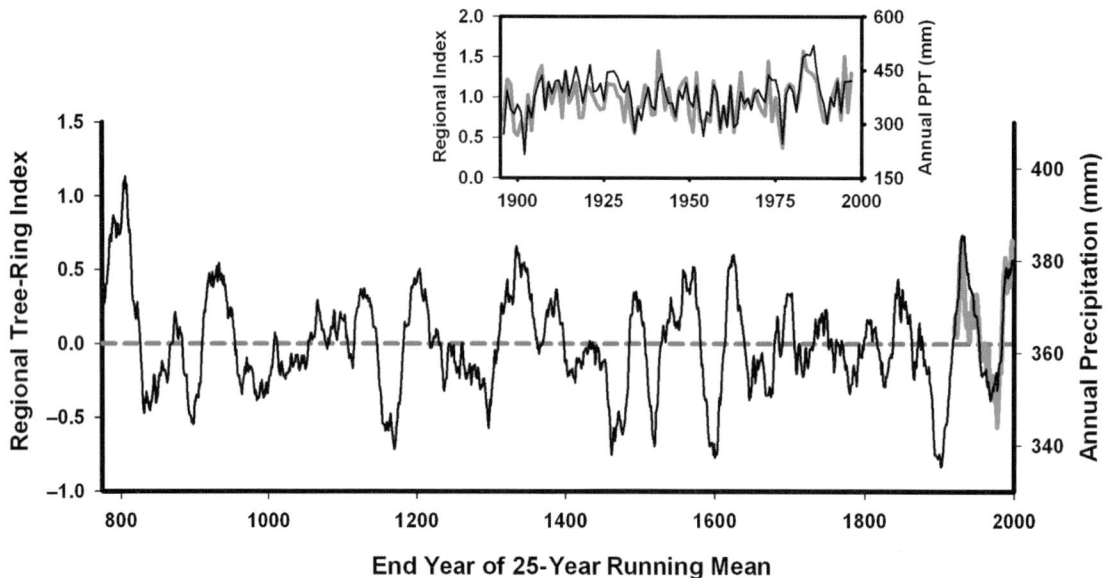

Fig. 7.1 Variation in annual precipitation (PPT) (25-year running mean) in the Upper Colorado River basin inferred from tree-ring records spanning the past 1225 years (black). Tree growth is represented as a regional standardized index based on z scores calculated from 11 of the oldest ring-width records in the basin (Meko et al. 2007). The gray trace shows a 25-year running mean of instrumentally measured precipitation during the past century. The inset shows the strong correlation between the composite tree-ring record (black) and measured annual precipitation in the region (gray). Note the decadal to centennial shifts in mean and variability, extended wet and dry periods, and continual evolution of the patterns in time. Modified from Fig. 1 in Jackson, S.T., Betancourt, J.L., Booth, R.K., and Gray, S.T. 2009a. Ecology and the ratchet of events: climate variability, niche dimensions, and species distributions. *Proceedings of the National Academy of Sciences* **106**, 19685–19692. Figure based on data from Meko et al. (2007).

and population size may be a legacy of a past climate event. For example, the 1950s drought in the American Southwest led to widespread mortality in conifer woodlands and forests, and the subsequent wet period led to a recruitment episode. Thus, many features of Southwestern ecosystems, ranging from population structure to fire susceptibility, are legacies of these climate events (Allen & Breshears 1998; Swetnam & Betancourt 1998). Similar legacy effects may be expected from the drought-induced beetle outbreaks that have recently devastated the Four Corners region (Breshears et al. 2005; Gitlin et al. 2006) and the temperature-related bark beetle outbreaks that are destroying extensive conifer forests in the central Rockies (Raffa et al. 2008).

Owing to legacy effects, the system state at a given time may be contingent not only upon time since the last major disturbance, but on climate conditions and events preceding and following that disturbance (Jackson et al. 2009a). For example, severe drought

events followed by extended wet periods have been implicated in recent range expansion and stand dominance of piñon pine (*P. edulis*) in northeastern Utah (Gray et al. 2006) and yellow birch (*Betula alleghaniensis*) in Upper Michigan (Booth et al. 2005). In the absence of the droughts, which evidently led to widespread mortality of incumbent trees, these species might not have expanded as widely and rapidly during the subsequent wet periods. Even in the absence of invading species, nonstationary climate variability may impart contingent effects on stand composition and ecosystem properties, whereby the latter cannot be understood (or predicted) without knowledge of the specific sequence of climatic events.

Although history cannot embody every possible realization of a dynamic or evolving system, the range of realizations can be maximized by observing the system as far back in time as possible, in as many different spatial settings as possible. Identification of system antiquity – its age of origin – can support such efforts.

Assessing the antiquity of ecosystems

All ecosystems have some finite time of origin. Ecologists traditionally view this as the time when a specific physical location became available for primary succession – the melting of an ice block to form a kettle lake, for example, or the retreat of a glacier or cooling of a lava flow to render a land surface capable of supporting terrestrial plants. But terrestrial ecosystems are defined as much by the functional properties of the dominant plant species as by the physical environment (Hooper et al. 2005), and paleoecological studies indicate that ecosystems at specific sites across the globe have undergone turnover (Figs. 7.2 and 7.3) (Jackson 2006). This ecosystem turnover results from changes in dominant species and prevailing disturbance regimes, driven by climatic changes as well as by human activities (particularly in the past three to four centuries). Thus, most terrestrial ecosystems are "secondary," insofar as species composition and ecosystem properties have undergone one or more allogenically driven turnover event since inception.

Secondary ecosystems bear legacies of preexisting ecosystems (e.g. soil biogeochemistry and carbon pools; populations of previous dominants), so the process of ecosystem turnover bears rough analogies with anagenetic (phyletic) evolution. Particular ecosystem realizations emerge as new species attain dominance, and dissolve as those dominants are replaced by new species. This process differs from secondary succession in that it is noncyclic and driven in most cases by environmental change.

The antiquity of specific vegetation patterns and disturbance regimes can provide clues to their overall resilience in the face of climate change. For example, ecosystems that have persisted *in situ* throughout the Holocene (the past 10 000 years) have experienced a wide range of climate changes and, as a first approximation, might be considered more resilient than ecosystems that have developed only in the past thousand years. Lodgepole pine (*Pinus contorta*) forests of the Rocky Mountains, ponderosa pine (*Pinus ponderosa*) forests of the southern Colorado Plateau, and subalpine balsam-fir (*Abies balsamea*) forests of New York and New England have undergone little compositional change in the past several thousand years (Fig. 7.3), despite considerable variation in temperature and precipitation (Jackson 2006). Even these compositionally stable forests have undergone climate-driven changes in fire regime and other properties (e.g. Spear et al.

1994; Swetnam & Betancourt 1998; Whitlock et al. 2003; Pierce et al. 2004). At a more local scale, jack pine (*Pinus banksiana*) forests of sandy outwash plains in Upper Michigan have persisted for a comparable period (Brubaker 1975), despite regional climate change that drove major forest changes on other sites (Davis et al. 1998; Booth et al. 2005).

In contrast, other ecosystems have developed within the last few decades or centuries (Fig. 7.3). These younger ecosystems typically arose from climate changes that were favorable for invasion or expansion of their dominant species. Relatively young ecosystems have experienced a narrower range of climate variability than older ecosystems nearby. Climate tolerances of young ecosystems may be as broad as those of older ecosystems, but the specific climates that prevailed in their locales were not capable of supporting them until relatively recently.

Ecosystem turnover and ecological novelties

Most regions have undergone some degree of ecosystem turnover during the Holocene (Fig. 7.3). In many regions, this turnover has consisted of changing vegetation composition via local to regional immigrations and extinctions, with little or no change in physiognomy or prevailing disturbance regime. Cases from the fossil record are legion, especially for the past 4000–5000 years (Figs. 7.2 and 7.3). These immigrations, and the accompanying declines of incumbent species, may have changed ecosystem properties by altering patch sizes, stand density, or litter and topsoil chemistry, but canopy, hydrological, and biogeochemical properties may have changed little (Jackson 2006).

Many regions and sites experienced more substantial ecosystem-level transformations during the Holocene, involving turnover in dominant species, vegetation physiognomy, and functional properties. These transformations, which range from replacement of one forest type by another (e.g. deciduous vs coniferous dominants) to alternation of forests with other vegetation types (woodland, savanna, prairie), become particularly apparent as the timescale expands to include the last 10 000 years. They are recorded over most of North America (Thompson & Anderson 2000; Baker et al. 2002; Williams et al. 2004). Some, however, are as recent as the past few hundred years (Fig. 7.3). In some cases, these changes represent shifts of ecotones along local to regional gradients (elevational, topographic,

(A)

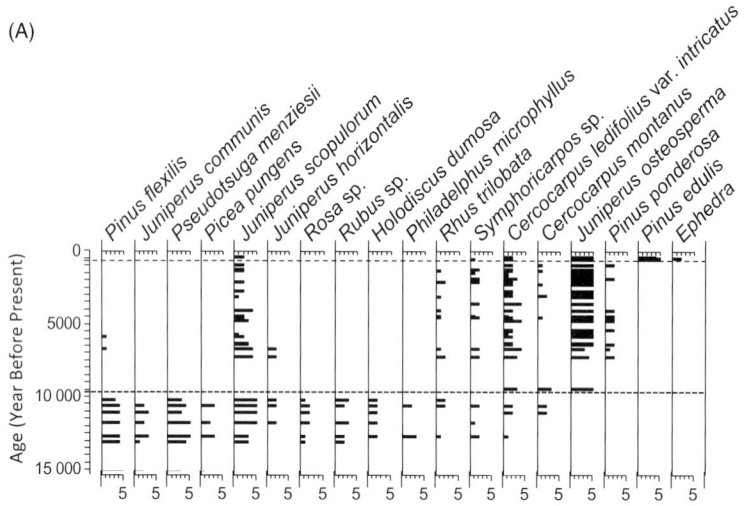

(B)   **Tannersville Bog, PA**

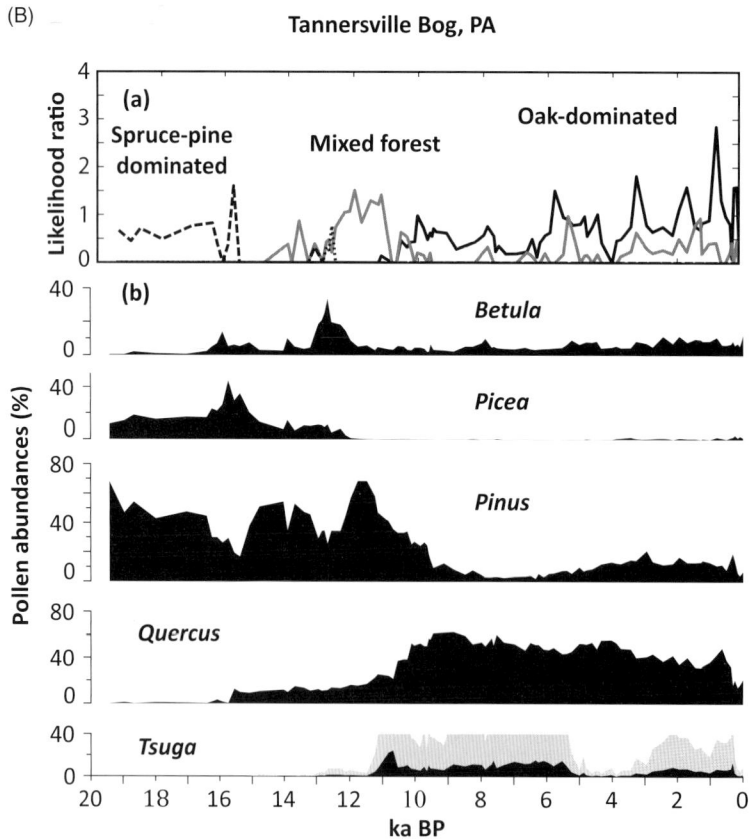

Fig. 7.2 Two representative paleoecological records showing emergence and turnover of ecosystems. (A). Example based on plant macrofossil abundances (1–5 relative scale) from a series of ^{14}C-dated woodrat (*Neotoma*) middens from Dutch John Mountain, northeastern Utah. The south-facing slope is occupied today by piñon/juniper woodland (*Pinus edulis* and *Juniperus osteosperma*), with scattered *Ephedra*, *Cercocarpus*, and *Juniperus scopulorum*. Piñon invaded and attained dominance 600–700 years ago (dashed horizontal line), marking the origin of the modern ecosystem. For the previous 8000 years, vegetation was structurally similar (i.e. evergreen conifer woodland), but dominated by Utah juniper. A major ecosystem transition (dotted horizontal line) from montane conifer forest/woodland (limber pine – *Pinus flexilis*; Douglas-fir – *Pseudotsuga menziesii*; blue spruce – *Picea pungens*) occurred ca. 10 000 years ago. From Jackson et al. (2005) and Gray et al. (2006). (B). Example utilizing pollen percentages from Tannersville Bog in northeastern Pennsylvania. The pollen data show turnover in forest dominants over the past 20 000 years (ka BP = 10^3 years before present). The top panel shows results of a Bayesian analysis of multivariate similarity of fossil pollen assemblages to modern pollen assemblages corresponding to specific modern ecosystems (dashed line: spruce-pine forest; gray line: mixed deciduous-conifer forest; dotted line: conifer woodland; solid black line: oak-dominated forest). Some fossil pollen assemblages show similarity to more than one modern ecosystem, indicating an ecotone, a regional mosaic, or forests lacking precise modern counterparts. The emergence of the modern ecosystem could be designated as the increase in hemlock (*Tsuga*) pollen 3000 years ago, or as the development of oak-dominated forest 10 000 years ago as oak (*Quercus*) replaced pine (*Pinus*) as the dominant pollen type. At a regional scale, forests may have undergone relatively little structural change in the past 10 000 years, although conifers have increased within the past few thousand years. Figure and analysis modified from Liu et al. (2010). Original pollen data from Watts (1979), obtained from the North American Pollen Database (NAPD) via http://www.neotomadb.org/.

edaphic), but in others, they represent ecotone movements or reorganizations at scales of hundreds of kilometers. These latter changes are generally congruent with large-scale climatic changes (Bartlein et al. 1998).

The late-glacial period between 17 000 and 10 000 years ago, when the continental ice sheets were in rapid retreat, is characterized over much of interior North America by forests and woodlands compositionally unlike any vegetation today (Fig. 7.4; Jackson & Williams 2004; Edwards et al. 2005; Williams & Jackson 2007). For example, forests of the upper Mississippi, Ohio, and Tennessee River valleys were dominated by a peculiar mix of spruce (*Picea*), oak (*Quercus*), hophornbeam (*Ostrya*), elm (*Ulmus*), birch (*Betula*), and black ash (*Fraxinus nigra*). Relative abundances of these taxa varied spatially, and in some areas, the canopy was either open in structure or the forests were widely interspersed with unforested patches. During part of this same period, Beringia (Alaska, northwest Canada, northeast Siberia) was vegetated by boreal deciduous forest, dominated by birch (*Betula*), larch (*Larix*), poplar (*Populus*), and willow (*Salix*). These novel ecosystems coincided with novel climates of the late-glacial period, when seasonal temperatures contrasted more sharply than at any time in the Holocene (Williams & Jackson 2007).

Recent analyses of future climate scenarios under greenhouse warming indicate that novel climates – new combinations of seasonal temperature and precipitation – are likely to arise over much of the world in the coming century, and existing climates will disappear in many regions (Saxon et al. 2005; Williams & Jackson 2007; Williams et al. 2007). Climate changes of the magnitude predicted under the most extreme scenarios (CO_2 tripling) may drive wholesale ecosystem replacements over all of North America, while more modest scenarios (CO_2 doubling) suggest ecosystem replacement over most of eastern North America and the American Southwest (Williams et al. 2007). Clearly, these scenarios will drive existing ecosystems far beyond the bounds of HRV measured at any temporal scale. Ecosystems will disappear, either gradually (via senescence and mortality of incumbents and colonization of species better suited to the new climate) or abruptly (via fires, pest/pathogen outbreaks, and weather extremes exceeding the physiological capacity of incumbent individuals).

Although climates capable of sustaining many existing ecosystems may shift to other parts of North America, some climates will disappear universally, and hence the ecosystems they control will undergo universal and irreversible termination or transformation. The spatial distribution of areas estimated at high risk differs among analyses, but ecosystems in the southeastern and northwestern United States appear particularly vulnerable (Saxon et al. 2005; Williams et al. 2007). At the same time, novel climates will arise in many regions, leading to ecosystems unlike any found today. Like disappearing climates, novel climates are predicted to be widespread.

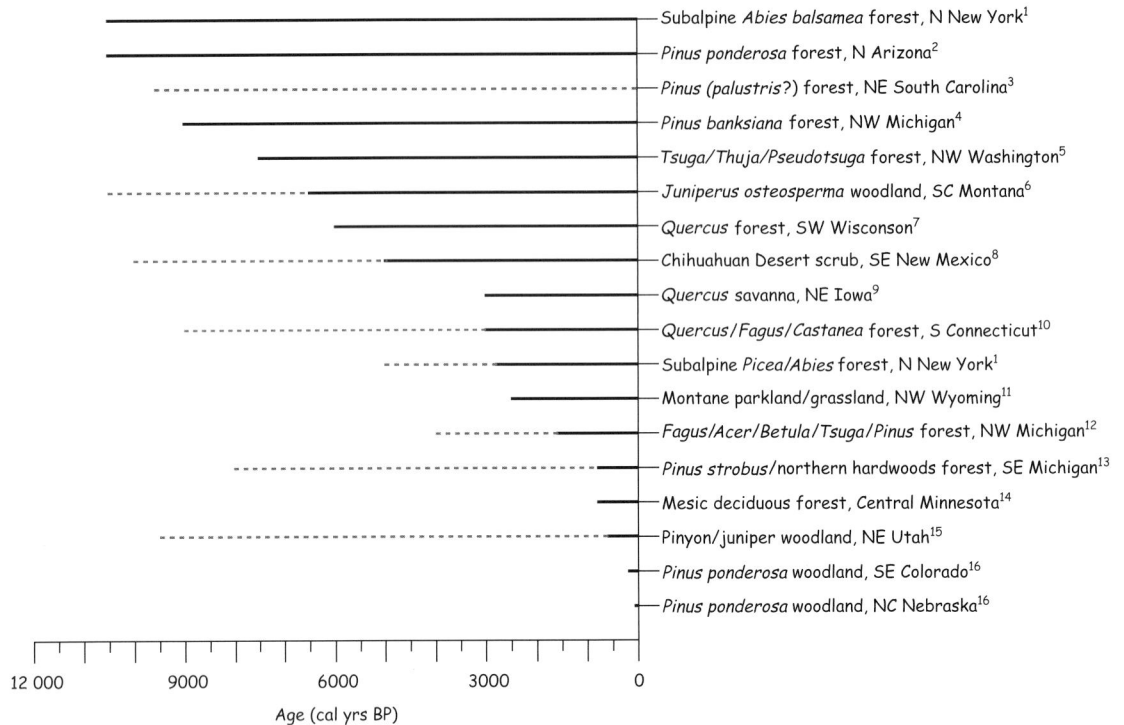

Fig. 7.3 Estimated antiquity of a variety of ecosystems from across the continental United States, inferred from paleoecological records (lake and peatland sediments, packrat middens, tree-ring demography). Solid lines show continuity through time of vegetation structure and dominant species. Dashed lines show continuity of vegetation structure and dominant functional types, in spite of turnover in dominant species. For example, subalpine *Picea/Abies* forests in northern New York arose approximately 2800 years ago with attainment of dominance by *Picea*, but these forests were preceded by forests dominated by *Abies balsamea*, which has similar physiognomy and canopy structure as *Picea/Abies* forest. Absence of dashed lines generally indicates a fundamental change in functional dominants and vegetation structure. For example, mesic deciduous forest in central Minnesota was preceded by *Quercus* savanna, while *Quercus* savanna in northeastern Iowa was preceded by tallgrass prairie. A dashed line is shown for *Pinus* forest in Northeast South Carolina because the species are unknown in the fossil record. [1]Jackson and Whitehead (1991); [2]Weng and Jackson (1999); [3]Hussey (1993); [4]Brubaker (1975); [5]Cwynar (1987); [6]Lyford et al. (2002); [7]Baker et al. (1992); [8]Betancourt et al. (2001); [9]Baker et al. (1996); [10]Davis (1969); [11]Lynch (1998); [12]Reeves (2006); [13]Booth and Jackson (2003); [14]Shuman et al. (2009); [15]Jackson et al. (2005), Gray et al. (2006); [16]Kaye et al. (2010); M.R. Lesser and S.T. Jackson, unpublished data.

Ecological novelties will arise not only because of novel climates but because rapid climate change and ecological responses will interact with invasive species, human land use, pests and pathogens, and other environmental stressors (Harris et al. 2006; Hobbs et al. 2006; Jackson & Hobbs 2009). Although ecological history provides no direct guide to the kinds of novel ecosystems that will arise, it reveals how ecosystems are periodically reorganized and can inform assessment of risks, vulnerabilities, and potential for adaptation (Jackson & Hobbs 2009; Dawson et al. 2011). Ecosystem turnover is a routine phenomenon at timescales of centuries to millennia (Figs. 7.2–7.4). Although not all the costs of these changes can be assessed in the currency of biodiversity or potential ecological goods and services, nearly all species living today are survivors of past ecological turnover. Although their survival suggests a substantial degree of natural "biodiversity resilience" in the face of environmental change (Dawson et al. 2011), it provides no

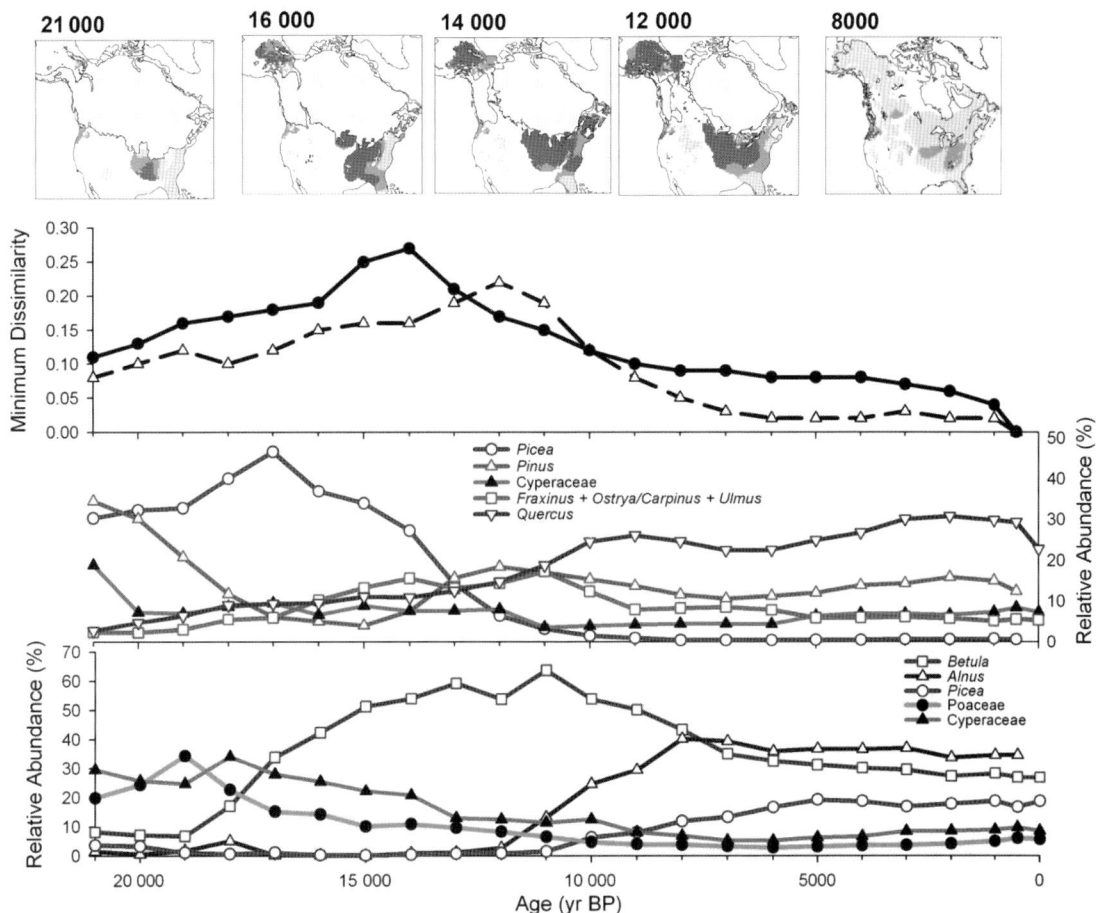

Fig. 7.4 Spatial, temporal, and taxonomic patterns of "no-analog" communities in North America during the last deglaciation. These communities were most common between 17 000 and 12 000 years ago (years BP). Red shading in maps is proportional to the multivariate dissimilarity between fossil pollen assemblages and their closest modern analogs. The second diagram shows that the peak no-analog period in eastern North America preceded the no-analog period in Alaska. The third and fourth diagrams show changing pollen percentages of dominant taxa in eastern North America and Alaska, respectively. Modified from Fig. 2 in Williams and Jackson (2007). Please refer to Colour Plate 1.

basis for complacency, given the projected rates of future environmental change and the compounding effects of multiple stressors.

Missing megafauna and the modern landscape

Grazing and browsing mammals can influence vegetation composition and structure, and human activities, particularly hunting, can influence their population sizes. Indeed, humans have driven animal populations to local, regional, and, in some cases, universal extinction. Documentation of missing faunal elements is motivating discussion of "rewilding" – introducing formerly occurring species or functional types in hope of restoring missing ecological functions (Oliveira-Santos & Fernandez 2010). These proposals often exceed the timescales of HRV baselines, extending restoration targets to include organisms and processes that disappeared thousands of years ago.

In temperate Europe, for example, bison (*Bison*), aurochs (*Bos*), deer (*Cervus*), horses *Equus*), and other large herbivores were widespread in the early Holocene (6000–10 000 years ago). Populations of these species diminished, and many disappeared during the "Neolithic Revolution" as human populations increased and cleared land for agriculture. Vera (2000) argued that grazing by these herbivores maintained a patchy landscape of open forests and meadows, and that closed-canopy forests are not natural components of the lowland European landscape. Various agencies and organizations in Europe are adopting or considering land-management practices based on this argument. However, Mitchell (2005) makes a compelling empirical case from paleoecological data that the composition and structure of pre-Neolithic forests were largely unaffected by herbivores.

To date, evidence for herbivore control of Holocene forest composition and structure is sparse in North America. Faison et al. (2006) present evidence for closed-canopy forests in the early Holocene of southern New England, and extensive studies provide no evidence for open forests or extensive unforested patches before European land clearance in that region (Foster & Motzkin 2003) as well as in other forested regions in eastern North America (Williams et al. 2004).

Until recently, North America was populated with a variety of now-extinct herbivores, including mammoths (*Mammuthus*), mastodons (*Mammut*), ground sloths (*Megalonyx*), horses (*Equus*), camels (*Camelops*), and giant beaver (*Castoroides*). These species, along with some top-level carnivores (dire wolf [*Canis dirus*], saber-tooth cat [*Smilodon*], short-faced bear [*Arctodus*]) all disappeared during the last deglaciation, between 15 600 and 11 500 years ago (Barnosky et al. 2004). The cause of these extinctions has been a source of controversy for more than a century. Most of the disappearances coincided with proliferation of the Clovis hunting culture, but they also occurred during a period of rapid climate change. It seems likely that human activities played a role, but climate-driven changes in habitat and resources may have amplified extinction vulnerability (Koch & Barnosky 2006).

Evidence is emerging that the decline of North American megaherbivores may have driven broadscale vegetation transformations (Gill et al. 2009). At sites in Indiana and New York, decline of *Sporormiella* (a dung-fungus marker for large herbivore activity) coincided with a rapid transition from spruce-dominated forest to deciduous forest, all during a time of relatively gradual climate change between 14 800 and 13 700 years ago (Robinson et al. 2005; Gill et al. 2009). At the Indiana site, this transition marks the appearance of "no-analog" communities and an increase in fire activity (Gill et al. 2009). Browsing by megaherbivores apparently dampened recruitment and productivity of hardwoods and reduced fuel loads, and their decline released hardwoods to dominate regional forests.

Recent proposals for "Pleistocene rewilding" – introduction of elephants, camelids, equids, and large carnivores from Africa and Asia into North America – argue that landscapes of North America have been ecologically depauperate for the past 13 000 years (Donlan et al. 2005; 2006). Elephants maintain open forests and patchy landscapes in Africa and southern Asia (Owen-Smith 1987; see Sinclair, Chapter 18, this book), and the recent findings of Gill et al. (2009) support the hypothesis that mastodons and other megaherbivores played a role in structuring Pleistocene ecosystems in North America. Furthermore, pre-Quaternary paleontological evidence suggests that, in the long run, megaherbivores are an inevitable ecological component of continental ecosystems (Smith et al. 2004). By this argument, North American ecosystems are ecologically incomplete (Donlan et al. 2006), and have been so since the late-glacial extinctions. However, an alternative perspective is that all ecosystems are contingent products of particular historical events, which in the case of postglacial North America happens to include the late-glacial megafaunal extinction events.

7.3 BEYOND BASELINES: THE EXTENDED HRV CONCEPT

The HRV concept attempts to incorporate a dynamic perspective into management of systems formerly treated as static. However, the historical perspectives reviewed here, ranging from the past few centuries to the past several millennia, indicate that our perceptions of ecosystems, including their variability, are strongly contingent upon the temporal scale at which we consider them. Ecosystems are evolving: as climate, human activities, herbivory, and other determinants change, so do ecosystems and their properties. From this perspective, HRV defined over the past few centuries appears, ironically, as a static approach – albeit one that encompasses an internal dynamic of disturbance and succession. To accommodate ongoing and future environmental changes, the HRV concept may itself need to evolve, looking backward deeper into time to better understand ecological dynamics, and forward to anticipate rates and directions of change that will be driven by natural and human-caused environmental change.

Applications of HRV already look backward, toward a baseline considered representative of the system state preceding disruption (Fig. 7.5). This baseline can often be extended further to encompass the entire period in which the ecosystem has existed – the extended HRV (eHRV) concept (Fig. 7.5). Such extension is possible, of course, only for ecosystems where historical, pale-

oecological, and paleoenvironmental records are available. But there are many such ecosystems, and many sources of information on their history. In general, eHRV assessment would have three primary objectives: (1) assessment of the range of environmental variation experienced by the ecosystem since its inception; (2) determination of the range of variation in ecosystem properties and disturbance regimes; and (3) identification of the environmental regime, particularly climatic, that has maintained the ecosystem. Separately or together, these three components of eHRV analysis would help gauge the sustainable limits of the system: How resilient has it been under past environmental

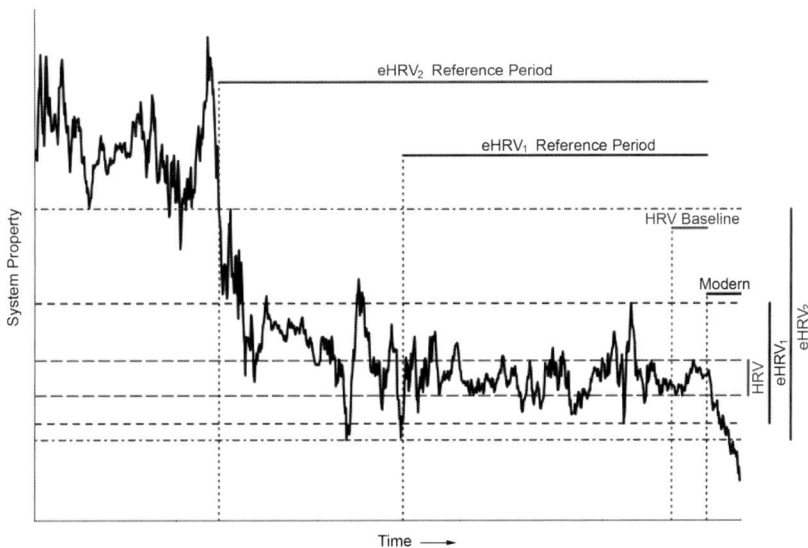

Fig. 7.5 The HRV baseline typically represents a small sample of the variability experienced by an ecosystem. A longer time course can reveal a broader range of system properties and system states. In this hypothetical time course, the y-axis could represent an intrinsic property of the ecosystem (e.g. productivity, biomass burning, nutrient flux, relative abundance of a key species) or an environmental variable relevant to those properties (e.g. seasonal temperature, annual or seasonal precipitation). The system property during the modern period (the post-Euro-American period in much of North America) might be well outside the HRV, which is the range of system properties observed during the HRV baseline period. Extending the HRV perspective backward in time can provide valuable information on system variability, resilience, and constraints. The vertical line from $eHRV_1$ Reference Period delineates the origination of the ecosystem (i.e. the time at which the dominant species and landscape configurations were attained). System properties often display a higher range of variation within the broader period since that event (i.e. the $eHRV_1$ reference period). Many ecosystems were preceded by ecosystems similar in structure and physiognomy but differing in relative abundance or dominance of key species. The line from e HRV_2 Reference Period delineates the origination of such an antecedent. Extending the reference period back to that event ($eHRV_2$) may encompass an even broader range of system properties and variables. Before the $eHRV_2$ reference period, the site, region, or landscape was occupied in this example by a fundamentally different ecosystem, possibly with properties well outside any conceivable range of variation. In some cases, only one relevant eHRV baseline may exist because current ecosystems were preceded by contrasting ecosystems (see Fig. 7.3). Climate change, land use, invasions, and other changes may force ecosystem properties into new trajectories (not shown), often far beyond the range of variation encountered by any previous historic state.

variations? How much variation in its fundamental properties has it experienced without being driven into a different system state? How much variability and change – both in terms of environment and system properties – can it sustain before it is set on an irreversible path to a different ecosystem? What are the boundaries of the system? What are the key thresholds for transition to a different system?

The range of environmental variation experienced by specific ecosystems can be assessed by examining paleoclimate and other paleoenvironmental records spanning the duration of those systems. In some cases, estimates of past annual or seasonal temperatures and precipitation can be obtained (e.g. tree-ring reconstructions, stable-isotope analyses); in other cases, the paleoclimatic information is available in relative or semiquantitative terms (e.g. lake levels, peatland water-table depths, dune activation and stabilization). Formal assessments have yet to be done in a comprehensive fashion, but this approach can provide indications of ecosystem resilience in the face of climate change. Ecosystems that have persisted with little change for most of the past 10 000 years (Fig. 7.3) have experienced considerable variation in temperature and precipitation in both eastern and western United States (Bartlein et al. 1998; Webb et al. 1998; Mayewski et al. 2004).

Paleoecological and paleoenvironmental records can provide information about the magnitude, frequency, and controls of temporal variation in a variety of ecosystem properties. Pollen and plant macrofossils reveal variations in composition, structure, and pattern of vegetation, even when species dominance or vegetation structure is constant. Sedimentary charcoal and fire-scar analyses indicate changing fire intensities and frequencies. Tree-ring demography records recruitment and mortality episodes as well as changing stand structure. Geochemical and biological data from lake and peatland sediments provide data on hydrological, erosional, and biogeochemical changes on surrounding uplands. Peatlands, lakes, and tree rings record carbon-flux rates. These methods have been applied widely to assess responses of ecosystem properties to past climate change, and could be applied systematically to assess eHRV for specific ecosystems.

Finally, identifying the environmental changes that led to inception of an ecosystem (Fig. 7.3) can also contribute to understanding its resilience to climate changes of the future (Jackson 2006). Specific ecosystems, characterized by their dominant species, distur-

bance regimes, and ecosystem properties, arise as a result of the development of specific climatic regimes, sometimes in concert with specific human activities (e.g. fires). Identification of the particular climate regimes and other factors that underpin particular ecosystems is important to determining vulnerability in the face of future climate change. Although this can be done by empirical modeling in modern climate space, such modeling may miss the climate variables and combinations most important to specific ecosystems (Jackson & Williams 2004; Williams et al. 2007; Jackson et al. 2009a). Determination of the date of origination of specific ecosystems, together with identification of particular macroclimatic regime changes that led to those originations, can provide an alternative approach. Once the specific climate realization that supports a specific ecosystem is known, the climatic thresholds at which it faces risk of collapse or conversion into a different ecosystem can be identified.

Application of eHRV requires some working definition of the ecosystem under consideration. Ecosystems can be defined many ways, but a useful approach is a continuum or hierarchy of specificity. At one end are the species composition, spatial heterogeneity, and disturbance regime of a site or region. Under this strictest definition, many ecosystems are likely to have relatively shallow antiquity because of the decadal to centennial nonstationarity of climate – the particular disturbance regime prevailing now or at the time of Euro-American occupation is unlikely to extend more than a few decades or centuries into the past (Fig. 7.1; Webb 1981; Millar & Woolfenden 1999). In certain regions, the spatial mosaic – the size and duration of patch elements, the frequencies at which they are generated, and the dominant processes generating them – may be sensitive to multidecadal to centennial climate variation.

The other end of the continuum consists of the formation or biome – the vegetation physiognomy that imparts many ecosystem properties. Formation-level properties, which are an aggregate of the elements of the prevailing spatial mosaic, are generally less sensitive to high-frequency climate variation, but they are subject to alteration by lower-frequency, high-magnitude climate change. Between this and the fine-scale, taxonomically precise pattern anchoring the other end of the hierarchy are several intermediates, including changing disturbance regimes and spatial mosaics with the same forest composition, changing mosaics under the same regional vegetation owing to differing elevational, edaphic, and aspect responses of species under moderately differing climates, and altered forest composition (i.e. different species assemblages) within the same biome or formation. Paleoecological records vary in sensitivity and precision owing to spatial, temporal, and taxonomic smoothing (Jackson 1997; Liu et al. 2010). Some studies must of necessity settle for formation-level inferences, while others can obtain more precise detail.

Many regions in the United States have sufficiently dense historical, paleoecological, and paleoclimatological data to attempt an eHRV assessment, and several recent summaries provide some foundations (Swetnam et al. 1999; Foster 2002; Gavin et al. 2007; Millar et al. 2007; Willard & Cronin 2007; Whitlock et al. 2008; Romme et al. 2009; Jackson et al. 2009b). Pursuing an integrated eHRV assessment is a scientific and social challenge that will require engagement among paleoecologists, ecologists, paleoclimatologists, and resource managers.

7.4 GEORGE WEBBER'S DILEMMA

Time presents a dilemma for conservationists and resource managers because things change, often irreversibly, along its axis. This problem has long been recognized in our culture, from Heraclitus to Thomas Wolfe (Wolfe 1940). Like George Webber, the protagonist of Wolfe's novel, we cannot go home again. We cannot restore ecosystems to some past baseline, static or dynamic, because ecosystems and their environmental controls have changed in the past 200 years and will continue to change indefinitely (Jackson & Hobbs 2009). We cannot return ecosystems to an early nineteenth-century state because we are no longer in the early nineteenth century. We can try to restore some ecosystem attributes – fire regimes, disturbance mosaics, stream discharge – that may be appropriate and sustainable in the early twenty-first century. Other ecosystem properties may be difficult or impossible to sustain, both because events of the past two centuries have left indelible imprints on ecosystems and because the environment that constrains and controls ecosystem properties has changed and will continue to change, particularly as we lurch our way into a greenhouse world.

HRV will continue to be valuable in the short term. Successful application will require consideration of pre-Columbian human activities, which will need to be assessed on a site-by-site or region-by-region basis using the best available information. Empirical, place-based knowledge should replace sweeping assumptions about pre-Columbian cultural parklands or untrammeled wilderness. HRV applications will also need to consider a broader array of climate variation than has been traditionally done. Environmental nonstationarity – past, present, and future – will need to become part of the HRV mind-set. The extended HRV concept may prove useful in dealing with nonstationarity and assessing ecosystem resilience and thresholds.

Ultimately, managers will need to shift focus toward asking whether ecosystems of interest will be sustainable under altered climate regimes. How far can these systems be pushed by environmental change before they collapse or change beyond the point of possible restoration? What desirable ecosystem properties and services can be preserved? Which are at risk of loss? Instead of managing for specific ecosystems, the focus may need to change toward identifying the ecological goods and services we value, defining the range of conditions under which they can be maintained, and assessing how evolving ecosystems can be managed to stay within that range (see Safford et al., Chapter 4, this book).

Meeting these challenges will require engagement of scientific and social perspectives. For decades, conservationists and resource managers have groped for objective, value-free baselines for management. The concept of "natural" has often served as a rallying point in this effort. The recognition that pre-Columbian human populations had widespread ecological effects, that the megafauna of North America has been impoverished for the past 10 000 years, and that climate variation and change represent ever-moving targets for ecosystem response suggests that an objective target may be elusive. The concept of natural need not be discarded, but its elasticity must be acknowledged. Scientists, managers, and policymakers will have to grapple more explicitly with values – ethical, aesthetic, and utilitarian – to identify appropriate and attainable goals in a rapidly changing world. The multiscaled perspectives provided by eHRV offer critical historical context for the value judgments that must be made.

ACKNOWLEDGMENTS

This chapter was inspired by discussions at a series of Project 3D workshops sponsored by the National

Commission on Science for Sustainable Forestry. I appreciate the opportunity, initiated by Bob Mitchell and nurtured by Norm Johnson, to think about variability, management, and policy in a long-term context. Discussions with Bob Mitchell, Norm Johnson, Connie Millar, Sally Duncan, Dan Binkley, Tom Spies, Fred Swanson, Brenda McComb, and others in the HRV working group were stimulating and challenging. Several others provided information or discussion at various junctures, including Mark Bush, Paige Newby, Richard Hobbs, Jeff Lockwood, Jack Williams, and Wally Woolfenden. Simon Brewer, Jennifer Andersen, and Mark Lesser assisted in preparing figures. Simon Brewer, Greg Hayward, Connie Millar, Dan Tinker, and two anonymous reviewers provided valuable critiques of the manuscript.

REFERENCES

Allen, C.D. (2002). Lots of lightning and plenty of people: an ecological history of fire in the upland Southwest. In *Fire, Native Peoples, and the Natural Landscape* (ed. T.R. Vale), pp. 143–193. Island Press, Washington, DC, USA.

Allen, C.D. & Breshears, D.D. (1998). Drought-induced shift of a forest-woodland ecotone: rapid landscape response to climate variation. *Proceedings of the National Academy of Sciences*, **95**, 14839–14842.

Baker, R.G., Bettis, E.A. III., Schwert, D., Horton, D.G., Chumbley, C.A., Gonzalez, L.A., & Reagan, M.K. (1996). Holocene paleoenvironments of northeast Iowa. *Ecological Monographs*, **66**, 203–234.

Baker, R.G., Maher, L.J., Chumbley, C.A., & van Zant, K.L. (1992). Patterns of Holocene environmental change in the Midwest. *Quaternary Research*, **37**, 379–389.

Baker, R.G., Bettis, E.A. III., Denniston, R.F., Gonzalez, L.A., Strickland, L.E., & Krieg, J.R. (2002). Holocene paleoenvironments in southeastern Minnesota: chasing the prairie-forest ecotone. *Palaeogeography, Palaeoclimatology, Palaeoecology*, **177**, 103–122.

Baker, W.R. (2002). Indians and fire in the Rocky Mountains: the wilderness hypothesis revisited. In *Fire, Native Peoples, and the Natural Landscape* (ed. T.R. Vale), pp. 41–76. Island Press, Washington, DC, USA.

Barnosky, A.D., Koch, P.L., Feranec, R.S., Wing, S.L., & Shabel, A.B. (2004). Assessing the causes of Late Pleistocene extinctions on the continents. *Science*, **306**, 70–75.

Bartlein, P.J., Anderson, K.H., Anderson, P.M., et al. (1998). Paleoclimate simulations for North America over the past 21000 years: features of the simulated climate and comparisons with paleoenvironmental data. *Quaternary Science Reviews*, **17**, 549–585.

Betancourt, J.L. & Van Devender, T.R. (1981). Holocene vegetation in Chaco Canyon, New Mexico. *Science*, **214**, 656–658.

Betancourt, J.L., Rylander, K.A., Peñalba, C., & McVickar, J. (2001). Late Quaternary vegetation history of Rough Canyon, south-central New Mexico, USA. *Palaeogeography, Palaeoclimatology, Palaeoecology*, **165**, 71–95.

Booth, R.K. & Jackson, S.T. (2003). A high-resolution record of Late Holocene moisture variability from a Michigan raised bog. *The Holocene*, **13**, 865–878.

Booth, R.K., Jackson, S.T., Forman, S.L., Kutzbach, J.E., Bettis, E.A. III., Kreig, J., & Wright, D.K. (2005). A severe centennial-scale drought in mid-continental North America 4200 years ago and apparent global linkages. *The Holocene*, **15**, 321–328.

Bormann, F.H. & Likens, G.E. (1979). *Pattern and Process in a Forested Ecosystem*. Springer-Verlag, New York, USA.

Breshears, D.D., Cobb, N.S., Rich, P.M., et al. (2005). Regional vegetation die-off in response to global-change-type drought. *Proceedings of the National Academy of Sciences*, **102**, 15144–15148.

Brubaker, L.B. (1975). Postglacial forest patterns associated with till and outwash in north-central upper Michigan. *Quaternary Research*, **5**, 499–527.

Connell, J.H. (1978). Diversity in tropical rain forests and coral reefs. *Science*, **199**, 1302–1310.

Cook, E.R., Seager, R., Cane, M.A., & Stahle, D.W. (2007). North American drought: reconstructions, causes, and consequences. *Earth-Science Reviews*, **81**, 93–134.

Cronon, W. (1983). *Changes in the Land: Indians, Colonists, and the Ecology of New England*. Hill & Wang, New York, USA.

Cwynar, L.C. (1987). Fire and the forest history of the North Cascade Range. *Ecology*, **68**, 791–802.

Davis, M.B. (1969). Climatic changes in southern Connecticut recorded by pollen deposition at Rogers Lake. *Ecology*, **50**, 409–422.

Davis, M.B., Calcote, R., Sugita, S., & Takahara, H. (1998). Patchy invasion and the origin of a hemlock-hardwoods forest mosaic. *Ecology*, **79**, 2641–2659.

Dawson, T.P., Jackson, S.T., House, J., Prentice, I.C., & Mace, G.M. (2011). Beyond predictions: biodiversity conservation under climate change. *Science*, **332**, 53–58.

Dean, J.S. (1988). Dendrochronology and paleoenvironmental reconstruction on the Colorado Plateau. In *The Anasazi in a Changing Environment* (ed. G.J. Gummerman), pp. 119–167. Cambridge University Press, Cambridge, UK.

Denevan, W. (1992). The pristine myth: the landscape of the Americas in 1492. *Annals of the Association of American Geographers*, **82**, 369–385.

Donlan, C.J., Berger, J., Bock, C.E., et al. (2005). Re-wilding North America. *Nature*, **436**, 913–914.

Donlan, C.J., Berger, J., Bock, C.E., et al. (2006). Pleistocene rewilding: an optimistic agenda for twenty-first century conservation. *American Naturalist*, **168**, 660–681.

Edwards, M.E., Brubaker, L.B., Lozhkin, A.V., & Anderson, P.M. (2005). Structurally novel biomes: a response to past warming in Beringia. *Ecology*, **86**, 1696–1703.

English, N.B., Betancourt, J.L., Dean, J.S., & Quade, J. (2001). Strontium isotopes reveal distant sources for architectural timber in Chaco Canyon, New Mexico. *Proceedings of the National Academy of Sciences*, **98**, 11891–11896.

Faison, E.K., Foster, D.R., Oswald, W.W., Hansen, B.C.S., & Doughty, E. (2006). Early Holocene openlands in southern New England. *Ecology*, **87**, 2537–2547.

Foster, D.R. (2002). Insights from historical geography to ecology and conservation: lessons from the New England landscape. *Journal of Biogeography*, **29**, 1269–1275.

Foster, D.R. & Motzkin, G. (2003). Interpreting and conserving the openland habitats of coastal New England: insights from landscape history. *Forest Ecology and Management*, **185**, 127–150.

Foster, D.R., Clayden, S., Orwig, D.A., Hall, B., & Barry, S. (2002). Oak, chestnut and fire: climatic and cultural controls of long-term forest dynamics in New England. *Journal of Biogeography*, **29**, 1359–1379.

Gavin, D.G., Hallett, D.J., Hu, F.S., et al. (2007). Forest fire and climate change in western North America: insights from sediment charcoal records. *Frontiers in Ecology and the Environment*, **5**, 499–506.

Gill, J.L., Williams, J.W., Jackson, S.T., Lininger, K.B., & Robinson, G.S. (2009). Pleistocene megafaunal collapse, novel plant communities, and enhanced fire regimes in North America. *Science*, **326**, 1100–1103.

Gitlin, A.R., Sthultz, C.M., Bowker, M.A., et al. (2006). Mortality gradients within and among dominant plant populations as barometers of ecosystem change during extreme drought. *Conservation Biology*, **20**, 1477–1486.

Gray, S.T., Jackson, S.T., & Betancourt, J.L. (2004). Tree-ring based reconstructions of interannual to decadal-scale precipitation variability for northeastern Utah since 1226 A.D. *Journal of the American Water Resources Association*, **40**, 947–960.

Gray, S.T., Betancourt, J.L., Jackson, S.T., & Eddy, R.G. (2006). Role of multidecadal climatic variability in a range extension of pinyon pine. *Ecology*, **87**, 1124–1130.

Harris, J.A., Hobbs, R.J., Higgs, E., & Aronson, J. (2006). Ecological restoration and global climate change. *Restoration Ecology*, **14**, 170–176.

Hobbs, R.J., Arico, S., Aronson, J., et al. (2006). Novel ecosystems: theoretical and management aspects of the new ecological world order. *Global Ecology and Biogeography*, **15**, 1–7.

Hooper, D.U., Chapin, F.S. III., Ewel, J.J., et al. (2005). Effects of biodiversity on ecosystem functioning: a consensus of current knowledge. *Ecological Monographs*, **75**, 3–35.

Hudson, C. (1997). *Knights of Spain, Warriors of the Sun: Hernando de Soto and the South's Ancient Chiefdoms*. University of Georgia Press, Athens, GA, USA.

Hunt, G.T. (1940). *The Wars of the Iroquois: A Study in Intertribal Trade Relations*. University of Wisconsin Press, Madison, WI, USA.

Hussey, T.C. (1993). *A 20 000-Year History of Vegetation and Climate at Clear Pond, Northeastern South Carolina*. MS thesis, University of Maine, Orono, ME, USA.

Jackson, S.T. (1997). Documenting natural and human-caused plant invasions using paleoecological methods. In *Assessment and Management of Plant Invasions* (ed. J.O. Luken and J.W. Thieret), pp. 37–55. Springer-Verlag, New York, USA.

Jackson, S.T. (2006). Vegetation, environment, and time: the origination and termination of ecosystems. *Journal of Vegetation Science*, **17**, 549–557.

Jackson, S.T. & Hobbs, R.J. (2009). Ecological restoration in the light of ecological history. *Science*, **325**, 567–569.

Jackson, S.T. & Whitehead, D.R. (1991). Holocene vegetation patterns in the Adirondack Mountains. *Ecology*, **72**, 641–653.

Jackson, S.T. & Williams, J.W. (2004). Modern analogs in Quaternary paleoecology: here today, gone yesterday, gone tomorrow? *Annual Review of Earth and Planetary Sciences*, **32**, 495–537.

Jackson, S.T., Betancourt, J.L., Lyford, M.E., Gray, S.T., & Rylander, K.A. (2005). A 40 000-year woodrat-midden record of vegetational and biogeographic dynamics in northeastern Utah. *Journal of Biogeography*, **32**, 1085–1106.

Jackson, S.T., Betancourt, J.L., Booth, R.K., & Gray, S.T. (2009a). Ecology and the ratchet of events: climate variability, niche dimensions, and species distributions. *Proceedings of the National Academy of Sciences*, **106**, 19685–19692.

Jackson, S.T., Gray, S.T., & Shuman, B.N. (2009b). Paleoecology and resource management in a dynamic landscape: case studies from the Rocky Mountain headwaters region (ed. G. Dietl and K.W. Flessa), pp. 61–80. Conservation paleobiology. Paleontological Society Papers 15.

Kaye, M.W., Woodhouse, C.A., & Jackson, S.T. (2010). Persistence and expansion of Ponderosa pine woodlands in the west-central Great Plains during the past two centuries. *Journal of Biogeography*, **37**, 1668–1683.

Keeley, J.E. (2002). Native American impacts on fire regimes of the California coastal ranges. *Journal of Biogeography*, **29**, 303–320.

Koch, P.L. & Barnosky, A.D. (2006). Late Quaternary extinctions: state of the debate. *Annual Review of Ecology, Evolution, and Systematics*, **37**, 215–250.

Liu, Y., Jackson, S.T., Brewer, S., & Williams, J.W. (2010). Assessing antiquity and turnover of terrestrial ecosystems in eastern North America using fossil pollen data: a preliminary study. IOP Conference Series: Earth and Environmental Science 9 012005. doi: 10.1088/1755-1315/9/1/012005.

Lyford, M.E., Betancourt, J.L., & Jackson, S.T. (2002). Holocene vegetation and climate history of the northern Bighorn

Basin, southern Montana. *Quaternary Research*, **58**, 171–181.

Lynch, E.A. (1998). Origin of a park-forest vegetation mosaic in the Wind River Range, Wyoming. *Ecology*, **79**, 1320–1338.

Mann, C.C. (2005). *1491: New Revelations of the Americas before Columbus*. Alfred A. Knopf, New York, USA.

Mayewski, P.A., Rohling, E.E., Stager, J.C., et al. (2004). Holocene climate variability. *Quaternary Research*, **62**, 243–255.

McAndrews, J.H. (1988). Human disturbance of North American forests and grasslands: the fossil pollen record. In *Vegetation History* (ed. B. Huntley and T. Webb III.), pp. 673–697. Kluwer, Dordrecht, the Netherlands.

Meko, D.M., Woodhouse, C.A., Baisan, C.A., Knight, T., Lukas, J.J., Hughes, M.K., & Salzer, M.W. (2007). Medieval drought in the upper Colorado River Basin. *Geophysical Research Letters*, **34**, L10705. doi: 10.1029/2007GL029988.

Milanich, J.T. & Hudson, C. (1993). *Hernando de Soto and the Indians of Florida*. University Press of Florida, Gainesville, FL, USA.

Millar, C.I. & Woolfenden, W.B. (1999). The role of climate change in interpreting historical variability. *Ecological Applications*, **9**, 1207–1216.

Millar, C.I., Stephenson, N.L., & Stephens, S.L. (2007). Climate change and forests of the future: managing in the face of uncertainty. *Ecological Applications*, **17**, 2145–2151.

Milly, P.C.D., Betancourt, J., Falkenmark, M., Hirsch, R.M., Kundzewicz, Z.W., Lettenmaier, D., & Stouffer, R.J. (2008). Stationarity is dead: whither water management? *Science*, **319**, 573–574.

Milner, G.R. (2004). *The Moundbuilders: Ancient Peoples of Eastern North America*. Thames and Hudson, London, UK.

Mitchell, F.G.H. (2005). How open were European primeval forests? Hypothesis testing using palaeoecological data. *Journal of Ecology*, **93**, 168–177.

Motzkin, G. & Foster, D.R. (2002). Grasslands, heathlands and shrublands in coastal New England: historical interpretations and approaches to conservation. *Journal of Biogeography*, **29**, 1569–1590.

Munoz, S.E. & Gajewski, K. (2010). Distinguishing prehistoric human influence on late-Holocene forests in southern Ontario, Canada. *The Holocene*, **20**, 967–981.

National Research Council. (2005). *The Geologic Record of Ecological Dynamics: Understanding the Biotic Consequences of Global Change*. National Academy Press, Washington, DC, USA.

Oliveira-Santos, L.G.R. & Fernandez, F.A.S. (2010). Pleistocene rewilding, Frankenstein ecosystems, and an alternative conservation agenda. *Conservation Biology*, **24**, 4–5.

Owen-Smith, N. (1987). Pleistocene extinctions: the pivotal role of megaherbivores. *Paleobiology*, **13**, 351–362.

Parker, A.J. (2002). Fire in Sierra Nevada forests: evaluating the ecological impact of burning by Native Americans. In *Fire, Native Peoples, and the Natural Landscape* (ed. T.R. Vale), pp. 233–267. Island Press, Washington, DC, USA.

Parkman, F. (1867). *The Jesuits in North America in the Seventeenth Century*. Little, Brown & Co., Boston, MA, USA.

Parshall, T. & Foster, D.R. (2002). Fire on the New England landscape: regional and temporal variation, cultural and environmental controls. *Journal of Biogeography*, **29**, 1305–1317.

Pauketat, T.R. (2009). *Cahokia: Ancient America's Great City on the Mississippi*. Viking, New York, USA.

Pickett, S.T.A. & White, P.S. (eds.). (1985). *The Ecology of Natural Disturbance and Patch Dynamics*. Academic Press, New York, USA.

Pierce, J.L., Meyer, G.A., & Jull, A.J.T. (2004). Fire-induced erosion and millennial-scale climate change in northern ponderosa pine forests. *Nature*, **432**, 87–90.

Raffa, K.F., Aukema, B.H., Bentz, B.J., Carroll, A.L., Hicke, J.A., Turner, M.G., & Romme, W.H. (2008). Cross-scale drivers of natural disturbances prone to anthropogenic amplification: the dynamics of bark beetle eruptions. *BioScience*, **58**, 501–517.

Reeves, K. (2006). *Holocene Vegetation Development of Hemlock-Hardwoods Forests in the Upper Peninsula of Michigan*. Master's thesis, University of Wyoming, Laramie, WY, USA.

Robinson, G.S., Burney, L., & Burney, D.A. (2005). Landscape paleoecology and megafaunal extinction in southeastern New York State. *Ecological Monographs*, **75**, 295–315.

Romme, W.H. & Knight, D.H. (1982). Landscape diversity – the concept applied to Yellowstone Park. *Bioscience*, **32**, 664–670.

Romme, W.H., Allen, C.D., Bailey, J.D., et al. (2009). Historical and modern disturbance regimes, stand structures, and landscape dynamics in piñon–juniper vegetation of the western United States. *Rangeland Ecology and Management*, **62**, 203–222.

Saxon, E., Baker, B., Hargrove, W., Hoffman, F., & Zganjar, C. (2005). Mapping environments at risk under different global climate change scenarios. *Ecology Letters*, **8**, 53–60.

Shimkin, D. (1986). Introduction of the horse. In *Handbook of North American Indians, Vol. 11: Great Basin* (ed. W. d'Azevedo), pp. 517–524. Smithsonian Institution Press, Washington, DC, USA.

Shuman, B., Henderson, A.K., Plank, C., Stefanova, I., & Ziegler, S.S. (2009). Woodland-to-forest transition during prolonged drought in Minnesota after ca. AD 1300. *Ecology*, **90**, 2792–2807.

Smith, B.D. (1989). Origins of agriculture in eastern North America. *Science*, **246**, 1566–1571.

Smith, F.A., Brown, J.H., Haskell, J., et al. (2004). Similarity of mammalian body size across the taxonomic hierarchy and across space and time. *American Naturalist*, **163**, 672–691.

Spear, R.W., Davis, M.B., & Shane, L.C.K. (1994). Late Quaternary history of low- and mid-elevation vegetation in the White Mountains of New Hampshire. *Ecological Monographs*, **64**, 85–109.

Swetnam, T.W. & Betancourt, J.L. (1998). Mesoscale disturbance and ecological response to decadal climatic variability in the American Southwest. *Journal of Climate*, **11**,
3128–3147.

Swetnam, T.W., Allen, C.D., & Betancourt, J.L. (1999). Applied
historical ecology: using the past to manage for the future.
Ecological Applications, **94**, 1189–1206.

Thompson, R.S. & Anderson, K.H. (2000). Biomes of western
North America at 18 000, 6000 and 0 1[4c] year BP recon-
structed from pollen and packrat midden data. *Quaternary
Science Reviews*, **27**, 555–584.

Vale, T.R. (2002). The pre-European landscape of the United
States: pristine or humanized? *Fire, Native Peoples, and the
Natural Landscape* (ed. T.R. Vale), pp. 1–39. Island Press,
Washington, DC, USA.

Vera, F.W.M. (2000). *Grazing Ecology and Forest History*. CABI
Publishing, Wallingford, UK.

Watson, T.A., Barnett, F.A., Gray, S.T., & Tootle, G.A. (2009).
Reconstructed streamflows for the headwaters of the Wind
River, Wyoming, United States. *Journal of the American
Water Resources Association*, **45**, 224–236.

Watts, W.A. (1979). Late Quaternary vegetation of Central
Appalachia and the New Jersey Coastal Plain. *Ecological
Monographs*, **49**, 427–469.

Webb, T. III. (1981). The past 11 000 years of vegetational
change in eastern North America. *BioScience*, **31**,
501–506.

Webb, T. III., Anderson, K.H., Bartlein, P.J., & Webb, R.S.
(1998). Late Quaternary climate change in eastern North
America: a comparison of pollen-derived estimates with
climate model results. *Quaternary Science Reviews*, **17**,
587–606.

Weng, C. & Jackson, S.T. (1999). Late-glacial and Holocene
vegetation and climate history of the Kaibab Plateau,
northern Arizona. *Palaeogeography, Palaeoclimatology, Palaeoecology*, **153**, 179–201.

Whitlock, C. & Knox, M.A. (2002). Prehistoric burning in the
Pacific Northwest: human versus climatic influences. In
Fire, Native Peoples, and the Natural Landscape (ed. T.R. Vale),
pp. 195–231. Island Press, Washington, DC, USA.

Whitlock, C., Shafer, S.L., & Marlon, J. (2003). The role of
climate and vegetation change in shaping past and future
fire regimes in the northwestern US and the implications
for ecosystem management. *Forest Ecology and Management*, **178**, 163–181.

Whitlock, C., Marlon, J., Briles, C., Brunelle, A., Long, C., &
Bartlein, P. (2008). Long-term relations among fire, fuel,
and climate in the northwestern US based on lake-sediment
studies. *International Journal of Wildland Fire*, **17**, 72–83.

Willard, D.A. & Cronin, T.M. (2007). Paleoecology and ecosystem restoration: case studies from Chesapeake Bay and
the Florida Everglades. *Frontiers in Ecology and the Environment*, **5**, 491–498.

Williams, J.W. & Jackson, S.T. (2007). Novel climates, no-
analog communities, and ecological surprises: past and
future. *Frontiers in Ecology and the Environment*, **5**, 475–482.

Williams, J.W., Shuman, B.N., Webb, T. III., Bartlein, P.J., &
Leduc, P.L. (2004). Late Quaternary vegetation dynamics
in North America: scaling from taxa to biomes. *Ecological
Monographs*, **74**, 309–334.

Williams, J.W., Jackson, S.T., & Kutzbach, J.E. (2007). Projected distributions of novel and disappearing climates by
2100 AD. *Proceedings of the National Academy of Sciences*,
104, 5738–5742.

Wolfe, T. (1940). *You Can't Go Home Again*. Harper-Collins,
New York, USA.

Woolfenden, W.B. (1996). Quaternary vegetation history. In
*Sierra Nevada Ecosystem Project: Final Report to Congress, Vol.
II: Assessments and Scientific Basis for Management Options*,
pp. 47–70. University of California, Centers for Water and
Wildland Resources, Davis, CA, USA. http://ceres.ca.gov/
snep/pubs/web/PDF/VII_C04.pdf.

Section 3

Modeling Historic Variation and Its Application for Understanding Future Variability

Robert E. Keane

USDA Forest Service, Rocky Mountain Research Station, Missoula Fire Sciences Laboratory, MT, USA

Although some may doubt its usefulness in a future with rapidly changing climates, exotic introductions, and increased human land use, the historical range of variation (HRV) of ecological landscape charactcristics provides a relatively useful reference point for evaluating the impacts of land-management activities. Unfortunately, comprehensive spatial and temporal data describing historical landscape conditions are rare for many areas, with most information being limited in geographic scope and relatively recent. The main problem facing many ecologists, scientists, and land managers is how to quantify the HRV of landscapes in a format that is scientifically credible, useful to land management, temporally deep, and spatially extensive, while still being relevant in today's changing world.

The best method for quantifying historical landscape conditions relies on a chronosequence or a series of maps or data layers from one landscape over many past time periods. However, temporally deep, spatially explicit

Historical Environmental Variation in Conservation and Natural Resource Management, First Edition. Edited by John A. Wiens, Gregory D. Hayward, Hugh D. Safford, and Catherine M. Giffen.

empirical chronosequences of landscape conditions are rare because aerial photography and satellite imagery were nonexistent before 1930, and paper maps of forest vegetation are scarce and inconsistent prior to 1900. Another method involves using digital maps from similar landscapes, taken from one or multiple time periods, and gathered across a geographic region to quantify the landscape HRV (Hessburg et al. 1999; 2000). This substitution of space for time assumes that all landscapes used to define HRV are similar in terms of environmental, disturbance, topography, and biological conditions. However, most landscapes are unique in terms of the biophysical environment and the manifestation of disturbance dynamics over time creates distinctive variations in landscape HRV characteristics because of differences in topography, orientation, wind direction, and many other microclimate, biotic, and edaphic characteristics (Keane et al. 2006).

In many situations, simulation modeling provides the only viable source for generating comprehensive HRV data. This third method involves simulating historical dynamics using landscape models to produce a chronosequence of simulated spatial data to use as reference. This approach assumes that succession and disturbance processes are simulated accurately in space and time. Many spatially explicit ecosystem simulation models are available for quantifying HRV patch dynamics (see Mladenoff & Baker 1999; Keane et al. 2004), but many are computationally intensive, difficult to parameterize and initialize, and complex in design, making them difficult to use across large regions over long time periods. Even with these limitations, simulation models often provide the only way to quantify HRV for many landscapes, and therefore, they are a critical tool for managing today's landscapes. Although spatial chronosequences are clearly preferable, simulated chronosequences provide a viable, and in some cases, the only, alternative to creating HRV data.

This section describes the use of simulation modeling to develop HRV chronosequences for land management. The first chapter deals with all the background, issues, and limitations of creating simulated HRV time series. Important topics include landscape size, selecting the most desirable model, and data parameterization issues. The next chapter provides examples of how simulated HRV time series can be used in natural resource management at various scales. Collectively, these chapters may provide the information needed to start an HRV project using a landscape simulation model to generate historical time series, which can then be used as a reference to compare management treatment alternatives.

REFERENCES

Hessburg, P.F., Smith, B.G. & Salter, R.B. (1999). A method for detecting ecologically significant change in forest spatial patterns. *Ecological Applications*, **9**, 1252–1272.

Hessburg, P.F., Smith, B.G., Salter, R.B., Ottmar, R.D. & Alvarado, E. (2000). Recent changes (1930's–1990's) in spatial patterns of interior northwest forests, USA. *Forest Ecology and Management*, **136**, 53–83.

Keane, R.E., Cary, G., Davies, I.D., et al. (2004). A classification of landscape fire succession models: spatially explicit models of fire and vegetation dynamic. *Ecological Modelling*, **256**, 3–27.

Keane, R.E., Holsinger, L. & Pratt, S. (2006). Simulating historical landscape dynamics using the landscape fire succession model LANDSUM version 4.0. General Technical Report RMRS-GTR-171CD. USDA Forest Service, Fort Collins, CO, USA.

Mladenoff, D.J. & Baker, W.L. (1999). *Spatial Modeling of Forest Landscape Change*. Cambridge University Press, Cambridge, UK.

Chapter 8

CREATING HISTORICAL RANGE OF VARIATION (HRV) TIME SERIES USING LANDSCAPE MODELING: OVERVIEW AND ISSUES

Robert E. Keane

USDA Forest Service, Rocky Mountain Research Station, Missoula, MT, USA

Historical Environmental Variation in Conservation and Natural Resource Management, First Edition. Edited by John A. Wiens, Gregory D. Hayward, Hugh D. Safford, and Catherine M. Giffen.

8.1   INTRODUCTION

Simulation modeling can be a powerful tool for generating information about historical range of variation (HRV) in landscape conditions. In this chapter, I will discuss several aspects of the use of simulation modeling to generate landscape HRV data, including (1) the advantages and disadvantages of using simulation, (2) a brief review of possible landscape models, and (3) the pitfalls and limitations of the simulation approach. This information provides a reference for planning, preparing, implementing, and interpreting HRV simulations and their results.

Before proceeding, it is probably best to define some major terms that will be used in this section. HRV time series refers to spatial or tabular data that represent landscape characteristics over time (e.g. percent coverage of a Douglas-fir vegetation type on a $500\,\text{km}^2$ landscape at 10-year intervals for 1000 years). While HRV can represent any ecosystem element at any scale, this section will only discuss HRV landscape dynamics and landscape structural and compositional characteristics. Size, arrangement, and pattern distribution of patches describe landscape structure, while landscape composition is often described by the relative abundance of ecosystem features across the spatial domain (e.g. percent area by cover types).

Modeling terminology can also be confusing, so it is important to define important terms as well. Model parameterization is the quantification of the major parameters required as input to the model. Parameters are static variables, such as smoke-emission factors or duff bulk densities, with values estimated by the user or model author. Parameters for some models are emergent properties (dynamically simulated output) for other models; fire-return interval, for example, is an input parameter in landscape succession model (LANDSUM) (Keane et al. 2006), but it is an explicitly simulated output variable in FIRESCAPE (Cary 1997). State variables are those dynamic variables explicitly simulated by a model, such as stand carbon or fuel loading. Model initialization involves quantifying state variables from plot data, geographic information system (GIS) maps, and previous simulation results to begin a simulation. Model execution refers to running the model to create output to analyze. The output contains predictions or estimations of response variables that, in this chapter, are the variables directly represented in the HRV time series.

How to create HRV time series

To understand the complexity involved in generating simulated HRV time series, it is important to know the general steps involved for the creation of a simulated HRV data set (Fig. 8.1):

1 State your objective. The most important step in the entire process is to succinctly state the objective of your HRV analysis so that you can properly select the best response variables and the most appropriate model.

2 Select response variables. The decision of which variables to include in the HRV time series will make model selection and parameterization easier, but a long list of response variables will limit the number of available models and complicate an HRV analysis.

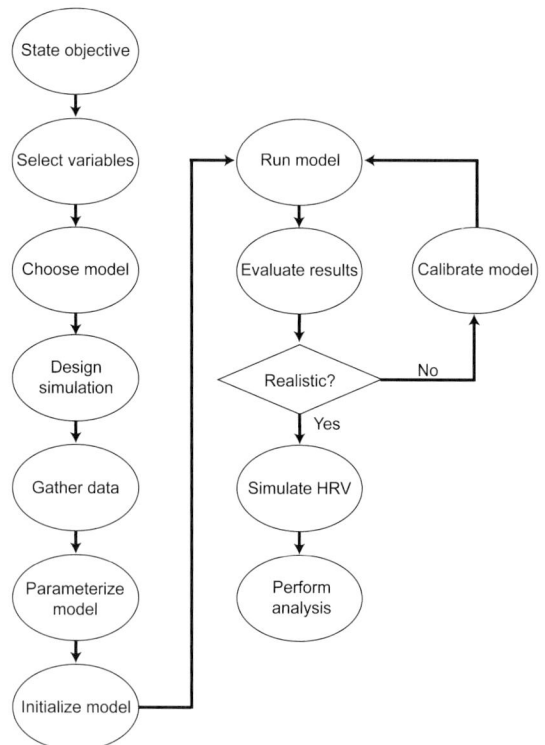

Fig. 8.1 The steps involved in creating HRV time series using landscape simulation modeling.

3 Choose your model. Selecting the right model is also an important step (Keane et al. 2009), so use experts to help decide the most appropriate model for a particular application; ensure that the input requirements for the model are available; find sufficient expertise to run the model; and make sure there are adequate computer resources to execute the model.

4 Design the simulation. Decide important simulation specifics such as the landscape extent (i.e. project area), pixel size, buffer width, length of the simulation, and output-reporting interval.

5 Gather data. The availability of data is critical to quantify parameters and starting conditions for the entire simulation area. If data are not available, literature searches, local experts, and statistical modeling can help quantify those missing parameters.

6 Parameterize model. Analyze collected data to quantify parameters for direct input into the model and be sure to document all sources of information used to quantify each parameter.

7 Initialize model. Many modelers use data that describe current conditions as a starting point in the model (e.g. a digital stand map classified to vegetation type could be used to initialize a landscape simulation).

8 Execute model. The model should be run again and again to be sure that all parameters and initial conditions are entered into the model correctly and to address all error statements and warnings.

9 Evaluate results. Always examine model outputs in detail to determine if the model is computing believable and realistic answers.

10 Calibrate model. Check simulation results against field data to ensure realism; if unrealistic predictions are generated, adjust parameters and initial conditions to get more reasonable answers, especially for those parameters with high uncertainty, such as those quantified from expert opinion and other relevant studies. (Repeat steps 8 and 9 until satisfied with results.)

11 Simulate HRV. The simulated HRV time series can be generated once the user is content with model behavior; these time series should be compared again against data to address validity.

12 Perform HRV analysis. Use the simulated HRV time series in land-management applications (e.g. change in similarity between simulated HRV time series and contemporary landscapes can be computed using Sorenson's index) (Keane et al. 2009). The analysis of HRV time series is the same whether it was created from simulation or empirical chronosequences.

8.2 SIMULATING HRV: THE GOOD AND THE BAD

Simulation modeling may provide the only vehicle for creating HRV time series with sufficient depth and detail for comprehensive land-management applications. However, simulation models should be used with great caution because all models are oversimplifications of reality and cannot represent the full range of landscape processes that influence HRV statistics. Understanding the strengths and weaknesses of the simulation approach is important to properly interpret and analyze HRV time series.

Advantages of the simulation approach

One good thing about simulation modeling is that it can produce spatially consistent HRV time series across relatively large areas. Although most simulation models may have limited accuracy, their real strength is that they have great precision: the models simulate disturbance and vegetation development processes in the same way across large regions and over long time spans. This helps to ensure that all ecosystems will be simulated at the same level of detail so that HRV time series are not biased toward certain ecosystems (providing they are represented in the same detail in model design).

The simulation approach also allows the generation of multiple time series reflecting alternate landscape histories, so that range and variation statistics can be computed across a wider envelope of historical and potential future conditions. Several climate-change scenarios, for example, could be simulated to generate a potential landscape-composition time series under future climates. For example, to predict future conditions, the succession simulation could include exotics to address their influence on historical departures and management actions, which can then be compared against the simulated HRV time series.

Another advantage is that simulation extrapolates limited field data consistently across large regions and over long time periods. Empirical data with inconsistent spatial or temporal coverage can parameterize landscape models, which can then simulate ecological processes to extrapolate parameter values across entire regions. A mountainous landscape, for example, may have limited fire-history data for a small portion of the

(a)

Simulated time series

(b)

**Difference across
fire scenarios**

Open tall Douglas fir
Open tall spruce/fir
Closed tall Douglas fir
Closed tall lodgepole pine
Closed tall spruce/fir

•••••• CC-HH
——— CC-HN
- - - - CC-HD

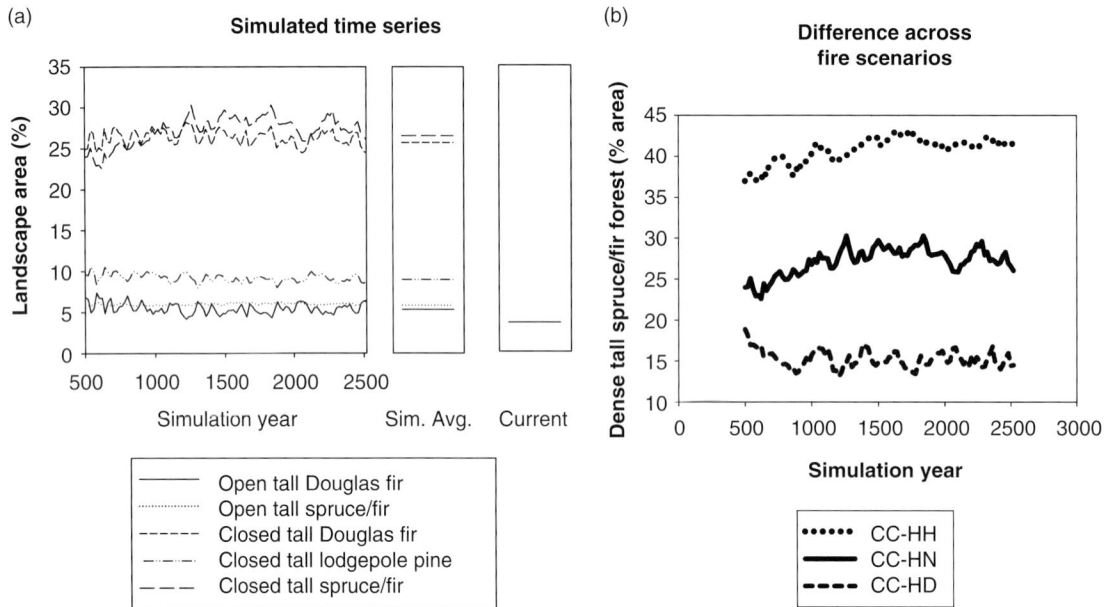

Fig. 8.2 An example of a simple five response variable (cover type extent) HRV time series for a mountainous subalpine Montana, USA landscape and its comparison to current conditions (from Keane et al. 2008).

area, but these data can parameterize fire history across the entire area based on biophysical factors.

Simulation models also generate many types of output to describe historical variation for a multitude of management objectives. For example, landscape models can generate maps of fire frequency and severity (Fig. 8.2) that can be compared with current fire atlases to estimate indices of departure in terms of fire regime (Keane et al. 2004). Fuel characteristics can be assigned to simulated vegetation types to generate historical chronosequences of fuels. The results can then be compared with current fuel maps to evaluate changes in fire hazard. Moreover, multiple output maps created by simulation can be merged to compute an integrated measure of HRV, as all output maps were created in the same context.

Landscape models can perform intermediate analyses to output additional HRV response variables. For example, landscape models can include wildlife-suitability statistical models to evaluate HRV of wildlife-habitat indexes. Models can also be modified to incorporate the latest research findings and computer technology, including improvements in model algorithms, parameters, initial conditions, and testing procedures, making them a highly adaptable tool for HRV generation.

Another advantage of simulation modeling is temporal depth. Creating deep temporal time series only requires additional years of simulation; deep temporal time series increases the number of observations, making analysis statistics more meaningful and powerful (Steele et al. 2006). Moreover, deep time series overcome issues of temporal autocorrelation by creating output at time intervals that are long enough to minimize correlation. Deep series are also long enough to ensure that the spatial properties of a fire regime are fully realized, which is especially difficult to evaluate in long fire-return-interval landscapes using empirically derived spatial chronosequences (Pratt et al. 2006).

Weaknesses of the HRV simulation approaches

Box and Draper (1987) said, "all models are wrong, but some are useful." This is the major drawback of all simulation approaches. It is impossible for a modeler to build a model that includes all the landscape and ecosystem processes that affect those variables selected to represent the HRV time series for several reasons. First, knowledge concerning the important processes and their linkages is often inadequate. Second, as model complexity increases, instability, computational requirements, input parameters, and interpretability increase to unrealistic levels. Lastly, the modeler's knowledge, experience, and ability often limit complex simulation designs. There is always a trade-off between model complexity and tenability. Empirical modeling approaches often yield the most accurate answers, but they are limited in scope, often data intensive, and incapable of directly simulating complex interactions. Mechanistic modeling approaches include greater detail in simulating ecological processes, making them more robust and comprehensive; these models, however, can be inaccurate, difficult to use, and somewhat unstable due to the increased complexity (Keane & Finney 2003).

There is often a lack of sufficient expertise and input parameter data to parameterize and execute the model. Land-management agencies with inadequate modeling expertise on staff may have to rely on outside experts to run, summarize, and interpret model output. A credible model used improperly is useless for HRV analysis, and model results that are interpreted inappropriately could also result in unwise management. It would be inappropriate, for example, to use a 1-km pixel resolution, landscape model to determine silvicultural thinning locations.

It is difficult to test and validate many landscape models because there are few spatial historical time series that are temporally deep and in the right context for comparison with model results (Keane & Finney 2003). Desirable characteristics for validation data include (1) compatible format and units with model output, (2) extensive time period representation, and (3) consistency across the entire landscape. Because comprehensive spatial historical data are rare, many modelers turn to the validation of a landscape model's modules or algorithms as a way to verify that the model is producing realistic results. Stage (1997) performed an extensive validation of succession parameters for the Columbia River Basin successional model (CRBSUM) landscape model (Keane et al. 1996a) using the forest vegetation simulator (FVS) stand growth model (Crookston & Dixon 2005) and found greater than 80% agreement. Keane et al. (2002) compared simulated fire area and pattern statistics from a 1000-year LANDSUM run to the historical fire atlas created by Rollins et al. (2001) and found >85% agreement with fire size but poor agreement with fire shape (<30%).

No single model will satisfy the varied demands of management, so compromises in simulation design must always be made. The best model might not be the most useful because (1) few people know how to run it, (2) there may be insufficient computer resources (software, hardware requirements) to run it, and (3) there may be insufficient field data for input parameter approximation. Most HRV simulation projects are designed around the people responsible for their completion. The following sections were written for them.

8.3 LANDSCAPE SIMULATION MODELS

Many types of landscape models generate HRV time series of landscape characteristics. One class of landscape models, called landscape fire succession models (LFSMs) (Keane et al. 2004), simulate the linked processes of fire and succession in a spatial domain. Although the complexity of spatial relationships of climate, vegetation, fire ignition, and fire spread may vary from model to model, all LFSMs, by definition, produce time-dependent, geo-referenced results in the form of digital maps or GIS layers. Keane et al. (2004) reviewed 44 LFSMs and then classified them into similar groups based on scale of application, simulation detail, and fire-modeling approaches. He et al. (2008) and He (2008) described and classified many forest landscape models that simulate natural and man-caused disturbance in forested landscapes. Baker (1989) examined several landscape change models and grouped them into whole, distributional, and spatial landscape models depending on the level of data aggregation. Mladenoff and Baker (1999) present details of some landscape models, and Gardner et al. (1999) provides a review of spatial fire spread and effects models. Perry and Millington (2008) also present a summary of disturbance and succession modeling efforts.

Several existing landscape models provide examples of the diverse approaches used to simulate landscape, climate, and fire dynamics. At the complex end, FIRE-BGC integrates the mechanistic FOREST-BGC biogeochemical model (Running & Coughlan 1988) with the FIRESUM gap-process model (Keane et al. 1989) to simulate climate-fire-vegetation dynamics in a spatial domain (Keane et al. 1996b). LANDIS is a diameter cohort process model that evaluates fire, climate, insects, windthrow, and harvest disturbance regimes and their effects on landscape pattern and structure

(He & Mladenoff 1999, Mladenoff 2004), with fire effects indirectly simulated at the stand level based on age-class structure, and succession simulated as a competitive process driven by species life history parameters. Roberts and Betz (1999) used life history parameters or vital attributes of Noble and Slatyer (1977) to drive succession in their polygon-based LANDSIM model, which simulates fire effects at the polygon level without a fire-spread model. The DISPATCH model of Baker (1999) stochastically simulates fire occurrence and spread based on the dynamically simulated weather, fuel loadings, and topographic setting, and then simulates subsequent forest succession as a change in cover type and stand age. Miller and Urban (1999) implemented a spatial application of fire in the Zelig gap-process model (Burton & Urban 1990) to assess the interaction of fire, climate, and pattern in Sierra Nevada forests. Any one of the models could be used to estimate HRV, but most are research models developed for exploring dynamic landscape interactions.

Most land managers find simple, parsimonious models preferable for HRV generation because they are easier to use, simpler to learn, quicker to execute, and more straightforward to interpret. More simplistic approaches include the SIMPPLLE model (Chew et al. 2003), which uses a multiple pathway, state-and-transition approach to simulate succession on landscape polygons and a stochastic approach to simulate fire. This same theme can be found in the FETM (Schaaf et al. 2004), VDDT (Kurz et al. 1999), and LANDSUM (Keane et al. 2006) models. Other similar approaches and models were used by Wimberly et al. (2000) for the Pacific Northwest and Keane et al. (1996a) for the interior Columbia River Basin.

8.4   SELECTING THE MOST APPROPRIATE MODEL

Selecting the most appropriate model for HRV generation is no trivial task since there are many models available and every model is different in design, input and output requirements, and detail of simulation. One important HRV model selection criterion is that the model should be spatially explicit, which means it should explicitly simulate spatial processes across a landscape (often called landscape or spatial models). The spatial simulation of ecosystem processes ensures that simulated landscape structural and composition variation includes those sources of variation due to

spatial factors. Karau and Keane (2007) found that output variation increased as the spatial extent decreased, indicating that without the explicit representation of spatial processes, stand-level models will tend to underestimate the range and variation of historical landscape dynamics, resulting in higher estimates of departure for current landscape conditions (Keane et al. 2009). A spatial model is vital for comprehensive HRV analysis because HRV is somewhat meaningless without a spatial context. Nonspatial landscape models, such as VDDT (Beukema & Kurz 1998), will underestimate variation in HRV response variables, thereby compromising the integrity of HRV time series.

The selected model should provide output that can be used directly in management HRV applications. If the model does not output fuels, for example, then estimating the HRV of fuel loadings will be difficult. Many managers overcome this problem by making the HRV variable in question an attribute of the entities or state variables that are simulated or by modifying the model to output the desired attributes. For example, fuel loadings can be assigned to each vegetation type (variable output by the model) to quantify wildland fuel HRV time series. Managers should evaluate the list of output variables from the models to determine which best fit their specific application.

The design, detail, and resolution of the simulation model are also important factors to evaluate when selecting a model for HRV application. The ecological processes integrated in the model, and their detail of representation, are critically important to insightful HRV generation. The greater the simulation complexity, the greater the variation will be in HRV response variables (Keane et al. 2009). Here are additional suggestions:

1 Select a model that simulates spatial processes at the right resolution. Spatial models that simulate fire spread at 30-m pixel size or less, for example, are the most appropriate for quantifying HRV for small landscapes.

2 Select a model that simulates the important processes that influence the selected response variables. Users should try to match model state and other simulated variables with desired output; if climate-change impacts must be assessed, then select a model that explicitly includes climate in its simulation.

3 Avoid models that require input parameters that are too difficult to quantify. Data input requirements for some models are so specific that they are difficult, if not impossible, to measure or estimate by managers. For example, maximum stomatal conductance as required by Biome-BGC (Thornton et al. 2002) is a parameter that few can accurately measure but is an important and sensitive parameter in the model.

The selection of the most appropriate model boils down to one single factor – the resources available to run the model. There must be knowledgeable, experienced people available to successfully run, or learn to run, the model and interpret model output. If there are few people available locally on staff, then it may help to look for other specialists, organizations, or businesses to perform these tasks. Computers used for HRV simulations must also be capable of executing the model in terms of hardware requirements (memory, disk space, processor speed), software necessities (ancillary analysis software, statistical packages), and network capacity (ability to transmit large amounts of data). Model design is probably the least-used selection factor for many managers because it requires considerable knowledge about simulation science and model specifications to effectively compare the vast array of models.

8.5 SIMULATION ISSUES AND LIMITATIONS

Simulation modeling is a complex process with many challenges in its implementation. A thorough knowledge of these limitations is critical for a comprehensive evaluation of HRV analysis results and implementations.

Parameterization

Perhaps the most important task in conducting HRV simulations is the accurate estimation of input parameters. There are often insufficient ecological data across an area for parameterization so data synthesis and modeling are often required to extrapolate existing data across the entire simulation landscape. Parameters can be approximated from many sources, and are listed here in order of preference:

1 Measurement. The actual measurement of input parameters across the simulation landscape produces the most credible simulation results; it is important to sample the full range of the parameter to minimize extrapolation into unsampled space.
2 Literature. A review of the literature to evaluate the parameter values measured on other landscapes by

other studies may provide a suitable alternative, but special care must be taken to ensure that the context of the measurement best fits the simulation landscape. Several literature syntheses of important model parameters are available (Korol 2001; Hessl et al. 2004).
3 Meta-modeling. Models can estimate parameters for other models. Stage et al. (1995), for example, used a growth and yield model (FVS) to estimate growth parameters for a gap model.
4 Iterative modeling. A parameter is approximated, input into the model, and then the model is executed to generate results that are compared to a reference to determine if the reference agrees with the simulated results. If not, changes to the parameter should be made incrementally until the results match the reference.
5 Expert opinion. A parameter is approximated by a group of experts in a systematic fashion, based on their past experience.
6 Default. Often, a modeler will prepare a sample model application for demonstration purposes, and use parameter values generated from the default input files.
7 Best guess. Sometimes, the only option is just to guess at the value based on past experience and consultation with modelers, and then use iterative modeling to calibrate the guess.

Technically, the author of each landscape model should conduct an extensive sensitivity analysis of their model's parameters to identify the importance of each so that detailed measurements or literature searches are not made on minor parameters, or best guesses are not used for the most important parameters.

Accurately estimated model parameters do not always ensure simulations with high accuracy or realism. There is always a disjunct between model design and model parameterization. Keane et al. (2006), for example, found fire regimes in mountainous settings were simulated incorrectly when fire-return-interval parameters were quantified from tree fire scars taken from topographic settings that experienced frequent fires (e.g. flat areas). The detail in the simulation model will never capture the detail of environmental factors that created the field evidence from which parameters are quantified. As a result, it is difficult to realistically simulate disturbance or vegetation development without building overly complex models that would be difficult to parameterize and inefficient to run for large landscapes over long simulation periods.

Autocorrelation

Most simulated time series are autocorrelated in space and time, as are real landscapes (Ives et al. 2003). Any part of a simulation landscape ultimately depends on the surrounding area (Turner 1987), and the condition of a landscape in one year greatly depends on the landscape composition and structure during the previous years (Reed et al. 1998). Moreover, the area covered by one vegetation type relates to the extent of all other vegetation types; an increase in one type must result in the corresponding decrease of one or more vegetation types on the landscape (Pratt et al. 2006). Autocorrelation may be considered part of HRV and therefore may not be important (Keane et al. 2011), but many statistical techniques require minimization of autocorrelation in response variables (Steele et al. 2006).

Temporal autocorrelation can be minimized by selecting a reporting interval that is long enough to reduce temporal autocorrelation but short enough to provide a sufficient number of observations to perform a valid statistical test and to adequately represent the variation. This reporting interval varies by landscape depending on fire frequency and succession transition times, but informal investigations suggest that 20–50 years would suffice for most landscapes (Pratt et al. 2006). A new set of statistical analysis tools may be needed to compute an index of departure that is both useful to land management and satisfies the assumptions of HRV analysis by minimizing direct effects from autocorrelation (Keane et al. 2011).

Scale

Using the same temporal and spatial extent to evaluate landscapes across large regions will introduce bias into the computation of HRV statistics because the scales of climate, vegetation, and disturbance interactions are inherently different across landscapes (Morgan et al. 1994). Karau and Keane (2007) found that simulated HRV chronosequences of landscapes smaller than 100 km^2 had increased variability in landscape composition due to the spatial dynamics of simulated disturbance processes. This is why HRV approaches are inappropriate when applied on small areas such as stands. A Douglas-fir stand historically dominated by ponderosa pine, for example, may appear to be outside HRV, but if it is within a 100-km^2 landscape composed primarily of ponderosa pine, it will certainly be within HRV. On the other hand, when evaluation landscapes are large (>500 km^2), it is often difficult to detect significant changes caused by ecosystem restoration or fuel treatments implemented on small areas (Keane et al. 2006).

There is an optimum landscape extent for HRV simulation, but this optimum depends on subtle differences in topography, climate, and vegetation across large regions. Karau and Keane (2007) computed optimal landscape sizes for flat and mountainous areas in central Utah under various fire-return intervals and simulation resolutions and found this optimum to differ by topography and geographical area (Table 8.1). This optimum also changes with simulation spatial resolution, which is important because coarser

Table 8.1 Optimal landscape sizes (km^2) for a mountainous and flat landscape for three different fire regimes – historical fire frequencies, half the historical fire frequencies, and double the historical frequencies. Simulations were done at four pixel resolutions (30, 90, 300, and 900 m) (see Keane et al. 2008 for more details).

Fire Frequency	Resolution			
	30 m	90 m	300 m	900 m
Mountainous landscape				
Half	87	103	104	111
Historical	90	99	105	108
Double	100	103	101	106
Flat landscape				
Half	107	73	72	81
Historical	84	104	97	98
Double	100	105	101	99

Note: There are no results for the 0.01-km^2 average fire size and the 900-m resolution simulation combination.

simulations require shorter execution times and less computer resources, making them more desirable for managers.

The input parameters more accurately represent the temporal depth of an HRV simulation than the length of the simulation (number of simulation years). Values given to parameters are usually derived, directly or indirectly, from field data, and these data often represent a small slice of time. Therefore, the simulated historical variation may not always represent the range of conditions needed to maintain healthy, resilient ecosystems. Given the age spans of some organisms used to quantify HRV (e.g. trees are often used to estimate fire histories), it is difficult to obtain comprehensive historical data over static climates and stable biophysical conditions. Therefore, the range and variation of most ecological characteristics tend to become greater as diverse climates are represented in the historical time series, resulting in a possible inflation of acceptable range and variation. The challenge, then, is to use an HRV time span supported by historical field data while also being representative of short-term future climate regimes.

Simulation complexity

The complexity of simulation models can have a major influence on both the creation of an HRV time series and the subsequent comparison of historical dynamics to current conditions. In general, output from complex mechanistic models tends to have higher variation than output from simplistic models. However, Keane et al. (2008) found that complex state-and-transition models with a large number of states had more elements to compare with current conditions and, as a result, the simulated variation of landscape elements was lower than in a simpler state-and-transition model, mainly because the large number of near-zero values for many states tended to lower departure estimates (Pratt et al. 2006). Departure estimation is best when succession pathway complexity is somewhat equal across all simulation landscapes and few states are rare on the landscape (Keane et al. 2011).

The design and detail of landscape models also affects HRV time series. Inadequate HRV simulation results will occur when critical processes are left out of the simulation model. For example, the lack of mountain pine beetle simulation in a landscape fire-succession model used to simulate HRV on lodgepole

pine landscapes may create historical time series that underestimate ranges and variation of lodgepole pine successional stages (Logan et al. 1998). Moreover, the detail of disturbance simulations also influences HRV simulated time series: simplistic cell-automata, fire-spread models, for example, may generate fire perimeters that are completely different when compared with perimeters simulated by complex vector spread algorithms (Keane et al. 2004).

An unfavorable side effect of HRV simulation modeling is that the range and variation of simulated response variables include the variation caused by uncertainty in model predictions. As mentioned before, it is nearly impossible to validate landscape models to quantify accuracy and error rates because of the lack of suitable spatial data. As a result, the simulated variation also includes undesirable and unquantifiable sources such as unintended stochasticity, model flaws, inadequate parameterization, and oversimplifications. It is important that modeling efforts have access to extensively sampled landscapes to create an extensive validation database that assesses unwanted sources of variation.

Model equilibration

The equilibrium models often used in HRV simulations may require long simulation times to come into equilibrium (Keane et al. 2006). Along with the input parameters, initial conditions often influence the time it takes for a model to reach equilibrium conditions. It is important to approximate the time to equilibrium for HRV simulations by graphically examining the time series and visually assessing when the initial conditions do not influence the simulated response variable. All HRV observations prior to this time should be removed from the time series, or the simulation should be executed again with output generated after this time. Pratt et al. (2006) estimated that at least 250 years were needed before the LANDSUM model came into equilibrium.

Using current landscape conditions to initialize historical landscapes can often be inefficient on two levels. First, the current landscape often contains too many patches, making simulation times longer and needlessly complex (Keane et al. 2006). Second, current landscapes are often so departed from historical conditions that it takes excessively long times to reach equilibrium. Moreover, exotic vegetation types that occur on the current landscape are often not represented in historical parameters, resulting in inappropriate HRV time series. Pratt et al. (2006) found that creating neutral initial landscapes composed of the most dominant vegetation type was the most efficient for HRV simulations.

Edge effects

One of the major problems in defining simulation landscapes is that the landscape edges create artificial boundaries across which spatial processes cannot traverse. Areas near the edge of the landscape, for example, have a limited number of surrounding pixels from which a fire can spread into them. Spatial processes, such as fires, cannot immigrate into the simulation landscape, resulting in decreased occurrence near landscape edges. This problem is exacerbated by biophysical factors that are directional vectors that act on many ecosystem processes such as fire, seed dispersal, and windthrow (e.g. wind direction and slope). Pixels near the direction from which the wind originates (upwind), for example, have a lower probability of burning than those downwind (Keane et al. 2002) (Fig. 8.3). Some modelers try to mitigate this problem by "wrapping" the landscape – fires that burn off one side of the landscape burn onto the landscape from the

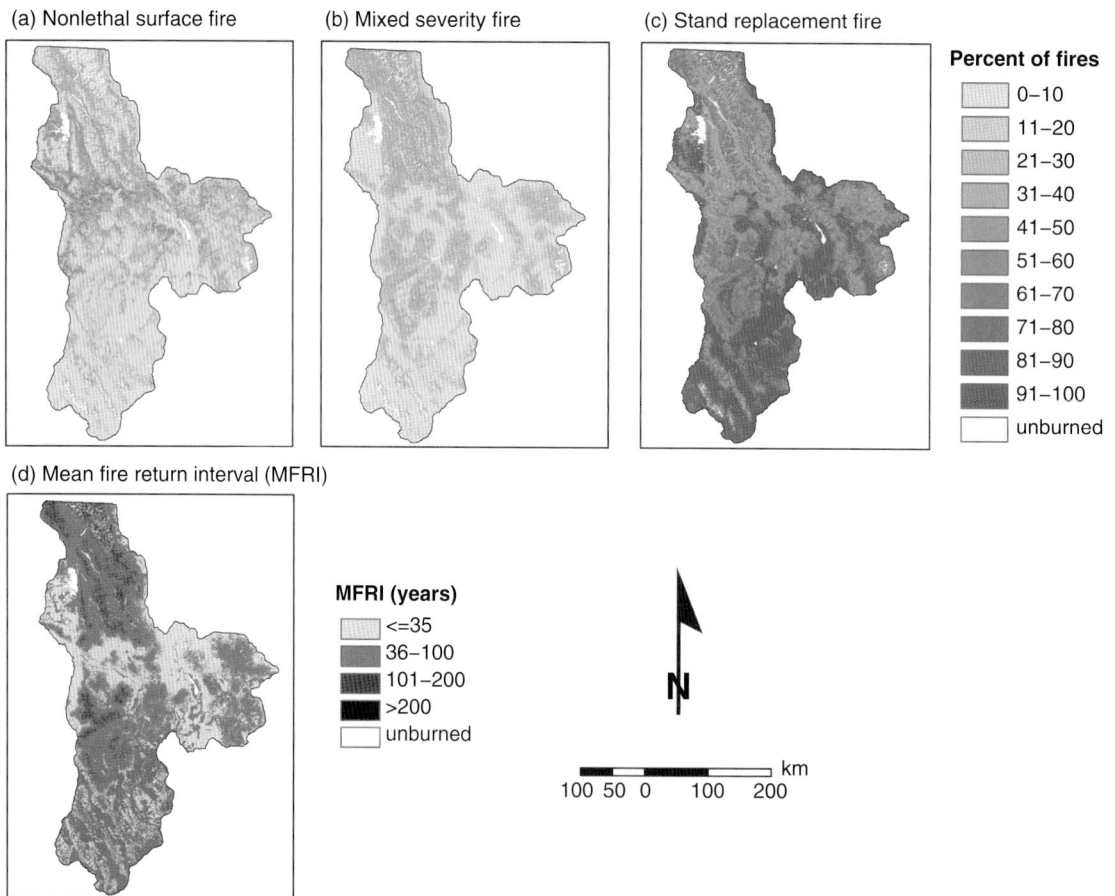

Fig. 8.3 Fire-regime maps produced by the LANDSUMv4 simulation model for the Northern Rockies region (from Pratt et al. 2006). Maps show the proportion of total fires that are (a) nonlethal surface fires, (b) mixed severity fires, and (c) stand-replacement fires. The mean fire-return interval (MFRI in years) is shown in the last map (d).

other side. This solution is inappropriate for many HRV simulations because topography, wind, and vegetation near one edge are not the same as on the other side of the landscape.

The best way to mitigate the edge effects is to surround the simulation landscape with a buffer (Fig. 8.3). To create a buffer, make the simulation landscape larger and then stratify the simulated output by the buffer and the context area. The LANDFIRE prototype project attached a 3-km buffer area around each 20 000-ha context area to create the simulation landscape (Pratt et al. 2006). The buffer should provide an adequate source for fires to immigrate from the windward and upslope side of the landscape. Each landscape is unique, so buffer width may differ for each setting. Modelers should always inspect fire-regime maps to determine if the buffer is large enough to minimize edge effects within the context landscape, keeping in mind that simulation time increases exponentially as landscapes get larger.

Landscape orientation and shape

The shape of the landscape is an important factor in the simulation of landscape dynamics. Fire frequencies in long, thin landscapes, for example, tend to be underestimated because simulated fires rarely reach their full size. Even with a large buffer, simulated fires spread quickly across thin context landscapes and burn only a fraction of its full size. Keane et al. (2002) found fire frequencies were approximately 20–40% less in long landscapes with high edge. Many managers like to use watersheds to define simulation landscapes, but watersheds are often long and linear with high edge, making them somewhat undesirable for HRV simulation. The best shapes are squares, rectangles, or circles that are large enough to contain both the buffer and context landscapes.

The directional orientation of the simulation landscape is also important. Long, thin landscapes that are oriented perpendicular to the wind direction will have far less burned area over time than the same landscapes oriented parallel to the wind (Keane et al. 2002). The orientation is especially important if watersheds define simulation landscapes because elongated watersheds that are positioned at right angles to the wind direction will tend to result in significantly less fire and narrow HRV time series.

Replication versus simulation time

Since most landscape models contain stochastic properties (i.e. probabilities are used to simulate some ecological processes), it is often suggested that the model be executed multiple times to quantify the variation of results due to the stochasticity inherent in the probability functions (Fig. 8.4). Repeated execution of the model is important in HRV simulations, especially in landscapes with rare disturbances, because it is possible that rare, large fires at the tail of the fire-size distribution may never be simulated. The inherent stochastic variation in simulated results should be considered part of the HRV and should probably not be averaged.

An alternative for many HRV simulations is simply to extend the simulation run, adding more simulated years to ensure adequate representation of disturbance regimes. Instead of simulating 10 replications of 1000 years, for example, one could simulate 10 000 years in one run. The length of the simulation also depends on the output interval, which is selected to minimize autocorrelation and to ensure adequate response variable sample size.

Simulation time versus real time

There is a common misconception that long-term simulation model HRV outputs are inappropriate because the simulation of fire and landscape dynamics occurred while unrealistically holding climate and fire regimes constant (Keane et al. 2006). This would be true if the objective of the fire modeling were to replicate historical fire events. However, the primary purpose of HRV modeling efforts is to describe variation in historical landscape dynamics, not to replicate them.

Simulation modeling allows the quantification of the entire range of landscape conditions by simulating the static historical fire regime for long time periods (e.g. thousands of years) to ensure all possible fire ignitions and burn patterns are represented in the HRV time series: long simulation periods ensure more fires are simulated on the landscape, resulting in better HRV estimates. In contrast, HRV time series from empirical historical records will tend to underestimate variation of landscape conditions because there are a limited number of fire events. The model input parameters represent the actual temporal context, while the simulation time represents the

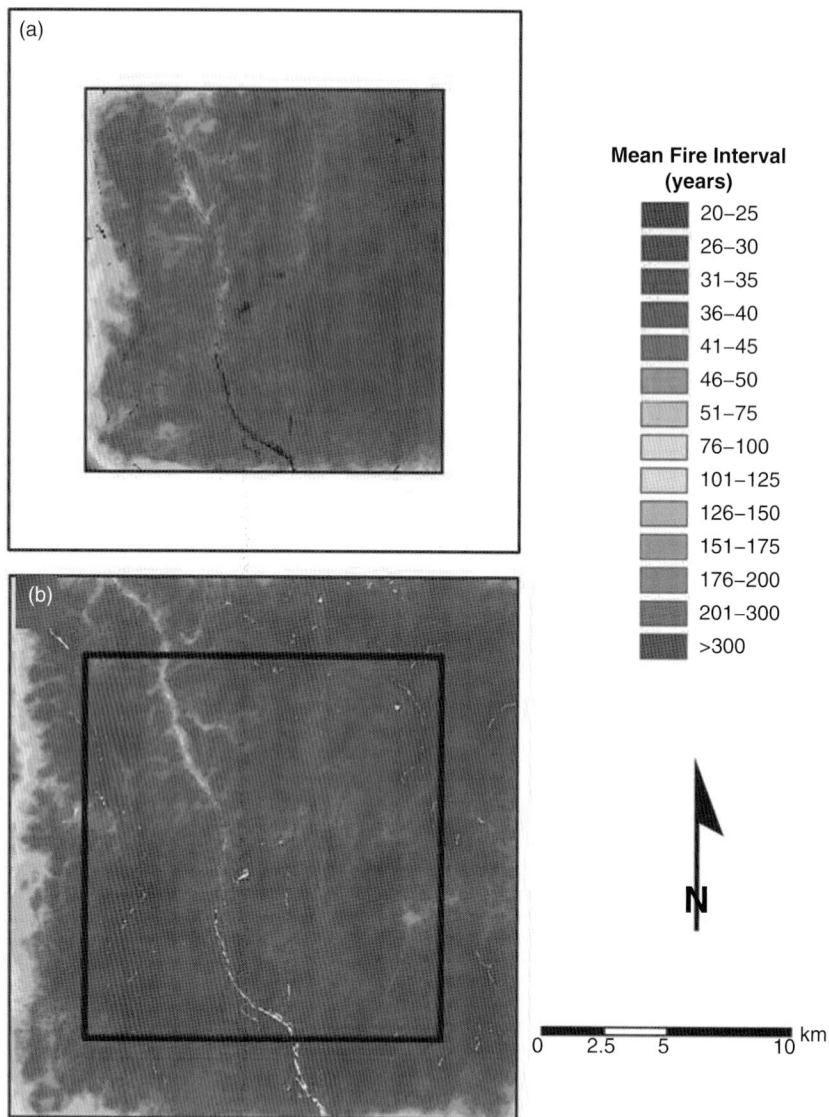

Fig. 8.4 An example of fire-regime simulation results showing the edge effect on fire frequency across the simulation landscape. (a) Fire frequency without a landscape buffer and (b) fire regime for the same landscape with a 3-km buffer. Please refer to Colour Plate 2.

length of time needed to adequately capture the range and variation of historical conditions. Because the time slice represented by model parameters often represents only four or five centuries, it may seem that only 500 years of simulation are needed (Fig. 8.5). However, the sampled fire events that occurred during this time represent only one unique sequence of the fire starts and growth that created the unique landscape compositions observed today. If these events happened at different times or in different areas, an entirely different set of landscape conditions would have resulted.

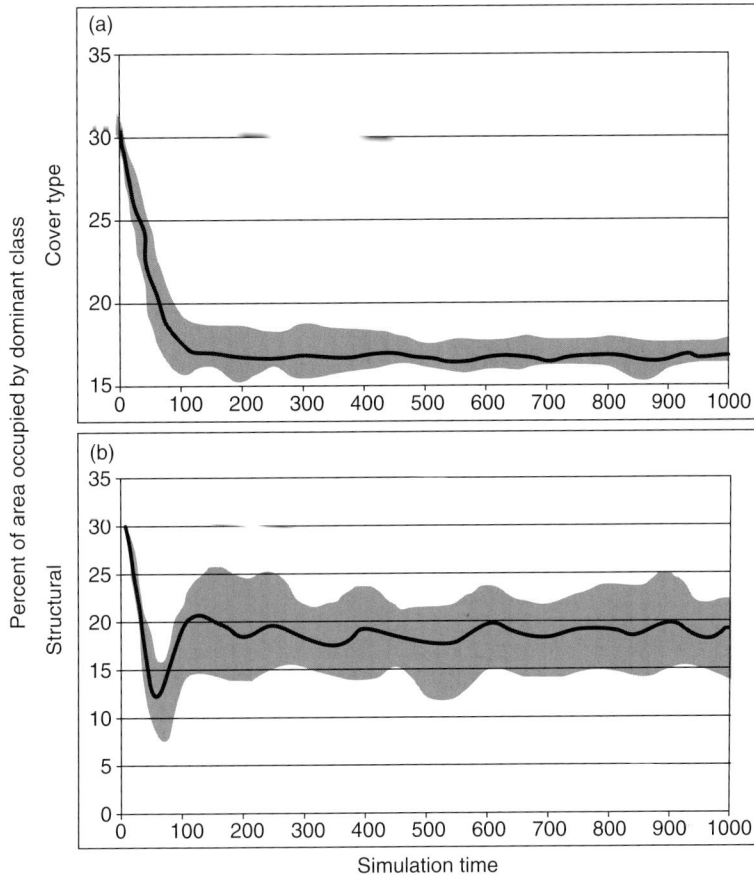

Fig. 8.5 Range of results for 20 LANDSUM model runs showing the inherent stochastic variation in predictions of percent landscape for (a) the dominant cover type and (b) structural stage.

8.6   CONCLUDING THOUGHTS

Much of HRV landscape modeling is about balancing realistic simulations of ecosystem dynamics with the often opposing goal of computational and logistical efficiency. This compromise becomes more important as simulation landscapes increase in size and complexity and as management issues expand in scope and scale. Simulation execution times tend to increase exponentially with increasing landscape size, but larger simulation landscapes are logistically simpler to prepare and produce better simulation results. There is no right way to simulate HRV, but there are many wrong ways. It is important that local experts review parameters, intermediate output, and results to ensure

realistic results. HRV simulations get easier with increased experience in simulation modeling.

Simulation modeling is a technology that will see more use in land management in the coming years. Costly and resource-intensive data surveys and field campaigns will become more difficult in the future with declining budgets, limited personnel, and dwindling expertise. And, as issues such as climate change become more complex and expand in geographic scope, the design and use of empirical approaches becomes more complicated and expensive. However, field data will continue to be absolutely essential to future simulation modeling efforts. The real challenges in the future will be to (1) design models that contain sufficient detail to simulate complex ecological interactions that are also useful to managers; (2) adapt field data inventory and monitoring projects to collect those data that can be used to build, parameterize, initialize, and validate models, as well as provide sufficient information to solve management problems; and (3) provide sufficient training and assistance to managers in using landscape models.

REFERENCES

Baker, W.L. (1989). A review of models of landscape change. *Landscape Ecology*, **2**, 111–133.

Baker, W.L. (1999). Spatial simulation of the effects of human and natural disturbance regimes on landscape structure. In *Spatial Modeling of Forest Landscape Change: Approaches and Applications* (ed. D.J. Mladenoff and W.L. Baker), pp. 277–308. Cambridge University Press, Cambridge, UK.

Beukema, S.J. & Kurz, W.A. (1998). *Vegetation Dynamics Development Tool: Users Guide, Version 3.0. 300.* ESSA Technologies, Vancouver, BC, Canada.

Box, G.E.P. & Draper, N.R. (1987). *Empirical Model-Building and Response Surfaces.* Wiley and Sons, New York, USA.

Burton, P.J. & Urban, D.L. (1990). An overview of ZELIG, a family of individual-based gap models simulating forest succession. In *Symposia Proceedings Vegetation Anagement: An Integrated Approach.* pp. 92–96. FRDA Report 109. Forestry Canada Pacific Forestry Centre, Victoria, BC, Canada.

Cary, G.J. (1997). FIRESCAPE: a model for simulation theoretical long-term fire regimes in topographically complex landscapes. In *Australian Bushfire Conference: Bushfire '97*, pp. 45–67. Australian Bushfire Association, Darwin, Australia.

Chew, J.D., Stalling, C., & Moeller, K. (2003). Integrating knowledge for simulating vegetation change at landscape scales. *Western Journal of Applied Forestry*, **19**, 102–108.

Crookston, N.L. & Dixon, G.E. (2005). The forest vegetation simulator: a review of its structure, content, and applications. *Computers and Electronics in Agriculture*, **49**, 60–80.

Gardner, R.H., Romme, W.H., & Turner, M.G. (1999). Predicting forest fire effects at landscape scales. In *Spatial Modeling of Forest Landscape Change: Approaches and Applications* (ed. D.J. Mladenoff and W.L. Baker), pp. 163–185. Cambridge University Press, Cambridge, UK.

He, H.S. (2008). Forest landscape models, definition, characterization, and classification. *Forest Ecology and Management*, **254**, 484–498.

He, H.S. & Mladenoff, D.J. (1999). Spatially explicit and stochastic simulation of forest-landscape fire disturbance and succession. *Ecology*, **80**, 81–99.

He, H.S., Keane, R.E., & Iverson, L. (2008). Forest landscape models, a tool for understanding the effect of the large-scale and long-term landscape processes. *Forest Ecology and Management*, **254**, 371–374.

Hessl, A.E., Milesi, C., White, M.A., Peterson, D.L., & Keane, R.E. (2004). Ecophysiological parameters for Pacific Northwest trees. General Technical Report PNW-GTR-618. USDA Forest Service, Portland, OR, USA.

Ives, A.R., Dennis, B., Cottingham, K.L., & Carpenter, S.R. (2003). Estimating community stability and ecological interactions from time series data. *Ecological Monographs*, **73**, 301–330.

Karau, E.C. & Keane, R.E. (2007). Determining landscape extent for succession and disturbance simluation modeling. *Landscape Ecology*, **22**, 993–1006.

Keane, R.E. & Finney, M.A. (2003). The simulation of landscape fire, climate, and ecosystem dynamics. In *Fire and Global Change in Temperate Ecosystems of the Western Americas* (ed. T.T. Veblen, W.L. Baker, G. Montenegro, and T.W. Swetnam), pp. 32–68. Springer-Verlag, New York, USA.

Keane, R.E., Arno, S.F., & Brown, J.K. (1989). FIRESUM: an ecological process model for fire succession in western conifer forests. General Technical Report INT-266. USDA Forest Service, Ogden, UT, USA.

Keane, R.E., Long, D.G., Menakis, J., Hann, W.J., & Bevins, C.D. (1996a). Simulating coarse-scale vegetation dynamics using the Columbia River Basin succession model: CRBSUM. General Technical Report RMRS-GTR-340. USDA Forest Service, Ogden, UT, USA.

Keane, R.E., Morgan, P., & Running, S.W. (1996b). FIRE-BGC: a mechanistic ecological process model for simulating fire succession on coniferous forest landscapes of the northern Rocky Mountains. Research Paper INT-RP-484. USDA Forest Service, Ogden, UT, USA.

Keane, R.E., Parsons, R., & Hessburg, P. (2002). Estimating historical range and variation of landscape patch dynamics: limitations of the simulation approach. *Ecological Modelling*, **151**, 29–49.

Keane, R.E., Cary, G., Davies, I.D., et al. (2004). A classification of landscape fire succession models: spatially explicit models of fire and vegetation dynamic. *Ecological Modelling*, **256**, 3–27.

Keane, R.E., Holsinger, L., & Pratt, S. (2006). Simulating historical landscape dynamics using the landscape fire succession model LANDSUM version 4.0. General Technical Report RMRS-GTR-171CD. USDA Forest Service, Fort Collins, CO, USA.

Keane, R.E., Holsinger, L., Parsons, R., & Gray, K. (2008). Climate change effects on historical range of variability of two large landscapes in western Montana, USA. *Forest Ecology and Management*, **254**, 274–289.

Keane, R.E., Hessburg, P.F., Landres, P.B., & Swanson, F.J. (2009). A review of the use of historical range and variation (HRV) in landscape management. *Forest Ecology and Management*, **258**, 1025–1037.

Keane, R.E., Holsinger, L., & Parsons, R.A. (2011). Evaluating indices that measure departure of current landscape composition from historical conditions. Research Paper RMRS-RP-83. USDA Forest Service, Fort Collins, CO, USA.

Korol, R.L. (2001). Physiological attributes of eleven North-west conifer species. General Technical Report RMRS-GTR-73, USDA Forest Service Rocky Mountain Research Station, Fort Collins, CO, USA.

Kurz, W.A., Beukema, S.J., Merzenich, J., Arbaugh, M., & Schilling, S. (1999). Long-range modeling of stochastic disturbances and management treatments using VDDT and TELSA. In *Proceedings of the Society of American Foresters 1999 National Convention*, pp. 349–355. Society of American Foresters, Portland, Oregon, USA.

Logan, J.A., White, P., Bentz, B.J., & Powell, J.A. (1998). Model analysis of spatial patterns in mountain pine beetle outbreaks. *Theoretical Population Biology*, **53**, 236–255.

Miller, C. & Urban, D.L. (1999). A model of surface fire, climate, and forest pattern in the Sierra Nevada, California. *Ecological Modelling*, **114**, 113–135.

Mladenoff, D.J. (2004). LANDIS and forest landscape models. *Ecological Modelling*, **180**, 7–19.

Mladenoff, D.J. & Baker, W.L. (1999). *Spatial Modeling of Forest Landscape Change*. Cambridge University Press, Cambridge, UK.

Morgan, P., Aplet, G.H., Haufler, J.B., Humphries, H.C., Moore, M.M., & Wilson, W.D. (1994). Historical range of variability: a useful tool for evaluating ecosystem change. *Journal of Sustainable Forestry*, **2**, 87–111.

Noble, I.R. & Slatyer, R.O. (1977). Post-fire succession of plants in Mediterranean ecosystems. In *Symposium on Environmental Consequences of Fire and Fuel Management in Mediterranean Ecosystems*, pp. 27–36. Palo Alto, CA, USA.

Perry, G.L.W. & Millington, J.D.A. (2008). Spatial modelling of succession-disturbance dynamics in forest ecosystems: concepts and examples. *Perspectives in Plant Ecology, Evolution and Systematics*, **9**, 191–210.

Pratt, S.D., Holsinger, L., & Keane, R.E. (2006). Modeling historical reference conditions for vegetation and fire regimes using simulation modeling. General Technical Report RMRS-GTR-175. USDA Forest Service, Fort Collins, CO, USA.

Reed, W.J., Larsen, C.P.S., Johnson, E.A., & MacDonald, G.M. (1998). Estimation of temporal variations in historical fire frequency from time-since-fire map data. *Forest Science*, **44**, 465–475.

Roberts, D.W. & Betz, D.W. (1999). Simulating landscape vegetation dynamics of Bryce Canyon National Park with the vital attributes/fuzzy systems model VAFS.LANDSIM. In *Spatial Modeling of Forest Landscape Change: Approaches and Applications* (ed. D.J. Mladenoff and W.L. Baker), pp. 99–123. Cambridge University Press, Cambridge, UK.

Rollins, M.G., Swetnam, T.W., & Morgan, P. (2001). Evaluating a century of fire patterns in two Rocky Mountain wilderness areas using digital fire atlases. *Canadian Journal of Forest Research*, **31**, 2107–2133.

Running, S.W. & Coughlan, J.C. (1988). A general model of forest ecosystem processes for regional applications: I. Hydrologic balance, canopy gas exchange and primary production processes. *Ecological Modelling*, **42**, 125–154.

Schaaf, M.D., Wiitala, M.A. Schreuder, M.D., & Weise, D.R. (2004). An evaluation of the economic tradeoffs of fuel treatment and fire suppression on the Angeles National Forest using the Fire Effects Tradeoff Model (FETM). In Proceedings of the II International Symposium on Fire Economics, Policy and Planning: A Global Vision, April 19–22, 2004, Córdoba, Spain. (ed. A Gonzales, technical coordinator), pp. 513–524. Gen. Tech. Rep. PSW-GTR-208. Albany, CA: Pacific Southwest Research Station, Forest Service, U.S. Department of Agriculture.

Stage, A.R. (1997). Using FVS to provide structural class attributes to a forest succession model CRBSUM. In *Forest Vegetation Simulator Conference*, pp. 137–147. USDA Forest Service, Fort Collins, CO, USA.

Stage, A.R., Hatch, C.R., Rice, D.L., Renner, D.W., Coble, J.J., & Korol, R. (1995). Calibrating a forest succession model with a single-tree growth model: an exercise in meta-modeling. Recent advances in forest mensuration and growth research, pp. 194–209. Danish Forest and Landscape Research Institute., Tampere, Finland.

Steele, B.M., Reddy, S.K., & Keane, R.E. (2006). A methodology for assessing departure of current plant communities from historical conditions over large landscapes. *Ecological Modelling*, **199**, 53–63.

Thornton, P.E., Law, B.E., Gholz, H.L., et al. (2002). Modeling and measuring the effects of disturbance history and climate on carbon and water budgets in evergreen needleleaf forests. *Agricultural and Forest Meteorology*, **113**, 185–222.

Turner, M. (1987). *Landscape Heterogeneity and Disturbance*. Springer-Verlag, New York, USA.

Wimberly, M.C., Spies, T.A., Long, C.J., & Whitlock, C. (2000). Simulating historical variability in the amount of old forest in the Oregon Coast Range. *Conservation Biology*, **14**, 167–180.

Chapter 9

MODELING HISTORICAL RANGE OF VARIABILITY AT A RANGE OF SCALES: AN EXAMPLE APPLICATION

Kevin McGarigal[1] *and William H. Romme*[2]

[1]University of Massachusetts, Amherst, MA, USA
[2]Colorado State University, Fort Collins, CO, USA

Historical Environmental Variation in Conservation and Natural Resource Management, First Edition. Edited by John A. Wiens, Gregory D. Hayward, Hugh D. Safford, and Catherine M. Giffen.
© 2012 John Wiley & Sons, Ltd. Published 2012 by John Wiley & Sons, Ltd.

9.1 INTRODUCTION

In this chapter, we present an application of a spatially explicit, landscape disturbance-succession simulation model (RMLands) developed collaboratively by scientists and resource managers to quantify the pre-1900 range of variation (hereafter referred to as the "historical range of variation" [HRV]) in landscape structure on the San Juan National Forest (SJNF) in southwestern Colorado. Our application was largely motivated by the need to provide a better *quantitative* description of HRV as a reference for comparison with current and potential future conditions to inform land-management planning, and as a supplement to a landscape condition analysis for the South Central Highlands Section in southwestern Colorado and Northern New Mexico (Romme et al. 2009). A critical management issue in the region is the presumed departure of the current landscape from its HRV and the resulting ecological impacts; in particular, the loss of ecosystem services (e.g. carbon sequestration, maintenance of native biodiversity). Here, we will only briefly review the methods employed in this particular application and report on only some of the major findings; a detailed description is provided elsewhere (http://www.umass.edu/landeco/research/rmlands/applications/hrv_sjnf/documents/sjnf.htm). Our purpose in this chapter is to illustrate the process of quantifying and interpreting HRV and highlight important considerations in the practical application of HRV assessment. Importantly, this application represents a collaboration between scientists (the authors) and a team of resource-management experts from the SJNF and Rocky Mountain Region (hereafter referred to as the "expert team").

9.2 METHODS

There are four essential steps to any HRV assessment, which we followed: (1) determine the spatial scale (i.e. geographic grain and extent) of the analysis, (2) define the temporal scale (i.e. temporal scope and resolution) of the analysis, (3) define the state variables of interest, and (4) define the approach for measuring the range of variation in the state variables. Below we describe each of these steps for the SJNF application.

Step 1. Define the spatial scale of the analysis

Choosing the appropriate spatial scale of the analysis is of paramount concern when describing HRV in landscape structure, especially if the description is to be quantitative. All quantitative measures of landscape structure vary with landscape scale – both grain and extent (Wu et al. 2003; Wu 2004; Wiens et al., Chapter 5, this book). In the context of HRV analysis, choosing an appropriate extent is particularly important (Keane et al. 2002; Karau & Keane 2007). In general, as the geographic extent of the landscape increases, temporal variability in landscape structure decreases – because larger landscapes are better able to incorporate and subsume the changes induced by disturbances than are smaller landscapes (Karau & Keane 2007). If the landscape is too big, however, it may span different physiographic regions, for which a single range of variation is ecologically meaningless. Conversely, if the landscape is too small, a single disturbance event may exceed the size of the landscape, causing the range of variation to be so large as to be rendered meaningless. The HRV concept is thus only meaningful at "intermediate" scales, and determining the intermediate scale or scales at which to apply HRV is challenging (Keane et al. 2009). Ideally, the landscape should represent a logical ecological unit in which the ecological patterns and processes are tightly coupled, resulting in measurable and ecologically meaningful fluctuations in landscape structure over time. And in general, the landscape should be considerably larger than the largest disturbance events – perhaps as much as a full order of magnitude larger.

Because this application was driven by the needs of the SJNF planning team, we defined the Forest as the overall geographic extent of the analysis. The SJNF environment is described in detail elsewhere (Romme et al. 2009), but briefly, it occupies 847 638 hectares in the approximate geographic center of the South Central Highlands Section of the southern Rocky Mountains Province in southwestern Colorado (Fig. 9.1). It is a mountainous landscape with complex physiography and climate. Seven major vegetation types of ecological and economic significance occur within the area, each with a unique ecological setting and history, as well as distinctive human impacts and changes since Euro-American settlement. At the lowest elevations (1900–2400 m), the vegetation is dominated by semidesert grasslands and savannahs and

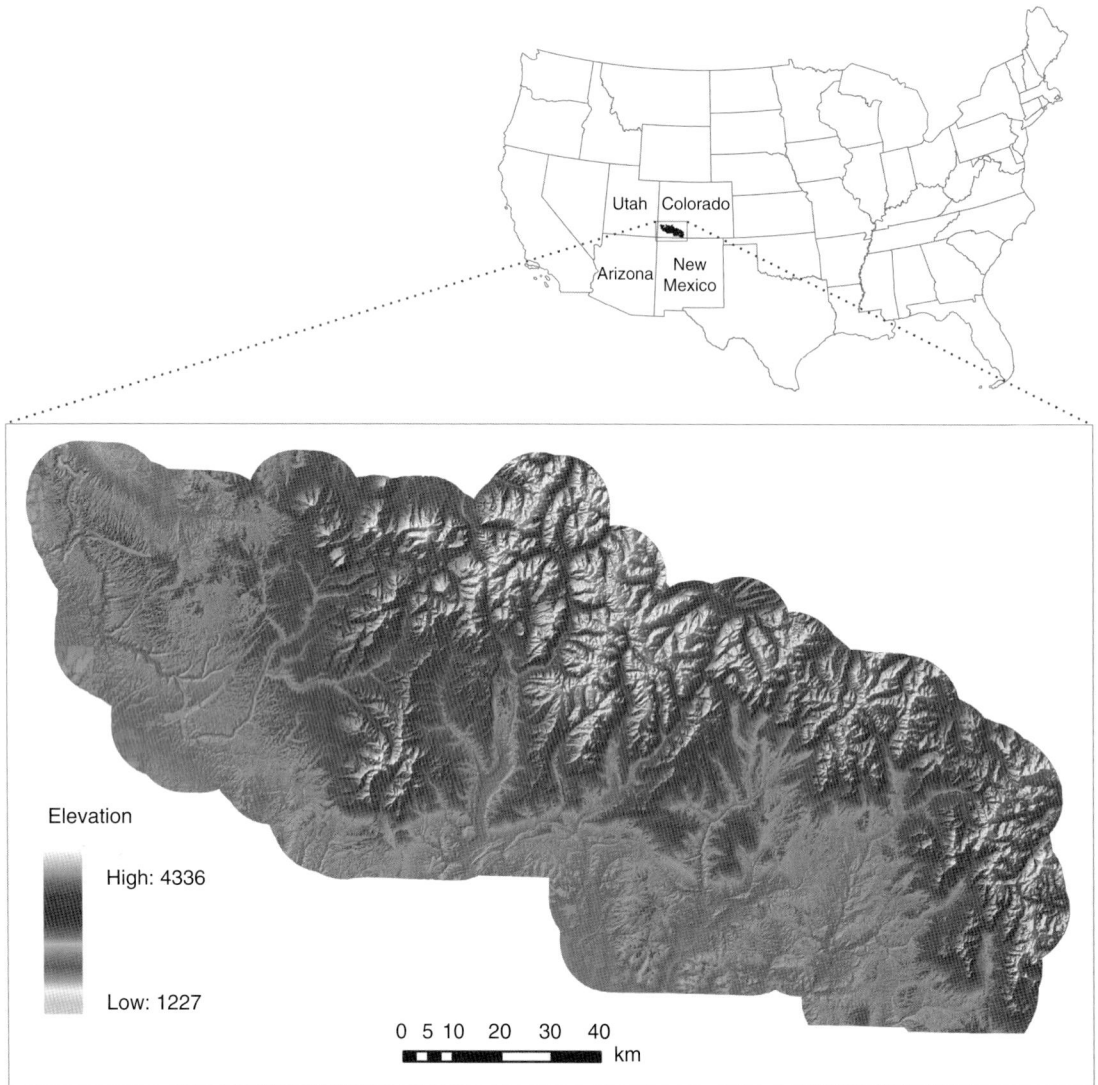

Fig. 9.1 San Juan National Forest (inclusive of a 10-km buffer) displayed as a topographic relief map in the Four Corners Region of the southwestern United States. The Forest encompasses 847 638 ha. Please refer to Colour Plate 3.

pinyon-juniper woodlands (primarily *Pinus edulis* and *Juniperus osteosperma*). At the foothills and on tops of broad plateaus and mesas, the vegetation ranges into ponderosa pine (*Pinus ponderosa*) forest interspersed with shrub-dominated stands (Petran chaparral dominated by *Quercus gambelii*). The middle slopes are covered by a mosaic of mixed conifers (*P. ponderosa*,

Psuedostuga menziesii, Abies concolor, Picea pungens) and quaking aspen stands (*Populus tremuloides*), broken by occasional meadows and grasslands. The highest elevations (2800–3800 m) contain extensive spruce-fir forests (primarily *Picea engelmannii* and *Abies lasiocarpa*), subalpine meadows, and treeless alpine communities on the highest peaks. Running through all these types are riparian woodlands and meadows along the borders of perennial rivers and streams.

Prior to European settlement, landscape dynamics were driven primarily by the patterns of wildfire, which varied dramatically with vegetation type. The median fire interval was only 10–20 years in the lower elevation ponderosa-pine type; 20–30 years in the dry mixed-conifer type; 50–100 years in the aspen type; and >100 years in the spruce-fir type (Romme et al. 2009). Many individual stands escaped fire for far longer than the median return interval and some burned at shorter intervals, creating a complex vegetation mosaic at the landscape scale. Under this "natural" disturbance regime, stand-replacement fires initiated stand development and maintained a coarse-grain mosaic of successional stages and cover types across the landscape, although other disturbance processes, such as landslides, floods, windthrow, insects, and disease also played a role on a finer scale. In particular, chronic, fine-scale processes (e.g. windthrow, insects, and pathogens) that kill individual trees or small groups of trees dominate the disturbance regime of individual stands in the later stages of development (Lertzman & Krebs 1991; Veblen et al. 1991; Roovers & Rebertus 1993).

To examine the effects of scale – specifically, landscape extent – on the measured dynamics in landscape structure, we subdivided the full landscape into three districts (corresponding to the existing US Forest Service Ranger Districts) of similar size and further selected a single watershed of comparable size nested within each District (Table 9.1). We simulated land-

Table 9.1 Landscape and sub-landscape extents for which the historical range of variation in landscape structure was estimated on the San Juan National Forest, Colorado.

Level[1]	Landscape/sub-landscape	Total area (ha)
1	San Juan National Forest	847638
2	Columbine District	308829
2	Pagosa District	282511
2	Dolores District	256298
3	Hermosa Creek Watershed	44103
3	Piedra River Watershed	39487
3	Narraguinnep Creek Watershed	31818

[1] Level 1 refers to the full landscape; level 2 sub-landscapes were nested within the level 1 landscape, and level 3 sub-landscapes were nested within the level 2 sub-landscapes (Fig. 9.1).

scape dynamics (see below) on the full SJNF landscape only. The resulting simulated landscapes were clipped to each of these sub-landscapes for purposes of quantifying HRV in landscape structure (see below). Our goal was not to examine a comprehensive range of scales, but rather to examine potentially realistic scales for planning and/or analysis.

Choosing a spatial resolution (or grain) is typically a matter of compromise (Wiens et al., Chapter 5, this book). For the SJNF application, we chose a grain (cell) size of 25 m, which allowed us to depict patches as small as 0.0625 ha (one cell). The raw input spatial data came from a variety of vector and raster data sources supplied by the SJNF. The principal land-cover data were composed of vector polygons derived from aerial-photo interpretation conducted by the SJNF. We converted all data to raster grids for use in RMLands. The 25-m grain size reflected a compromise between our desire to accurately represent linear landscape elements such as roads and streams that potentially serve as fragmenting features on the one hand, and our desire to emphasize coarse vegetation patterns in a computationally efficient manner on the other. In addition, a 25-m cell size allowed us to represent depth-of-edge distances at a 25-m resolution, which gave us sufficient flexibility to quantify potential edge effects. Overall, a 25-m cell size seemed a reasonable compromise given these considerations, while at the same time it seemed sufficient to our task of assessing habitat at an ecologically relevant scale for a suite of indicator species (see previously referenced website for additional information). Owing to differences in the original mapping resolution among patch types, we also coarsened the patch mosaic by increasing the minimum mapping unit to 0.5 ha (eight cells) to maintain consistency in the spatial resolution of the patch mosaic among patch types. We did not resample the grid at this coarser resolution; rather, we dissolved away any patch less than eight cells in size using a nearest-neighbor-replacement algorithm.

Step 2. Define the temporal scale of the analysis

Choosing the appropriate temporal scale of the HRV analysis is equally important. A crucial part of describing HRV is selecting the reference time period used to characterize system dynamics. There is no single, inherently optimal period for all applications. In general, the relevance of HRV is lost if too long a time period is used, because conditions such as climate (the major driver of disturbance regimes) and species composition may not be stable enough to express a meaningful range of variation. Similarly, the HRV concept is inappropriate if too short a time period is used, because the landscape may not have had enough time to cycle through one or more rotations of the characteristic disturbance regime and express a stable range of variation. Ultimately, for the HRV concept to be meaningful, a historical period must exist for which there was (or was assumed to be) a relatively stable range of variation in the state variables of interest (see below). The period must be long enough for the system to exhibit a stable range of variation and not so long that the system no longer exhibits dynamic equilibrium conditions (Morgan et al. 1994).

For the SJNF application, we chose the period from about 1300 to the late 1800s, representing the period from Ancestral Puebloan abandonment to Euro-American settlement, as the reference period. This period is often referred to as the *period of indigenous settlement*, in contrast to the period of Euro-American settlement that began in the mid to late 1800s (Romme et al. 2009). The period of several centuries prior to 1900 represents a time when broadscale climatic conditions were generally similar to those of today but Euro-American settlers had not yet introduced the sweeping ecological changes that now have greatly altered many Rocky Mountain landscapes through fire suppression, grazing, road-building, timber cutting, recreation, and other activities. It cannot be emphasized too strongly that the chosen reference period was *not* a time of stasis, climatically, ecologically, or culturally. For example, the "Little Ice Age" occurred during this time, and there were small shifts in the position of the upper timberline and in the elevational breadth of the forest zone on the middle slopes of the mountains. Local human inhabitants obtained horses and new technology and were affected by disease and displacement of other tribes brought about by European colonization farther to the east. Nevertheless, compared with some other periods in history, the period from about 1300 to the late 1800s was a time of relatively consistent environmental and cultural conditions in the region, and a time for which we have a reasonable amount of specific information to enable us to model the system (Swetnam & Betancourt 1990).

As with spatial resolution, choosing a temporal resolution (or grain) is typically a matter of compromise (Wiens et al., Chapter 5, this book). For the SJNF application, we chose a resolution of 10 years. This resolution reflected a compromise between our desire to accurately represent fire and insect/pathogen disturbances (and their climate drivers) on the one hand and our interest in relatively coarse vegetation successional stages (emphasizing forested cover types) and computational efficiency on the other hand. The choice of a 10-year temporal resolution meant that each time step in the simulation represented a 10-year period, and it also meant that the drivers of disturbance (and succession) processes operated to effect change at the same resolution. Consequently, drivers such as climate varied at a 10-year resolution, despite the fact that climate clearly operates annually. This represented a trade-off in the model. To parameterize the climate parameter in the model for the SJNF application, we summarized the historical climate data for 10-year intervals differently for each disturbance process. For example, the climate parameter for wildfire disturbance was based on the observed mean Palmer Drought Severity Index for each 10-year period, whereas the climate parameter for spruce beetle (*Dendroctonus rufipennis*) disturbance was drawn from a random normal distribution for each 10-year period.

Step 3. Define the state variables of interest

Another important part of describing HRV is selecting the state variable or variables to use to characterize the system. While there are myriad possible ecological state variables, in the context of HRV, state variables are typically measures of landscape structure, representing either landscape *composition*, the number and/or types of landscape elements and their areal extent, or landscape *configuration*, the spatial character and arrangement, position and/or orientation of landscape elements. The choice of landscape metrics is not a trivial one, as it requires a deep understanding of the metrics and their limitations. The selected metrics should reflect the objectives of the analysis and be appropriate, given the scope and limitations of the available data.

In most HRV applications, the analysis is limited to measures of landscape composition, which tend to be less sensitive to landscape definition issues associated with spatial scale (Step 1). The most common application, including ours, involves a categorical classification of vegetation into discrete cover types (or ecological communities) and seral stages (or stand condition classes). In this case, the percentage of the landscape in each patch type (defined here as a combination of cover type and seral stage) or the percentage of a particular cover type in each seral stage serve as state variables for describing the composition of the landscape at any point in time. The emphasis on landscape composition is justifiable because in most cases, our knowledge of the spatial character of historical disturbance regimes and landscape conditions is weak. However, because spatial configuration can have important ecological consequences, we believe it is prudent to examine configuration metrics as well and adjust the level of uncertainty in the results accordingly.

For the SJNF application, we selected a broad suite of landscape metrics as state variables, including both landscape composition and configuration metrics deemed to provide a universally strong and consistent characterization of the dominant landscape-structure gradients (Neel et al. 2004; Cushman et al. 2008); see Table 9.2 for a list of the landscape metrics (see McGarigal et al. 2002 for complete descriptions of each metric). As noted above, we defined the landscape based on a combination of cover type and seral stage, resulting in 57 patch types or classes. We applied the metrics at the *landscape level* to describe the structure of the entire landscape patch mosaic (i.e. the collection of all patches representing different cover types and seral stages); however, the metrics can also be applied at the *class level* to describe the structure of each patch type (i.e. unique combination of cover type and seral stage). Some of the configuration metrics are functional metrics, requiring application-specific parameterization. The expert team parameterized these metrics for this application.

To simplify the results and facilitate interpretation for management, we reclassified the landscape mosaic defined by the original 57 patch types (i.e. each unique combination of cover type and seral stage) into a smaller set of 21 aggregated patch types or classes and rescaled the spatial resolution of the landscape by increasing the minimum mapping unit from 0.065 ha (one cell) to 0.5 ha (eight cells). Five of these new patch types (e.g. water, barren) were static (i.e. unchanging over the course of the simulation), and two lacked data on current seral stages, leaving 14 dynamic patch types (Table 9.2). These new classes represented "habitats of special interest" to management; for example, old-growth forests and low elevation fire-maintained open-canopy conifer forest. Full definitions of these

classes are provided elsewhere (http://www.umass.edu/landeco/research/rmlands/applications/hrv_sjnf/documents/fragstats_analyses.htm). For the purposes of this chapter, we report only the state variables corresponding to the reclassified and rescaled landscape.

Step 4. Define the approach for measuring HRV in the state variables

Lastly, a method for quantifying HRV in the selected state variables (given the designated spatial and temporal scale of analysis) must be determined. This step is burdened with the greatest number of practical and technical challenges, as there are numerous alternatives (Keane et al. 2009). Here, we provide a brief description of the approach we developed and implemented for the SJNF application, with some considerations to our choices.

Step 4.1 Simulate landscape dynamics

Model development

We elected to develop and apply a spatially explicit landscape disturbance-succession model to simulate the dynamics in landscape structure. An empirical approach based solely on field data was not deemed a viable option in this case, as the empirical record, while allowing us to characterize certain aspects of the historical disturbance regime (e.g. disturbance return intervals for the major cover types), did not allow us to quantify landscape structure (e.g. seral-stage distribution of major cover types). Instead, we used the empirical record to guide model parameterization and calibration (see below). To meet our needs, we developed the Rocky Mountain Landscape Simulator (RMLands), a grid-based, spatially explicit, stochastic landscape simulation model designed to simulate disturbance and succession processes affecting the structure and dynamics of Rocky Mountain landscapes. RMLands simulates two key processes: succession and disturbance, implemented sequentially within 10-year time steps for a user-specified period of time. Succession is implemented using a stochastic state-based transition approach in which vegetation cover types transition probabilistically between discrete states (seral stages). Natural disturbances include wildfire and a variety of insects/pathogens, including pinyon

Table 9.2 Range of variation in landscape structure under the simulated HRV disturbance scenario on the San Juan National Forest, Colorado, and the degree of departure of the current landscape from the simulated range of variation (see text for details). Landscape structure is based on the reclassified and rescaled landscape in which cover types and seral stages (stand conditions) were aggregated into a smaller set of habitats of special interest at a minimum mapping unit of 0.5 ha.

Landscape metric	Current landscape		Percentiles of simulated distribution							CV[1]	HRV departure index[2]
	Metric value	Percentile of HRV	0	5	25	50	75	95	100		
Landscape composition[3]											
Pinyon-Juniper woodlands: Early seral	0.32	79	0.04	0.10	0.16	0.23	0.29	0.40	0.44	127	16
Pinyon-Juniper woodlands: Mid seral	0.55	0	0.66	0.85	1.21	1.58	1.96	2.61	3.00	111	100
Pinyon-Juniper woodlands: Late seral	1.89	100	0.02	0.04	0.42	0.65	0.95	1.39	1.77	206	100
Low-elevation conifer: Early seral	0.28	84	0.06	0.11	0.17	0.22	0.26	0.34	0.42	108	34
Low-elevation conifer: Mid seral	6.94	100	0.78	0.89	1.00	1.09	1.18	1.29	1.42	37	100
Low-elevation conifer: Late seral	14.68	100	4.53	5.66	6.70	7.70	8.76	10.67	11.77	65	100
Low-elevation conifer: Fire-maintained	0.00	0	8.96	10.13	12.09	13.07	14.17	15.23	16.49	39	100
High-elevation conifer: Early seral	1.15	0	1.03	1.91	2.75	3.49	4.62	5.99	9.71	117	99
High-elevation conifer: Mid seral	6.84	89	2.24	2.82	4.25	5.31	6.16	7.39	10.26	86	56
High-elevation conifer: Late seral	29.66	93	17.12	19.56	22.89	24.69	26.84	29.93	33.64	42	73
Aspen-dominated stands: Early seral	0.74	5	0.14	0.75	1.44	2.20	3.23	4.93	7.15	190	80
Aspen-dominated stands: Mid seral	2.62	3	1.86	2.80	4.11	5.07	5.99	7.74	8.99	98	86
Aspen-dominated stands: Late seral	2.08	99	0.87	1.14	1.43	1.59	1.74	1.94	2.18	51	96
Sagebrush-dominated stands	1.87	35	1.46	1.59	1.78	1.99	2.17	2.38	2.66	40	0
Simpson's diversity index	0.85	0	0.84	0.86	0.87	0.88	0.89	0.90	0.91	4	98
Simpson's evenness index	0.89	1	0.88	0.90	0.92	0.93	0.93	0.94	0.95	4	96
Landscape configuration[4]											
Patch density	2.58	0	4.46	5.06	5.56	5.73	5.88	6.08	6.36	18	100
Edge density	49.06	0	68.37	72.95	77.67	79.52	81.11	83.78	87.87	14	100
Mean patch size	38.78	100	15.73	16.45	17.01	17.46	17.98	19.75	22.43	19	100
Area-weighted mean patch size	7003.54	61	1669.76	2386.65	3942.81	5829.44	8694.65	15302.77	20047.35	222	0
Correlation length	3187.30	85	1532.91	1758.93	2135.71	2512.67	2897.22	3710.63	4489.62	78	40
Mean shape index	1.75	0	1.89	1.90	1.91	1.92	1.94	1.97	2.00	4	100
Area-weighted mean shape index	8.29	22	6.76	7.56	8.41	9.22	10.38	13.38	16.33	63	11
Mean core area	30.66	100	11.50	12.12	12.69	13.11	13.48	14.99	16.95	22	100
Area-weighted mean core area	6072.22	61	1417.56	2007.64	3323.31	4953.43	7397.30	13152.72	17251.84	225	0
Mean core area index	34.76	0	42.46	44.27	46.11	47.53	48.77	50.62	53.31	13	100
Area-weighted mean core area index	79.05	100	72.45	73.31	74.25	74.81	75.41	76.32	77.59	4	100

Mean proximity index	1312.91	68	412.67	547.87	791.80	1102.79	1423.13	2274.96	2993.21	157	0
Area-weighted mean proximity index	7070.70	82	1038.60	1872.03	3072.08	4180.54	5686.80	11464.06	21804.46	229	27
Contrast-weighted edge density	19.46	0	24.37	25.30	25.93	26.43	27.01	27.97	28.66	10	100
Total edge contrast index	38.74	100	31.03	31.89	32.35	32.85	33.44	34.37	36.20	8	100
Contagion	56.41	100	48.81	49.59	50.64	51.49	52.35	53.55	55.26	8	100
Interspersion and juxtaposition index	74.90	81	68.56	70.63	72.35	73.63	74.64	75.76	77.35	7	23

Summary indices[5]:

Landscape Composition Departure Index	80
Landscape Configuration Departure Index	65
Landscape Structure Departure Index	72

[1] CV = coefficient of variation in the simulated distribution, computed as the difference between the 5th and 95th percentiles divided by the median and multiplied by 100 to convert to a percentage.

[2] HRV departure index represents the degree of departure of the current landscape condition from the historical range of variation and is given here specifically as the degree of departure from the 25 to 75 percentile range of variation, where a 0 represents no departure (i.e. within the 25–75 percentiles of variation) and 100 represents complete departure (i.e. cutside the 0–100 percentiles of variation); see important caveats in the text (Section 9.4).

[3] Landscape composition represents the distribution of area among patch types (in this case, aggregated combinations of cover types and seral stages (stand conditions). Only dynamic patch types (i.e. those that change in area over time in response to disturbance and succession) are included here; static patch types (i.e. those that we treated as constant over time, such as water and barren) are excluded since they cannot exhibit any "departure." The patch types included here represent a smaller set of classes created by reclassifying (and rescaling) cover types and seral stages (see text for details). In addition, landscape composition includes two diversity metrics (Simpson's diversity and Simpson's evenness) that are derived from the distribution of area among patch types.

[4] Landscape configuration represents the spatial character, distribution, and arrangement of patches (across all patch types). The landscape metrics listed here are described in McGarigal et al. (2002).

[5] Landscape composition departure index = mean departure index across composition metrics; landscape configuration departure index = mean departure index across configuration metrics; landscape structure departure index = mean of the landscape composition and configuration departure indices.

decline (pinyon ips beetle, *Ips confusus*, and black stain root rot, *Leptographium wagneri*), mountain pine beetle (*Dendroctonus ponderosae*), Douglas-fir beetle (*Dendroctonus pseudotsugae*), spruce beetle, and spruce budworm (*Choristoneura occidentalis*). Briefly, each natural disturbance is modeled as a stochastic process consisting of the following key components:

• Climate – Climate plays a significant role in determining the spatiotemporal characteristics of the disturbance regime. Climate is specified as a time-varying global parameter for each disturbance process that optionally effects initiation, spread, and mortality of all disturbances within a time step. Climate can be specified as constant with a user-specified level of temporal variability, a trend over time (with variability), or as a user-defined trajectory – perhaps reflecting the climate conditions during a specific reference period. Essentially, the climate parameter acts as a dial to modify the probability of initiation, spread, and mortality of each disturbance processes.

• Initiation – Disturbance events are initiated at the cell level. Each cell has a probability of initiation in each time step that is a function of its susceptibility to disturbance based on its current state (e.g. cover type, seral stage, stand age, time since last disturbance, aspect, topographic position).

• Spread – Once initiated, disturbance spreads to adjacent cells in a probabilistic fashion. Each cell has a probability of spread that is a function of its susceptibility to disturbance, modified by its relative topographic position, orientation to prevailing winds, and the influence of potential barriers to spread (e.g. streams). The spread of a single disturbance is further modified to reflect variable weather conditions associated with the disturbance event itself. This event modifier affects final size of the disturbance and is a user-defined size distribution. In addition, there is an optional provision for the "spotting" of disturbances during spread so that disturbances are not limited only to contiguous spread.

• Mortality – Following spread, each cell is evaluated to determine the magnitude of ecological effect of the disturbance. Each cell can exhibit high or low mortality of the dominant plants. High mortality occurs when all or nearly all (>75%) of the dominant plant individuals are killed. Cells are aggregated into vegetation patches for purposes of determining mortality response, where patches are defined as spatially contiguous cells having the same cell attributes (e.g. identical disturbance history, age).

• Transition – Following mortality determination, each high-mortality vegetation patch is evaluated for potential immediate transition to a new seral stage (state). Transition pathways and rates of transition between states are defined uniquely for each cover type and are conditional on several attributes at the patch level. Hence, this is a conditional Markovian chain-like process, whereby the transitions are conditioned on antecedent information (e.g. time since last disturbance) and other state variables (e.g. elevation). These disturbance-induced transitions are differentiated from the successional transitions that occur at the beginning of each time step in response to gradual growth and development of vegetation over time. Low-mortality vegetation patches potentially transition to new seral stages during the succession phase of the simulation.

Model parameterization

We parameterized the model based on our best estimate of how disturbance and succession operated in the landscape during the historical reference period. Model parameterization involved developing a state-based model for each cover type and defining succession and disturbance-induced transition pathways and the factors influencing those pathways. The definition of state-based models was largely based on expert knowledge, but several local and regional empirical studies were drawn upon to parameterize the disturbance regimes (e.g. to specify target mean fire-return intervals for each cover type). Much of this parameterization was derived from information presented in Romme et al. (2009). In addition, we made use of extant data on historical climate (e.g. Palmer Drought Severity Index), local meteorological conditions (e.g. wind directions), and historical fire records (e.g. size distribution from National Fire Database) to help parameterize the major disturbance regimes. Additional details of model parameterization can be found at http://www.umass.edu/landeco/research/rmlands/applications/hrv_sjnf/documents/sjnf.htm.

Model calibration

After we verified that the model was implemented correctly and produced behaviors consistent with expectations – a stage of modeling we refer to as "verification"

– we calibrated the model for our application. Model calibration is an essential but sometimes overlooked or misused step in HRV modeling that involves adjusting certain model parameters to achieve specified outputs. It is critically important that the state variables of interest (Step 3) *not* be used to calibrate the model. For example, if the state variable of interest is the seral-stage distribution of each cover type, then it is inappropriate to establish targets for these variables and then calibrate the model to achieve these targets – no insights can be gained from this circular approach. Instead, we contend that model calibration should focus on the disturbance regime and not vegetation response, for two reasons. First, disturbance is the principal driver of vegetation change: disturbance is the cause, vegetation is the effect. We are interested in the question, Given the historical disturbance regime (cause), what is the vegetation dynamic (effect)? Second, our empirical understanding of landscape dynamics under historical reference conditions is largely based on characteristics of the disturbance regime, so we should ensure that our model is simulating the disturbance regime correctly and let the vegetation response be an emergent property of the model.

We established objective targets for mean disturbance return interval at the scale of an individual cell of a given vegetation type (equivalent to disturbance-rotation period for that vegetation type on the landscape) for each disturbance agent and each cover type and then calibrated the model by adjusting certain parameters (e.g. frequency of initiations, Weibull hazard function shape parameter) and rerunning the model until we realized return intervals within ±10% of our targets. In addition, we established subjective targets for a few other aspects of each disturbance regime, such as the frequency of major insect outbreaks, based on more limited empirical data and expert opinion, and calibrated the model accordingly.

Model execution

We used RMLands to simulate vegetation patterns over an 800-year period under the reference period disturbance regime. Although the historical reference period was roughly 600 years long, we simulated landscape dynamics for a longer period, which we determined during preliminary trials was long enough to capture the stable range of variation in the state variables. It did not matter that the simulation period exceeded the

historical reference period since the disturbance regime was treated as stationary; that is, the parameters that defined the disturbance regime did not change during the simulation. All that mattered was that the state variables achieved a stable range of variation – which was essentially guaranteed by the stationary model parameters over a sufficiently long simulation (in this case, a period representing roughly twice the length of the longest disturbance rotation period). In addition, we ran the simulation five times to capture stochastic variation among landscape trajectories, although a single simulation five times as long would have provided a similar result.

Step 4.2 Quantify state variables

We used FRAGSTATS (McGarigal et al. 2002) to quantify the state variables associated with landscape structure (Table 9.2). Briefly, FRAGSTATS is a program for quantifying the spatial structure of categorical patch mosaics. Here, we used FRAGSTATS to quantify composition and configuration of the landscape (based on the reclassified and rescaled landscape, see above) at each time step of the simulation.

Step 4.3 Summarize HRV

There are numerous statistical measures available for characterizing HRV in a state variable based on simulated landscape dynamics, including mean, range, quantiles, standard deviation, coefficient of variation (CV), skewness, frequency, spatial arrangement, and size and shape distributions. Although the absolute range of a state variable would seem to be a logical choice among measures, the range, when used alone, may not be appropriate because rare, extreme events define these bounds. Quantiles provide a more robust measure as they fully describe the empirical distribution without involving any underlying assumptions about the shape of the distribution and the use of, say, the 5–95th percentiles to describe the range of variation is insensitive to the extremes. Although sometimes used, measures of central tendency such as the mean and median also are inappropriate descriptors of HRV when used alone as they do not explicitly express variation – a fundamental aspect of the HRV concept.

We summarized the 0, 5, 25, 50, 75, 95, and 100th percentiles of the distribution in each state variable (i.e. landscape structure metrics) over an 800-year simulation period after eliminating the first 100 years of simulation to allow the current landscape to equilibrate to the historical disturbance regime – something referred to as the model equilibration period. Because we were interested in characterizing vegetation dynamics under reference-period conditions, it was necessary first to transform the current landscape (i.e. starting point of our simulations) into one in balance with the reference-period disturbance regime. Preliminary analyses suggested that most measured landscape attributes equilibrated within roughly 100 years, and some did so in less time. Thus, for our purposes, we simply ignored the first 100 years of the simulation for reporting the HRV in landscape structure. In addition, we combined the five simulations (after removing the first 100 years from each simulation) and treated each time step as a single independent observation of the landscape, resulting in 350 total observations (5 runs times 70 snapshots at 10-year intervals).

To examine the effect of scale (landscape extent) on HRV, we computed the percentiles of the distribution in each state variable at each scale (Table 9.1), as above. We computed a 90% nonparametric CV in each landscape metric for each landscape. We defined the 90% CV as the difference between the 5th and 95th percentiles of the metric distribution, divided by the median (50th percentile) and multiplied by 100 to convert to a percentage. This measure is similar to a conventional CV, but instead of using the standard deviation (which assumes normal distribution) as a measure of variance, we used the 5th–95th percentile range (which involves no distribution assumptions), and instead of using the mean (which is most appropriate for a symmetrical distribution) to represent the central tendency, we used the median (which is more appropriate for asymmetrical distributions). This provided a relative measure of the range of variation in each landscape metric that was comparable across metrics and landscape extents. Finally, we computed the average CV across landscape composition metrics and landscape configuration metrics (Table 9.2) and regressed these mean CVs against landscape extent using nonlinear regression based on a power function.

Step 4.4 Quantify current departure

Lastly, we compared the structure of the current landscape to the simulated HRV in landscape structure to determine the degree of "departure" of the current landscape. For our purposes, the "current" condition refers to the landscape in 2003 after the 2002 Missionary Ridge fire, which burned ca. 20 000 ha in the central portion of the SJNF. A detailed description of how we derived our "departure index" is provided elsewhere (http://www.umass.edu/landeco/research/rmlands/applications/hrv_sjnf/documents/methods.htm), but briefly, for each state variable (Table 9.2), it involved comparing the current landscape to the simulated HRV in terms of percentiles of the HRV distribution. If the current landscape was within the 25–75th percentiles of the HRV distribution, the departure index was set to zero, and if the current landscape was outside the 0–100th percentiles of the HRV distribution, the departure index was set to 100. If the current landscape was between the 0 and 25th or between the 75 and 100th percentiles of the HRV distribution, the departure index was scaled from 0 to 100, with larger values indicating greater departure. The choice of the 25–75th percentiles (i.e. the center 50% of the HRV distribution) to represent "no departure" was arbitrary.

We summarized the HRV departure indices as follows. First, we computed a *landscape composition departure index* as the mean departure index across patch types (combinations of cover types and seral stages) based on the PLAND metric (percentage of the landscape in each patch type; Table 9.2). This summary index measures the degree of current departure in terms of the proportional distribution of area among patch types. Second, we computed a *landscape configuration departure index* as the mean departure index across the landscape configuration metrics (Table 9.2). This summary index measures the degree of current departure in terms of the spatial pattern of the landscape mosaic. Lastly, we combined the landscape composition and configuration departure indices into an overall *landscape structure departure index* by taking the average of the two component indices.

9.3   SCOPE AND LIMITATIONS

HRV analyses can only be understood and interpreted within the scope and limitations of the approach used. For the SJNF application, the most important considerations are as follows:

First, our approach was designed to simulate vegetation dynamics for the period from about 1300 to the

late 1800s, stretching from Ancestral Puebloan abandonment to Euro-American settlement. As noted
above, this period represents a time when broadscale
climatic conditions were generally similar to those of
today but Euro-American settlers had not yet introduced sweeping ecological changes, a time of relatively
consistent (though not static) environmental and cultural conditions in the region, and a time for which we
have a reasonable amount of specific information to
enable us to model the system. Nevertheless, it is a
somewhat arbitrary reference period.

Second, our approach relied on the use of computer
models, and while it is important to recognize the
many advantages of models, it is critical to understand
that models are abstract and simplified representations
of reality. Thus, our results should not be interpreted
as "golden." Rather, they should help identify the most
influential factors driving landscape change, identify
critical empirical information needs, identify interesting system behavior (e.g. thresholds), identify the limits
of our understanding, and help us to explore "what if"
scenarios. Moreover, models are only as good as the
data used to parameterize them. To the extent possible,
we utilized local empirical data; however, we also drew
on relevant scientific studies, often from other geographic areas, and relied heavily on expert opinion.
Thus, our results should not be viewed as definitive, but
rather as an informed estimate of the HRV based on
our current scientific understanding. Moreover,
because of our partial reliance on expert opinion to
parameterize the model, the emergent properties cited
above are only partially emerging from ecological processes, and are partially an artifact of assumptions
made by experts. Moreover, readers should be aware
that there are many different approaches for modeling
disturbance and succession (Keane et al. 2002, 2009)
and each approach has its own set of assumptions and
requirements for parameterization. For example, some
models (e.g. LANDIS; He & Mladenoff 1999) use a
"process-based" approach to succession, in which the
mechanisms that govern succession are represented
explicitly, as opposed to the "pathway-based" approach
used in RMLands, in which succession is implemented
as simple conditional transitions between discrete
states.

Third, our approach focused on upland vegetation
types, largely for pragmatic reasons. Vegetation patterns and dynamics of riparian and aquatic vegetation
are more complex, more variable, and more difficult to
model in a straightforward fashion than are patterns
and dynamics of upland vegetation. We note, however,
that the patterns and ecological processes of surrounding upland vegetation have profound influences on
aquatic ecosystems; thus, our results for terrestrial
vegetation provide a partial basis for future assessments of aquatic HRV.

Lastly, our approach focused on the effects of two
major natural disturbances: fire and insects/diseases.
Other kinds of natural disturbances also occur, but the
impacts of these other disturbances tend to be localized
in time or space and have far less impact on vegetation
patterns over broad spatial and temporal scales than
do fire and insects. Thus, we believe that our results
provide a useful reference for management.

9.4 MAJOR FINDINGS

HRV

It is widely accepted that during the reference period,
natural-disturbance processes operated to maintain a
complex vegetation mosaic of successional stages and
cover types. It is less clear whether this vegetation
mosaic was stable in structure (composition and configuration) or the degree to which it varied over time.
With our simulations, we sought to quantify the range
of variation in landscape structure during the reference period to help ascertain the degree of dynamism
in landscape structure and to provide a benchmark for
comparison with alternative future land-management
scenarios. To this end, our simulations produced
several important findings:

First, and most importantly given our objectives, we
were able to derive estimates of the HRV for a broad
suite of state variables (Table 9.2). For example, one of
the patch types of management concern was high
elevation late-seral conifer forest, a type of "old-
growth" forest that has received much attention in the
Southern Rocky Mountain region because of the
special wildlife habitat and aesthetic qualities that it
affords. This patch type consists of sparse to dense
stands of a mixture of conifer tree species, including *P.
engelmannii*, *A. lasiocarpa*, *Abies lasiocarpa var arizonica*,
A. concolor, *Pseudotsuga menziesii*, *P. pungens*, *Pinus
flexilis*, and *Pinus strobiformis*, although commonly
only one or two of these species will be present in a
stand. This type is found at middle to higher elevations
(2200–3800 m), although at the lower elevations, it is
restricted to cooler and moister sites (e.g. north-facing

slopes). In addition, this patch type consists of stands in later stages of development, characterized by an abundance of canopy gaps, a multilayered canopy consisting of multiple-age cohorts or an uneven age structure, large live trees, an abundance of snags and coarse woody debris, and a patchy to well-developed understory. We estimated that the HRV (represented as the 5th–95th percentile range of variation in the simulated landscape) in the extent of this patch type was roughly 20–30% of the landscape. These estimates, however, are of the percentage of the "landscape" composed of this patch type, not the percentage of high-elevation conifer forest in this seral stage; the latter would be a much higher percentage.

Another patch type of management concern was low elevation fire-maintained open-canopy conifer forest. This patch type develops when low-mortality fire burns a stand in the later stages of development, and it is maintained by periodic but frequent low-mortality fires. This patch type consists of stands characterized by moderate to dense ground cover of grasses, forbs, and sometimes patchy low shrubs (primarily *Q. gambelii*), and a low density of large trees (primarily *P. ponderosa*) of varying size classes with a patchy distribution and open canopy; it is found principally at lower to middle elevations (2000–2800 m). We estimated that the HRV in the extent of this patch type was roughly 10–15% of the landscape.

Second, the vegetation mosaic was remarkably variable in structure over time. Although the landscape could be characterized as a shifting mosaic of successional stages and cover types, it was not a steady-state shifting mosaic (sensu Bormann & Likens 1979) – the composition of the mosaic was not constant. Although the vegetation mosaic was not in a steady-state equilibrium, the mosaic was generally in dynamic (or *bounded*) equilibrium (sensu Turner et al. 1993). That is to say, while landscape structure varied over time, it generally fluctuated within bounds about a stable mean (http://www.umass.edu/landeco/research/rmlands/applications/hrv_sjnf/figures/mts_fig_equilibrium.gif). For example, Fig. 9.2 depicts the proportional abundance of seral stages within the high-elevation conifer forest (top figure) and low-elevation conifer forest (bottom figure). In both cases, relative abundance of the seral stages appears to fluctuate within relatively stable bounds, although in the case of the high-elevation forest the variation was somewhat episodic, and in the case of low-elevation forest the variation did not stabilize until after about

100 years. This dynamic equilibrium behavior was essential to our objective of describing the range of variation in landscape structure, because the concept of a "range of variation" implies that the range is stable.

Third, most metrics achieved a stable, bounded equilibrium within a 100 to 300-year period, although it took twice that length of time to verify that the range of variation was in fact stable. Not surprisingly, the period required for equilibration in landscape composition varied among cover types. In general, cover types that experienced shorter disturbance return intervals and/or faster rates of succession equilibrated in the shortest period (e.g. low-elevation conifer forest; Fig. 9.2 bottom). Some cover types were already in apparent equilibrium (i.e. the current landscape condition appeared to be within the simulated HRV) but took much longer to verify that they were in equilibrium due to longer disturbance return intervals and/or slower rates of succession (e.g. high-elevation conifer forest; Fig. 9.2 top).

Lastly, the range of variation in landscape structure cannot be expressed by a single metric, at least not effectively, because the metrics associated with different aspects of landscape composition and configuration exhibit varying degrees and patterns of variation. Surprisingly, landscape-composition metrics exhibited greater variation (on average, based on the CV) than landscape-configuration metrics (Table 9.2). Thus, while the composition of the vegetation mosaic fluctuated markedly over time, the spatial pattern of the mosaic was more stable. The variation in configuration was principally associated with changes in the size and continuity of large patches in the landscape. This suggests that large, severe disturbance events, those that occurred relatively infrequently but that substantially altered the seral-stage distribution and created a coarse-grained mosaic of vegetation patches, were disproportionately important in regulating the dynamism in landscape structure. In contrast, the relatively frequent small disturbances had little impact on overall landscape pattern or change through time.

Current departure

One of the principal purposes of gaining a better quantitative understanding of the historical reference period is to know whether recent human activities have caused landscapes to move outside their historical range of

High Elevation Conifer

Low Elevation Conifer

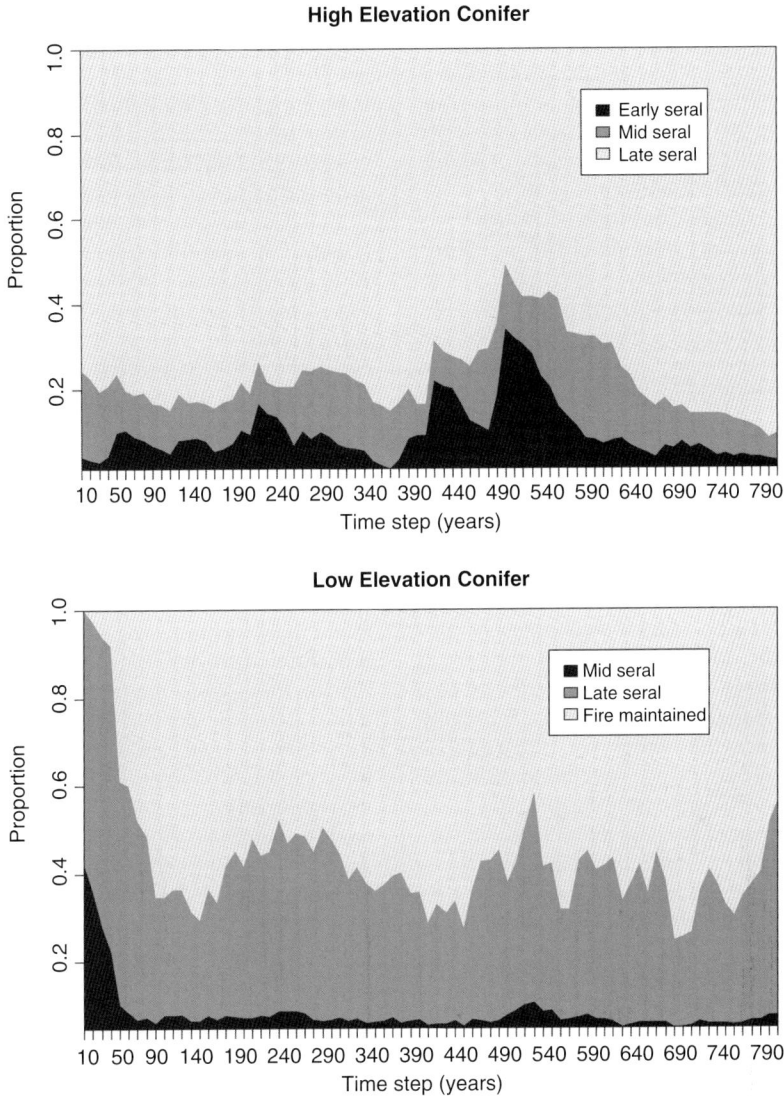

Fig. 9.2 Percent of high-elevation conifer forest (top) and low elevation conifer forest (bottom) composed of each seral stage over an 800-year simulated landscape trajectory under the historical reference period conditions on the San Juan National Forest, Colorado, USA. Note, the low-elevation forest: early seral stage is not shown because it comprised <1% for the duration of the simulation.

variation. To this end, we derived an index of departure to assess the degree to which the current landscape deviates from the simulated HRV and made the following key findings (but see important caveats below):

First, the current landscape structure appears to deviate substantially from simulated HRV, although the level of "departure" varies among landscape metrics (Table 9.2) and varies spatially across the forest in relation to differences among cover types. Many characteristics appear far outside the HRV. Indeed, almost half of the landscape metrics (7/16 composition metrics and 10/19 configuration metrics) have 100%

departure indices (Table 9.2). A large departure index can result from either too much or too little of a given landscape variable; for example, the index for mid-seral pinyon-juniper woodlands is 100 because the current landscape is at the 0th percentile, whereas the index for late-seral pinyon-juniper woodlands is also 100, but this is because the current landscape is at the 100th percentile (Table 9.2).

Second, in general, the current landscape has fewer, larger, more extensive and less isolated patches with less edge habitat than existed under the simulated HRV. The larger patches tend to be geometrically less complex and contain proportionately more core area than existed under the simulated HRV. Due to lack of extensive disturbance, large patches of late-seral conifer forest and an almost total absence of early-seral forest dominate the current high-elevation landscape. Low-elevation forests, especially ponderosa pine, also contain an overabundance of stands in the late-seral stages, but the most notable departure is the complete absence of stands in the fire-maintained open-canopy condition. Overall, the current landscape is more contiguous and less structurally diverse than ever existed under the simulated HRV. This can be interpreted as a more homogenous landscape, in which the lack of any extensive disturbances during the past 100 years has led to large, mostly late-seral patches with low contrast due to the paucity of younger seral stages. This landscape condition appears to be largely a legacy of the last century of land-management practices, in particular fire exclusion. However, the generally benign climate of the twentieth century also was a significant reason for the lack of large, stand-replacing disturbances, either by fire or spruce beetle.

Due to the above vegetation conditions and patterns, the current landscape appears to deviate substantially from the HRV in susceptibility to at least four of the simulated insect/pathogen disturbances. The current landscape appears especially vulnerable to pine beetle, spruce budworm, and spruce beetle outbreaks. In addition, given the direct link between vegetation patterns and wildlife habitat, it is not surprising that the current landscape appears to favor some species and not others. In particular, species associated with high-elevation, late-seral forest interior, such as American pine marten (*Martes americana*), are the principal beneficiaries of the current landscape departure. Conversely, species associated with edges and interspersion of cover type and seral stages, such as elk (*Cervus elaphus nelsoni*) and olive-sided flycatcher (*Contopus*

borealis) are disadvantaged in the current landscape. The paucity of disturbances over the past century has left the current landscape rather deprived of edge habitat and has reduced the overall interspersion and juxtaposition of the vegetation mosaic, with negative consequences on habitat capability for these two indicator species.

Managers need to be cognizant of an important limitation when interpreting these HRV departure results. Although it is clear that the current landscape structure is not within the modeled range of variation, the magnitude of departure is less clear due to inconsistencies in the spatial resolution and accuracy of the initial cover-type map. Specifically, the fine-grained heterogeneity in vegetation created by the disturbance processes in RMLands was probably not comparably represented in the initial input map of current vegetation cover types and seral stages. We took precautions to safeguard against reaching spurious conclusions in this regard. First, we eliminated the finest heterogeneity by rescaling the vegetation maps to a 0.5-ha minimum mapping unit and then evaluated HRV departure on these rescaled vegetation maps in addition to the original high-resolution maps (not presented here). Second, in our interpretation of HRV departure, we emphasized several area-weighted landscape metrics that are insensitive to variations affecting very small patches. Nevertheless, several landscape configuration metrics sensitive to fine-grained heterogeneity were incorporated into the overall configuration departure index. Consequently, while we feel confident in concluding that the current landscape structure is well outside the modeled range of variation, it is important to be aware that our reported HRV-departure indices, except for the seral-stage departure index and landscape-composition departure index, are probably biased high.

In addition, any conclusions regarding HRV departure depend on an accurate mapping of seral stages in the current landscape. In particular, we lack reliable age and seral stage data for most nonforested types (e.g. mountain shrublands, mesic sagebrush, pinyon-juniper woodlands). Consequently, our initial assignment of nonforest stands to condition classes (seral stages) was based on interpolation from sparse data or on a random assignment based on seral-stage distributions estimated by local experts. In either case, we are not confident that our current condition estimates are accurate. Hence, until more complete data on current stand age and condition are available, the HRV-departure results for these cover types must be viewed with extreme caution.

Lastly, our simulations indicate that returning the landscape structure to a condition that falls within the simulated HRV would likely be a difficult and long-term undertaking, if it were deemed desirable. We deduced this from the time it took the current landscape to equilibrate to the reference-period disturbance regime. We can infer that if management activities were designed to emulate natural disturbance processes, then it would take a length of time equal to the equilibration period to return the landscape to its HRV. In our simulations, most landscape structure metrics equilibrated within 100–300 years. It is possible, however, that aggressive management could shorten the recovery to the HRV. It must be emphasized, however, that this does not imply that our management goal should be to recreate all of the ecological conditions and dynamics of the reference period. Complete achievement of such a goal would be impossible, given the climatic, cultural, and ecological changes that have occurred in the last century. Moreover, the extent and intensity of disturbance required to emulate the natural disturbance regime would be unacceptable socially, economically, and politically.

Effects of scale

The pattern detected in any ecological mosaic is a function of scale (Wiens 1989). In this regard, our simulations produced several important findings:

First, we examined landscape-structure dynamics at two spatial resolutions: (1) 0.0625-ha (25-m cell size) minimum mapping unit, and (2) 0.5-ha minimum mapping unit. Despite our expectations, the results were largely insensitive to spatial resolution. It was apparent that small patches had a trivial impact on most configuration metrics and virtually no impact on the metrics selected for interpretation (i.e. area-weighted metrics). This is not to say that the fine-grained patterns of heterogeneity are unimportant ecologically, only that at the scale of the large landscape extents we examined (tens to hundreds of thousands of hectares), the quantitative importance of the fine-grained patterns was dwarfed by the coarse-grained patterns created by the larger patches.

Second, we established the temporal extent of our simulations based on our desire to capture and describe a stable range of variation in landscape structure. In

general, a minimum of 100–300 years is needed to capture the full range of variation in most metrics, and twice that long to confirm that the range is stable. Thus, a management strategy designed to emulate the natural disturbance regime would take 100–300 years to see the landscape fluctuate through its full range of conditions. This is a humbling thought, given that most professional careers last no more than 30 years – a blip on the scale of these landscape dynamics – and that most policies are geared toward 10- to 20-year planning horizons.

Third, when we examined progressively smaller spatial units of the entire simulated landscape (Table 9.1), temporal variability increased (as would be expected), and there was an apparent threshold in the relationship between landscape extent and temporal variability (Fig. 9.3). Specifically, the magnitude of variation in landscape structure increased only modestly as the landscape extent decreased from the Forest scale (847 638 ha) to the District scale (average = 282 546 ha), but increased dramatically as the landscape extent decreased to the watershed scale (average = 38 469 ha). We interpret this to mean that at the District extent (and larger), the landscape is large enough to fully incorporate the disturbance regime and exhibit stable dynamic behavior. We conclude that under the simulated disturbance regime, characterizing HRV is best done at the District or Forest scale.

Fourth, each of the sub-landscapes we examined was unique in its absolute range of variation in landscape structure (results not presented here), demonstrating that no two landscapes in this mountainous region are identical; each has unique characteristics of topography, vegetation, and so on, that affect its dynamical behavior. Most of these differences can be attributed to differences in landscape composition. The challenge to managers is in deciding whether to give explicit recognition to these differences when establishing management direction or to subsume these difference at the Forest level on pragmatic grounds. We believe that it is probably sufficient to characterize HRV at the Forest scale for purposes of general communication, but that it would be wise if possible to use the District-specific HRV results when setting management targets.

Lastly, despite the importance of landscape extent and context on the measured range of variation, the degree of departure of the current landscape from the simulated HRV was relatively invariant to scale and context. Thus, all of the sub-landscapes we

Landscape composition

$y = 22742x^{-0.371}$
$R^2 = 0.7$

Landscape configuration

$y = 866.03x^{-0.272}$
$R^2 = 0.85$

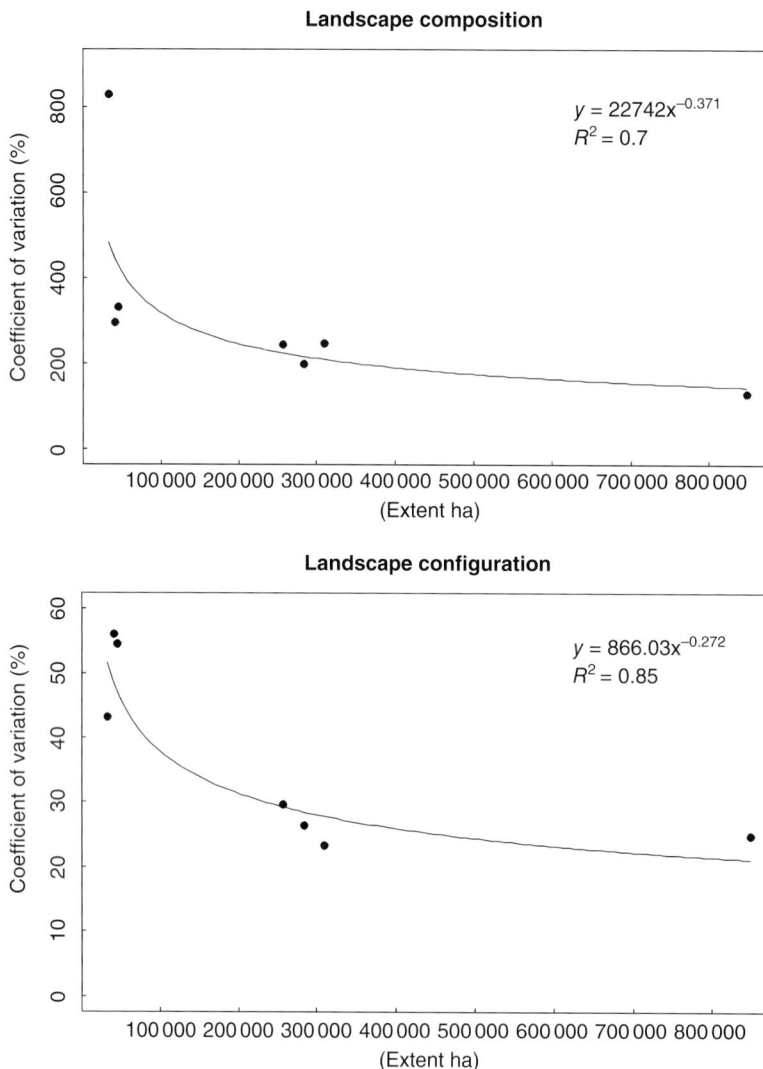

Fig. 9.3 Coefficient of variation in landscape composition (top) and landscape configuration (bottom) in relation to landscape extent under the simulated historical reference period conditions on the San Juan National Forest, Colorado, USA.

examined appear to be roughly equally outside their simulated HRV.

9.5 CONCLUSION

In conclusion, modeling HRV allowed us to demonstrate what may be generally understood by experts but rarely quantified in practice: that the landscape is highly dynamic and best conceptualized as a shifting mosaic of vegetation states over time. Further, given the variable disturbance regime across the landscape, the period over which the landscape may exhibit dynamic equilibrium conditions may vary from one to several centuries; thus, over a relatively short observational period of say a single professional career of 30 years, we should not expect the landscape to exhibit its full range of variation. Moreover, we learned that quantifying HRV requires a multivariate approach, as no single landscape metric (or state variable) can adequately characterize landscape structure and dynamics. Importantly, simulation modeling allowed us to expand our assessment of HRV to include a variety of landscape configuration metrics that deal with spatial structure of the landscape in addition to simple landscape composition metrics. In addition to quantifying HRV, simulation modeling allowed us to quantify current landscape departure and establish that the current landscape appears to deviate substantially from simulated HRV in many aspects. In particular, our results indicate that the current landscape lacks the coarse-grained spatial heterogeneity in vegetation structure that was likely characteristic of the historical reference period, and that recovery of this landscape structure may be a long process. More importantly, we learned that while "departure" is a simple concept, quantifying it is fraught with numerous challenges, not the least of which is defining a meaningful index of departure given the limitations of extant data. Lastly, our results clearly indicate that scale and landscape context matter a great deal to HRV assessment. Specifically, both landscape and temporal extent strongly influence quantitative measures of HRV, there may be an intermediate scale (in space and time) for which HRV is most meaningful, and because landscapes are idiosyncratic, transferring HRV results from one landscape to another is problematic. Many of the lessons learned from this application derive from the use of simulation modeling. We believe this application demonstrates the vital role of simulation modeling in historical ecology.

REFERENCES

Bormann, F.H. & Likens, G.E. (1979). *Pattern and Process in a Forested Ecosystem*. Springer-Verlag, New York, USA.

Cushman, S.A., McGarigal, K., & Neel, M. (2008). Parsimony in landscape metrics: strength, universality, and consistency. *Ecological Indicators*, **8**, 691–703.

He, H.S. & Mladenoff, D.J. (1999). Spatially explicit and stochastic simulation of forest-landscape fire disturbance and succession. *Ecology*, **80**, 81–99.

Karau, E.C. & Keane, R.E. (2007). Determining landscape extent for succession and disturbance simulation modeling. *Landscape Ecology*, **22**, 993–1006.

Keane, R.E., Parsons, R.A., & Hessburg, P.F. (2002). Estimating historical range and variation of landscape patch dynamics: limitations of the simulation approach. *Ecological Modeling*, **151**, 29–49.

Keane, R.E., Hessburg, P.F., Landres, P.B., & Swanson, F.J. (2009). The use of historical range and variability (HRV) in landscape management. *Forest Ecology and Management*, **258**, 1025–1037.

Lertzman, K.P. & Krebs, C.J. (1991). Gap-phase structure of a subalpine old-growth forest. *Canadian Journal of Forest Research*, **21**, 1730–1741.

McGarigal, K., Cushman, S.A., Neel, M.C., & Ene, E. (2002). FRAGSTATS: Spatial Pattern Analysis Program for Categorical Maps. Computer software program produced by the authors at the University of Massachusetts, Amherst. http://www.umass.edu/landeco/research/fragstats/fragstats.html.

Morgan, P., Aplet, G.H., Haufler, J.B., Humphries, H.C., Moore, M.M., & Wilson, W.D. (1994). Historical range of variability: a useful tool for evaluating ecosystem change. *Journal of Sustainable Forestry*, **2**, 87–111.

Neel, M.C., McGarigal, K., & Cushman, S.A. (2004). Behavior of class-level landscape metrics across gradients of class aggregation and area. *Landscape Ecology*, **19**, 435–455.

Romme, W.H., Floyd, L., & Hanna, D. (2009). *Historical Range of Variability and Current Landscape Condition Analysis: South Central Highlands Section, Southwestern Colorado & Northwestern New Mexico*. Colorado Forest Restoration Institute, Colorado State University, Fort Collins, CO, USA.

Roovers, L.M. & Rebertus, A.J. (1993). Stand dynamics and conservation of an old-growth Engelmann spruce-subalpine fir forest in Colorado. *Natural Areas Journal*, **13**, 256–267.

Swetnam, T.W. & Betancourt, J.L. (1990). Fire-southern oscillation relations in the southwestern United States. *Science*, **249**, 1017–1020.

Turner, M.G., Romme, W.H., Gardner, R.H., O'Neil, R.V., & Kratz, T.K. (1993). A revised concept of landscape equilibrium: disturbance and stability on scaled landscapes. *Landscape Ecology*, **8**, 213–227.

Veblen, T.T., Hadley, K.S., & Reid, M.S. (1991). Disturbance and stand development of a Colorado subalpine forest. *Journal of Biogeography*, **18**, 707–716.

Wiens, J.A. (1989). Spatial scaling in ecology. *Functional Ecology*, **3**, 385–397.

Wu, J. (2004). Effects of changing scale on landscape pattern analysis: scaling relations. *Landscape Ecology*, **19**, 125–138.

Wu, J., Shen, W., Sun, W., & Tueller, P.T. (2003). Empirical patterns of the effects of changing scale on landscape metrics. *Landscape Ecology*, **17**, 761–782.

Section 4

Case Studies of Applications

Gregory D. Hayward

USDA Forest Service, Alaska Region, Anchorage, AK, USA

More than 50 years ago, Charles Cooper (1960) described the relationship between fire, grazing, climate change, and patterns of forest structure in the ponderosa pine (*Pinus ponderosa*) forests of the southwest. Predating modern programs of ecological restoration by over half a century, Cooper's synthesis arose from an understanding of historical ecology spanning multiple temporal scales and integrating several ecological processes. Today, the "southwest ponderosa pine story" may represent the most-referenced example of the successful application of historical ecology to forest management in North America (e.g. Swetnam et al. 1999; Covington 2000). In his monograph, Cooper prophetically reviewed the

challenges of restoration experienced by land managers today. For instance, Cooper struggled to envision an economically feasible approach to remove dense pine regeneration and restore forest structures compatible with the frequent fire regime described in his historic analysis. Equally compelling was the political struggle he portrayed in a lengthy quote from a report from the Secretary of Agriculture, which ended with "It has been alleged that the fires set in the early days by the Indians and the first settlers were beneficial. As a matter of fact, these early fires were enormously destructive" (USDA 1910, as cited by Cooper 1960). Cooper described long-running debates regarding the consequences of frequent fires on forest productivity

Historical Environmental Variation in Conservation and Natural Resource Management, First Edition. Edited by John A. Wiens, Gregory D. Hayward, Hugh D. Safford, and Catherine M. Giffen.

and the potential conversion of forests to nonforest vegetation resulting from uncharacteristic fire frequency and intensity.

Application of historical ecology 50 years ago led to dramatic changes in the understanding of interactions between management and ecosystem dynamics. Cooper, along with some of his management contemporaries, recognized the need to be aware of ecological dynamics at temporal scales appropriate to the environment of concern. They realized that the ecological changes observed during an individual's career may not represent the most common dynamics. Land managers instead must predict the future of systems that operate at temporal scales beyond individual human experience. Historical evidence, and therefore the formal development of historical ecology, was critical.

As managers have gained experience in applying historical ecology and historical range of variation (HRV) concepts to land management over the past half-century, the benefits have become clear. In this section, we examine a series of eight case studies demonstrating the value of historical ecology in resource management and conservation. These examples span the space-time continuum from short-term, local restoration efforts (Chapter 15) to broad management plans that may consider environmental history over 10 000 years or more (Chapters 10, 11, and 17). The stories, told by both practitioners and research scientists, demonstrate the power of partnerships and the challenge of joining professionals from different cultures and different reward systems. Several chapters suggest that ideas from historical ecology or HRV assessments were instrumental in bringing together groups who previously found little common ground for collaboration (Chapters 12 and 15).

One of the striking observations from this set of case studies relates to the tremendous range of management settings that benefit from insights provided by historical ecology. In Chapter 14, the rather mundane documentation of range contraction led to specific management decisions for yellow-legged frog, while a more nuanced appreciation for wetland hydrology pointed the way to water management strategies to expand the range of wood storks. Chapters 10, 11, 15, and 16 examine management of dominant ecological processes from flooding to fire to plant dispersal based on insights that often become apparent only through careful analysis of temporal patterns spanning centu-

ries rather than the 1–4 years of traditional research programs.

By design, this assortment of case studies includes examples of management efforts guided by substantial HRV assessments requiring enormous time and resources *and* cases where the application of historical ecology developed as an indirect consequence of confronting difficult issues of temporal dynamics. In Chapter 15, the value of comprehending historical patterns of flood dynamics grew from short-term monitoring of species establishment after a management action. In contrast, forest managers spent years synthesizing historical data on fire dynamics to establish guidelines for broadscale forest management in Chapters 11, 12, and 13.

This section opens and closes with synthetic chapters. Chapter 10 provides a comprehensive example illustrating the rich array of environmental features that may be evaluated from a thorough HRV assessment. The closing chapter (Chapter 17) ties the previous examples together with a series of four vignettes that demonstrate how to apply historical knowledge using a structured framework to make land-management decisions. We suggest that this framework can help to transform an academic understanding of historical ecology or a management-oriented HRV assessment into insights that motivate critical management decisions. Application of this framework can facilitate effective restoration, rigorous effects analysis, efficient research and monitoring, and insightful planning that will become critical to land management and conservation in the face of climate change.

As you examine this selection of case studies, we encourage you to consider how the management or conservation decisions made in each example might have differed if knowledge of temporal dynamics resulting from historical ecology was not available.

REFERENCES

Cooper, C.F. (1960). Changes in vegetation, structure, and growth of southwestern pine forests since white settlement. *Ecological Monographs*, **30**, 129–164.

Covington, W.W. (2000). Helping western forests heal. *Nature*, **408**, 135–136.

Swetnam, T.W., Allen, C.D., & Betancourt, J.L. (1999). Applied historical ecology: using the past to manage for the future. *Ecological Applications*, **9**, 1189–1206.

USDA. (1910). *Annual Report*. US Department of Agriculture, Washington, DC, USA.

Chapter 10

REGIONAL APPLICATION OF HISTORICAL ECOLOGY AT ECOLOGICALLY DEFINED SCALES: FOREST ECOSYSTEMS IN THE COLORADO FRONT RANGE

Thomas T. Veblen,[1] *William H. Romme,*[2] *and Claudia Regan*[3]

[1]University of Colorado, Boulder, CO, USA
[2]Colorado State University, Fort Collins, CO, USA
[3]USDA Forest Service, Rocky Mountain Region, Lakewood, CO, USA

Historical Environmental Variation in Conservation and Natural Resource Management, First Edition. Edited by John A. Wiens, Gregory D. Hayward, Hugh D. Safford, and Catherine M. Giffen.
© 2012 John Wiley & Sons, Ltd. Published 2012 by John Wiley & Sons, Ltd.

10.1 INTRODUCTION

The purpose of a historical range of variation (HRV) assessment can be described most accurately and simply as providing a comprehensive understanding of the range of critical ecological processes and conditions that have characterized particular ecosystems over specified time periods and under varying degrees of human influences (Veblen & Donnegan 2005). There are many potential uses of the information contained in an HRV assessment (see Chapters 1 and 2, this book), but three applications stressed in the HRV assessments produced for the National Forests in the Rocky Mountain Region (USDA Forest Service Region 2; Veblen & Donnegan 2005; Romme et al. 2009) are preservation of biodiversity, maintenance of resilience in the context of climate change, and evaluation of the effects of past management practices on ecosystem services. The goal is to synthesize and evaluate information on historical ecology to inform discussions of potential management goals and strategies in a collaborative context. This chapter illustrates how the understanding of HRV can directly influence management of forest ecosystems and lead to major changes in management direction. Changes in management result when the HRV clearly synthesizes the understanding of ecosystem processes and the socio-political system is able to embrace the new scientific understanding and apply it to meet management goals.

In this chapter, we summarize some applications of HRV assessments produced for National Forests in the Rocky Mountain Region. Our aim is to describe how knowledge of historical ecology can improve resource planning and management strategies when rigorously considered along with social perspectives in developing goals and strategies in ecosystem-based resource management (sensu Christensen et al. 1996). In particular, we evaluate how goals of ecological restoration may or may not be compatible with mitigation of risks of wildfire and insect outbreaks in different ecosystem types. There are important contrasting social interests in these two sets of goals – restoration and mitigation – and identification of desired conditions related to both goals requires a sound understanding of historical ecology. It is especially critical to understand how current forest-landscape conditions have been shaped by the interplay between land-use practices, such as fire exclusion, and climatic variability. This knowledge, in turn, is essential for determining how much effort would be required to achieve and sustain restoration and mitigation goals in different ecosystems.

In the late 1990s, Region 2 of the USDA Forest Service collaborated with several forest-ecology research groups to produce HRV assessments for each National Forest in the Region. Each contracting group consisted of university-based researchers who had a long history of conducting research in the geographic areas included in each HRV assessment. Although reports on National Forests located in Wyoming and southwestern Colorado were completed (Dillon et al. 2005; Meyer et al. 2005; Kulakowski & Veblen 2006; Romme et al. 2009), this chapter focuses on HRV assessments for National Forests located in the Colorado Front Range (Veblen & Donnegan 2005). The HRV assessments of National Forests in the Rocky Mountain Region conducted in the early 2000s conceptually reflect the paradigm of ecosystem-based management accepted by researchers and resource managers at the time. Under this paradigm, management focuses on sustaining ecosystems while also sustaining yields of particular forest products and uses. Implementation of ecosystem management requires a thorough understanding of the natural processes that shape and maintain ecosystems and, in particular, how those processes may be affected by specific land-use practices (Christensen et al. 1996).

Public perception and the national policy context

In response to a recent trend of increasing area burned by wildfires in the western United States, the federal agencies responsible for wildland fire management developed the National Fire Plan in 2000, and in 2003, Congress passed the Healthy Forests Restoration Act. The current interagency federal wildfire policy is represented by a National Cohesive Wildland Fire Management Strategy (USDA and USDI 2011), which contains a variety of provisions to support hazardous fuel reduction and forest-restoration projects. Part of the rationale for the current national strategy is summarized as follows:

> Declining vegetative health across the national landscape has contributed to the increasing risk of catastrophic wildfire which threatens landscapes and communities. These factors – including weather variability, fire exclusion, spread of insects and diseases and non-native species, changing land use, fragmentation, and

urbanization – pose a significant challenge to establishing and maintaining healthy, resilient landscapes and communities.

A key premise of the recent succession of national-policy documents on wildfire policy is the belief that many decades of fire suppression have resulted in unnatural accumulations of fuel and created the potential for extreme fire events in ecosystem types that formerly did not experience high-severity fire events (Covington 2000). Resource professionals and fire-history researchers debate the relative importance of twentieth-century fire exclusion on current fuel characteristics for different ecosystem types; it is now widely recognized that the goals of fire hazard reduction and ecological restoration, while critical in certain ecosystems, are not compatible in other ecosystem types (Veblen 2003; Schoennagel et al. 2004). As a result, area-specific knowledge is required on the historical

ecology of fire and its effects in particular ecosystems and in specific places to balance these potentially compatible or divergent management goals. In the following section, we use the example of the Colorado Front Range to illustrate how historical ecology has been used to evaluate and revise default expectations derived from the underlying premises of the 2000 National Fire Plan and the 2003 Healthy Forest Restoration Act (Table 10.1).

Fire mitigation and ecological restoration in the front range

Regional fire mitigation and restoration policy context

Given the rapid residential growth in the wildland–urban interface during the last half of the twentieth

Table 10.1 A comparison of ecological assumptions and the management implications of those assumptions, based on "default expectations" prior to historical studies in the Colorado Front Range versus insights and implications derived from recent historical-ecology research. We emphasize that our depiction of "default expectations" does not characterize the views of all managers or policymakers in the region; many individuals questioned conventional wisdom even before rigorous historical research was conducted. However, each of the ideas listed as a "default expectation" was held by some influential managers and policymakers, and these ideas were clearly revealed in strategic policy statements and programs such as the National Fire Plan of 2000 and the Healthy Forest Restoration Act of 2003.

Default expectation prior to historical ecology studies	Management implications of default expectation	Revised paradigm derived from historical research	Management implications of revised paradigm
Frequent low-severity surface fires were dominant throughout the forests of the Front Range prior to 1900; high-severity fires were uncommon or rare	Fire regimes have changed substantially since 1900 throughout the Front Range	High-severity or mixed-severity fires occurred at variable intervals in most forests of the Front Range . . . frequent low-severity surface fires characterized only the lowest elevations	Fire regimes have changed substantially since 1900 only at the lowest elevations of the Front Range
Pre-1900 fire frequency, extent, and severity were controlled prlmarily by ignitions and fuels conditions	Managers can reduce fire risks effectively by reducing fuel loads, for example, through thinning or low-intensity prescribed burning	Pre-1900 fire frequency, extent, and severity were controlled primarily by climate . . . variation in fuels controlled fire only at the lowest elevations	Fuel reduction programs (e.g. thinning or low-intensity burning) may reduce fire risks at the lowest elevations, but will be less effective elsewhere in the Front Range
Fire-return intervals have become dramatically longer since 1900, primarily because of fire suppression		Fire-return intervals have lengthened since 1900, but this is due primarily to less frequent occurrence of climatic conditions conducive to large fires	

Continued

Table 10.1 *Continued*

Default expectation prior to historical ecology studies	Management implications of default expectation	Revised paradigm derived from historical research	Management implications of revised paradigm
Low tree densities and fuel loads characterized most forests of the Front Range prior to 1900	Thinning of forest stands represents both fuel reduction/ fire mitigation and ecological restoration throughout the Front Range	Tree densities and fuel conditions were highly variable prior to 1900, and high-density stands were common	Thinning and fuel reduction do not represent ecological restoration except at the lowest elevations
Tree densities and fuel loads have increased dramatically in most forests since 1900, primarily because of fire suppression		Only small increases in tree density have occurred in most forests, with the greatest increases at the lowest elevations and little or no change at higher elevations	
The likelihood of high-severity fire is greater today than before 1900 because of increasing fuels and tree densities	We can reduce fire risk by manipulating fuels across the landscape	Large, high-severity fires occurred prior to 1900 whenever climate conditions were favorable; the same is true today	We can reduce fire risk only at local scales, for example, in the immediate vicinity of vulnerable structures
Recent extensive and severe bark beetle outbreaks are occurring primarily because of unnaturally dense and unhealthy forest conditions	More intensive forest management, for example, thinning and harvesting, can slow the current outbreak and reduce the risk of future outbreaks	The recent bark beetle outbreaks are occurring primarily because of the climate changes of the past quarter-century	The current outbreak cannot be controlled by standard forestry practices . . . continuing climate change portends additional eruptive activity by native or introduced insects
Extensive and severe insect outbreaks create unusual fuel conditions that lead to destructive forest fires	Prompt removal of beetle-killed trees will reduce fire risk	No strong connection exists between insect-caused tree mortality and subsequent fire occurrence or severity	Removal of beetle-killed trees is justified for safety purposes, for example, along roads and trails, but will not reduce fire risk
Uncharacteristically severe fires could result from the combination of unnaturally dense stands and insect-caused mortality, leading to loss of structures and degradation of watersheds	More intense forest management, especially thinning, harvesting, and prescribed low-intensity burning, can reduce losses throughout the Front Range	High-severity fires are normal in most Front Range forests, and these forests can be expected to recover naturally from future fires – unless conditions deteriorate due to climate change	Houses and other vulnerable infrastructure are at high risk of fire damage because of the inherent nature of Front Range forests as well as current and projected climate conditions . . . risks can be reduced (but not eliminated) by localized fuel reduction and attention to "Firewise" guidelines

century, the Colorado Front Range figures prominently in recent federal, state, and local initiatives aimed at mitigating fire hazard and restoring ecosystems. The most significant recent coalitions of multigovernment and nongovernment stakeholders have been The Front Range Fuels Treatment Partnership Roundtable created in 2002 and the Colorado Front Range Landscape Restoration Project (CFLRP) funded by the national Collaborative Forest Landscape Restoration Program in 2010. These coalitions provide the primary venues for collaborative planning and implementation of fuels mitigation and ecological restoration projects in the Front Range in parts of Arapaho and Roosevelt and Pike and San Isabel National Forests. In 2009, the US Congress established the Collaborative Forest Landscape Restoration Program for the purpose of encouraging the collaborative, science-based ecosystem restoration of priority landscapes (USDA Forest Service 2011). In this context, ecological restoration is defined as:

> . . . the process of assisting the recovery of resilience and adaptive capacity of ecosystems that have been degraded, damaged, or destroyed. Restoration focuses on establishing the composition, structure, pattern, and ecological processes necessary to make terrestrial and aquatic ecosystems sustainable, resilient, and healthy under current and future conditions.

In its report, *Living with Fire: Protecting Communities and Restoring Forests*, the Front Range Roundtable (2006) identified restoration of ponderosa pine (*Pinus ponderosa*) forests in the lower montane life zone (1800–2400 m) as its highest ecologically based priority. The goal of this restoration is to return the ecological structures and processes associated with these forests to their HRV. Because ecosystem structure and processes naturally change over time, the Roundtable recommended that areas be restored to reflect a historical *range* of ecological structures and processes rather than targeting a single snapshot in time. The CFLRP adopted a restoration focus on lower montane ponderosa pine forests based in part on an expert-opinion analysis conducted on behalf of the Roundtable (Kaufmann et al. 2006), which in turn was derived from information summarized in the HRV assessment

of Front Range National Forests (Veblen & Donnegan 2005), peer-reviewed publications, preliminary results from ongoing research projects, and field experiences of each author. Under the current CFLRP restoration initiative in the Front Range, the conclusions from this expert-opinion analysis help to define desired conditions in stand and landscape-scale stand structures that are consistent with historical fire regimes. Vegetation treatments and subsequent monitoring focus on parameters such as tree sizes, densities, and spatial patterns, as well as non-tree openings so that a more complex mosaic of forest structures will be attained across a large landscape area.

The relevance of rocky mountain HRV assessments to development and implementation of fuels treatment and ecological restoration projects

An important goal of conducting an HRV assessment is to frame questions in ways that link interpretations of ecological research to information needs of resource managers and stakeholders. In the context of fire mitigation and ecological restoration in the Front Range, three key questions were addressed in the Front Range HRV Assessment (Veblen & Donnegan 2005):
1. How did historical fire frequency, extent, and severity vary across and within the major forest types?
2. How have twentieth-century policies of fire exclusion altered historical fire regimes across and within the different forest types?
3. What have been the effects of climate variability on historical fire regimes of each major forest type?

The Front Range HRV Assessment (Veblen & Donnegan 2005) focused on changes in disturbance patterns and processes as they influence forest conditions such as structure (tree ages, sizes), successional status, and relative dominance by different tree species, extent of forest cover types, and tree condition. The Assessment considered a range of spatial scales including (1) a stand level (~0.01 to a few km²) defined by units that are relatively homogeneous; (2) a landscape scale (a few km² to ~100 km²), which included a substantial range of community types and structures; and (3) a broad regional scale (>100 km²). Knowledge of past patterns was uneven across spatial scales. For example, historical conditions were often well documented at a stand scale, but effects of past fires were typically known with less certainty at broader spatial scales.

Likewise, knowledge of past landscape patterns varied over the reference period. Tree-ring data on past fire activity were sparsely available beginning in ~1200 AD (Brown et al. 1999) but were not sufficiently abundant to allow landscape-scale analyses until ~1600 AD (Veblen et al. 2000; Sibold et al. 2006).

Information sources used in the Front Range HRV assessment included a broad range of data sources from microfossils to documentary descriptions of landscape conditions and land-use practices (see Chapter 3, this book). The sources of information that proved most useful were tree-ring reconstructions of fire activity and climate variability, nineteenth-century Department of Interior surveys of vegetation conditions and land-use practices conducted in Timber Reserves (i.e. precursors to National Forests), and historical photographs from the 1880s onward. The tree-ring fire histories (Goldblum & Veblen 1992; Mast et al. 1998; Brown et al. 1999; Kaufmann et al. 2000; Veblen et al. 2000; Ehle & Baker 2003; Sherriff & Veblen 2006, 2007) and reconstructions suggesting defoliator (budworm) and bark beetle events (Hadley & Veblen 1993; Eisenhart & Veblen 2000) lead to especially important insights.

Historic fire regimes in the front range and the focus on ponderosa pine ecosystems

Broadscale vegetation patterns and historical fire regimes

In northern and central Colorado (~39 to 41°N) the Front Range rises from the plains to the continental divide. Increasing moisture availability with elevation is reflected by an elevation gradient of life zones and vegetation patterns (Fig. 10.1). Ponderosa pine is the dominant tree in the Lower Montane Zone, with Douglas fir (*Pseudotsuga menziesii*) also present or codominant in locally mesic locations. The Upper Montane Zone is also characterized by ponderosa pine, which forms nearly pure stands in places. However, several other tree species are common and may be codominant in places, including Douglas fir, lodgepole pine (*Pinus contorta*), limber pine (*Pinus flexilis*), and aspen (*Populus tremuloides*), and (in the south) white fir (*Abies concolor*). Engelmann spruce (*Picea engelmannii*) and subalpine fir (*Abies lasiocarpa*) are sometimes minor components of stands in the higher parts of the Upper Montane Zone. Above the Upper Montane Zone is the

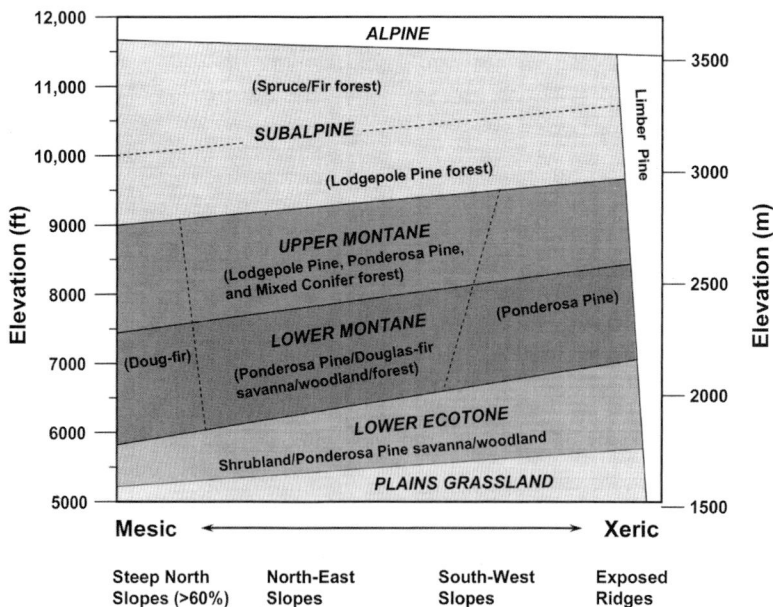

Fig. 10.1 Major vegetation zones in the Colorado Front Range. (Source: Kaufmann et al. 2006).

Subalpine Zone, dominated by lodgepole pine, Engelmann spruce, and subalpine fir.

Along this complex elevation gradient, historical fire regimes differed in frequency, extent, type (surface vs crown fire), and severity. Tree-ring fire and forest-structure data permit characterization of fire regimes during the historical reference period from about 1600 to 1900 AD. This period includes a significant time predating permanent Euro-American settlement in about the 1850s and the initiation of organized fire suppression activities in the early 1900s (Veblen & Donnegan 2005). Historical fires burned in a complex fashion in response to variation in weather, fuels, and topography so that individual fires burned with variable severity. Moreover, given the changes in fire conditions among fire events, the same point in the landscape might experience variable fire conditions, ranging from high severity in one event to low severity in other events. Despite the spatial and temporal variability in fire severity, common fire patterns experienced within certain forest types provide the basis for developing a classification of fire regimes that forms a continuum of fire-regime types in Front Range forest ecosystems (Kaufmann et al. 2006). Frequent low-severity fire regimes represent surface fires recurring about <35 years within a stand (~100 ha) that kill young seedling/saplings and maintain open low-density stands of large fire-resistant trees. At the other end of the spectrum, infrequent high-severity fire regimes are characterized by stand-replacing fires that recur after 100+ years within a stand. Mixed-severity (or variable-severity) fire regimes, in contrast, represent mosaics of low- to high-severity both within stands and across topographically complex landscapes.

Historical fire regimes in the subalpine zone

Historical evidence suggests that few if any opportunities for ecological restoration occur within the Subalpine Zone except in areas of mining, road building, or inappropriate timber harvest methods. These forests were shaped primarily by infrequent, extensive, stand-replacing fires (Veblen 2003; Buechling & Baker 2004; Sibold et al. 2006). Surface fires affected only small percentages of the surface area of these forests. As a consequence, fire exclusion has not led to forest conditions requiring restoration. For example, in a study area of 30 000 ha of subalpine forest, surface fires affected only 1–3% of the surface area of the lodgepole

pine forests (Sibold et al. 2006). Furthermore, the surface fires that did occur in lodgepole pine forests did not have a significant effect on tree mortality or tree establishment (Sibold et al. 2007).

Despite consistent findings in the peer-reviewed literature that fire regimes in subalpine forests in Colorado and southern Wyoming are overwhelmingly dominated by infrequent stand-replacing events (Romme & Knight 1981; Kipfmueller & Baker 2000), the existence of a widespread fire-regime type of frequent and widespread surface fires in Colorado lodgepole pine forests was repeatedly asserted during the development of historical fire-regime classes as part of the Fire Regime Condition Classification of LANDFIRE in the early 2000s. Lodgepole pine forests of the Front Range do not show evidence of departures from their historical conditions that could be explained by exclusion of surface fires. Dense tree densities historically were the norm in lodgepole pine forests in the Front Range and should not be confused with alleged impacts of suppression of surface fires (Veblen 2003; Sibold et al. 2006). Any management attempt to create a landscape of open woodlands of lodgepole pine would not be consistent with the historical ecology of this ecosystem type and should not be presented as an appropriate goal of ecological restoration.

Pure spruce-fir forests in the Front Range are characterized by historical fire regimes with the long-fire free intervals of more than a hundred to many hundreds of years (Buechling & Baker 2004; Sibold et al. 2006). The fires that structured these forests were primarily stand-replacing crown fires in which most or all trees were killed over 1000s of hectares. Although fire suppression activities of the twentieth century may have reduced the number of fires in some land-management units, any resulting departures from historical ecological condition or fuel-defined fire hazard in spruce-fir forests were minimal because of the long fire-free intervals natural to this forest type. Likewise, tree-ring reconstructions identify extensive spruce beetle outbreaks in this forest type during the nineteenth century, long before any putative effects of fire exclusion (Eisenhart & Veblen 2000).

Although there are currently no proposals to initiate high-severity fire regimes in the subalpine zone based on restoration objectives, HRV information has informed critical discussions of fire-hazard mitigation and resource planning more generally in these ecosystems (Veblen & Donnegan 2005). For example, the high proportion of stands originating in the mid- to late-1800s is strongly linked to severe drought that occurred synchronously over large areas of the Rocky Mountain region in response to variability in hemispheric and global-scale drivers of climate variability (Sibold & Veblen 2006; Schoennagel et al. 2007). It follows that, with warming and drying conditions in the future, we should expect large synchronous fires over extensive areas of the Subalpine Zone in Colorado similar to the large fire events of 2002. Likewise, the historical ecology of lodgepole pine forests supports the interpretation that vast areas of forests recovering from climate-driven nineteenth-century fires were in age classes susceptible to mountain pine beetle attack when the current outbreak began in the mid-1990s. Thus, climate variability rather than any effects of fire exclusion is the primary contributor to regional forest susceptibility to mountain pine beetle outbreaks (Romme et al. 2009). Analogously, tree-ring reconstructions of spruce beetle outbreaks in northwestern Colorado document widespread outbreaks in the nineteenth century and earlier that are likely associated with natural climate variability and clearly not linked to any past land-use practices (Veblen et al. 1991; Eisenhart & Veblen 2000). The importance of extensive but relatively infrequent wildfires and bark beetle outbreaks in the Subalpine Zone that are controlled primarily by climatic variation, however, does not imply an absence of management options for resource managers. For example, the markedly reduced probability of wildfire spread in young (<100 years old) conifer stands is strong justification for fire-use policies in areas remote from infrastructure where wildfires are allowed to burn large areas in National Forests and National Parks in Colorado. Historical research strongly supports managers' perceptions that young conifer-dominated as well as the aspen-dominated post-fire stands are important buffers to subsequent fire and bark beetle spread (Bebi et al. 2003; Bigler et al. 2005).

The historical research summarized above is now widely known to resource practitioners and is central to discussions of potential management responses to hazards and risks of wildfire and insect outbreaks in the Subalpine Zone of Colorado (Table 10.1). The mountain pine beetle outbreak that began in 1996 and infested 1.6 million hectares in Colorado and Wyoming provides a good example (USDA Forest Service, Forest Health Protection and its Partners 2010). Initially, managers and the media asserted that decades of fire suppression and a paucity of forest harvesting caused

the epidemic. There were calls for aggressive management actions, including broadscale fuel treatments (Denver Post 2002; Best 2007). Evaluation of the existing historical ecological research on these forests, however, supported a more limited management response with a focus on removal of hazard trees and strategic fuels reduction near settlements and high-value infrastructure such as powerlines and communication facilities (Kaufmann et al. 2008; Romme et al. 2009).

Historical fire regimes in ponderosa pine ecosystems

In comparison with the Subalpine Zone, more frequent historical fires followed by more effective fire exclusion in the twentieth century in the ponderosa pine forests of the Montane Zone makes it a more appropriate target for restoration projects (Kaufmann et al. 2006). In many dry ponderosa pine ecosystems of the western United States, twentieth-century exclusion of formerly frequent low-severity fires is believed to have resulted in dramatic increases in tree density, a shift to potential crown fires, and increases in forest insect outbreaks (Covington & Moore 1994; Covington 2000). For Southwest ponderosa pine ecosystems, a formalized restoration framework (Friederici 2003) aims to restore frequent low-severity fires and related forest conditions in order to reduce the risk of crown fires in these ecosystems. As recently as the early 2000s, the Southwest model was the default expectation among most forest managers for ponderosa pine forests in the Colorado Front Range (Table 10.1), despite published research that identified higher severity stand-replacing or partially stand-replacing fires as important components of the pre-fire-exclusion fire regimes of these ecosystems in Colorado (Veblen & Lorenz 1986; Goldblum & Veblen 1992). We suggest that the Southwest ponderosa model was widely and uncritically accepted for Front Range ponderosa pine ecosystems for three reasons: (1) the convergence of ecological restoration with a reduction in crown-fire risk was attractive to both managers and the general public; (2) the low-severity restoration framework had been persuasively supported through outreach to the general public, managers, and national decision makers; and (3) research identifying the limited applicability of the low-severity model to Front Range ponderosa pine forests had not been effectively synthesized into an alternative restoration framework.

The Front Range HRV Assessment (Veblen & Donnegan 2005) and recent publications conclude that the low-severity model is not appropriate for most ponderosa pine forests in the Front Range (Sherriff & Veblen 2007; Schoennagel et al. 2011). The ideas and evidence in this research have been summarized into a variable-severity fire-regime model for ponderosa pine ecosystems presented as a coherent alternative to the low-severity model (Table 10.2; Baker et al. 2007). In this model, natural fires vary in severity and frequency, sometimes burning at low severity in surface fuels and sometimes burning as high-severity fires in the crowns of trees, or as a mixture of both fire types. In the variable-severity model, most of the landscape historically experienced or is capable of supporting higher severity fires that kill high percentages (e.g. >40%) of canopy trees in small to moderate-sized patches (e.g. 1 to >100 ha). A key distinction from the low-severity model is that moderate to high-severity fires create significant opportunities for the regeneration of shade-intolerant tree species by creating large canopy openings. Thus, cohort structures reflect the regeneration opportunities that follow fires.

While analysis based on expert opinion (Kaufmann et al. 2006) was an important qualitative guide, managers needed more spatially explicit criteria for design of ecologically based restoration. Spatial modeling based on widely distributed stand-scale reconstructions of fire frequency and fire effects permitted the mapping of probability surfaces for different historical fire-regime types (Sherriff & Veblen 2006, 2007). The primary goal was to map areas that historically had been characterized by frequent low-severity fires (mean fire-return interval of <30 years) for the period 1700–1860 AD. Low-severity fires were identified by a fire-scar date that was followed by little or no post-fire tree establishment, in contrast to conspicuous pulses of establishment following moderate and severe fires. Low-severity fires were further identified by the high percentages of trees that survived the fire event. For a 60875-ha area in the ponderosa pine zone of the eastern slope of the Front Range, the results indicated that <20% of the ponderosa pine zone had a historical fire regime (pre-1915) of relatively frequent low-severity fires (Fig. 10.2). More than 80% of the ponderosa pine zone experienced a mixed or variable-severity fire regime in which some fire events killed high proportions of canopy trees and typically resulted in dense

Table 10.2 Comparison of two models of fire and forest structure in ponderosa pine and ponderosa pine–Douglas fir forests (after Baker et al. 2007).

Low-severity historical fire-regime model	Variable-severity historical fire regime-model
Old-growth trees dominant	Old-growth patches, common but patches of other ages occur
Low-severity surface fires only	Variable fire severity: mixture of low-, moderate-, and high-severity
Trees widely spaced, tree density low	Tree density variable but including some large dense patches
Low-severity fires kill few canopy trees	Moderate and high-severity fires kill canopy trees in groups or over large (multiple hectares) areas
Frequent surface fires:	Surface fires
Kill most small trees	Kill some small trees, leaving some patches
Prevent fuel buildup	Have varied effects on fuels
	May enhance tree regeneration
High-severity fires are absent	High-severity fires commonly create open areas suitable for tree regeneration
Fire exclusion leads to:	Fire exclusion leads to:
High rate of tree regeneration and woody encroachment	Fewer opportunities for tree regeneration
Fuel buildup	Varied fuel effects but not greatly different from historic conditions
Uncharacteristic high-severity fires	Decrease in natural high-severity fires

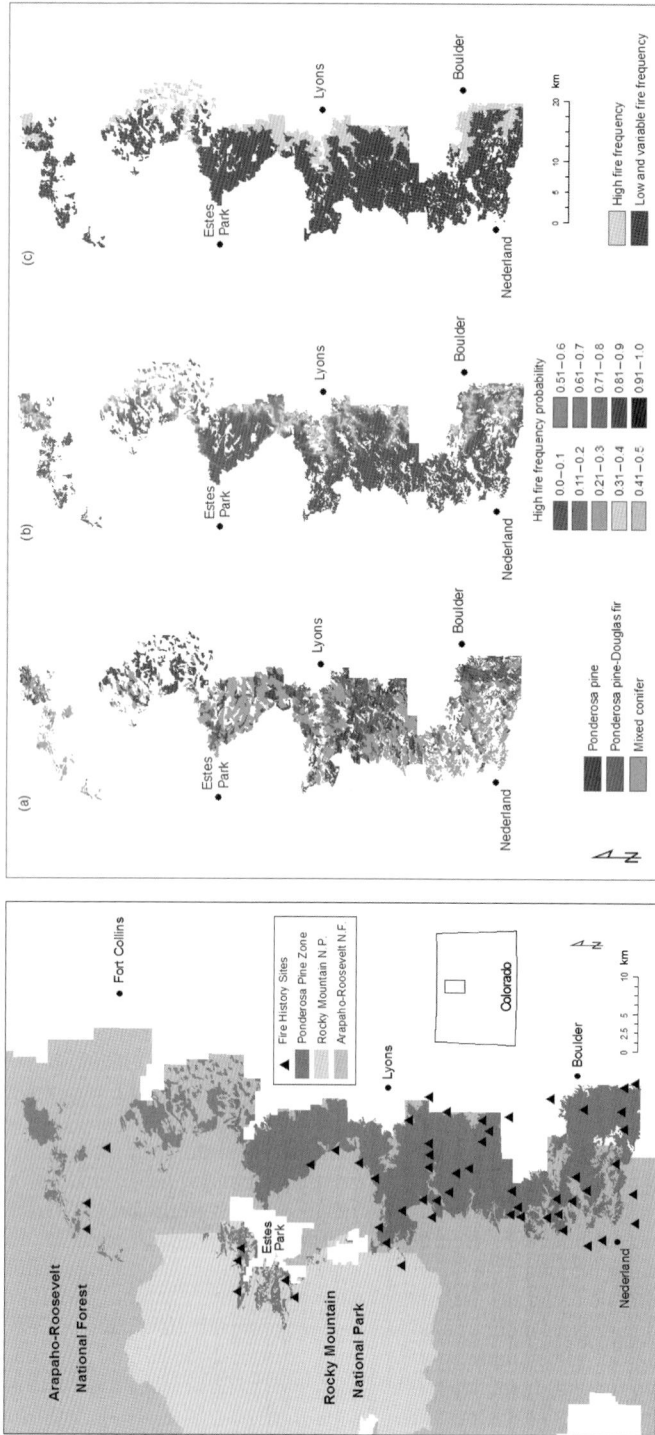

Fig. 10.2 Location map (left panel) showing sites sampled for fire history in the ponderosa pine zone of the northern Colorado Front Range from which historical fire regimes were reconstructed (dark gray Study Area) in Arapaho-Roosevelt National Forest. Map (right panel) showing areas of historical fire regimes of high fire frequency (light gray) versus areas of low and variable fire frequency (including high-severity fire; dark gray). (Source: Sherriff and Veblen 2007.) Additional maps are included in Fig. 10.2 in the color insert. Please refer to Colour Plate 4.

post-fire cohorts dominating the modern landscape (Sherriff & Veblen 2007). Low elevation was the primary predictor of location of low-severity fire regime.

How has twentieth-century fire exclusion altered historical fire regimes in the ponderosa pine zone of the front range?

Given that only the Lower Montane Zone within the ponderosa pine zone of the Front Range was characterized by relatively frequent historical low-severity fires, it follows that modern fire exclusion would result in the greatest departure from historical fire frequency for these low elevations. Although fire exclusion is reflected by declines in the number of tree-ring fire scars during the twentieth century throughout the Upper Montane Zone as well (Veblen et al. 2000), the ecological consequences of fire exclusion are measurably distinct in the Upper versus Lower Montane Zones (Sherriff & Veblen 2006). Studies using tree ages (Veblen & Lorenz 1986; Sherriff & Veblen 2006) and repeat historical photography (Veblen & Lorenz 1991; Mast et al. 1997) in the Lower Montane Zone (~1870–2200 m) indicate that formerly sparse ponderosa pine stands increased in density during the twentieth century and that nearly treeless grasslands have been invaded by trees (Fig. 10.3).

In contrast, at mid- to upper elevations (~2440–2750 m), only small percentages of living ponderosa pine established in the post-1920 fire exclusion period (Sherriff & Veblen 2007; Schoennagel et al. 2011). Tree and stand ages primarily reflect stand initiation following mid- to late-nineteenth-century fires rather than reduced fire frequency during the twentieth century. Repeat photography from the late 1800s to ~1980s (Veblen & Lorenz 1991) show that stands were relatively dense and supported high-severity burns long before any possible effects of fire exclusion (Fig. 10.4). Historical aerial photograph comparisons from ~1940 to 1999 (Platt & Schoennagel 2009) show only minor tree encroachment into open grasslands during the twentieth-century period of fire exclusion. Furthermore, tree ages do not support the common hypothesis that Douglas fir has invaded previously pure stands of ponderosa pine during the post-1920 fire exclusion period in the Mid- to Upper Montane Zone. Where young (i.e. post-1920) Douglas fir are abundant, older Douglas fir are also abundant.

The effects of climate variability on fire regimes of the front range ponderosa pine zone and the implications for future wildfires under climate change

Tree-ring fire-history studies in the northern Front Range ponderosa pine zone reveal that fires were more frequent and extensive during the second half of the nineteenth century than the previous ~100 years (Veblen et al. 2000; Sherriff & Veblen 2008). Although many of these fires were intentionally set during the Euro-American settlement period, high-severity fire events that created regeneration opportunities in the middle to upper elevation zone of ponderosa pine are strongly associated with severe droughts during the nineteenth century (Sherriff & Veblen 2008; Schoennagel et al. 2011). These episodes of widespread fire and drought are correlated with subcontinental scale patterns of synchronous wildfire activity across western North America that have been linked to variability in major climate drivers such as El Niño Southern Oscillation (ENSO) and mid- to high-latitude sea-surface temperatures (Sherriff & Veblen 2008).

Given the differences in fire frequency between the higher and lower elevations within the ponderosa pine zone, it is not surprising that the relationship of fire activity to interannual climatic variability also differs by elevation (Sherriff & Veblen 2008). Below ~2100 m, major fire years follow moister conditions by two years that favored fine-fuel accumulation before dry fire years. Thus, fire spread appears to have been at least partially limited by the lack of abundant fine fuels in most years. In contrast, in the Mid- and Upper Montane Zone (~2200–2800 m), the positive influence of moister years disappears and fire depends solely on drought during the fire year. Limitation of spread of historical fires by insufficient fuel quantities or continuity in the Lower Montane Zone implies that fuel reductions are likely to be an effective mitigation strategy in this environment. In contrast, in the Upper Montane Zone fuel desiccation was the primary limitation to historical fire occurrence. This suggests that reduction of fire potential would require more drastic fuels removal. These differences in sensitivity of fire regime to climate variability are consistent with the recognition that Lower and Upper Montane ponderosa pine forests require substantially different fire mitigation and ecological restoration frameworks.

Knowledge of how past climate variability affected wildfire activity and its ecological consequences in

Fig. 10.3 Views to the northwest of the western part of the city of Boulder and the Lower Montane Zone in 1905 (above) and 1988 (below) showing an increase in abundance of ponderosa pine on slopes that formerly were relatively treeless grasslands. The upper left of the scene shows a relatively dramatic shift from open woodlands to dense ponderosa pine forest while the upper right shows only modest encroachment of ponderosa pine into grasslands. (Source: Veblen and Lorenz 1991.)

Fig. 10.4 Views to the southeast of Sugarloaf Mountain in the Upper Montane Zone of Boulder County in 1905 (above) and 1984 (below) showing dense forest that burned in a high-severity fire that occurred in the late 19th century and by the 1980s had recovered to a dense stand dominated by ponderosa pine and Douglas-fir. Snow is present in the 1905 photograph. The 1905 photograph depicts relatively dense stands of ponderosa pine and Douglas-fir in the background at a time that pre-dates any fire suppression. (Source: Veblen and Lorenz 1991.)

ponderosa pine forests is critical for assessing the likely impacts of current and future climate change in these ecosystems. Increases in area burned by wildfires in the western United States since the 1980s are correlated with broadscale warming and lengthening of fire seasons (Westerling et al. 2006). Steep warming trends across Colorado in the last 50 years (Ray et al. 2008), in combination with continued annual and supra-annual variability in precipitation, will likely continue to be conducive to years of widespread wildfires in Front Range forests. The widespread burning in Upper Montane ponderosa pine forests (Sherriff & Veblen 2008) and in subalpine ecosystems (Sibold et al. 2006) well documented during warm-dry episodes of the nineteenth century is an analog for wildfire activity under continued warming and droughts in the early twenty-first century. While historical research documents the resiliency of Front Range forests to climate-induced increases in wildfire as well as budworm and bark beetle outbreaks (Veblen & Donnegan 2005), it also implies an urgent need for the development of adaptation policies to reduce the vulnerability of humans and infrastructure to these expected large-scale ecological disturbances. In the case of the ponderosa pine ecosystems, adaptation needs involve not only common-sense fuels treatments in close proximity to homes but also more politically difficult changes in land-use policies to restrict residential development in inherently fire-prone environments.

10.2 CHALLENGES OF INTEGRATING HRV ASSESSMENTS INTO DIALOGUES ABOUT RESOURCE-MANAGEMENT ISSUES

There is a strong congruence between the objectives of the Rocky Mountain HRV assessments (Veblen & Donnegan 2005; Romme et al. 2009) and the role of HRV in restoration management as stated by Forest Service leadership (USDA 2011). In practice, however, there are many challenges to appropriate consideration of HRV in particular management projects or programs. Unarticulated assumptions about the purpose and application of HRV information may impede effective communication. Scientists and managers are both responsible for clearly stating their underlying assumptions about the application of HRV to land management. Scientists need to emphasize that

HRV assessments inform management by describing processes of ecological change. Furthermore, they must clearly acknowledge that application of HRV does not require that management attempt to return landscapes to a single and unchanging past condition. Scientists also need to carefully inform themselves about the information needs of managers and, interactively with managers, define appropriate objectives for HRV assessments tailored to particular management units.

Unarticulated assumptions may lead either to misapplication of HRV information or to resistance to its incorporation into decision making. Frequently, managers assume that the sole purpose of an HRV assessment is to quantify past reference conditions to set management targets. This assumption is often reflected in an inappropriate emphasis on quantifying static stand conditions such as densities of trees by size class and landscape patterns such as patch size and proportions of areas in different seral stages. A singular focus on static past reference conditions may lead to underappreciating the importance of temporal and spatial variation in ecological conditions and processes. Stakeholders may assume that management designed to create conditions similar to a reference period should take priority over other potential management goals. Such an assumption can create resistance to any use of information from an HRV assessment because of a perceived incompatibility with certain ecosystem services and products. The ecological understanding provided by an HRV should be used in making resource-management decisions that assure compatibility and sustainability of multiple uses, including resource extraction and delivery of ecosystem services such as wildlife habitat, recreational opportunities, watershed protection, and species preservation.

Early and frequent communication between HRV scientists and resource managers is important in increasing the likelihood that the HRV results will be relevant to management concerns. This was particularly challenging in the case of some of the HRV assessments produced for the Rocky Mountain Region due to tight schedules and the limited capacity of both research and management personnel. Sufficient investment of time (both agency staff time as well as scientist time) is essential to sustaining dialogues among researchers, resource practitioners, and stakeholders to ensure that the HRV process meets the changing needs of adaptive management.

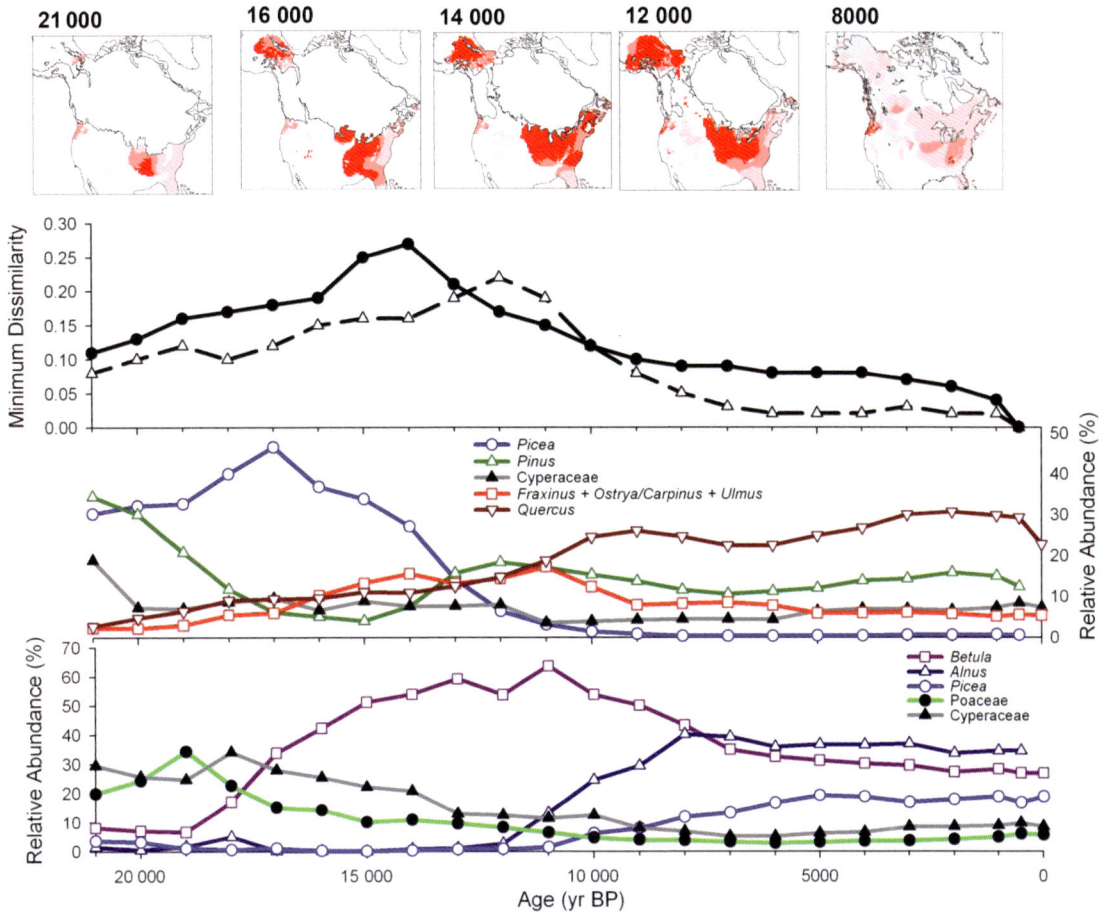

Plate 1 (Figure 7.4) Spatial, temporal, and taxonomic patterns of "no-analog" communities in North America during the last deglaciation. These communities were most common between 17 000 and 12 000 years ago (years BP). Red shading in maps is proportional to the multivariate dissimilarity between fossil pollen assemblages and their closest modern analogs. The second diagram shows that the peak no-analog period in eastern North America preceded the no-analog period in Alaska. The third and fourth diagrams show changing pollen percentages of dominant taxa in eastern North America and Alaska, respectively. Modified from Fig. 2 in Williams and Jackson (2007).

Historical Environmental Variation in Conservation and Natural Resource Management, First Edition. Edited by John A. Wiens, Gregory D. Hayward, Hugh D. Safford, and Catherine M. Giffen.
© 2012 John Wiley & Sons, Ltd. Published 2012 by John Wiley & Sons, Ltd.

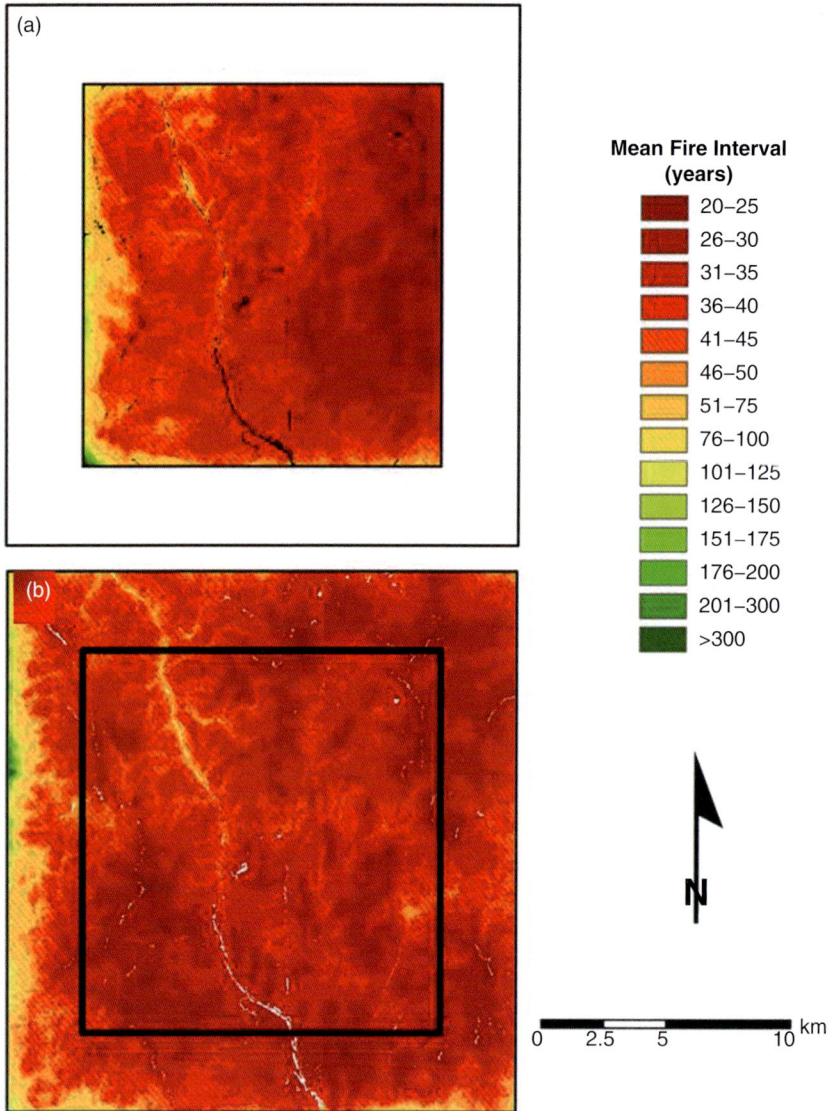

Mean Fire Interval (years)

■	20–25
■	26–30
■	31–35
■	36–40
■	41–45
■	46–50
■	51–75
■	76–100
■	101–125
■	126–150
■	151–175
■	176–200
■	201–300
■	>300

Plate 2 (Figure 8.4) An example of fire-regime simulation results showing the edge effect on fire frequency across the simulation landscape. (a) Fire frequency without a landscape buffer and (b) fire regime for the same landscape with a 3-km buffer.

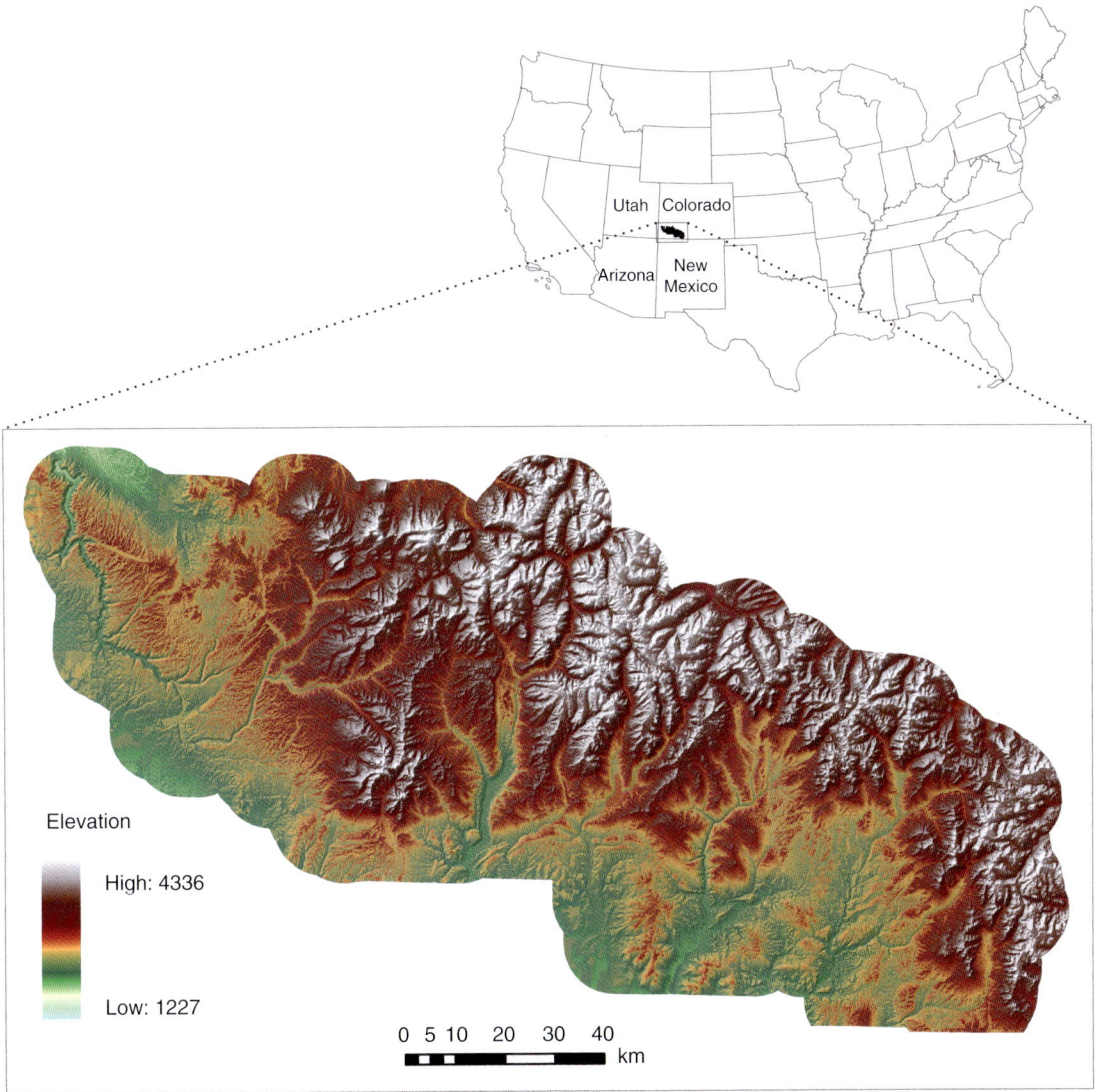

Plate 3 (Figure 9.1) San Juan National Forest (inclusive of a 10-km buffer) displayed as a topographic relief map in the Four Corners Region of the southwestern United States. The Forest encompasses 847 638 ha.

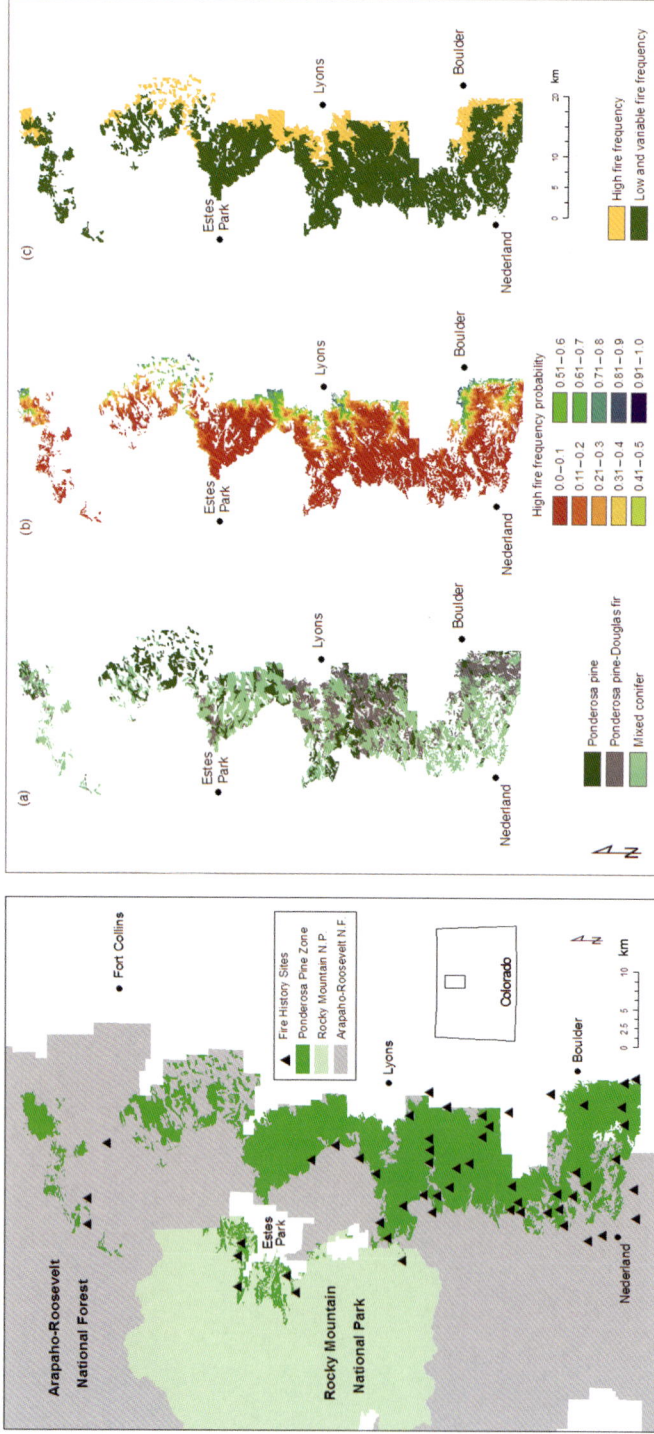

Plate 4 (Figure 10.2) Location map (left panel) showing sites sampled for fire history in the ponderosa pine zone of the northern Colorado Front Range from which historical fire regimes were reconstructed (dark gray Study Area) in Arapaho-Roosevelt National Forest. The panels on the right are maps showing (a) forest cover types of ponderosa pine forest, ponderosa pine-Douglas-fir forest, and mixed conifer forest of Douglas-fir and ponderosa pine with lodgepole pine; (b) the probability that the historical fire regime was one of high fire frequency (and low fire severity); and (c) areas of historical fire regimes of high fire frequency (yellow) versus areas of low and variable fire frequency (including high-severity fire; green). From Sherriff and Veblen (2007).

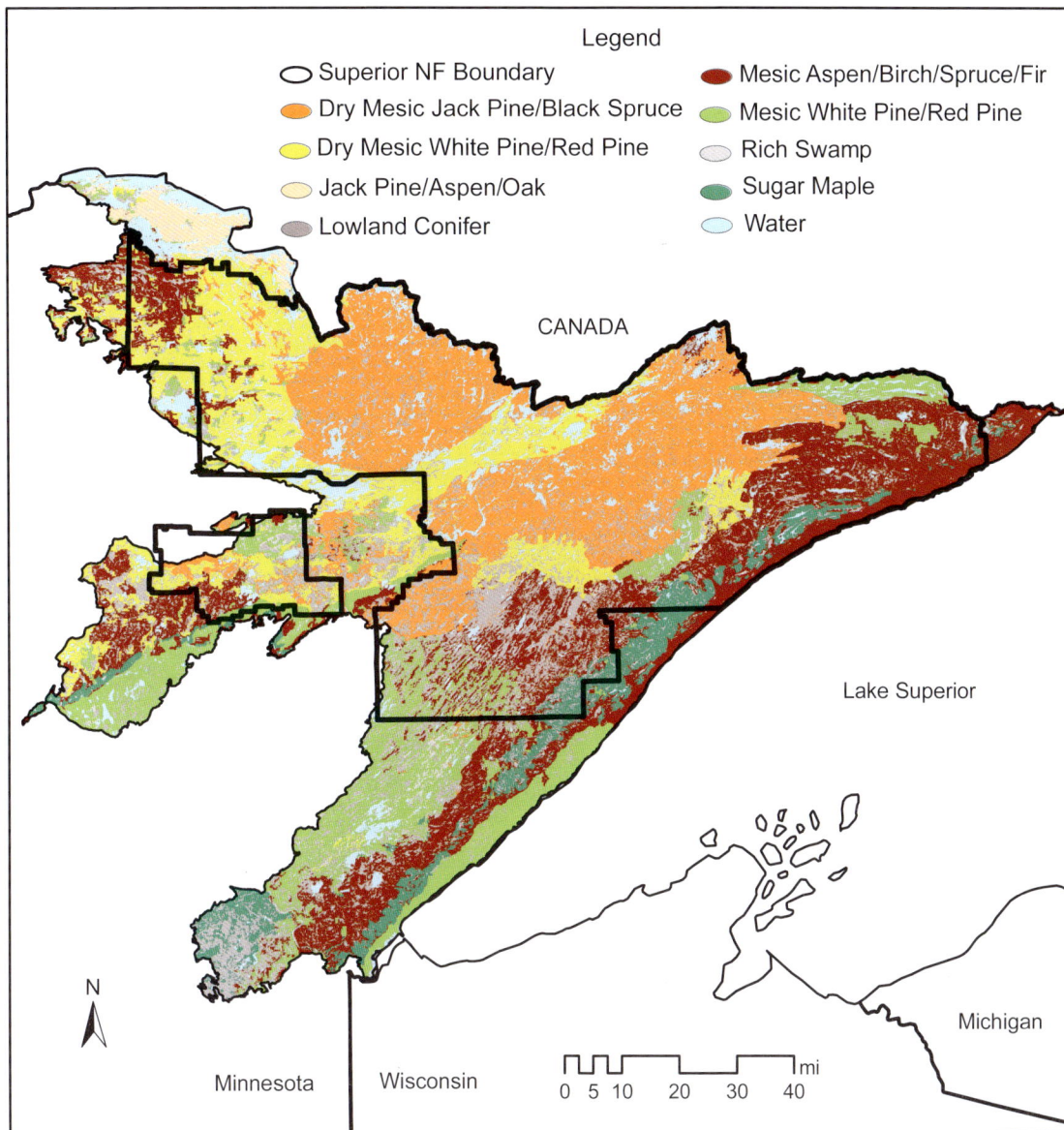

Plate 5 (Figure 12.2) Landscape Ecosystems on the Superior National Forest within the Superior Uplands Section used to define ecological units for Forest planning and implementation.

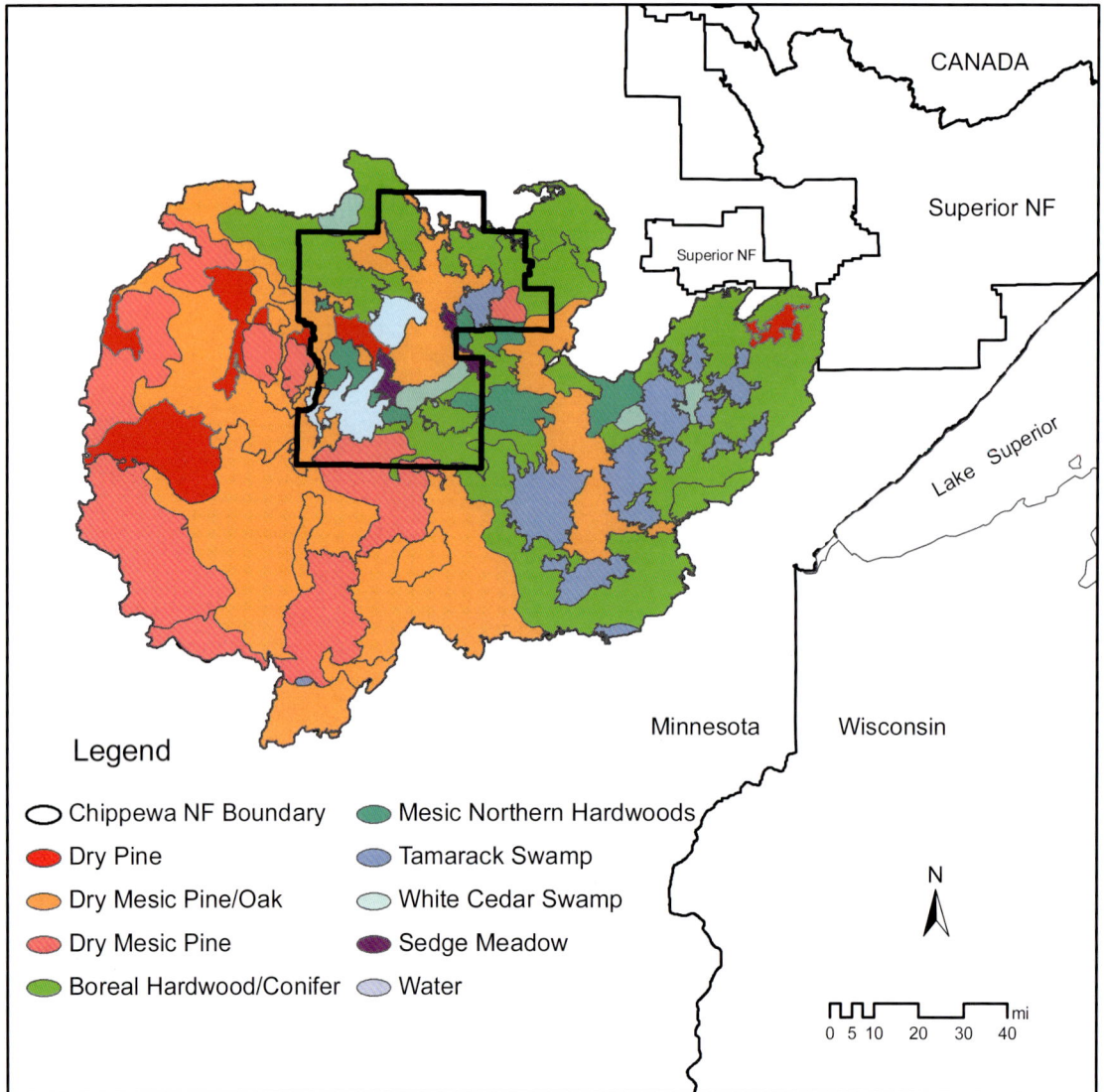

Plate 6 (Figure 12.3) Landscape Ecosystems on the Chippewa National Forest within the Minnesota Drift and Lake Plains Section used to define ecological units for Forest planning and implementation.

Plate 7 (Figure 17.2) Three ecosystems developed as case studies in a framework for applying the historical range of variation concept to ecosystem management. (a) Lodgepole pine forests in Yellowstone National Park, showing rapid forest regeneration in areas burned in the large 1988 fires; (b) mixed-conifer-aspen forests in the San Juan Mountains of southwestern Colorado; (c) piñon-juniper woodlands in Mesa Verde National Park, Colorado, showing an area on the ridgetop where a very old woodland burned in 2000, and an area on the adjacent slope that was missed by the 2000 fire and has not burned in many centuries. *Photos by W.H. Romme.*

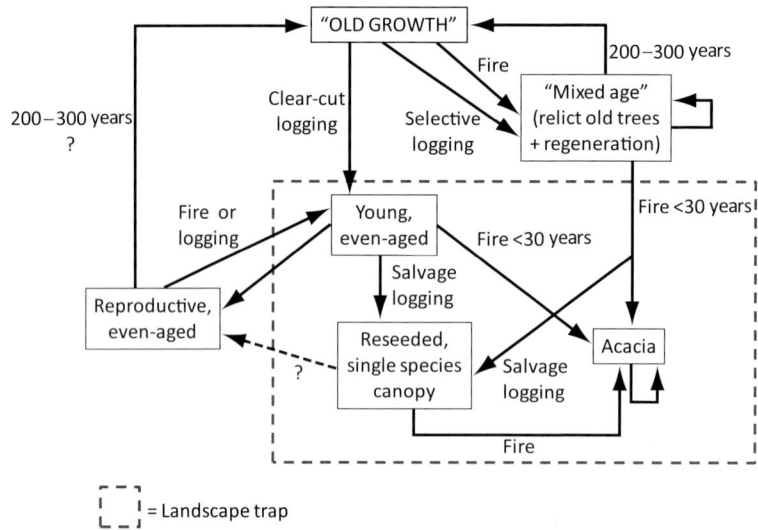

Plate 8 (Figure 19.1) Conceptual model of the interplay between forest age, wildfire, clear-cut logging, and salvage logging in montane ash forests. The hatched line encompasses a landscape fire trap in which there are negative feedbacks between natural and human disturbances that preclude the development of older, less fire-prone stands.

10.3   CONCLUSIONS

Our case study of fire mitigation and ecological restoration in the Colorado Front Range illustrates how historical ecology changed perspectives of managers and the public as they developed objectives for landscape-scale restoration. At a broad scale, understanding the historical role of fire across the elevation gradient of forest ecosystems demonstrated that recognizing differences among forest types was critical to developing sustainable management options. The clear differences in fire regimes exposed by historical analysis illuminated the choice to focus on lower elevations of the ponderosa pine zone as the most appropriate target for restoration. Understanding the histories of wildfire and insect outbreaks in subalpine forests demonstrated that few opportunities for mitigation and restoration occur in these forests because of the disturbance regimes inherent to these higher elevation ecosystem types. In the Subalpine Zone, both fires and bark beetle outbreaks are mainly controlled by climatic variability and kill most large trees across vast areas at infrequent intervals. The frequency and extent of these disturbances show little evidence of being modified by past land-use practices, including fire exclusion. Significant mitigation of the potential for fire and insect disturbances in the future would require unprecedented alteration of vegetation attributes over large areas; actions that would not be economically or socially acceptable. However, knowledge of HRV of subalpine forest ecosystems is vitally important for public education and outreach initiatives in support of adaptation strategies to minimize societal vulnerability to the inevitable large-scale, severe fires and insect epidemics in these habitats.

In contrast, knowledge of historical ecology has been instrumental in developing site-specific plans for restoration and fuel-mitigation projects in the ponderosa pine zone. The Rocky Mountain HRV assessments informed discussions of how and where to apply vegetation treatments for restoration and mitigation. Furthermore, the understanding of historical ecology helped managers and the public frame expectations for future forest conditions. The assessments were not used to establish a template describing a static set of conditions for future forests, but rather they guided planning and data gathering necessary for project-level restoration treatments. The HRV assessments were particularly useful in gaining social acceptance of

management initiatives. Success was realized when planners carefully distinguished between outcomes consistent with restoration versus outcomes that primarily would achieve fire-mitigation goals.

While HRV assessments are increasingly being embraced by decision makers in the Front Range, the challenge of applying historical ecology to plan for social adaptation to the ecological consequences of climate change remains. Understanding historical ecology and its strong control by climate variability will help bound societal expectations for what is and what is not achievable or sustainable forest management.

REFERENCES

Baker, W.L., Veblen, T.T., & Sherriff, R.L. (2007). Fire, fuels and restoration of ponderosa pine-Douglas-fir forests in the Rocky Mountains, USA. *Journal of Biogeography*, **34**, 251–269.

Bebi, P., Kulakowski, D., & Veblen, T.T. (2003). Interactions between fire and spruce beetle in a subalpine Rocky Mountain forest landscape. *Ecology*, **84**, 362–371.

Best, A. (2007). The spreading epidemic of pine bark beetles. Colorado Central Magazine (December). http://cozine.com/2007-december/the-spreading-epidemic-of-pine-bark-beetles/.

Bigler, C., Kulakowski, D., & Veblen, T.T. (2005). Multiple disturbance interactions and drought influence fire severity in Rocky Mountain subalpine forests. *Ecology*, **86**, 3018–3029.

Brown, P.M., Kaufmann, M.R., & Shepperd, W.D. (1999). Long-term landscape patterns of past fire events in a montane ponderosa pine forest of central Colorado. *Landscape Ecology*, **14**, 513–532.

Buechling, A. & Baker, W.L. (2004). A fire history from tree rings in a high-elevation forest of Rocky Mountain National Park. *Canadian Journal of Forest Research*, **34**, 1259–1273.

Christensen, N.L., Bartuska, A.M., Brown, J.H., et al. (1996). The report of the Ecological Society of America committee on the scientific basis for ecosystem management. *Ecological Applications*, **6**, 665–691.

Covington, W.W. (2000). Helping western forests heal. *Nature*, **408**, 135–136.

Covington, W.W. & Moore, M.M. (1994). Southwestern ponderosa forest structure: changes since Euro-American settlement. *Journal of Forestry*, **92**, 39–47.

Denver Post. (2002). Groups challenge Routt logging plan. Environmentalists: effort to fight beetles won't work. May 31, 2002:B-03.

Dillon, G.K., Knight, D.H., & Meyer, C.B. (2005). Historic range of variability for upland vegetation in the Medicine

Bow National Forest, Wyoming. General Technical Report RMRS-GTR-139. USDA Forest Service, Fort Collins, CO, USA.

Ehle, D.S. & Baker, W.L. (2003). Disturbance and stand dynamics in ponderosa pine forests in Rocky Mountain National Park, USA. *Ecological Monographs*, **73**, 543–566.

Eisenhart, K. & Veblen, T.T. (2000). Dendrochronological detection of spruce bark beetle outbreaks in northwestern Colorado. *Canadian Journal of Forest Research*, **30**, 1788–1798.

Friederici, P. (ed.). (2003). *Ecological Restoration of Southwestern Ponderosa Pine Forests*. Island Press, Washington, DC, USA.

Front Range Roundtable. (2006). Living with fire: protecting communities and restoring forests. www.frftp.org/roundtable/report.pdf.

Goldblum, D. & Veblen, T.T. (1992). Fire history of a ponderosa pine–Douglas-fir forest in the Colorado Front Range. *Physical Geography*, **13**, 133–148.

Hadley, K.S. & Veblen, T.T. (1993). Stand response to western spruce budworm and Douglas-fir bark beetle outbreaks, Colorado Front Range. *Canadian Journal of Forest Research*, **23**, 479–491.

Kaufmann, M.R., Regan, C.M., & Brown, P.M. (2000). Heterogeneity in ponderosa pine/Douglas-fir forests: age and size structure in unlogged and logged landscapes of central Colorado. *Canadian Journal of Forest Research*, **30**, 698–711.

Kaufmann, M.R., Veblen, T.T., & Romme, W.H. (2006). Historical fire regimes in ponderosa pine forests of the Colorado Front Range, and recommendations for ecological restoration and fuels management. Findings of the Ecology Workgroup, Front Range Fuels Treatment Partnership Roundtable. http://www.frftp.org/roundtable/pipo.pdf.

Kaufmann, M.R., Aplet, G.H., Babler, M., et al. (2008). The status of our scientific understanding of lodgepole pine and mountain pine beetles – a focus on forest ecology and potential fire behavior. GFI report 2008-2. The Nature Conservancy, Arlington, VA, USA.

Kipfmueller, K.F. & Baker, W.L. (2000). A fire history of a subalpine forest in south-eastern Wyoming, USA. *Journal of Biogeography*, **27**, 71–85.

Kulakowski, D. & Veblen, T.T. (2006). *Historical Range of Variability for Forest Vegetation of the Grand Mesa National Forest, Colorado*. p. 84. USDA Forest Service and Colorado Forest Restoration Institute, Fort Collins, CO, USA. http://warnercnr.colostate.edu/publications-and-reports/.

Mast, J.N., Veblen, T.T., & Hodgson, M.E. (1997). Tree invasion within a pine/grassland ecotone: an approach with historic aerial photography and GIS modeling. *Forest Ecology and Management*, **93**, 187–194.

Mast, J.N., Veblen, T.T., & Linhart, Y.B. (1998). Disturbance and climatic influences on age structure of ponderosa pine at the pine/grassland ecotone, Colorado Front Range. *Journal of Biogeography*, **25**, 743–755.

Meyer, C.B., Knight, D.H., & Dillon, G.K. (2005). Historic range of variability for upland vegetation in the Bighorn

National Forest, Wyoming. General Technical Report RMRS-GTR-140. USDA Forest Service, Fort Collins, CO, USA.

Platt, R.V. & Schoennagel, T. (2009). An object-oriented approach to assessing changes in tree cover in the Colorado Front Range, 1938–1999. *Forest Ecology and Management*, **258**, 1342–1349.

Ray, A.J., Barsugli, J.J., & Averyt, K.B. (2008). Climate change in Colorado. Western Water Assessment Conservation Board. http://wwa.colorado.edu/climate_change/ClimateChangeReportFull.pdf.

Romme, W.H. & Knight, D.H. (1981). Fire frequency and subalpine forest succession along a topographic gradient in Wyoming. *Ecology*, **62**, 319–326.

Romme, W.H., Floyd, M.L., & Hanna, D. (2009). *Historical Range of Variability and Current Landscape Condition Analysis: South Central Highlands Section, Southwestern Colorado & Northwestern New Mexico*. USDA Forest Service and Colorado Forest Restoration Institute, Fort Collins, CO, USA.

Schoennagel, T., Veblen, T.T., & Romme, W.H. (2004). The interaction of fire, fuels and climate across Rocky Mountain forests. *BioScience*, **54**, 661–676.

Schoennagel, T., Veblen, T.T., Kulakowski, D., & Holz, A. (2007). Multidecadal climate variability and interactions among Pacific and Atlantic sea surface temperature anomalies affect subalpine fire occurrence, western Colorado (USA). *Ecology*, **88**, 2891–2902.

Schoennagel, T., Sherriff, R.L., & Veblen, T.T. (2011). Fire history and tree recruitment in the upper montane zone of the Colorado Front Range: implications for forest restoration. *Ecological Applications*, **21**, 2210–2222.

Sherriff, R.L. & Veblen, T.T. (2006). Ecological effects of changes in fire regimes in *Pinus ponderosa* ecosystems in the Colorado Front Range. *Journal of Vegetation Science*, **17**, 705–718.

Sherriff, R.L. & Veblen, T.T. (2007). A spatially explicit reconstruction of historical fire occurrence in the ponderosa pine zone of the Colorado Front Range. *Ecosystems*, **9**, 1342–1347.

Sherriff, R.L. & Veblen, T.T. (2008). Variability in fire-climate relationships in ponderosa pine forests in the Colorado Front Range. *International Journal of Wildland Fire*, **17**, 50–59.

Sibold, J.S. & Veblen, T.T. (2006). Relationships of subalpine forest fires in the Colorado Front Range with interannual and multidecadal-scale climatic variation. *Journal of Biogeography*, **33**, 833–842.

Sibold, J.S., Veblen, T.T., Chipko, K., Lawson, L., Mathis, E., & Scott, J. (2007). Influences of surface fire, mountain pine beetle, and blowdown on lodgepole pine stand development in the northern Colorado Front Range. *Ecological Applications*, **17**, 1638–1655.

Sibold, J.S., Veblen, T.T., & Gonzales, M.E. (2006). Spatial and temporal variation in historic fire regimes in subalpine forests across the Colorado Front Range in Rocky Mountain

National Park, Colorado, USA. *Journal of Biogeography*, **32**, 631–647.

USDA Forest Service, Forest Health Protection and its Partners. (2010). USDA Forest Service, Rocky Mountain Region – Aerial Survey Data.. http://www.fs.fed.us/r2/resources/fhm/aerialsurvey/.

USDA. (2011). *US Forest Service Ecological Restoration*. USDA Forest Service, Washington DC, USA. http://www.fs.fed.us/restoration.

USDA and USDI. (2011). *A national Cohesive Wildland Fire Management Strategy*. USDA and USDI, Washington, DC, USA. http://www.forestsandrangelands.gov/index.shtml.

Veblen, T.T. (2003). Historic range of variability of mountain forest ecosystems: concepts and applications. *The Forestry Chronicle*, **79**, 223–226.

Veblen, T.T. & Donnegan, J.A. (2005). *Historical Range of Variability for Forest Vegetation of the National Forests of the Colorado Front Range*. USDA Forest Service and Colorado Forest Restoration Institute, Fort Collins, CO, USA. http://www.fs.fed.us/r2/projects/scp/tea/HRVFrontRange.pdf.

Veblen, T.T. & Lorenz, D.C. (1986). Anthropogenic disturbance and recovery patterns in montane forests, Colorado Front Range. *Physical Geography*, **7**, 1–24.

Veblen, T.T. & Lorenz, D.C. (1991). *The Colorado Front Range: A Century of Ecological Change*. University of Utah Press, Salt Lake City, UT, USA.

Veblen, T.T., Hadley, K.S., Reid, M.S., & Rebertus, A.J. (1991). The response of subalpine forests to spruce beetle outbreak in Colorado. *Ecology*, **72**, 213–231.

Veblen, T.T., Kitzberger, T., & Donnegan, J. (2000). Climatic and human influences on fire regimes in ponderosa pine forests in the Colorado Front Range. *Ecological Applications*, **10**, 1178–1195.

Westerling, A.L., Hidalgo, H.G., Cayan, D.R., & Swetnam, T.W. (2006). Warming and earlier spring increase western U.S. forest wildfire activity. *Science*, **313**, 940–943.

Chapter 11

INCORPORATING CONCEPTS OF HISTORICAL RANGE OF VARIATION IN ECOSYSTEM-BASED MANAGEMENT OF BRITISH COLUMBIA'S COASTAL TEMPERATE RAINFOREST

Andy MacKinnon[1] and Sari C. Saunders[2]

[1]BC Ministry of Natural Resource Operations, Coast Area Operations, Victoria, BC, Canada
[2]BC Ministry of Natural Resource Operations, Coast Area Operations, Nanaimo, BC, Canada

Historical Environmental Variation in Conservation and Natural Resource Management, First Edition. Edited by John A. Wiens, Gregory D. Hayward, Hugh D. Safford, and Catherine M. Giffen.

11.1 INTRODUCTION

Together with adjacent southeast Alaska, the Central and North Coasts (CNC) of British Columbia, Canada, sustain approximately one-quarter of the world's unlogged, coastal temperate rainforest (Schoonmaker et al. 1997). As a result of tension, particularly in the late 1990s, surrounding resource development in the CNC (also known as the "Great Bear Rainforest"), novel coalitions were created within and among environmental groups, the forest industry, and coastal First Nations (aboriginal groups) (Price et al. 2009). A new planning process was initiated for the area (Fig. 11.1) which resulted in the creation of new protected areas and the development of a new management system – Ecosystem-Based Management (EBM). The stated goals of EBM are to manage human activities to ensure the coexistence of healthy, fully functioning ecosystems and human communities (Coast Information Team 2004).

This chapter describes how historical range of variation (HRV) is utilized in EBM on British Columbia's coast. We discuss the challenges of using HRV to manage a remote landscape where the dominant disturbance regime consists of infrequent and small-scale gap dynamics nested within more infrequent, stand-replacing events, and when effects of projected climate change are uncertain.

11.2 ECOLOGICAL CHARACTERISTICS OF CNC OF BRITISH COLUMBIA

The CNC geographic area corresponds closely to the temperate Perhumid Rainforest (Alaback 1996), with considerable precipitation in all seasons; greater than 15% of annual total precipitation may occur from June through August (Redmond & Taylor 1997). The area is characterized by high annual precipitation (3–5 m), an ocean-moderated climate, and often steep, mountainous terrain. Forests are dominated by conifers, especially western hemlock (*Tsuga heterophylla*) and western red cedar (*Thuja plicata*). Sitka spruce (*Picea sitchensis*) occurs along floodplains and exposed coasts. At higher elevations, common species are mountain hemlock (*Tsuga mertensiana*), yellow cedar (*Chamaecyparis nootkatensis*), and amabilis fir (*Abies amabilis*). There are significant populations of large carnivores such as wolves (*Canis lupus*) and omnivores such as bears: grizzly bears (*Ursus arctos* ssp. *horribilis*), black

bears (*Ursus americanus*), and a genetic white variant of the black bear (*U. americanus* ssp. *kermodei*) sometimes referred to as the "spirit bear." Rivers and streams have populations of up to five species of anadromous salmon (*Oncorhynchus* spp.). For more, see Pojar (2003).

The natural disturbance regimes, long-lived tree species, and slow decay rates of standing and downed dead material produce a landscape dominated by old-growth coniferous forests and structurally complex stands (MacKinnon 2003) containing huge volumes of living and dead material (BC Ministry of Forests 2001). These productive stands are interspersed with forested and nonforested bogs and small lakes and ponds that cover one-half to three-fourths of lower elevations (Banner et al. 1986), becoming more common in the west and north of the region.

11.3 NATURAL DISTURBANCE IN BRITISH COLUMBIA'S CNC

Natural disturbance over much of coastal British Columbia is rare and small in extent (Daniels & Gray 2006) with the old growth seral stage characterized by disturbances resulting from death of one or a few trees at a time (fine-scale gap dynamics; e.g. Lertzman et al. 1996). Fires are historically infrequent, with return intervals of >1000 years (Lertzman et al. 2002). Certain site types experience fires only every several thousand years (Gavin et al. 2003) due to low lightning ignition rates and high fuel moisture. Pearson (2010) examined a combination of air photos and geographic information system (GIS) forest-cover databases for unlogged areas covering 1.5 million hectares of British Columbia's Central Coast and concluded that forests >140 years old made up approximately 97% of the land base. Stand-replacing disturbances had a return interval of 4440 years, with geomorphic processes affecting 1.4% and fire (restricted geographically) affecting 1.3% of forest area within the 140 years. Wind was relatively unimportant as a disturbance agent at this larger spatial extent, though a few areas of Haida Gwaii and northern Vancouver Island have been highlighted as being susceptible to stand-replacing wind events. This contrasts with the importance of large-scale wind disturbance identified in southeast Alaska (Kramer et al. 2001) and may, perhaps, be explained by the presence of an offshore "barrier" of islands along most of British Columbia's coast.

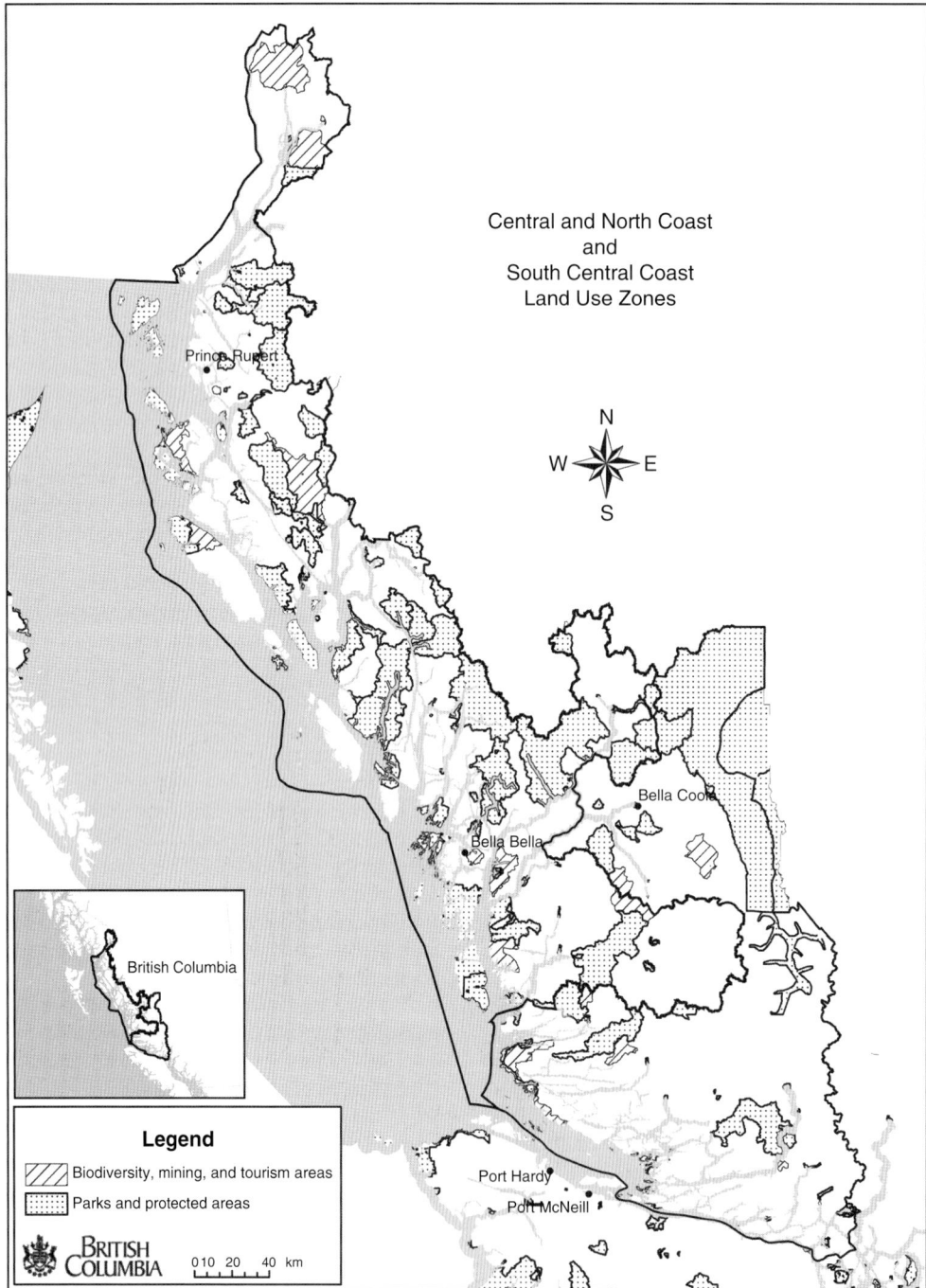

Fig. 11.1 Land-use zones on the Central and North Coast (CNC) and South Central Coast (SCC) of British Columbia, Canada. There are three land-use zones within the EBM planning area: Protected Areas; Biodiversity, Mining and Tourism Areas (BMTAs); and Ecosystem-based Management Operating Areas. BMTAs were designed to contribute to the conservation of species, ecosystems, and seral stage diversity by being located adjacent to existing Conservancies and other types of Protected Areas and by limiting the land uses within the zones. Commercial timber harvesting and commercial hydroelectric power projects are prohibited within BMTAs.

In the CNC forests, stand ages often exceed ages of the oldest trees (Daniels & Gray 2006), and >80% of the landscape may exist in old-seral condition. Price and Daust (2003) used forest inventories and other data to estimate natural disturbance intervals and resulting age-class distributions for combinations of fine-scale ecosystems within physiographic units across the CNC. They estimated that, for example, 95–98% of upland forest in the wettest, hypermaritime area, and 61–96% of the upland forest in the drier Inner Coast, was >250 years old. Because the CNC area is predominantly old-growth forest, representation (landscape-level) and retention (within-stand) targets under EBM are defined in terms of old forest.

11.4 REGIONAL PLANNING AND EBM

In British Columbia, regional or subregional Land and Resource Management Plans (LRMPs) provide goals for allocation and use of resources over broad geographic regions. The plans indicate spatial land-use zones, resource-management objectives for these zones, objectives for economic and social development, and strategies for implementation of the plan. The planning process, historically led by the provincial government, is complex due to the number of resource values and stakeholders that may be affected over the large areas of application.

A fundamentally different planning process took place in the CNC relative to other regions of the province. By 2001, First Nations on the coast and the provincial government agreed to interact as equal parties in the planning process via Government-to-Government discussions. The Governments came to consensus on an approach for planning in the CNC, including the definition, principles, and goals of EBM. EBM was defined as, "…an adaptive approach to managing human activities that seeks to ensure the coexistence of healthy, fully functioning ecosystems and human communities…" (Coast Information Team 2004). An international group of scientists, the Coast Information Team (CIT) (hereafter the "team"), assembled to support planning efforts and ensure the best available science-informed EBM. The team conducted a number of spatial analyses and produced a series of reports, including an EBM Handbook (Coast Information Team 2004). Among other items, the team recommended employing HRV of ecosystem elements and ecological processes as a basis for maintaining ecological integ-

rity. The Government-to-Government commitment to full implementation of EBM by March 31, 2009 upheld the understanding that HRV be a key benchmark in ecological management. The ability of these management strategies to achieve their intended objectives with respect to biodiversity conservation was to be evaluated through processes of adaptive management.

11.5 HRV CONCEPTS AND BRITISH COLUMBIA'S BIODIVERSITY GUIDEBOOK

Incorporating natural disturbance into forest management is not a new concept in British Columbia. The province (95 million hectares) is divided into administrative Landscape Units, each encompassing one or more watersheds and ranging in size from approximately 50000–75000 ha. Within each Landscape Unit, a certain percentage of each biogeoclimatic variant – an area of homogeneous macroclimate and dominant, overstory vegetation (Pojar et al. 1987) – is set aside as old-growth reserves. The old-growth reserves are made up of Old Growth Management Areas, large areas established at the Landscape Unit scale, and smaller Wildlife Tree Reserves, established as each cutblock is logged. The area to be set aside as Old Growth Management Areas and Wildlife Tree Reserves is determined, in part, by HRV; managers determine a natural distribution of seral stages and retain at least half the natural proportion of mature and old forest – the latter defined as >250 years for productive, low-elevation coastal ecosystems.

The province has been divided into five Natural Disturbance Types based on frequency and type of disturbance (Biodiversity Guidebook; BC Ministry of Forests & BC Ministry of Environment 1995):
• Natural Disturbance Type 1 – ecosystems with rare stand-initiating events;
• Natural Disturbance Type 2 – ecosystems with infrequent stand-initiating events;
• Natural Disturbance Type 3 – ecosystems with frequent stand-initiating events;
• Natural Disturbance Type 4 – ecosystems with frequent stand-maintaining fires; and
• Natural Disturbance Type 5 – alpine tundra and subalpine parkland (no disturbance regimes specified).

For each Natural Disturbance Type, a certain percentage of old forest must be left unlogged, with a higher amount to be retained in landscapes naturally dominated by old forests. Though the percentage to be left was based on concepts of natural disturbance, it was considerably below HRV of amounts of old forest in these landscapes, to reduce impact on timber supply. Most of coastal British Columbia is classed as Natural Disturbance Type 1 with a mean return interval for stand-initiating disturbances of 250 years; guidelines indicate a minimum of 13–19% of the forest in a Landscape Unit should be maintained as "old" forest, based on the expected seral-stage distribution under natural disturbance and the relative "biodiversity emphasis" of the Landscape Unit. This percentage is of the biogeoclimatic variant, not of each ecosystem type ("site series"; Pojar et al. 1987) within that variant. The Guidebook did, however, recommend that rare ecosystems be maintained unlogged at higher proportions than they occurred in a Landscape Unit and that other ecosystems be maintained in proportion to the amounts they occurred in a Landscape Unit.

11.6 INCORPORATING NATURAL DISTURBANCE INTO EBM

Data on the frequency, spatial extent, and intensity of disturbance (and thus historical range of amounts of forest in different seral stages) were limited for development of the Biodiversity Guidebook. Difficulties of quantification and geographic delineation of natural disturbance patterns likely reduced the utility of the concept for management applications (e.g. Hilbert & Wiensczyk 2007) and emphasized the need for regionally specific information when defining characteristic patterns of a landscape (e.g. DeLong 2007). The intent in developing EBM for British Columbia's CNC was to use data specific to natural disturbances of the CNC for estimating HRV of amounts of old forest, and to develop a foundation for evaluation of ecological risk with habitat loss.

The use of natural disturbance data and concepts of HRV were incorporated into EBM of forested systems for the CNC through consideration of (1) how much old forest to retain in managed EBM areas (representation targets); (2) what types of ecological units to use when applying representation targets; (3) how to incorporate concepts of risk into representation levels; and (4) what spatial scales of protection and representation are important for maintaining ecological integrity. Consideration of HRV also influenced the development of legal objectives associated with sus-

taining aquatic health and the ecological integrity of the upland–riparian interface ("hydroriparian guidelines"). Here we focus on the role of HRV in objectives for maintenance of terrestrial landscape and stand-level biodiversity. Forests and forestry were the focus for planning and implementation of EBM in the CNC because forestry is the primary agent of anthropogenic change in this region.

The amount of old forest naturally occurring in different ecosystems (the HRV for old forest) suggests the conditions to which species in those systems should be adapted and under which ecological processes might be expected to be fully functioning. The CIT considered the degree to which amounts of old forest could deviate from this HRV benchmark without significant risk to ecological integrity in managed EBM areas (land outside the Land-Use zones depicted in Fig. 11.1). Literature analysis by Dykstra (2004) suggested that thresholds in population persistence – primarily for vertebrates, and also for some invertebrates and plant species – occurred at approximately 70% and 30% of *original* habitat, that is, the area of forest >250 years old under a natural disturbance regime. When >70% of original habitat remained, there appeared to be little effect on population persistence; as amount of original habitat remaining dropped below 30%, there was an increased rate of extirpation of local populations. The CIT scientific panel recommended retaining more than 70% of original habitat (i.e. 70% HRV) as a "low-risk" strategy and defined "less than 30% of original habitat" (i.e. 30% HRV) as a "high-risk" strategy for conservation. Thresholds were to be applied to quantitative estimates of *mean* HRV (hereafter just HRV) of old forest first developed by Price and Daust (2003) for these coastal landscapes. These thresholds have also been used in evaluating the risk associated with various EBM implementation options by Holt and Sutherland (2004).

Further research (Price et al. 2007) concluded that "low risk" could be defined as retaining more than 60% of *total* habitat, and "high risk" could be defined as retaining 30% or less of *total* habitat – not in terms of *original* habitat, as defined above. These definitions of "high risk" and "low risk" have not yet been incorporated into operational implementation of EBM. Recent work focused on habitat needs for individual focal species (mountain goats [*Oreamnos americanus*], grizzly bears [*U. arctos horribilis*], northern goshawks [*Accipiter gentilis*], tailed frogs [*Ascaphus truei*], and marbled murrelets [*Brachyramphus marmoratus*]). Still, scientists and managers continue to grapple with the lack of data specific to the CNC on species' responses to habitat loss and the sometimes conflicting empirical and modeled relationships found for other regions (e.g. Lindenmayer & Luck 2005).

The CIT also recognized the importance of retaining the type of ecosystems characteristic to the CNC; consideration of HRV was applied to the choice of conservation units. The representation targets were to be applied to ecosystems called "site series." These ecological units are described for British Columbia in terms of their soil conditions (nutrient and moisture) and vegetation communities within larger climatic units (e.g. Banner et al. 1993). This represents a fundamental difference from other parts of British Columbia, where reserved old-growth forests are calculated as a certain percentage of a variant – a climatically homogenous unit (see above). Elsewhere in British Columbia, it might be possible – under current legislation and policy – to protect one forest type (e.g. western red cedar – amabilis fir – blueberry) and cut another (e.g. amabilis fir – western red cedar – foamflower) as long as they are both in the same broader climatic unit. Under EBM, each individual ecosystem type must be protected, recognizing these are important conservation elements in their own right. Ecosystem mapping is, however, currently available for <20% of the EBM area, making representation by these units difficult to achieve. Targets have been applied using "site series surrogates" developed from site productivity and forest inventory data for the region. These "surrogates" crosswalk poorly for many of the actual site series, but still provide a better level of representation than simply using the variants.

The Coast Information Team (2004) recommended implementing representation targets across a range of spatial scales for long-term maintenance of ecological integrity. The panel suggested that the initial (70% of old forest HRV), low-risk target be applied over very large areas (millions of hectares); for example, over the entire CNC Plan area, or a First Nations traditional territory. Within these large areas, however, individual watersheds or groups of watersheds might be managed to a higher risk level – but never to the "high-risk" definition of less than 30% of HRV remaining in an ecosystem. Representation was to be measured over several scales, allowing more flexibility at finer (watershed) scales as long as the subregional totals were at least 70% HRV. Because of the predominance of fine-scale disturbance dynamics and rare, stand-replacing

events on this landscape, this broad extent was considered ecologically appropriate for establishment of risk thresholds in habitat representation (Price et al. 2009). Precedence for this type of multiscale recommendation in planning was set in 1995 by the Clayoquot Sound Scientific Panel for Clayoquot Sound on the west coast of Vancouver Island (CSSP 1995).

In the CNC, implementation of flexible retention targets at spatial extents smaller than the subregion has been operationally difficult to date, and the recommendation that 70% of HRV be maintained in each ecosystem type (or surrogate ecosystem) has yet to be implemented. The current legislated orders indicate that representation across individual Landscape Units should sum to 50% of HRV for old-seral stages of individual ecosystems across the combined planning boundaries of the CNC and the EBM planning unit adjacent to the south (South Central Coast [SCC]; Fig. 11.1).

11.7 CHALLENGES IN APPLYING CONCEPTS OF HRV TO EBM IN THE CNC

Applying a management model informed by historic range of variation in ecosystem representation and disturbance regime to a landscape such as British Columbia's CNC presents a number of challenges. Simply quantifying the amounts and distributions of ecosystems characteristic of the region is difficult. Ecosystem mapping is costly in terms of time and resources; the work requires air photo expertise and field checking (Ecosystems Working Group 1998) across large tracts of inhospitable terrain. Implementation of multiscale targets is hampered by the geographic bias of existing mapping – most ecosystem mapping is in areas of interest to forest companies, generally the more productive forest types. Thus, it is difficult to determine, in total or spatially, areas of each ecosystem to which to apply HRV targets. Although the use of surrogate ecological units to represent site series (the desired implementation unit) may adequately represent the range of productivity of the forested land base to reduce ecological risk, use of these surrogates may result in biased representation of some ecosystems, in terms of total area and spatial placement.

Similarly, quantifying natural disturbance accurately across this region is a complex task. When the disturbance regime is dominated by infrequent, small-scale events and large-scale events that occur only every few thousand years, determining the nature, frequency, intensity, and extent of these events requires analysis of detailed, stand-level data over millions of hectares. Confidence in the details of the reference conditions that have been characterized is limited and no similar, alternate landscape exists for data substitution (Hessburg et al. 1999).

Even overcoming these conceptual hurdles, and assuming relevant and accurate information is in hand, application of HRV targets for old forest retention is difficult in managed areas with such infrequent natural disturbance. For example, in ecosystems where 96–99% of unlogged areas are old-growth forest, any logging will quickly move the seral structure outside of that window and below the 70% of HRV currently hypothesized to represent a high-risk threshold. Accordingly, application of HRV concepts to old forest representation in the CNC is premised not on managing within HRV, but instead on managing within defined classes of ecological risk that often fall outside of HRV.

Implementation of EBM at spatial and temporal scales that are ecologically meaningful is also operationally difficult. The characteristic, low-intensity disturbance regime requires attention to stand-level features and retention of structure within harvested areas. Current EBM implementation calls for additional (beyond ecosystem-level representation), stand-level retention of at least 15% of area of each cutblock; however, this target was not derived from HRV considerations and is not discussed further here. The nesting of a gap-phase dynamic within the regime of larger, geomorphic events requires thoughtful application of the multiscale representation targets that reflects an understanding of the variability in hydrogeomorphic function across this landscape. For example, Price and Daust (2003) illustrated the ecologically significant impact of spatial extent and analysis unit on quantified return intervals and the interpretation of drivers of disturbance regime. Their estimates of HRV for amounts of old forest varied for timber analysis (administrative) units versus bioclimatic units (variants) versus combinations of ecosystems (site series) within hydroriparian subregions (broader climatic-physiographic units), the latter providing the most homogeneous units with respect to disturbance frequency and variation over the analytical time frame. Managing in landscapes that experience "competing" disturbance types over different scales requires that

both social and ecological trade-offs be evaluated (Seymour et al. 2002). Potential for shifts in disturbance regime with projected changes in climate increases the difficulty of deciding on appropriate temporal and spatial extents for managing this landscape.

Minimizing conflict between ecological and social (or community-based) objectives of EBM is challenging. While a natural-disturbance model may be useful in sustaining healthy ecosystems, it has limited direct utility in ensuring healthy human communities. Healthy communities depend, to a certain extent, on healthy ecosystems. Protecting forests under EBM reduces timber supply and may reduce employment related to forestry in CNC. On the other hand, these protected forests may allow development of new economic models in the CNC area.

11.8 IMPLICATIONS OF CLIMATE CHANGE FOR USE OF HRV WITHIN EBM

Projected climate change is a source of uncertainty for application of HRV in the CNC landscape. Shifts in climatic envelopes (Hamann & Wang 2006) and differential adaptive (Miyamoto et al. 2010) and dispersal capabilities of plant populations will result in altered distributions of individual species and new ecological communities (e.g. Hansen et al. 2001). Some emissions scenarios projected for the Central Coastal of British Columbia suggest significant loss of climatic conditions for upper elevation (Mountain Hemlock) forests within 15 years and continued loss of these systems for at least the 75 years evaluated (Hamann & Wang 2006). Variability in ecosystem types and their distributions may thus be highly dynamic over a single or a few forest-management rotation period(s). Flexibility may be required in implementation of representation targets for ecosystems as climatic conditions shift across the landscape.

Changes in temperature and precipitation are likely to affect disturbance regimes, including the type, timing, frequency, extent, and intensity of occurrences (Dale et al. 2001). Shifts in mean annual temperature and precipitation are anticipated to alter hydrologic regimes across British Columbia (Pike et al. 2008a). In addition, altered timing and number of extreme climate events would alter hydrological resources and disturbance regimes. For example, projections for increased storm frequency and intensity would increase the impacts of windthrow, tree breakage, and flooding (Pike et al. 2008b). These storms may also increase rates of landslides, currently a primary stand-replacing disturbance in the CNC. Prolonged rainfall associated with high-intensity precipitation events triggers the majority of landslides in coastal British Columbia. Modeling of current and projected conditions (for the south coast) support predictions of increased landslide frequency (Jakob & Lambert 2009). Thus, with climate change in the CNC landscape, we would expect shifts in the amounts of forest among seral stages. Current representation targets for "old" forest may require revision as predictions for changes in disturbance regime and bioclimatic modeling are refined and validated.

With climate change, the spatial context – including appropriate resolution and extent – for development of representation targets will require reevaluation (Keane et al. 2009). Although this highlights the additional challenge of reconciling temporal scales when using HRV to manage within political and operational time frames, Keane et al. (2009) note that existing vegetation communities do reflect variation in past climate, and projections of historical conditions may be more reliable than management based on less certain future conditions. The ability to accept uncertainty in resource management outcomes, to define and use risk thresholds, and to incorporate active adaptive management become that much more critical.

New protected areas established under EBM and current efforts at identifying landscape-level old forest representation may allow ecosystem adaption under climate change and reduce the risk of failing to achieve coarse-filter conservation objectives. Pojar (2010) suggests that:
• Intact, functional, natural ecosystems and diverse, complex systems are likely to be more resilient.
• Management needs to focus attention on species that are ecologically critical, such as keystone species, ecosystem engineers, top carnivores, and abundant herbivores.
• Management ought to reorient conservation efforts "towards: (1) maintaining well-functioning, resilient ecosystems of sometimes novel composition that continue to deliver ecosystem services; and (2) maximising the diversity of native species and ecosystems" (Pojar 2010; see also Campbell et al. 2009).

11.9 LESSONS FOR THE FUTURE FROM EBM IN THE CNC

The stated goals of EBM are to manage human activities to ensure the coexistence of healthy, fully functioning ecosystems and human communities. EBM incorporates elements of HRV in its management regime. The HRV-based decisions to apply coarse-filter representation of ecosystems and to leave a percentage of each ecosystem as old forest are relatively objective, science-informed approaches for conserving biodiversity that can be agreed upon by all parties in the planning process.

Adaptive management was adopted as a central tenet to the implementation of EBM and remains a challenge in itself (Tyler et al. 2009). The period of full implementation of EBM only began in 2009, and much remains to be accomplished. The hope is that ongoing incorporation of new data, analytical techniques, and emerging theory into legislated EBM will ensure that management of the CNC continues to be science-informed. Efforts to establish multiscale research in the CNC – through both retrospective studies and new, experimental activities and monitoring (e.g. Fenger et al. 2009) – should improve our understanding of the utility of HRV and its application to representation planning and retention targets over time.

Implementing EBM in British Columbia's CNC includes the use of HRV concepts in several key and novel areas, including:
• setting coarse-filter, ecosystem representation objectives for forested ecosystems; that is, considering "how much total area is enough?"
• setting representation goals that may fall outside HRV in the context of hypothesized thresholds of risk to ecological integrity; that is, considering "what is the nature of the relationship between representation and ecological integrity or persistence?"
• choosing ecological units to which representation objectives should apply
• developing representation objectives across a range of spatial scales
• setting representation targets to maintain functioning hydroriparian ecosystems
• evaluating the role of stand-level retention in meeting both stand- and landscape-level biodiversity objectives assessing the quality of water and health of stream systems.

Incorporating HRV concepts into resource management has been challenging within the CNC landscape. Inventory is limited, the terrain is inaccessible, and the disturbance regime is difficult to study at this extent. The primary disturbances affecting stand dynamics are infrequent and small in scale. These are nested within stand-replacing disturbances that occur only every few 1000 years. Quantifying the relationships among current disturbance regimes, climate, and stand and landscape-level structure is complex. The use of HRV concepts to achieve ecological goals for the CNC is further complicated by the uncertainty surrounding climate projections and the implications these have for changes in disturbance patterns and suitable conditions for existing vegetation species. EBM implementation is in its early stages on British Columbia's CNC area. A program of adaptive management and monitoring will be critical to evaluating whether goals established through this large experiment are ultimately met.

ACKNOWLEDGMENTS

We would like to thank the editor and three anonymous reviewers for their helpful comments, and Todd Davis for producing Fig. 11.1.

REFERENCES

Alaback, P.B. (1996). Biodiversity patterns in relation to climate: the coastal temperate rainforests of North America. In *High Latitude Rainforests and Associated Ecosystems of the West Coast of the Americas: Climate, Hydrology, Ecology and Conservation* (ed. R.G. Lawford, P.B. Alaback, and E. Fuentes), pp. 105–133. Springer-Verlag, New York, USA.

Banner, A., Pojar, J., & Trowbridge, R. (1986). Representative wetland types of the northern part of the Pacific Oceanic Wetland Region. Research Report 85008-PR. BC Ministry of Forests, Victoria, BC, Canada.

Banner, A., MacKenzie, W., Haeussler, S., Thomson, S., Pojar, J., & Trowbridge, R.L. (1993). *A Field Guide to Site Identification and Interpretation for the Prince Rupert Forest Region.* Land Management Handbook Number 26, Parts 1 and 2. BC Ministry of Forests, Victoria, BC, Canada. http://www.for.gov.bc.ca/hfd/pubs/Docs/Lmh/Lmh26.htm.

BC Ministry of Forests. (2001). *Mensuration Data from the Provincial Ecology Program.* Working Paper 62/2001. BC Ministry of Forests, Victoria, BC, Canada. http://www.for.gov.bc.ca/hfd/pubs/Docs/Wp/Wp62.htm.

BC Ministry of Forests & BC Ministry of Environment. (1995). Biodiversity guidebook: forest practices code of British Columbia. Victoria, BC, Canada. http://www.for.gov.bc.ca/tasb/legsregs/fpc/fpcguide/biodiv/biotoc.htm.

Campbell, E.M., Saunders, S., Coates, K.D., et al. (2009). Ecological resilience and complexity: a theoretical framework for understanding and managing British Columbia's forest ecosystems in a changing climate. Technical Report 55. BC Ministry of Forests, Victoria, BC, Canada.

Clayoquot Sound Scientific Panel (CSSP). (1995). Sustainable ecosystem management in Clayoquot Sound: planning and practices. http://archive.ilmb.gov.bc.ca/slrp/lrmp/nanaimo/clayoquot_sound/archive/reports/panel.html.

Coast Information Team. (2004). Ecosystem-based management planning handbook. Coast Information Team, Victoria, BC, Canada. http://www.citbc.org/c-ebm-hdbk-fin-22mar04.pdf.

Dale, V.H., Joyce, L.A., McNulty, S., et al. (2001). Climate change and forest disturbances. *BioScience*, **51**, 723–734.

Daniels, L.D. & Gray, R.W. (2006). Disturbance regimes in coastal British Columbia. *BC Journal of Ecosystems and Management*, **7**, 44–56.

DeLong, S.C. (2007). Implementation of natural disturbance-based management in northern British Columbia. *The Forestry Chronicle*, **83**, 338–346.

Dykstra, P. (2004). Thresholds in habitat supply: a review of the literature. Scientific Background Report to Coast Information Team. Victoria, BC, Canada. http://ilmbwww.gov.bc.ca/citbc/b-Thresh-Dykstra-Sep04.pdf.

Ecosystems Working Group. (1998). *Standard for Terrestrial Ecosystem Mapping in British Columbia.* Resource Inventory Committee, Victoria, BC, Canada. http://archive.ilmb.gov.bc.ca/risc/pubs/teecolo/tem/indextem.htm.

Fenger, M., Howard, S., Loo, S., & Holt, R.F. (2009). Adaptive management in experimental watersheds – research design. Final report to the EBM Working Group. http://archive.ilmb.gov.bc.ca/slrp/lrmp/nanaimo/cencoast/ebmwg_docs/am04a_final_report.pdf.

Gavin, D.G., Brubaker, L.B., & Lertzman, K.P. (2003). Holocene fire history of a coastal temperate rain forest based on soil charcoal radiocarbon dates. *Ecology*, **84**, 186–201.

Hamann, A. & Wang, T. (2006). Potential effects of climate change on ecosystem and tree species distribution in British Columbia. *Ecology*, **87**, 2773–2786.

Hansen, A.J., Neilson, R.P., Dale, V.H., et al. (2001). Global change in forests: responses of species, communities, and biomes. *BioScience*, **51**, 765–779.

Hessburg, P.F., Smith, B.G., & Salter, R.B. (1999). Detecting change in forest spatial patterns from reference conditions. *Ecological Applications*, **9**, 1232–1252.

Hilbert, J. & Wiensczyk, A. (2007). Old-growth definitions and management: a literature review. *BC Journal of Ecosystems and Management*, **8**:15–31. http://www.forrex.org/publications/jem/ISS39/vol8_no1_art2.pdf.

Holt, R.F. & Sutherland, G. (2004). Central coast coarse filter ecosystem trends risk assessment – base case. Report to Coast Information Team. Victoria, BC, Canada. http://www.citbc.org/c-cencst-cf-base-31Mar04.pdf.

Jakob, M. & Lambert, S. (2009). Climate change effects on landslides along the southwest coast of British Columbia. *Geomorphology*, **107**, 275–284.

Keane, R.E., Hessburg, P.F., Landres, P.B., & Swanson, F.J. (2009). The use of historical range and variability (HRV) in landscape management. *Forest Ecology and Management*, **258**, 1025–1037.

Kramer, M.G., Hansen, A.J., Taper, M.L., & Kissinger, E.J. (2001). Abiotic controls on long-term windthrow disturbance and temperate rain forest dynamics in southeast Alaska. *Ecology*, **82**, 2749–2768.

Lertzman, K.P., Sutherland, G.D., Inselberg, A., & Saunders, S.C. (1996). Canopy gaps and the landscape mosaic in a temperate rain forest. *Ecology*, **77**, 1254–1270.

Lertzman, K.P., Gavin, D.G., Hallet, D.J., Brubaker, L.B., Lepofsky, D., & Mathewes, R. (2002). Long-term fire regime estimated from soil charcoal in coastal temperate rainforests. *Ecology and Society*, **6**, 5. http://www.ecologyandsociety.org/vol6/iss2/art5.

Lindenmayer, D.B. & Luck, G. (2005). Synthesis: thresholds in conservation and management. *Biological Conservation*, **124**, 351–354.

MacKinnon, A. (2003). West coast, temperate, old-growth forests. *The Forestry Chronicle*, **79**, 475–484.

Miyamoto, Y., Griesbauer, H.P., & Green, D.S. (2010). Growth responses of three coexisting conifer species to climate across wide geographic and climate ranges in Yukon and British Columbia. *Forest Ecology and Management*, **259**, 514–523.

Pearson, A.F. (2010). Natural and logging disturbances in the temperate rain forests of the Central Coast, British Columbia. *Canadian Journal of Forest Research*, **40**, 1970–1984.

Pike, R.G., Spittlehouse, D.L., Bennett, K.E., Egginton, V.N., Tschaplinski, P.J., Murdock, T.Q., & Werner, A.T. (2008a). Climate change and watershed hydrology: Part 1 – Recent and projected changes in British Columbia. *Streamline Watershed Management Bulletin*, **11**, 1–8.

Pike, R.G., Spittlehouse, D.L., Bennett, K.E., Egginton, V.N., Tschaplinski, P.J., Murdock, T.Q., & Werner, A.T. (2008b). Climate change and watershed hydrology: part II – hydrologic implications for British Columbia. *Streamline Watershed Management Bulletin*, **11**, 8–13.

Pojar, J. (2003). Biodiversity in the CIT region. Scientific background report. Coast Information Team, Victoria, BC, Canada. (http://ilmbwww.gov.bc.ca/citbc/b-Biodiv-CITReg-02Apr04.pdf).

Pojar, J. (2010). A new climate for conservation – nature, carbon and climate change in British Columbia. Report prepared for Working Group on Biodiversity, Forests and Climate. http://www.forestethics.org/new-climate-for-conservation-report.

Pojar, J., Klinka, K., & Meidinger, D.V. (1987). Biogeoclimatic ecosystem classification in British Columbia. *Forest Ecology and Management*, **22**, 119–154.

Price, K. & Daust, D. (2003). The frequency of stand-replacing natural disturbance in the CIT area. Scientific background report. Coast Information Team, Victoria, BC, Canada. http://ilmbwww.gov.bc.ca/citbc/b-DistFreq-PriceDaust-Oct03.pdf.

Price, K., Holt, R., & Kremsater, L. (2007). Representative forest targets: informing threshold refinement with science. http://www.forrex.org/program/con_bio/PDF/Workshops/Forest_Workshop/representation_paper.pdf.

Price, K., Roburn, A., & MacKinnon, A. (2009). Ecosystem-based management in the Great Bear Rainforest. *Forest Ecology and Management*, **258**, 495–503.

Redmond, K. & Taylor, G. (1997). Climate of the coastal temperate rain forest. In *The Rain Forests of Home* (ed. P.K. Schoonmaker, B. von Hagen, and E.C. Wolf), pp. 25–41. Island Press, Washington, DC, USA.

Schoonmaker, P.K., von Hagen, B., & Wolf, E.C. (1997). Introduction. In *The Rain Forests of Home* (ed. P.K. Schoonmaker, B. von Hagen, and E.C. Wolf), pp. 1–6. Island Press, Washington, DC, USA.

Seymour, R.S., White, A.S., & DeMaynadier, P.G. (2002). Natural disturbance regimes in northeastern North America – evaluating silvicultural systems using natural scales and frequencies. *Forest Ecology and Management*, **155**, 357–367.

Tyler, S., Daust, D., Price, K., Soto, C., & Overstall, R. (2009). Adaptive management framework for the central and north coasts of British Columbia. Final report to the EBM Working Group. http://archive.ilmb.gov.bc.ca/slrp/lrmp/nanaimo/cencoast/ebmwg_docs/AMF_overview_v4_20090129.pdf.

Chapter 12

INCORPORATING HRV IN MINNESOTA NATIONAL FOREST LAND AND RESOURCE MANAGEMENT PLANS: A PRACTITIONER'S STORY

Mary Shedd,[1] *Jim Gallagher,*[2] *Michael Jiménez,*[1] *and Duane Lula*[3]

[1]USDA Forest Service, Superior National Forest, Duluth, MN, USA
[2]USDA Forest Service, Chippewa National Forest (retired), Cass Lake, MN, USA
[3]USDA Forest Service, Superior National Forest (retired), Duluth, MN, USA

Historical Environmental Variation in Conservation and Natural Resource Management, First Edition. Edited by John A. Wiens, Gregory D. Hayward, Hugh D. Safford, and Catherine M. Giffen.
© 2012 John Wiley & Sons, Ltd. Published 2012 by John Wiley & Sons, Ltd.

12.1 INTRODUCTION

Minnesota's two national forests, the Chippewa National Forest and the Superior National Forest, applied an understanding of the historical range of variation (HRV) of Minnesota's northern forest ecosystems as a planning tool in the joint revision of our Forest Land and Resource Management Plans (Forest Plans) in 2004 (USDA 2004a). HRV was used as a reference and framework (not a target) for developing and analyzing impacts of vegetation management alternatives that would, as the stated goals of the forest plans describe, "sustain the health, diversity, and productivity of the national forests to meet the needs of present and future generations." In this chapter, we describe a "practitioner's story" as an early example of application of HRV concepts for broadscale management of forest ecosystems. After briefly describing the ecological and socioeconomic setting of the Forests, we examine:

• Why we used our understanding of HRV as a planning tool
• How we defined HRV
• How we collaborated with partners in the development of HRV
• How we characterized elements of HRV
• How we used HRV to develop and analyze forest-management alternatives.

We close by reflecting on lessons learned, including how our perspectives have changed with the deeper understanding of historical ecology that has developed. Nearly a decade after initiating our effort to use HRV and after 6 years of implementing Forest Plans influenced by our understanding of HRV, we are in a position to look back and ask "Was this a successful use of HRV?" and "What would we do differently?"

12.2 ECOLOGICAL AND SOCIOECONOMIC SETTING

Ecological setting

Climate, landscape, and vegetation communities

The Chippewa and Superior National Forests fall in the Laurentian Mixed Forest Province of the national hierarchical ecological classification system (USDA 1993) (Fig. 12.1). This province spans a large portion of northern Minnesota, Wisconsin, and Michigan, with a total of about 9 million hectares in the northern third of Minnesota. In Minnesota, this Province is characterized by its moist and cool-temperate climate and by "northwoods" or boreal-hardwood transition zone vegetation communities. The landscape is composed of conifer forests, mixed hardwood/conifer forests, conifer bogs, and swamps. The forests are dominated by white pine (*Pinus strobus*), jack pine (*Pinus banksiana*), red pine (*Pinus resinosa*), trembling aspen (*Populus tremuloides*), paper birch (*Betula papyrifera*), black spruce (*Picea mariana*), white spruce (*Picea glauca*), balsam fir (*Abies balsamifera*), tamarack (*Larix laricina*), white cedar (*Thuja occidentalis*), and sugar maple (*Acer saccharum*). Glaciers covered the entire area some 10 000 years ago and had a profound effect on surface features. The topography ranges from rugged terrain with thin soils and bedrock outcropping to hilly or undulating plains with deeper soils. Carved out by the glaciers, and as much a defining feature of the landscape as the forests, are the ubiquitous numerous large and small lakes, streams, peatlands, and array of other wetlands.

Disturbance

The forest of the Laurentian Mixed Forest Province that exists today developed as a result of the influence of both natural and human processes. Prior to European settlement in the 1800s, natural disturbance processes and the cultural activities of Native Americans shaped the composition and structure of the ecosystems. Fire, wind, insects and disease, and some flooding, together with natural plant community succession, were the primary influences on vegetation, wildlife, and aquatic systems. Native Americans introduced local disturbance associated with settlements and, more broadly, disturbance to manage the land. Dominant among these was intentional and unintentional fire. For example, fires were intentionally set to enhance game habitat and blueberry crops. Similar to other regions of North America, anthropogenic fire was likely both additive and compensatory to fire that resulted from lightning.

Human influence on vegetation increased in both intensity and extent with the introduction of extensive logging and subsequent large, severe, slash-fueled fires during the European-settlement era (late nineteenth to early twentieth centuries). This led to extraordinary

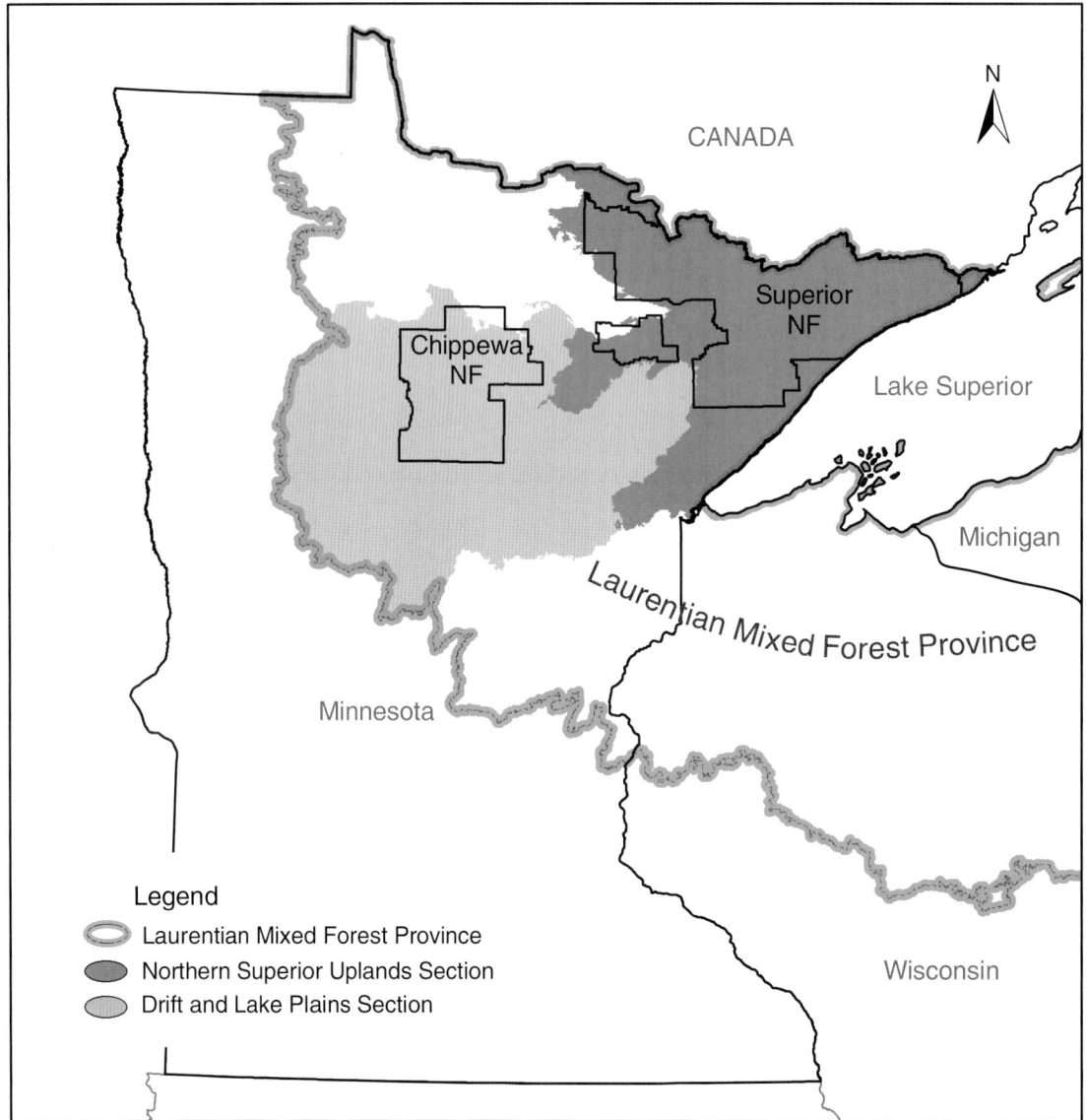

Fig. 12.1 Ecological land units identified on the Chippewa and Superior National Forests in Minnesota, showing their location within the National Hierarchical Framework of Ecological Units. The Northern Superior Uplands and the Minnesota Drift and Lake Plains sections represent lower levels of the hierarchy than the Laurentian Mixed Forest Province.

changes in forest ecosystems. Both forest composition and landscape structure have become dramatically more homogeneous. Dominance shifted from late-successional or mid-seral conifer (red and white pine, spruce) to early successional hardwood species (aspen). Conifers now comprise 30–40% of tree density, com-

pared with 60–70% during the pre-Euro settlement era. Following this pulse of severe disturbance during the settlement period, even-aged timber harvest replaced fire as the primary disturbance factor. Although natural disturbances remain powerful agents of forest change, management practices –

including changing land uses – now play a stronger role than natural processes in shaping landscape composition, structure, and pattern.

Social setting

Minnesota's National Forests are a patchwork of land ownerships. Of the 1.4 million hectares within the Superior's boundary, 37% is state, private, and county-owned. Of the 0.65 million hectares within the Chippewa boundary, 58% is state, private, county, or tribally owned. Across most of northern Minnesota, including the national forests, Ojibwe bands hold treaty authority rights to hunt, fish, and gather on state and federal land. Both Forests are home to numerous small towns (<6000 inhabitants) and rural residences. People who live on in-holdings within the broad boundaries of the National Forests or the hundreds of thousands who annually visit the federal lands depend on the Forests for their aesthetic, spiritual, consumptive, commodity, recreational, and scientific uses and values. The mix of ownerships is a crucial consideration in planning for ecological and socioeconomic sustainability since ecosystems are not defined by, but span, legal and administrative boundaries.

Economic setting

For the last 100 years, the economy of the region has varied in vigor from thriving to struggling, anchored primarily in the values and uses of the natural resources of the area. These are often referred to as the "Three T's: Timber, Tourism, and Taconite" (a mined iron-rich or *ferrous* rock). On the Superior, potential development of nonferrous mining is rapidly expanding, but multiple-use management for forest products (saw-timber and pulp) and high-quality recreational settings (tourism) contribute most strongly to the economy. Biomass for energy production is now an emerging demand and also has potential to contribute to economies. The unpredictable nature of the natural resource-based economy is apparent in the pattern of the timber market during the past few decades. In the 1990s through the early part of this century, forest products experienced high demand and supported a vigorous economy; however, since 2008, the timber and pulp market has been depressed. Despite the fluctuations in demand and market strength, it is difficult

to predict, but reasonable to forecast, that the area's natural resources will continue to be the drivers of local economies.

Integrating ecological, social, and economic drivers

The premise driving Chippewa and Superior Forest Plans was that desired conditions of forest vegetation, reflected by goals and objectives within the Plans, tie together the social, ecological, and economic aspects of managing the forest for the public good. For example, scenic beauty, clean water, and productive fish and wildlife populations in forested settings contribute to the economic well-being of communities by drawing tourists into the area. The results of vegetation management provide settings that tourists enjoy and return to year after year. Forests are important in the cultural and economic well-being of Native Americans. The same forest management, while moving toward restoration of desired conditions, can reduce wildfire risk to communities and offer a sustainable level of forest timber products as outputs of managing for those conditions.

12.3 WHY WE USED OUR UNDERSTANDING OF HRV AS A PLANNING TOOL

The Chippewa and Superior's decision to use HRV as a planning tool was based on the need to craft management direction to support sustainability of the forests over the long term. In 1997, when the two National Forests began jointly revising Forest Plans, the application of HRV concepts was becoming an accepted but new approach in land management on public and private lands (Christensen et al. 1996). Support for the approach stemmed primarily from its potential to inform decisions that ultimately could improve the ecological sustainability of land management. However, the focus on sustainability recognized the inseparable relationship between ecological and socioeconomic sustainability. In the socioeconomic context, sustainability is represented by the intersection of a variety of values, products, and services provided by the Forest ecosystems. Specifically, this decision rested on two related basic assumptions: (1) the closer ecological processes operate to their expected bounds of variation, the greater the confidence that diversity and productivity will be sustained for future generations, and (2) in the absence of conditions defined by the historical record, adequate ecological representation of historical systems should provide ecosystem integrity and productivity for future generations.

A second compelling reason for using the concept of HRV related to collaboration and the potential to integrate land-use planning among diverse landowners. Adjacent land managers had also recognized the potential for HRV to address ecological, social, and economic sustainability on their lands. HRV offered a framework to help join together many different management approaches across the complex ownership pattern of northern Minnesota. Partners, including the Superior, Chippewa, state, and county land managers, the forest products industry, academia, and special interest groups in the larger ecological landscape of Minnesota developed HRV assessments as they began land-management planning. These managers intended to use HRV to inform voluntary management guides to achieve forest composition and age objectives across shared landscapes. The partner group was part of the Minnesota Forests Resource Council (http://www.frc.state.mn.us), a group appointed by the governor to develop statewide, sustainable forest-resource policies and practices. Concurrent planning by the Minnesota Forests Resources Council and the National Forests using common ecological and social-economic assessments provided an opportunity to pursue collaborative, complementary goals that spanned administrative or political boundaries.

12.4   HOW WE DEFINED HRV

While there are a number of relatively concise definitions of HRV, ours was not one of them. Early in our planning process, we recognized a challenge of gaining public, or even internal, acceptance and understanding of our proposed use of HRV (see Box 12.1). A simple definition of HRV did not provide the breadth of understanding necessary to engage the public or planners. Rather, it was important to elaborate on the assumptions and limitations of using HRV. It was important to acknowledge the uncertainties associated with understanding HRV – limitations resulting from inadequate data and the challenge associated with HRV assumptions. A prime example of the latter is climate change. If, as we hypothesized at the time, HRV

Box 12.1 HRV definition used to describe the concept in the environmental impact statement for the Superior and Chippewa Forest Plans (USDA 2004b).

Ecological systems are dynamic in nature; however, they have historically changed sufficiently slowly that there was apparent continuity in landscape processes across multiple species life cycles and human generations. The composition, structure, and processes of ecosystems fluctuate over time. In order to address ecosystem sustainability, these fluctuations must be interpreted in light of the natural and historical variation of the landscape (USDA 1999). This variation is predictable when it happens under a relatively stable set of physical and climatic conditions, disturbance regimes, and natural succession processes (when a system regenerates without human influences). The term "historical range of variation" (HRV) is used to describe these fluctuations. The importance of understanding the concept of HRV lies in the premise that ecosystems operating within HRV are more resilient after disturbances; and therefore the effects of disturbances, either human-induced or natural, are more predictable. Conversely, ecosystems operating outside of HRV tend to be affected by disturbances in ways that are much different than those conditions under which plants, soils, animals, and other ecosystem elements evolved. Disturbance effects on landscapes functioning outside of RNV become much less predictable; and the risk of losing resiliency and compositional, structural, or functional elements of ecosystems increases (USDA 1999). Understanding the historical range of variability of ecosystem composition, structure, and processes can help identify what elements of the ecosystem may need special consideration in forest management.

provides insights regarding vegetation dynamics over time, would those predictions hold if climate changed? Could managers continue to achieve desired conditions under a changing climate?

Describing HRV as a tool to aid planners required explicit communication of limitations. For example, we were careful to communicate that evaluation of HRV did not suggest a desire to return ecosystems to historical conditions. Returning to earlier conditions was not desirable because it was not socially or ecologically possible and because doing so would not meet the public's demands for multiple use of the National Forests. Forest users who make a living from or enjoy outcomes and outputs of current vegetation (such as scenic recreational settings, forest products, wildlife populations, sense of place), expressed a common concern that using HRV implied an intention to manage toward historical conditions.

Fears regarding the application of HRV remained common until the Superior and Chippewa Forest Plans were signed. The Forest Service emphasized in communications with the public that the conditions described by HRV were not a target and returning ecosystems to pre-Euro settlement conditions was not a goal; however, several factors led to the disconnect between attempted communication and public perceptions. First, a portion of the public liked current forest conditions – they felt a sense of place in the conditions they experienced daily. This raised a concern that if HRV was a "target" condition, their forest environment would change. Others believed that our characterization of past conditions was not adequately documented. They suggested that scientists did not understand the impacts Native Americans had on historical conditions. Some believed that future climate change invalidated using HRV.

Ultimately, acceptance or tolerance for application of HRV grew from an understanding that the final Forest Plans did not rely on HRV as a target for vegetation conditions. In fact, while HRV informed management direction, HRV and historic conditions are not mentioned in the Forest Plans.

12.5 HOW WE COLLABORATED WITH PARTNERS IN THE DEVELOPMENT OF HRV

Collaboration framework

A hallmark of Forest Plan revision in northern Minnesota was close collaboration between the US Forest Service and Minnesota Forests Resources Council (MFRC) (http://www.frc.state.mn.us). The Council promotes sustainable forest resource policies and practices statewide by coordinating land-management activities. More than just a political body, the MFRC represents a forum where diverse interests discuss and resolve issues regarding the management of Minne-

sota's forests. MFRC helped depoliticize forestry in Minnesota by emphasizing collaboration and science. Members of the Council include representatives of commercial logging contractors, conservation organizations, county land departments, environmental organizations, the forest products industry, game management organizations, labor organizations, the Minnesota Department of Natural Resources, the Minnesota Indian Affairs Council, nonindustrial private forest landowners, research and higher education, the resort and tourism industry, secondary wood products manufacturers, and the US Forest Service.

In 1998, MFRC initiated efforts to develop landscape plans to coordinate, implement, and monitor cross-ownership forest management for Minnesota. At the beginning of these efforts in the Northeastern (Superior) and North Central (Chippewa) Landscape Regions, the MFRC chartered a panel of experts to define aspects of the HRV for forested communities in northern Minnesota based on a synthesis of historical ecology. The expressed purpose was to provide a context for all landowners for (1) analyzing forest-management alternatives, (2) making land-management allocation decisions, (3) describing desired future conditions, and (4) defining management area prescriptions. As part of MFRC, the National Forests participated in this effort.

12.6 HOW WE CHARACTERIZED HRV

An expert panel of scientists, working for MFRC, developed an HRV assessment for forested regions in Minnesota. As described in Chapters 3 and 5, initiating an HRV assessment requires explicit definition of the temporal and spatial extent of the analysis. The charge of the team was to identify the appropriate temporal and spatial extent and resolution to defining major forest and nonforest ecosystems; determine disturbance regimes for the ecosystems; and quantify the extent and spatial distribution of native plant communities and the seral stages that formerly existed on the landscape (Minnesota Forest Resources Council 1999, 2000).

Time frame

The expert panel defined the period from 1600 to 1900 AD as most appropriate to characterize forested landscapes of Minnesota. The range of forest conditions during this time period was thought to most closely represent the climate, natural processes, and disturbances under which the current forest ecosystems and the accompanying biological diversity of northern Minnesota developed.

Delineation of forest ecosystems

A next step in characterizing HRV was to clearly delineate the spatial extent of analysis. The expert panel agreed that the appropriate spatial scale for characterizing HRV was the ecoregion within the National Hierarchical Framework of Ecological Units identified as the "Section" (Cleland et al. 1997). For the Superior National Forest and the mixed ownerships within its boundary, the forest conditions within the Northern Superior Uplands (Fig. 12.1) from 1600 to 1900 AD provide the appropriate characterization of HRV. For all lands within the Chippewa National Forest boundary, the forest conditions within the Minnesota Drift and Lake Plains (Fig. 12.1) during that same time period provide the appropriate characterization of HRV. Within the Sections, further ecological subdivisions, based largely on soils, native plant communities, and disturbance regimes, were delineated and coined "Landscape Ecosystems" in the Forest Plans (Figs. 12.2 and 12.3).

Elements of HRV

Adequately describing conditions of existing ecosystems is challenging due to limitations of data and the difficult process of placing indistinct ecological units into specific categories. Characterizing conditions from hundreds of years ago is substantially more difficult. There are numerous, complex, and interconnected elements that could and – for the best scientific understanding of HRV – should be characterized, from vegetation seral stages, composition and spatial patterns, to disturbances, to species abundance and distribution, to aquatic ecosystem function, to carbon and nutrient cycling, to soils, and to microbes. The expert panel was charged with identifying the elements of HRV that would be most feasible and useful to characterize. These were the key elements:
• Vegetation composition: extent of major forest stand types (such as aspen or spruce-fir) and percentage of individual species of trees across the delineated ecosystems.

• Vegetation structure or growth stage: extent of forest in successional stages, described in terms of extent of different forest age classes.
• Historical ecosystem function: extent and frequency of major disturbance types (e.g. fire, wind) and natural succession. Information on forest spatial patterns was emerging as we finalized Plans and was generally characterized. Information on other elements of ecosystem function, such as insect and disease outbreaks, flooding, soil nutrient cycling, carbon cycling, and wildlife-caused disturbances were summarized but not quantified.

Interpretations of numerous data sources and science literature contributed to the expert panel's quantitative and qualitative characterization of the key elements above. These sources included late 1800s to early 1900s Government Land Office Public Land Survey bearing tree notes that recorded tree species along survey lines; mapped pre-Euro settlement vegetation; lake and swamp pollen cores that allowed estimate of tree species distribution and abundance over time extending back over 9000 years; climate data dating back over 100 years; wildfire history from burn scars and current vegetation patterns providing records for the past couple centuries; current and historical silvicultural practices; historical data on insect and disease outbreaks over the past 100 years; Forest Inventory and Analysis of forest type, extent, growth, mortality, and removals in Minnesota's forests; Landsat satellite images classified into vegetation types; and terrestrial ecological unit inventories, including soil and landform characterization and approximately 4500 vegetation relevé plots of existing vegetation. The expert panel also considered socioeconomic historical information related to timing and extent of timber harvest, Native American uses of the land, wildlife and fish harvest, agriculture, and mining. (Detailed metadata for most of these resources can be found at the Minnesota Forest Resources Council [MFRC]'s Interagency Information Cooperative web site at http://iic.gis.umn.edu.)

State and transition models were developed for each delineated ecosystem. The models predicted extent of vegetation growth stages in each successional pathway based on prevalent tree species, disturbance return intervals, and the kinds of disturbances that dominated each ecosystem during the reference period. Figure 12.4 is an example from this effort for the Dry Pine "Landscape Ecosystem" on the Chippewa National Forest. This figure illustrates three primary distur-

Legend

○ Superior NF Boundary   ● Mesic Aspen/Birch/Spruce/Fir

◐ Dry Mesic Jack Pine/Black Spruce   ▒ Mesic White Pine/Red Pine

◌ Dry Mesic White Pine/Red Pine   ○ Rich Swamp

○ Jack Pine/Aspen/Oak   ◐ Sugar Maple

◐ Lowland Conifer   ○ Water

CANADA

Lake Superior

N

Michigan

Minnesota   Wisconsin

0 5 10   20   30   40 mi

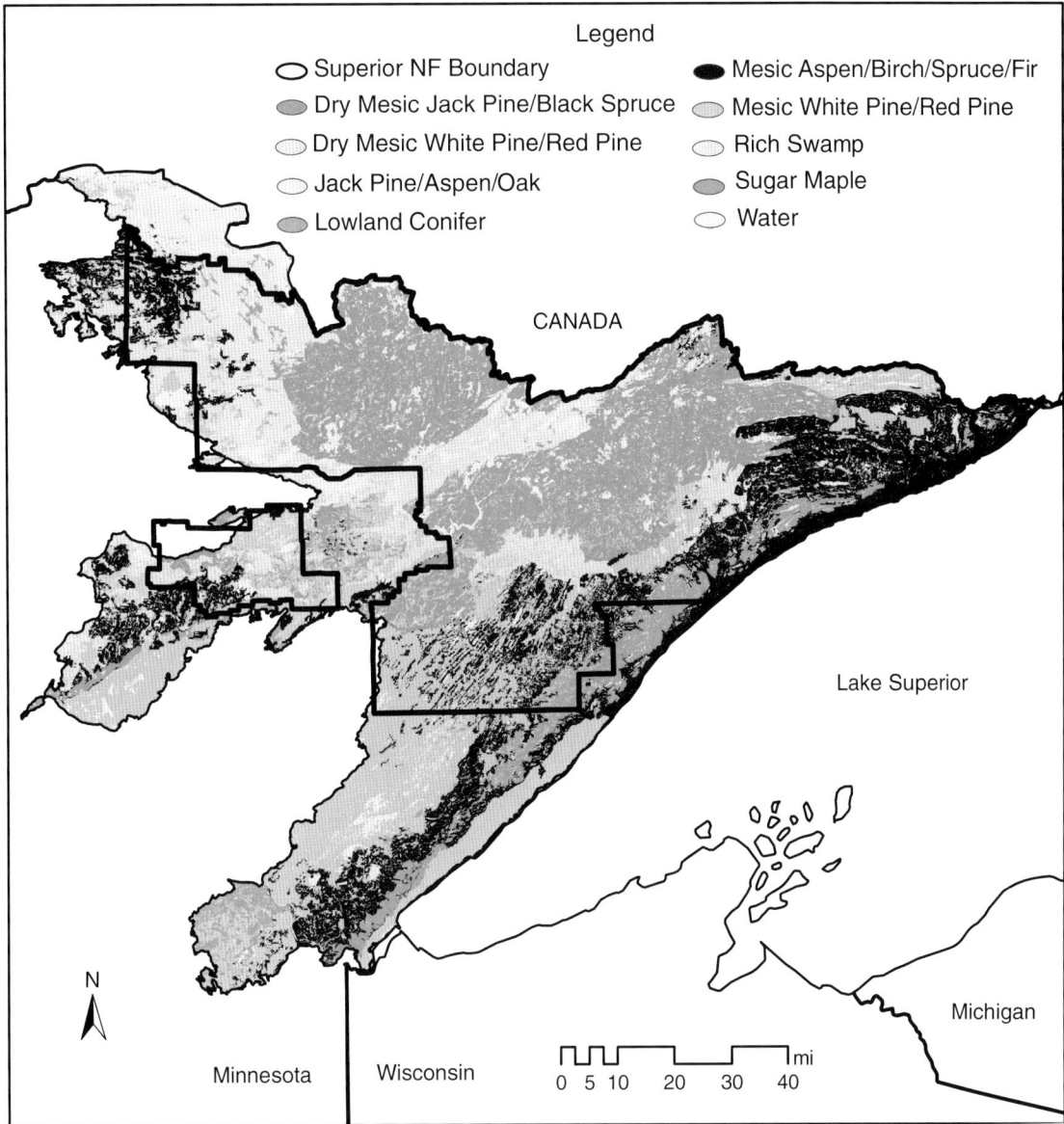

Fig. 12.2 Landscape Ecosystems on the Superior National Forest within the Superior Uplands Section used to define ecological units for Forest planning and implementation. Please refer to Colour Plate 5.

bance regimes (wind, stand-replacing fire, and ground fire) and successional pathways in the absence of disturbances for this historically jack pine-dominated system resulting in five, vegetation-growth stages.

Once reference conditions and successional pathways were modeled and calculated, these ranges of

percentages were used to quantify variation in spatial extent of vegetation by section (Fig. 12.5).

The expert panel's HRV assessment represented ecological inferences developed using the expertise of the panel, the evidence of ecological patterns in the available data, and interpretation of results from the state

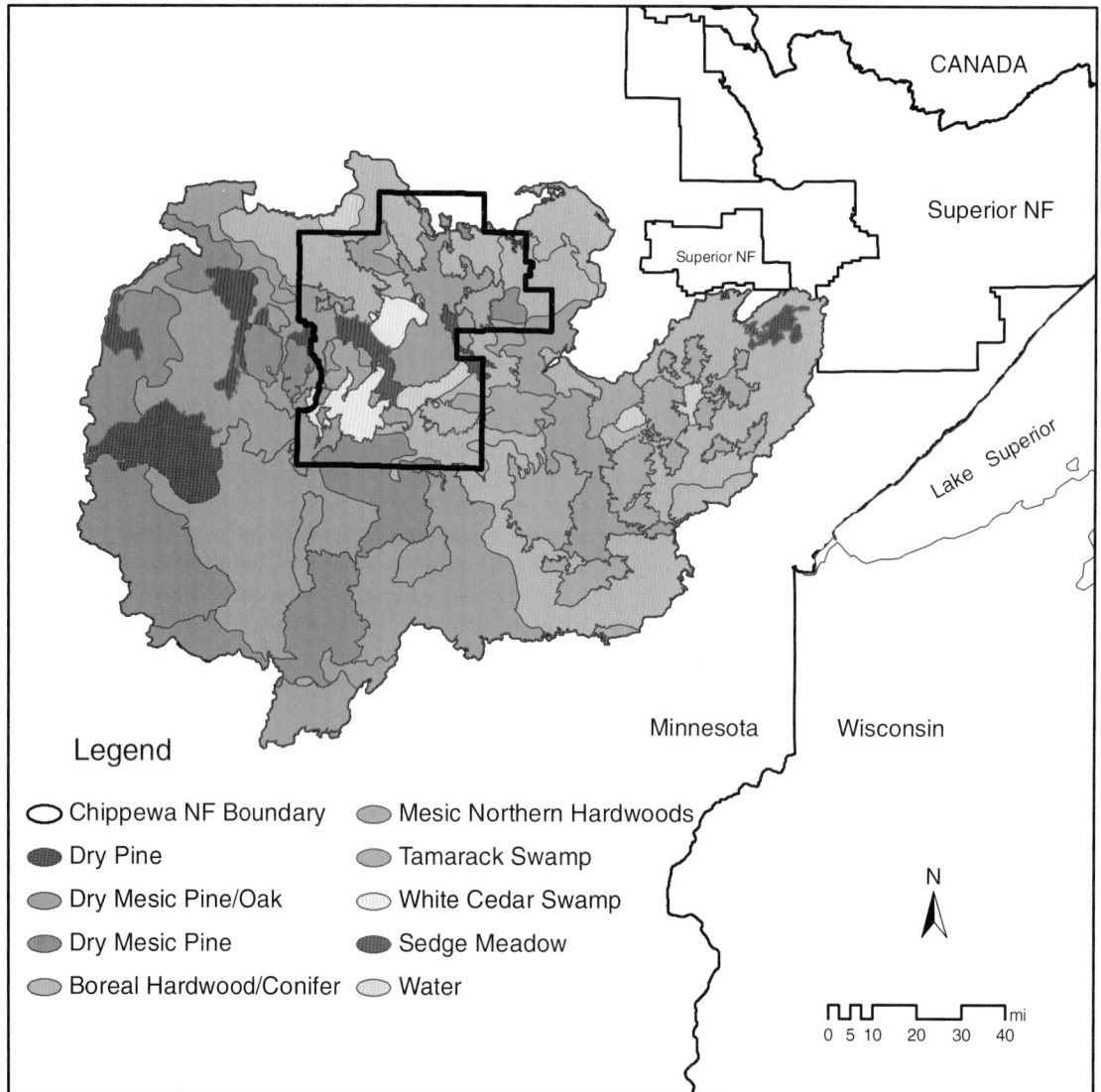

Fig. 12.3 Landscape Ecosystems on the Chippewa National Forest within the Minnesota Drift and Lake Plains Section used to define ecological units for Forest planning and implementation. Please refer to Colour Plate 6.

and transition modeling. Similar processes are described in Dillon et al. (2003), Veblen and Donnegan (2005), and others. In general, the panel acknowledged that each science-based source had a host of limitations that required careful interpretation of patterns illustrated in the data. Factoring in these limitations and resulting assumptions is required for understanding the interpretation limits of any HRV assessment.

Examples of several challenges encountered while developing the HRV assessment provide some perspective on the product. One significant gap concerned historical conditions of aquatic ecosystems. In the absence of any dependable synthesis, the team made the assumption that managing for sustainable forest vegetation conditions would result in sustainable aquatic ecosystems. Another gap was reliable information on

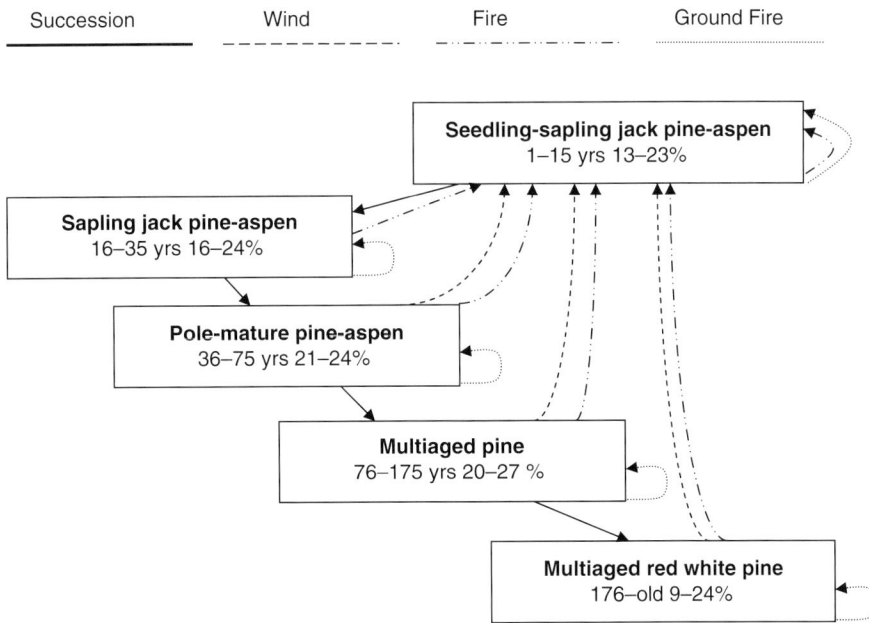

Fig. 12.4 Example of successsional pathways for the Dry Pine Landscape Ecosystem in the Minnesota Drift and Lake Plains Section for the Chippewa National Forest. Pathways illustrate the consequences of disturbance and recovery and were developed by a team of scientists from multiple organizations.

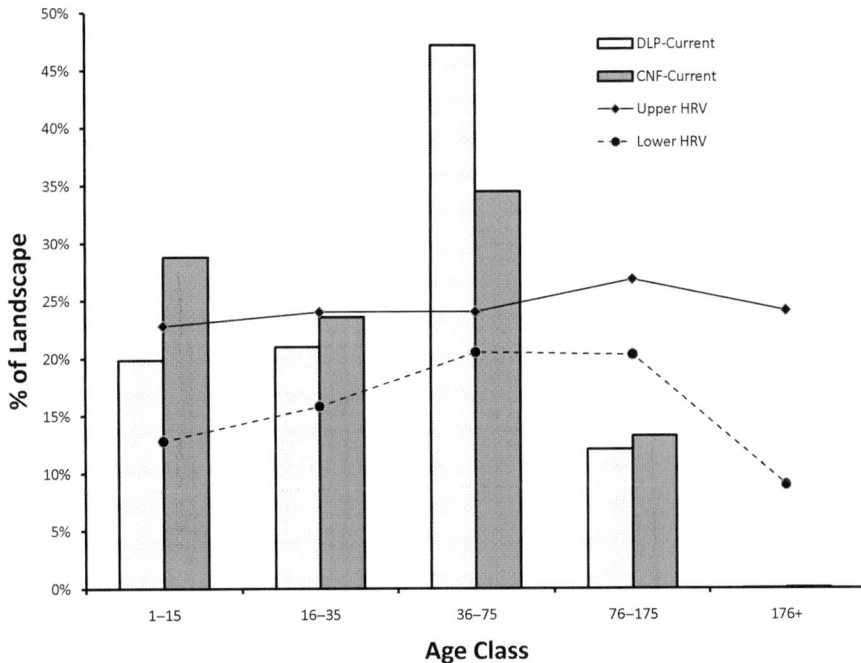

Fig. 12.5 Age-class distribution of forests in the Dry Pine Landscape Ecosystem in 2004 compared to distribution under Historical Range of Variation (HRV) estimated through evaluation of historical dynamics by the science team for the Drift and Lake Plains Section on the Chippewa National Forest.

historical conditions for non-tree vegetation such as shrubs, forbs, lichens, or bryophytes. Therefore, to manage for diverse non-tree vegetation, we relied on characterizations of current understory vegetation and non-forest conditions from studies in examples of relatively undisturbed native vegetation. Historical information on animals, particularly population abundance and dynamics, was largely absent and syntheses unavailable. To guide planners, we relied on habitat modeling and assumed that species response to habitat had not changed since the reference period. We could make assumptions about above- or below-ground biomass (live or dead); however, we did not quantify or model variation of biomass. Finally, given the general nature of Forest Plan decisions, much of this information (e.g. changes in lichen and bryophytes) played little role in development of Forest Plan alternatives; however, the synthesis is available for more, fine-scale project planning and analysis.

12.7   HRV AND DEVELOPMENT OF MANAGEMENT ALTERNATIVES

Appendix G of the Final Environmental Impact Statement for Forest Plan Revision reviews the process used to incorporate an understanding of HRV into development of management alternatives. In this section, we briefly summarize those planning steps.

Step 1: Build a collaborative planning framework with other landowners and forest stakeholders

The collaborative planning framework developed through the MFRC was integral to the process of revising plans. Though each landowner has distinct goals, legal guidance, and unique roles to play in managing shared landscapes, members of MFRC identified and built on shared goals to craft common desired conditions. These included maintaining resilient, functioning forest ecosystems, and ensuring that forests provide a range of products, services, and values that contribute to economic stability, environmental quality, social satisfaction, and community well-being.

Collaborative planning processes outside the MFRC framework were equally important and available to nonmember landowners, Native American bands, and stakeholders. Some of these processes are required under law, policy and regulations, including Executive Order 13175 (November 6, 2000) which recognizes the sovereignty of federally recognized Native American tribes and the special government-to-government relationship with the United States government; the National Environmental Policy Act of 1970 (NEPA); and Forest and National Forest Management Act of 1976 (NFMA). NEPA and NFMA and their federal implementing regulations ensure that the public can participate in and comment on revision. Our general approach emphasized early, frequent, open, and useful public involvement.

Step 2: Characterize the HRV and existing conditions

The characterization of the HRV has been described in sections above and was accomplished by the expert panel composed of university and agency scientists and managers. Over the course of 3–4 years, the key products of the panel's work used in plan revision included:

• Characterizations and estimates of HRV in forest structure and composition organized by ecological Sections (Frelich 1998, 1999, 2000).

• Assessment of changes in landscape spatial patterns from pre-European settlement to present (Wolter & White 2002).

• Current and historical forest conditions and trends in the North Central Landscape Region (Minnesota Forest Resources Council 2000) and Northeastern Landscape Region (Minnesota Forest Resources Council 1999). These two assessments included characterizations of current and historical conditions for forest composition, structure, and spatial patterns as well as narrative descriptions of other ecological elements that were less quantifiable (e.g. wildlife populations, species at risk, water quality, and riparian areas). The assessments also summarized conditions and trends for social and economic elements such as land use, ownership patterns, and parcel size, road densities, demographics, employment, forest products industry, tourism industry, and recreation uses and facilities. These are available at http://www.frc.state.mn.us/resources_documents_landscape.html.

Step 3: Develop and analyze management alternatives

Building on the understanding developed in Step 2, the team quantified current and historical ecological conditions for subdivisions of the Sections called "Landscape Ecosystems" for National Forest System lands only. This information helped managers develop alternatives for vegetation management. Peoples' interests in forest conditions were complex and conflicting, so alternatives provided decision makers and the public with a way to assess the consequences, outcomes, and outputs of differing ways of managing the forest. By comparing past and present ecological conditions with an eye toward retaining representation of dominant elements of the system and within the context of the public's interests, we developed a range of seven alternatives for each National Forest. The alternatives were framed in what the US Forest Service calls desired future conditions. The planning horizon included consideration of changes for up to 100 years; a time frame sufficient to achieve goals for some conditions with extreme transition periods. Desired future conditions in the Dry Pine Landscape Ecosystem illustrating vegetation composition (forest-stand type percentage) and structure (vegetative growth stage or age class) are outlined in Tables 12.1 and 12.2.

For the Dry Pine Landscape Ecosystem, one can recognize management themes of each alternative (Tables 12.1 and 12.2). On each Forest, the themes ranged from an emphasis in Alternative C on high amounts of younger hardwood forest in small (<16 ha) patches supporting significant harvest levels, to high amounts of older conifer forest in larger patches (Alternative D) emphasizing development of certain aesthetic and biodiversity values. Correspondingly, Alternative C favored habitat for early-successional species like white-tail deer (*Odocoileus virginianus*) while Alternative D favored habitat for old-forest late-successional species like the northern goshawk (*Accipiter gentilis*). Alternative F approximated the average ecological conditions described in the HRV assessments – this feature was made clear to the public during dialogue regarding the plan alternatives.

The environmental effects analysis followed standard agency practices required by the National Environmental Policy Act. The HRV assessment facilitated this analysis and communication of the results to the public. The analysis focused on comparison of current environmental conditions with historical conditions defined by the HRV assessment and the future conditions predicted under each alternative. Quantitative comparisons were developed for vegetation composition (forest stand types) and structure (forest age

Table 12.1 Stand diversity objectives for Dry Pine Landscape Ecosystem (LE) by Alternative (percent of the LE dominated by a forest type).

Forest types	Existing 2001	HRV Midpoint	Alternatives (100 year)						
			A	B	C	D	E[1]	F[2]	G
	(%)	%	(%)	(%)	(%)	(%)	(%)	(%)	(%)
Jack pine	29	69	Not by LE	67	48	50	53	69	61
Red pine	39	22		24	40	44	35	22	30
Aspen	22	2		2	5	0	3	2	2
Oak	3	2		2	2	2	2	2	2
Spruce/fir	1	2		2	3	3	2	2	2
Paper birch	3	1		1	1	1	3	1	1
Northern hardwoods	1	1		1	1	1	1	1	1
White pine	1	1		1	1	1	1	1	1
Total	100		0	100	101	102	100	100	100

[1] Alternative E was the selected alternative for Forest Plan Revision.
[2] Alternative F percentages represent midpoint of HRV percentage for forest types.

Table 12.2 Age-class objectives for Dry Pine Landscape Ecosystem (LE) by Alternative (percent of the LE in an age class). Note that the age classes here match those of the forest development stages shown in Fig. 12.4.

Age class	HRV	Existing 2001	Alternatives						
			A	B	C	D	E[1]	F[2]	G
	(%)	(%)	(%)	(%)	(%)	(%)	(%)	(%)	(%)
1–15	13–24	28	Not by LE	14	20	10	14	17.8	15
16–35	16–24	26		18	26	13	19	19.9	20
36–75	21–24	36		37	34	25	38	22.3	40
76–175	20–27	10		16	14	27	23	23.6	15
175+	9–24	<1		15	7	25	6	16.6	10
Total	100	100	0	100	100	100	100	100	100

[1]Alternative E was the selected alternative for Forest Plan Revision.
[2]Alternative F percentages represent midpoint of HRV percentage for forest types.

classes) and for approximately 30 different wildlife habitat types. We qualitatively examined changes in spatial patterns, within-stand tree species diversity, and disturbances (fire, insect, and disease). We also quantitatively and qualitatively analyzed spatial patterns such as patch size, connectivity, interior forest, and edge habitats. The environmental consequences then informed social and economic consequences of the alternatives. For example, we could use predictions of future vegetation to project economic outcomes and outputs associated with timber harvest, fish and wildlife harvest, and tourism.

Step 4: Revised forest plans

Several years of collaboration with the Forests Resources Council and an array of public organizations produced seven alternative management scenarios for the Chippewa and Superior National Forests. Agency leaders chose an alternative (E) that provides a compromise and good mix of conditions to balance diverse – and frequently conflicting – interests and guide management over the next few decades. The theme of the chosen alternative directs managers to use active management to provide a mix of young to old-growth conifer, and hardwood forests with a strong emphasis on diverse wildlife habitats. Under this alternative, forest composition, structure, and spatial patterns do not reflect historical conditions (as described in the HRV assessment), but management will empha-

size actions that move forests incrementally toward conditions thought to be most common in the past.

Despite the focus on HRV during plan development and analysis, HRV is not mentioned in the Forest Plan for either the Chippewa or Superior National Forests. However, the influence of historical ecology is clear in the statements of desired conditions and measureable objectives in the Forest Plans (see Box 12.2).

In addition to the forest-wide direction, the objectives for particular ecosystem types reflect the influence of historical ecology. Measurable objectives provided concrete goals for the first two decades of management, with reference to the long-term (100-year) goal. Tables 12.3 and 12.4 are Forest Plan excerpts of management objectives for the Dry Pine. Figure 12.4 and Tables 12.1 and 12.2 illustrate the analogous historical conditions.

Objectives for tree species diversity ("within-stand diversity" based on total percentage of trees, not total area of forest type) and wildlife habitat (types representing major biological communities affected by management) were also developed through an understanding of historical conditions described in the HRV assessment.

The preceding examples illustrate the influence of historical ecology on management direction described in the final Forest Plans for both the Chippewa and Superior National Forests. However, as emphasized early in this chapter, the scope of planning within northern Minnesota and the role of historical ecology in land-management planning was much broader.

Box 12.2 Historical ecology in forest plans.

The application of historical ecology on the Superior and Chippewa National Forests played a critical role in defining desired conditions and the restoration objectives that grew from the vision for the future of the Forests. The following excerpts from the Forest Plans are examples (Chapter 2 in USDA 2004a). While these examples illustrate desired conditions and restoration goals for vegetation composition and structure, the understanding of historical conditions also resulted in desired conditions and restoration objectives related directly to spatial pattern and ecological processes.

Desired Condition

Vegetation (live and dead) is present in amounts, distributions, and characteristics that are representative of the spectrum of environmental conditions that would have resulted from the natural cycles, processes, and disturbances under which current forest ecosystems and their accompanying biological diversity evolved. The ecosystem composition, structure, and process representation considers time frames, a variety of landscape scales, and current biological and physical environments. Resource conditions exist that minimize undesirable occurrences of non-native, invasive species.

Vegetation Composition and Structure:

Desired Condition

Vegetation conditions that have been degraded or greatly diminished in quality or extent on the landscape by past land use are restored to conditions more representative of native vegetation communities. These conditions, in ecologically and socially appropriate areas, result from gradually re-establishing

a. Old forest and old-growth forest age classes and vegetative growth stages, while providing for a full array of forest age classes and vegetative growth stages.
b. Uneven-aged and multiaged forests with a variety of tree ages and different vegetation layers (heights) within the same community, while also providing for even-aged forests.
c. The full range of successional stages in non-forested lands such as bogs, fens, grasses, and shrublands.
d. Diverse mixes of trees, shrubs, herbs, mosses, lichens, and fungi species . . . that are more representative of native vegetation communities.
e. Diverse structure in native vegetation communities that have been harvested, salvaged, prescribed burned, or have undergone natural disturbance. Structural diversity components will be provided by small patches of forest (reserve islands), scattered or clumped standing, mature and older live trees; dead trees; and coarse woody debris (down logs and branches).

Table 12.3 Vegetation composition objectives for the Dry Pine Landscape Ecosystem, Chippewa National Forest, MN.

Forest type	Existing (2003)		Objectives		
			Decade 1	Decade 2	Long-term 100-year goal
Upland forest type	Acres	(%)	(%)	(%)	(%)
Jack pine	3300	27	35	41	53
Red pine	4900	41	39	37	35
White pine	200	1	2	2	1
Spruce/fir	200	1	1	2	2
Oak	400	3	3	3	2
Northern hardwoods	100	1	1	1	1
Aspen	2700	23	16	12	3
Paper birch	300	2	2	2	3
Total	12 100	100	100	100	100

Table 12.4 Vegetation age-class objectives for the Dry Pine Landscape Ecosystem, Chippewa National Forest, MN.

Age class	Existing (2003)		Objectives		
			Decade 1	Decade 2	Long-term 100-year goal
Upland forest type	Acres	(%)	(%)	(%)	(%)
0–9	1800	14	12	10	9
10–39	5000	40	45	45	28
40–79	4700	37	24	28	34
80–179	1100	8	19	17	22
180+	0	0	0	0	6
Total	12 500	100	100	100	100

Plans for federal lands were developed concurrently, collaboratively, and consistently with the MFRC (Minnesota Forest Resources Council 2003a,b). Thus, despite a patchwork of ownership and a broad diversity of management goals, a general management framework was developed and implemented using a common assessment of HRV to unite the collaborators.

12.8 LESSONS LEARNED ON THE USE OF HRV IN FOREST MANAGEMENT

More than 10 years have passed since the Minnesota National Forests and partners began land-management planning using HRV as a conceptual focus. With 6 years of operational experience "under our belts," managers from the two Forests can evaluate successes and challenges associated with implementing the plan. This section summarizes some of our key observations. We reflect on and share lessons learned that may be of interest to others considering incorporating HRV into planning.

Response to change

Change to forests is certain – whether from climate, social, land use, or other processes and stressors. An important legacy of the Minnesota planning effort is a planning framework that facilitates adjustments in response to changes in the physical, biotic, and socio-economic environments. The framework of broadscale objectives for vegetation defined in the Chippewa and Superior Forest Plans establishes management direc-

tion for these National Forests that we consider robust to uncertainty, especially uncertainty related to climate change.

The direction established in the Forest Plans provides the opportunity to alter management in response to changing conditions, such as shifts in species range or invasion by non-native species. This planning framework has been tested almost from the beginning as we have looked for ways to respond to change. For example, the impending spread of emerald ash borer (*Agrilus planipennis*) into Northern Minnesota is expected to significantly reduce the abundance and distribution of ash species (*Fraxinus* spp.) on the Chippewa and Superior National Forests. While the Forest Plans did not predict or provide specific management guidance to address this, managers (including partners and members of the public) are identifying and beginning to implement adaptive management options to address this issue, while remaining consistent with the general direction of plans. An appreciation of HRV facilitated these discussions – the potential for rapid change was characterized in the HRV document.

The framework established an understanding that the range of historical conditions described in the HRV assessment allows for a bounded continuum of change, rather than seeking a single condition to achieve and maintain healthy and resilient ecosystems. We recognize that future climatic and ecological conditions are uncertain, but the Forest Plan's objectives, referenced to HRV, establish a range of possible ecological conditions consistent with a range of scenarios, including alternative trajectories of climate change and invasive species.

What would we do differently? During the next planning cycle, we would continue to incorporate HRV concepts because of the important ways it provides understanding of ecosystem health and resiliency. However, we would be more explicit about potential future changes to forests, especially those related to climate change. This likely would include a more thorough ecosystem vulnerability assessment to identify the most urgent and significant threats to ecosystem integrity. In our management planning, we used HRV to identify elements that were greatly diminished from those expected under HRV, but we did not follow through with assessment of what impact that had on ecosystems or the impact that diminishment had on ecosystem and socioeconomic resiliency. For example, though tamarack has significantly declined on the landscape, forests retain function and people have adapted to the loss. What is important is understanding those threats both through the lens of HRV and of predictions of probable impacts so that managers and public have dialogue about management priorities.

Public and internal acceptance of HRV as a planning tool

At the time Forest Plans were finalized, and as they have been implemented, there has been public and internal acceptance for the use of HRV. This support contrasts with the skepticism and confusion that characterized early discussions of HRV. Many were reluctant to use HRV or even adamantly opposed to considering an approach linked to historical ecology. Arguably, the greatest concern was that HRV would define an ecological target which many people assumed to be both ecologically impossible and socially unacceptable. Other concerns included the paucity of data to adequately characterize HRV; the perception that climate change negated the applicability of HRV; fears that elements of HRV (forest composition and age structure) defined in an assessment would lack the resolution necessary to define ecosystem characteristics in the past; and doubts that specialists understood how to use the assessment to improve management.

The transformation of perspective from distrustful or skeptical to supportive or even excited about the application of historical ecology was remarkable. Two factors were responsible for the change in perception. First, once the plan alternatives began to take form, participants recognized that HRV did not define narrow ecological targets. Where goals pointed toward ecological conditions similar to reference conditions, these were planned in long-term incremental steps providing the opportunity to consider the social and economic consequences with opportunity for adjustment. Second, participants recognized that HRV provided a feasible and practical model for assessing sustainability and no alternative model was available. Through clear communication of potential biases and limitations in understanding of HRV, leadership established the legitimacy of historical ecology as the best available science for the current planning process.

The switch in public perception of HRV from initial rejection by some, to support, raises the question – could this new science tool have been introduced more effectively? In retrospect, we suggest that the arm-wrestling over and misunderstanding of HRV may have been necessary and very productive. Through rigorous discussions, the public and agencies developed a common understanding of the concept and similar understanding of the strengths and weaknesses of the tool.

What would we do differently? Starting over, we might have begun with more effective education regarding the application of HRV to land management by being more explicit about the goals of using HRV and the limits of its application. We would more directly explain the use of HRV in a changing climate. Furthermore, we would place an even greater emphasis on how an understanding of HRV intersects with social-economic interests. Because of the changing and deeper understanding of applications of HRV in the last 10 years, and in light of the increased awareness of climate change, it is likely that we would have a more rigorous, and ultimately productive dialogue than we had 10 years ago.

Collaborative planning

Collaborative planning was a vital part of the Forest Planning effort, and the application of HRV played a critical role in successful integration among diverse interests. Both the Forest Service and partners of the MFRC are extremely proud of the model for land-management planning that developed from the effort. Beyond the application of HRV, collaboration improved monitoring, inventory, education, research, and specific management projects that developed from the partnership to tackle a suite of social, economic, and natural resource issues. But it was the early success represented by wrestling over application of HRV that facilitated successful collaboration. It helped establish a foundation from which all the partners, with our diverse interests and responsibilities, could promote sustained use and enjoyment of Minnesota's forest resources.

What would we do differently? In the future, collaborative planning will become, in our view, more critical than ever. In past planning, we and partners of the MFRC established science-based management plans to address public's needs and interests in forest resources. Because of the uncertainties of climate change – as well as social change and adaptability – during the next era of planning, our science foundation may be less certain and may involve a whole suite of different approaches in addition to HRV. We will need to more effectively foster a public debate and dialogue about management priorities. A difficult set of questions must be addressed. For instance, where should limited resources be directed? Should management actions work to resist or promote resilience or to intentionally accommodate change and adaptation of ecosystems and their benefits? Should different landowners take on different management priorities? These are questions for creative and experienced managers and researchers to jointly consider as they engage public stakeholders in future planning.

Forest plan implementation

Developing an effective assessment of HRV for managers is difficult and always plagued by incomplete data, difficult interpretation of diverse input, and the challenge of synthesizing complex ecological patterns to form a reasonable picture of past variation. The HRV assessment supporting land-management planning in Minnesota was no different. For example, lack of data on historical conditions of aquatic systems was one of the greatest gaps. Despite this, proactive management to protect and restore wetlands and riparian areas is successfully being implemented through management of surrounding forests combined with site-specific protections. Lack of quantified data on historical conditions of above-ground biomass, shrub-forb layer, or some elements of ecosystem function (e.g. insect and disease outbreaks) has not prevented proactive man-

agement of these elements. The Forest Plan's narrative-style desired future descriptions and management objectives to move elements toward HRV conditions encourages interdisciplinary resource management specialists to consider new and emerging science and social-economic issues rather than rely on "cookbook" solutions.

What would we do differently? Although we would argue that a *perfect* understanding of HRV is not necessary for its use in forest planning and management, we would continue to develop information on aquatic systems, wildlife populations, and soil and nutrient cycling and other resources as equally important elements of ecosystem condition. This would be valuable from two perspectives. First, it would bring a stronger science basis to planning for sustainability and public benefits of these elements. Second, it would represent evidence and direct communication to public and managers of the importance of these elements. For many, the focus on forest composition, structure, and pattern suggests less concern for non-forest resources compared to forests. Though the plans were explicit about the interconnection between forests and water, wildlife, and soil, a bias toward forest commodities was perceived by some people. Other than "beefing up" HRV characterization for other elements, we see no driving need to change the plan's narrative-style management direction because of the flexibility it provides to change strategies and implementation tactics or management practices.

12.9 CONCLUSION

In the most general sense, and despite struggles and uncertainty over how to objectively incorporate historical ecology into land-management planning, we endorse the focused application of HRV assessments. Looking both backward to the start of planning and forward to the changing forests, benefits associated with focused planning and clearer understanding of ecosystems far outweigh challenges and costs. This is especially true when planning is understood to be an iterative and adaptive process. Despite clear gaps in understanding of both current and historical conditions, we and our partners must make management decisions – the insights from HRV provided increased confidence in the efficacy of management direction in light of the acknowledged uncertainty.

REFERENCES

Christensen, N.L., Bartuska, A.M., Brown, J.H., et al. (1996). The report of the Ecological Society of America on the scientific basis for ecosystem management. *Ecological Applications*, **6**(3), 665–691.

Cleland, D.T., Avers, P.E., McNab, W.H., Jensen, M.E., Bailey, R.G., King, T., & Russell, W.E. (1997). National hierarchical framework of ecological units. In *Ecosystem Management Applications for Sustainable Forest and Wildlife Resources* (ed. M.S. Boyce and A. Haney), pp. 181–200. Yale University Press, New Haven, CT, USA.

Dillon, G.K., Knight, D.H., & Meyer, C.B. (2003). *Historic Variability for Upland Vegetation in the Medicine Bow National Forest, Wyoming*. Report submitted to the Medicine Bow National Forest and the Division of Renewable Resources, Rocky Mountain Region. USDA Forest Service, Lakewood, CO, USA. http://www.fs.fed.us/r2/projects/scp/tea/medbownationalforesthrv.pdf.

Frelich, L.E. (1998). Natural disturbance and variability of forested ecosystems in northern Minnesota. Report prepared for the Minnesota Forest Resource Council and the National Forests in Minnesota, St. Paul, MN, USA.

Frelich, L.E. (1999). Range of natural variability in forest structure for the northern Superior Uplands. Report prepared for the Minnesota Forest Resources Council and the National Forests in Minnesota, St. Paul, MN, USA.

Frelich, L.E. (2000). Natural range of variability estimates for vegetation growth stages of Minnesota's drift and lake plains. Report prepared for the Minnesota Forest Resources Council and the National Forests in Minnesota, St. Paul, MN, USA.

Minnesota Forest Resources Council. (1999). Minnesota northeast regional landscape: current conditions and trends assessment. Report LT-0799. Minnesota Forest Resources Council, St. Paul, MN, USA. http://www.frc.state.mn.us/documents/council/landscape/NE%20Landscape/MFRC_CurrentConditions&Trends_NE_MN_1999_Report.pdf.

Minnesota Forest Resources Council. (2000). Minnesota north central landscape current conditions and trends assessment. Report LT-0500. Minnesota Forest Resources Council, St. Paul, MN, USA. http://www.frc.state.mn.us/documents/council/landscape/NC%20Landscape/MFRC_CurrentConditions&Trends_NC_MN_2000-05-01_Report.pdf.

Minnesota Forest Resources Council. (2003a). Recommended desired outcomes, goals, and strategies: northeast landscape region. Report LP-0303a. Minnesota Forest Resources Council, St. Paul, MN USA. http://www.frc.state.mn.us/documents/council/landscape/NE%20Landscape/MFRC_ForestResource%20Management_LandscapePlan_NE_MN_2003-03-01_Plan.pdf.

Minnesota Forest Resources Council. (2003b). *Recommended Desired Outcomes, Goals, and Strategies: North Central Landscape Region*. Minnesota Forest Resources Council, St. Paul, MN, USA. http://www.frc.state.mn.us/documents/council/landscape/NC%20Landscape/MFRC_ForestResourceManagement_LandscapePlan_NC_MN_2003-03-01_Plan.pdf.

USDA. (1993). *National Hierarchical Framework of Ecological Units (ECOMAP)*. USDA Forest Service, Washington, DC, USA.

USDA. (1999). Sustaining the people's lands: recommendations for stewardship of the national forests and grasslands into the next century. Report by the USDA Committee of Scientists. USDA, Washington, DC, USA. http://www.fs.fed.us/news/news_archived/science/cosfrnt.pdf.

USDA. (2004a). *Land and Resource Management Plans for the Chippewa and Superior National Forests*. USDA Forest Service, Milwaukee, WI, USA. http://www.fs.fed.us/r9.

USDA. (2004b). *Final Environmental Impact Statement for the Chippewa and Superior National Forest Plans*. USDA Forest Service, Milwaukee, WI, USA. http://www.fs.usda.gov/Internet/FSE_DOCUMENTS/fsm91_048471.pdf.

Veblen, T.T. & Donnegan, J.A. (2005). *Historical Range of Variability for Forest Vegetation of the National Forests of the Colorado Front Range*. USDA Forest Service, Golden, CO, USA.

Wolter, P.T. & White, M.A. (2002). Recent forest cover type transitions and landscape structural changes in northeast Minnesota, USA. *Landscape Ecology*, **17**, 133–155.

Chapter 13

APPLYING HISTORICAL FIRE-REGIME CONCEPTS TO FOREST MANAGEMENT IN THE WESTERN UNITED STATES: THREE CASE STUDIES

Thomas E. DeMeo,[1] Frederick J. Swanson,[2] Edward B. Smith,[3] Steven C. Buttrick,[4] Jane Kertis,[5] Jeanne Rice,[6] Christopher D. Ringo,[7] Amy Waltz,[8] Chris Zanger,[8] Cheryl A. Friesen,[9] and John H. Cissel[10]

[1]USDA Forest Service, Pacific Northwest Regional Office, Portland, OR, USA
[2]USDA Forest Service, Pacific Northwest Research Station, Corvallis, OR, USA
[3]The Nature Conservancy, Flagstaff, AZ, USA
[4]The Nature Conservancy, Portland, OR, USA
[5]USDA Forest Service, Siuslaw National Forest, Corvallis, OR, USA
[6]USDA Forest Service, Mt. Hood National Forest, Sandy, OR, USA
[7]TetraTech CES, Mt. Vernon, WA, USA
[8]The Nature Conservancy, Bend, OR, USA
[9]USDA Forest Service, Central Cascades Adaptive Management Partnership, Springfield, OR, USA
[10]Joint Fire Science Program, Boise, ID, USA

Historical Environmental Variation in Conservation and Natural Resource Management, First Edition. Edited by John A. Wiens, Gregory D. Hayward, Hugh D. Safford, and Catherine M. Giffen.
© 2012 John Wiley & Sons, Ltd. Published 2012 by John Wiley & Sons, Ltd.

13.1   INTRODUCTION

Studies of fire history have long been a core component of our understanding of fire regimes. Fire regimes – frequency, severity, size, and seasonality – provide a conceptual framework for restoring fire as an ecological process, and are the basis for management applications or strategies that utilize the natural or historical range of variation (HRV) concept (Hann et al. 1997; Barrett et al. 2006; Keane et al. 2009). Throughout this book, HRV has been defined as "the variation of ecological characteristics and processes over scales of time and space that are appropriate for a given management application" (Romme et al., Chapter 1, this book). Wong and Iverson (2004) were more specific, defining HRV (referring to it as the "range of natural variability") as "the temporal and spatial distribution of ecological processes and structures prior to European settlement of North America." The issues of how spatial and temporal scales, reference period, and data availability and accuracy affect HRV are discussed in other chapters of this book. Here we present several examples of real-world applications to illustrate the practical value of fire-history studies, as well as the challenges and concerns of doing so.

First, however, consider the gradient of fire regimes, ranging from frequent, low-severity fires (such as in ponderosa pine [*Pinus ponderosa*] ecosystems) to frequent, high-severity fires (in grasslands), to mixed-severity regimes (as in mixed-conifer forests across the American West), to the longer-interval regimes where fires are uncommon but are of high severity when they do occur (e.g. in some lodgepole pine [*Pinus contorta*] ecosystems and in moister western hemlock [*Tsuga heterophylla*] forests). These fire-regime gradients have profound ecological and management implications. One obvious implication arises from comparisons of current and past fire regimes in a given ecosystem. If we assume that past fire regimes were functioning in a resilient, sustainable manner, comparisons with the past can provide a measure of how well current ecosystems are functioning (i.e. are fire frequency and severity congruent with the HRV?). Lack of approximate congruence between past (or, potentially, future) patterns and current patterns could lead to fires with adverse outcomes on water, soil, wildlife, and biodiversity, attributes whose current expressions on the landscape are due in large part to their history.

Some fire regimes of forest ecosystems in the western United States are currently experiencing a departure from the HRV, for a variety of reasons. In some cases, a century or more of timber extraction, grazing, and fire suppression has led to greater tree density and more extensive mid-seral, closed-canopy forest relative to the historical range of conditions (Peterson et al. 2005). Changed forest structure has multiple ecological impacts, including changed disturbance regimes that can feed back into profound and sudden changes in fundamental ecosystem attributes such as species composition, forest structure, soils, nutrient cycling, and wildlife habitat. In other cases, where current rates of anthropogenic fire are much higher than under past conditions, similarly profound ecosystem changes can result. An example occurs in southern California, where high fire frequency has denuded mountain slopes of their shrub cover and led to high rates of soil erosion and the invasion of exotic plants (Stephenson & Calcarone 1999).

In this chapter, we present three case studies illustrating how historical ecology and HRV can be used to guide resource management and restoration in fire-prone forest ecosystems. The first two case studies – from Oregon and Arizona – involve application of fire-regime condition class (FRCC) methodology, providing a simple, easily understandable technique to measure current departure from historical conditions in forest structure and fire regime. The third case study – also from Oregon – describes a collaborative landscape-management process in which management plans and activities are derived directly from studies of HRV. In the three studies, common themes include the fundamental value of historical ecological data to contemporary and future resource management; the importance of using timely, easily understood assessment methods; and the need to implement findings in a collaborative framework.

13.2   CASE STUDY 1: FRCC AND THE NORTHWEST FIRE LEARNING NETWORK (NFLN): USING HRV AND STAKEHOLDER VALUES TO HELP PRIORITIZE AND FACILITATE FOREST RESTORATION

FRCC

FRCC (Hann et al. 2004; Barrett et al. 2006; Barrett et al. in press) is a quantitative system for identifying degree of ecological departure from the HRV in fire frequency, severity, and seral stages in a given area.

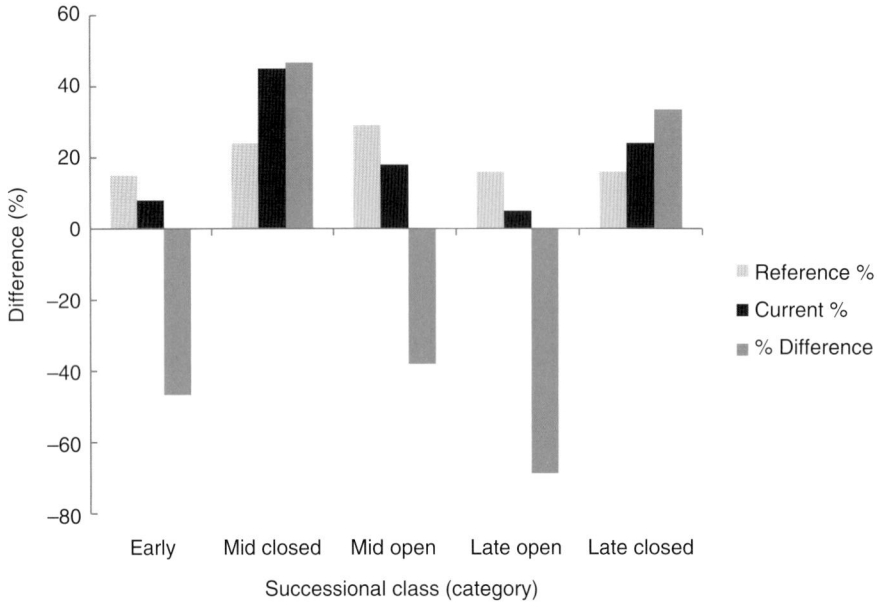

Fig. 13.1 Fire-regime condition class (FRCC) percent departure: successional class (defined by their canopy cover and size) reference and current percent in the upper Deschutes River basin (Oregon, USA) landscape, and the percent difference between the reference and current values. A Sörensen's Index-based similarity measure is used to determine the percent departure of the entire landscape. (See Hann et al. 2004.)

Hundreds of field practitioners have been trained in the method, and modeled FRCC is a key data layer in the national LANDFIRE spatial database (LANDFIRE 2010). The Healthy Forests Restoration Act (HFRA) of 2003 specifically directed Federal land-management agencies to use FRCC as a planning metric, and the method has been used in many national and regional restoration and fuel-management prioritization exercises (Hann & Bunnell 2001; Provencher et al. 2008, Rice et al., unpublished report).

This measure of a landscape's ecological departure from HRV consists of two parts: succession-class departure and fire-regime departure. The succession-class departure is a measure of the difference between the current mix of seral stages on a landscape and the historical mix. For each vegetation type, landscapes are organized into different seral stages defined by their successional status, canopy cover, and dominant species size (a mappable surrogate for age). Estimates of the proportions of the historical landscape occupied by each seral stage are modeled using nonequilibrium, aspatial, state-and-transition modeling, usually with

the Vegetation Dynamics Development Tool (VDDT) (ESSA 2007). The average outcome of multicentury Monte Carlo simulations within VDDT is taken as the "reference" state, and this average outcome is compared with the contemporary mix of seral stages on the landscape, derived from field observation or (more commonly) from maps or satellite images. Sorenson's Index (a simple similarity index) is then used to quantify the level (percent) of "departure" from the reference state (see Hann et al. 2004 for specifics; Fig. 13.1). Similarly, the fire-regime departure compares historical and current measures of fire frequency and severity by means of a simple percent departure metric (Hann et al. 2004). In the end, the departure percentages for both succession-class and fire-regime departures are categorized into "Condition Classes," where Condition Class (CC) 1 represents current values that are 0–33% different from the reference values (CC1 is often defined as being "within HRV"), CC2 includes values that are >33–66% different ("moderately departed"), and CC3 represents values >67% different ("highly departed"). The key strength of FRCC is its relative simplicity and ease of communication. It has moved landscape-ecology assessments from a rather small group of expert practitioners in academia and research to a broad audience of users. The use of FRCC in LAND-FIRE, in implementing HFRA, and in reporting accomplishments has helped institutionalize the method. Moving HRV concepts to this much broader array of practitioners and interests is no small achievement in technology transfer. Map displays of the three Condition Classes are an obvious and straightforward presentation of restoration need. Another strength is the consistency and availability of FRCC map data, which exist nationwide for all land ownerships.

Like all methods that try to simplify complex ideas into an easily digestible format, FRCC has shortcomings. Determining the historical reference conditions for FRCC applications relies heavily on VDDT. The average final outcome of the VDDT runs is used as the set of reference conditions for FRCC, rather than some measure of the variability in reference conditions. This reliance on the mean values from the state-and-transition models is probably the most criticized feature of FRCC (see Chapter 4, this book). In addition, expert judgment must often fill data gaps in the VDDT input parameters, especially in vegetation types for which there is not a rich history of HRV information. For example, in their tree-ring studies, Swetnam and Brown (2010) found that forest ecosystems with shorter fire-return intervals (which have been heavily studied) correlated fairly well with FRCC results, but those with longer intervals (which have been much less studied) did not. Another problem in the use of FRCC is its sensitivity to scale, but this is a problem common to all applications of historical data (see Chapter 5, this book). Even modest changes in the area analyzed can lead to significant changes in FRCC outcome. FRCC methodology was developed using relatively few structural stage classes (five), relatively uniform assumptions about the time period of the range of variation, and a focus on overstory structural conditions that tends to ignore species composition or the importance of understory fuels to fire behavior (Keane et al. 2009, H. Safford, pers. comm.).

Applying FRCC to restoration planning in oregon, USA

Changes in forest structure in central Oregon have occurred since Euro-American settlement due to fire exclusion, grazing, and timber management (Agee 1993). In forests historically adapted to frequent fire, changes in forest structure have led to changes in fire behavior, resulting in increased fire risk to human communities as well as a decline in or loss of plant and animal species adapted to fire regimes (Hessburg et al. 2005). Current forests may lack key elements of sustainable and resilient systems, posing a risk to their long-term persistence.

To help address these issues, federal agencies have treated thousands of hectares of forests and woodlands over the past 10 years through application of prescribed fire, thinning, and wildland fire to reduce fuel loads and/or modify fire behavior (National Fire Plan, http://199.134.225.81/index.htm), but the level of work has not matched the need (Reinhardt et al. 2008). Underinvestment in the use of fire and forest restoration, lack of or limitations in wood-product processing infrastructure, market conditions, fragmented ownership patterns, and long-standing disagreements over how forests should be managed continue to impede progress in addressing this problem.

The Nature Conservancy analyzed LANDFIRE National data for Oregon to examine the scope of Oregon forest departure from their modeled HRV (Macdonald et al. 2006) to better understand the scale of the problem. The analysis focused only on those forests and woodlands that are adapted to frequent fire (forests and woodlands with low- or mixed-severity fire regimes – Fire Regimes I and III) and today have moderately or highly altered species composition and structure (Condition Classes 2 and 3) as a result of fire exclusion and other factors. Out of the approximately 12.1 million hectares of forest and woodland in Oregon, over 6 million hectares are in Fire Regimes I and III and in Condition Classes 2 or 3; that is, moderately to highly departed from their HRV for forest structure, fire regime, or both. Public land, mostly Forest Service and Bureau of Land Management, accounted for almost 3.8 million hectares of this area.

To treat these 3.8 million hectares over the next 20 years, over 192 000 ha would need to be treated annually, not including treatments needed to maintain those areas that are in CC1 (i.e. considered to be within the historical range). Current treatment levels would need to more than double to close the gap between identified treatment needs and current efforts. With an increase in treatment rate comes a need to locate treatments strategically to maximize their effectiveness, minimize adverse impacts, protect important natural resources, and resume natural processes.

The NFLN and the Upper Deschutes River Basin

The NFLN was established in 2005 by The Nature Conservancy in cooperation with the Forest Service and Bureau of Land Management (see http://www.conservationgateway.org/topic/fire-learning-network). The Network's goal is to facilitate forest restoration by addressing barriers to restoring forests and fire resiliency on federal, state, and private landscapes. In Oregon and Washington, the NFLN has focused its efforts on several large, multi-jurisdictional landscapes. Here we consider the Upper Deschutes River Basin, an 800 000-ha landscape managed principally by the Forest Service, but with areas managed by private, state, and local government landowners as well.

In the Upper Deschutes Basin, the NFLN prioritized areas for restoration using a two-step process, first collecting and summarizing spatially explicit ecological information and then incorporating input from local stakeholders. In the first step, an NFLN technical team set out to develop an assessment of current forest conditions across the Basin with regard to their historical sustainability and to prioritize forests and stands in need of restoration. Using satellite imagery classified according to Forest Inventory and Analysis (FIA) ground plots, the team mapped vegetation structure across the Basin and assigned forest stands to FRCC succession classes. The FRCC Mapping Tool (Johnson & Tirmenstein 2007) characterized current percent departure (and CC) from the published FRCC reference conditions for central Oregon. The NFLN used these data to create a prioritized "Treatable Forest Stands" map, where high abundances of mid-successional stage, closed-canopy stands in ponderosa pine forests were found. The mid-seral closed class is often one of the most departed successional stages in currently fire-suppressed forests that once supported frequent surface fires (Fig. 13.1). Active management in these stands has the potential to shift forest structure to a mid-succession, open-canopy condition better aligned with HRV.

FRCC is a tool to help identify and prioritize stands for restoration, but it does not address legal challenges brought about by conflicting values and stakeholder distrust. To address this, NFLN held two stakeholder workshops in 2009 to develop a list of attributes valued by participants and spatial layers representing these values on the landscape. Participants in each workshop were asked what they valued most in the Upper Deschutes Basin and, in their opinion, what landscape attributes were most vulnerable to severe wildfires and other uncharacteristic changes (i.e. disturbance events outside the HRV that did not occur prior to Euro-American settlement; Hann et al. 2004). The resulting list of values was collaboratively refined, collapsing values that were redundant and translating values to attributes that could be spatially represented. The participants in each workshop agreed on the top values. Examples of these values included riparian areas and wetlands, threatened and endangered species, old-growth ponderosa pine and dry mixed-conifer forest, water quality, wildlife habitat, high-value recreation, and scenic quality/viewsheds. From the two workshops, spatial data were compiled to represent each value. Combined value scores (referring to the number of values that occupy the same pixel) were mapped. These results were then filtered by the Treatable Stands restoration priority map developed in step one. Areas where multiple stakeholder values and restoration needs overlapped were visually identified as high-value areas.

Step One identified 89 000 ha with restoration needs based on FRCC departure for fire regime, succession class departure, or both. Values overlaid on these restoration areas (Step Two) can help prioritize areas where stand-structure restoration might happen first. For example, although only 2 ha on the Upper Deschutes landscape registered all values identified at both workshops, 14 000 ha captured seven or more values. Based on the workshop results, the NFLN anticipates a large constituency of stakeholders will support restoration treatments on these areas with minimized legal challenges, facilitating more rapid and focused restoration activities within this important landscape.

13.3 CASE STUDY 2: COMPARISON OF HISTORICAL VERSUS CURRENT FIRE FREQUENCY AND SEVERITY IN ARIZONA

In 2005, the United States Forest Service entered into a cost-share agreement with The Nature Conservancy in Arizona to produce descriptions of HRV and vegetation-succession models for major vegetation types of the Southwest Region (Arizona and New

Mexico). This assessment summarized empirical studies for 11 major vegetation types, identifying departure from HRV and thus departure from the range of structure, composition, and disturbances. This information shapes regional priorities for ecological restoration and fuel-treatment activities and helps shape ecological goals for forest-plan revisions across the Region's 11 National Forests. HRV descriptions and successional models can be found at TNC (2011).

These descriptions include landscape vegetation and species composition, age- and size-class structure, and other attributes of patch or stand structure. The latter includes patch size and spatial distribution, vegetation density, and canopy closure and/or cover. HRV descriptions also include descriptions of the natural processes shaping potential vegetation through time, including succession or patch dynamics, climatic fluctuations, fire regimes, hydrological processes, nutrient cycling, predator–prey interactions, herbivory, insects and pathogens, windthrow, avalanche, and erosion.

The variability and rate of change of these characteristics are as important to understanding HRV as is the magnitude of their change. Thus, defining the time period for which the HRV is described is important, as is defining the influence of humans on changes in ecosystem characteristics. Longer time periods often reveal a greater range of variation. Several authors have noted that contemporary patterns of vegetation and their dynamic processes developed in the Southwest during the early Holocene, around 11 000–8000 years ago (Allen 2002). Due to limitations in the availability of recorded data from tree rings, pollen, and charcoal, we limited our description of the HRV (unless otherwise noted) to the period from 1000 AD to 1880, when significant European settlement in the region began. We refer to that portion of the HRV resulting from conditions after 1880 as the postsettlement or current period. To explore differences between presettlement (historical conditions) and postsettlement (current conditions), we used published values for ecosystem disturbances to build quantitative succession models for both.

Composition, structure, and function of ecosystems

Anthropogenic disturbance has led to major changes in ponderosa pine forest structure and function. With the introduction of grazing animals at various times

during the nineteenth century, and active fire suppression throughout the twentieth century and into the twenty-first century, low-severity and frequent surface fires have been replaced with high-severity and infrequent crown fires (see Chapter 1, this book). Although the effects of these large, stand-replacing fires are variable, several fires have led to long-term changes from forested systems to grasslands, shrublands, and areas of dense pine regeneration (Savage & Mast 2005). Areas that have not yet burned have a higher density of young trees, changing the quality of habitat for wildlife and humans, and increasing the probability of stand-replacing fire (Covington & Moore 1994; Allen 2002).

Modeling outputs and trade-offs

As in the previous case study, the VDDT was used to develop both historical (pre-1880) and current (1880 to present) period models of vegetation change for several potential natural vegetation types of the Southwest. The VDDT software allows the user to model succession as a series of vegetation states differing in structure, composition, and cover, and to specify the amount of time it takes to move from one vegetation state to another in the absence of disturbance. Various disturbance agents (e.g. surface fires, stand-replacing fires, grazing, and insect outbreaks) affecting the transition of vegetation between states can then be incorporated as probabilistic functions. By varying the types and rates of disturbance across the landscape, the effects of different management treatments on future vegetation can be investigated. For the ponderosa pine model, we included the effects of fire, insects, and precipitation; we display the modeling outputs as proportions of ecological states or successional stages (Fig. 13.2).

In historical times, ponderosa pine occurred primarily as late-seral open forests dominated by large trees, with relatively small patches of younger trees (Covington & Moore 1994). The VDDT modeling process uses this type of information to generate reference conditions (ESSA Technologies Ltd. 2007). Fire suppression and the resulting uncharacteristic stand-replacing fires generated a novel state on the landscape, the so-called "anomalous grassland" (Savage & Mast 2005; Kuenzi et al. 2008). This novel state has experienced complete or nearly complete overstory tree removal and, in many cases, soil sterilization that favors exotic

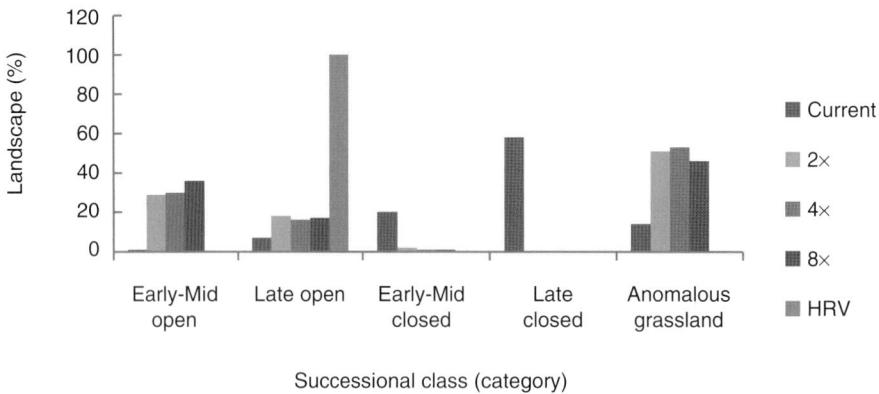

Fig. 13.2 Changes in ponderosa pine (*Pinus ponderosa*) forest condition under 100-year scenarios of varying stand-replacing fire (SRF) frequency. The current condition is compared with scenarios of increasing fire frequency (two, four, and eight times the current frequency; 2×, 4×, and 8×, respectively) and with the historical range of variation (HRV). With a warming climate, increases in fire frequency are anticipated. Data displays such as this allow planners to prepare for future scenarios, such as increases in a novel (anomalous) grassland state (last successional class displayed on *x*-axis). See text for further details.

species of grasses and forbs. At many sites, this novel state has been shown to persist for decades, suggesting an alternative stable state outside the HRV. Recently, succession models have explored scenarios of management over small to large areas (Provencher et al. 2008). We used our current-conditions model and changed the frequency of stand-replacing fires to give a simplistic indication of possible ecological trajectories for Southwest ponderosa pine under a warming climate. As a surrogate for increasing temperatures and drought, we can model increased frequency of stand-replacing fire across the landscape and compare modeling results to current and historical conditions (Fig. 13.2). Under current conditions, stand-replacing fire occurs about once every 100 years. When stand-replacing fire occurs as frequently as once every 50 years, about 50% of the landscape moves to the novel, anomalous grassland state. Other predicted changes include an expansion in early- and mid-open classes and losses in area within the closed-canopy condition (Fig. 13.2).

This methodology is a useful tool to describe potential landscape changes. By gaining some understanding of what these changes may be, management can be adjusted to help maintain landscape resiliency in the face of the more frequent fires expected in a warming climate.

13.4 CASE STUDY 3: INCORPORATING HISTORICAL ECOLOGY AND CLIMATE CHANGE INTO LAND MANAGEMENT: BLUE RIVER LANDSCAPE PLAN, WILLAMETTE NATIONAL FOREST, OREGON

The Blue River Landscape Plan and Administrative Study area, in the Willamette National Forest of western Oregon's Cascade Range, applies historical ecology information to the planning and conduct of active management on National Forest lands (Cissel et al. 1999). Management is planned in the context of general objectives of the Northwest Forest Plan (NWFP) (USDA and USDI 1994) to protect terrestrial and aquatic species, and in response to the Plan's specific charge to the Central Cascades Adaptive Management Area (AMA). This AMA, incorporating the Blue River Plan area, explores "approaches for integrating forest and stream management objectives and implications of natural disturbance regimes." The terrain is steep, ranging from 350 m to 1600 m elevation; the climate is wet (mean annual precipitation is ca. 2500 mm in lower elevation areas) and ranges from rain-dominated at low elevations to snow-dominated at upper elevations. The highly productive, dominantly conifer forest cover is in the western hemlock zone at low and middle elevations and the Pacific silver fir (*Abies amabilis*) zone at upper elevations (Franklin & Dyrness 1988). Timber harvest began in the 1950s and continues today, albeit at a much slower pace than during the 1950s through the 1980s. Most unharvested stands in the area date from wildfires in the 1500s and 1800s.

Fire history in the study area was interpreted from tree-establishment and fire-scar dates, using analysis of tree rings up to approximately 800 years old (Weisberg 1998). Archival and paleoecological records were also used. Site fire frequency ranged from 50 years to more than 500 years, and severity ranged from light ground fire to stand-replacement disturbance. About 25% of the native-forest area was clear-cut, and plantations were established in 1950–1990 in patches averaging ca. 15 ha in size. Additional areas of mature forest (80 to 200 years old) experienced light thinning during this period.

The management plan was guided by historical wildfire-regime information and considerations for watershed processes and species of special interest (Cissel et al. 1999). Historical fire regimes were used to develop three prescriptions for forest-stand management using cutting frequencies of 100, 180, and 260 years, with respective levels of live tree-canopy retention at 50, 30, and 15%. The geographic distribution of historical wildfire regimes was used to designate the spatial patterns of cuts, which were also designed in part to minimize further fragmentation of older forest created by earlier clear-cutting (Franklin & Forman 1987). Prescribed fire was selected to create snags and to maintain fire as a keystone process in the forest landscape. Some reserve areas for selected species and riparian reserves were designated, although much less extensively than in standard NWFP land allocations.

In keeping with the adaptive management purpose of this planning area, prescriptions were developed and implemented to examine effects of management practices. Monitoring, modeling, and other assessments of the practices and their effects on the environment have been conducted (Swanson et al. 2003; Gray & Miller 2006). The first two timber sales were executed as planned and environmental effects matched those expected. However, the overall rate of forest harvest has been much less than anticipated for both the NWFP and the Blue River areas because of public pressure against cutting of unharvested forests (Moeur et al. 2005; Rapp 2008).

The Blue River landscape-planning process considered climate variability. Historical variation in fire occurrence at the centennial scale (Weisberg & Swanson 2003) raises questions about ecological effects of a forest-harvest program in an area with century-long periods of extensive fire separated by two centuries of little fire. Because the nature of the future climate and its ecological impacts are uncertain, interdisciplinary deliberations on management implications of future climate change have not yet led to specific changes in the landscape plan. These deliberations instead led to several questions and considerations:

1. Do areas of persistent cold-air drainage provide thermal refuges from climate warming? Recent research (Daly et al. 2009; Tepley 2010) documents substantial topographic effects on microclimate and fire regimes, suggesting that management to address climate change should be nuanced with respect to topography and air drainage.

2. Will the Cascades forest system exhibit abrupt and dramatic threshold responses to climate change? Some high-latitude ecosystems appear to be profoundly affected by climate change, both directly and indirectly (by fire, insects, etc.), but the Oregon Cascade forests do not, yet. Will they? Perhaps the limited influence of freezing temperatures in this maritime climate will limit some potential forms of threshold behavior in hydrologic systems and invertebrate communities observed in areas of colder climate (Williams & Liebhold 2002). It will be important to distinguish the effects of climate change from system change in response to other factors, including legacies of past land use and natural disturbance (see Chapter 5, this book).

3. What does the experience of developing and implementing early stages of the Blue River Landscape Plan say about alternative approaches to landscape management? First, it has been possible to design and implement some features of an HRV-based landscape-management approach that includes timber harvest. Second, there is value in blending approaches; in fact, the Blue River Landscape Plan is a blend of reserves derived from species-conservation approaches with a dynamic landscape-HRV approach. Third, climate change will likely bring new management imperatives, which will likely drive a new landscape-management approach aimed at fostering resilience of ecosystems in a changing environment. Finally, there will invariably be public opposition to new ideas in landscape management, so education and public involvement will be critical to success (Shindler & Mallon 2009).

4. Are current Federal land-management practices of fire suppression and harvest, limited mainly to thinning in plantations, creating landscapes vulnerable to climate-change impacts? The answer may depend on the historical fire regime. Fire suppression in forests historically characterized by frequent fire may lead to development of dense, stressed stands vulnerable to the added stress of climate change. On the other hand, thinning plantations in forests with historically low fire frequency, using approaches that maintain and enhance natural heterogeneity of stand and landscape composition and structure, may reduce stress imposed by climate change.

What does the future hold for continued implementation of the Blue River Landscape Plan and Study and for related learning opportunities? As for research, the H.J. Andrews Experimental Forest Long-Term Ecological Research program, located within the Blue River Plan area, can be expected to persist and provide insights to critical questions about landscape dynamics in a changing climate (Franklin et al. 1990). On the land-management and policy fronts, the Blue River Study has been hampered by near elimination of support for the AMA program at the NWFP scale and near cessation of logging of previously unharvested forest, which was expected to be a major source of wood products under at least the first few decades of the NWFP and Blue River Landscape Plan.

Nevertheless, collaboration of the Willamette National Forest and its long-standing partners in the science community carries the Blue River landscape work forward. These circumstances illustrate the need for long-term, institutional support for research and planning for managing landscapes that historical ecology tells us are inherently dynamic.

13.5 CONCLUSIONS

Although HRV does not provide an absolute target or goal for ecosystem restoration in itself, a knowledge of local to regional variability helps to understand current and potential future conditions in relation to the past. Use of HRV can provide a foundation for building quantitative models representing how ecosystems function. Input parameters of successional models can be changed to build a variety of plausible scenarios, allowing land managers and stakeholders to compare results of different treatment strategies and changes in

disturbance frequency. Use of these tools can help us think strategically about goal setting and can inform a collaborative dialogue about the scale and scope of actions necessary to achieve those goals.

In this chapter we explored three practical applications of this process. The first two case studies (in Oregon and Arizona) compared current versus historical vegetation conditions using state-and-transition modeling and FRCC methodology, based on simple comparisons of forest structure between modeled reference conditions and current vegetation maps. Although FRCC methodology has some important limitations, it has found widespread use in many western landscapes in identifying restoration needs, due principally to its simplicity and its widespread availability.

Although significant effort and funding has been directed toward restoring Federally administered forest lands in the western United States with thinning and prescribed burning, it has become obvious that the areas treated are insufficient to improve conditions across many watersheds and regions. National, regional, and local FRCC assessments underline the extent of the fire-suppression problem and make clear the need for greater use of unplanned ignitions ("wildland fire use") in meeting restoration goals (Miller & Landres 2004; Doane et al. 2006). FRCC brings an understanding of landscape ecology to multiple levels of land-management agencies and to practitioners who would not otherwise be exposed to this perspective. The value of this in shaping future goals for managing fire regimes and biodiversity should not be underestimated.

The third case study (Blue River) employed more traditional fire-history methods to illustrate the application of the historical fire regimes in selecting the pattern of timber harvest rotations, and also to assess implications of climate change. This case study showed the value of blending assessment approaches and how stakeholder values and activism can complicate assessment and planning.

These case studies provide several important messages about applying HRV concepts to management. First, in a management context, relatively simple, extensive methods, such as FRCC assessment, are needed to communicate HRV concepts to a broad audience. They are also needed to make timely assessments in a continually changing planning horizon. This simplicity and timeliness comes at a cost of accuracy and a heavy reliance on state-and-transition modeling. The latter involves multiple assumptions, which may be difficult to verify.

Second, although the relevance of HRV has been questioned as changes in the climate move us into new domains of variation, Keane et al. (2009) point out that HRV parameters are good placeholders for initial conversations on goal-setting in the near term. When comparing several plausible climate-projection models, the outputs show more variability and uncertainty than exists from HRV estimates of the past (Stainforth et al. 2005; Keane et al. 2009). We know that past-forested landscapes exhibited a level of resilience that maintained forest structure and composition for hundreds of years, so using HRV conditions as a starting point for goal-setting – along with modifications for anticipated climate changes – should foster resilience as the changes are manifested.

Third, in predicting climate-change effects on forests, an assessment of the HRV is essential to understanding how these systems rebounded from disturbance and how they will continue to respond to future changes. The Blue River case study shows how HRV concepts can be used to answer questions about the effect of a warming climate on management options. Until credible and useful estimates of a future range of variation are available, HRV provides a powerful tool to help determine, illustrate, and communicate management options.

Finally, all three case studies indicate that information alone is never sufficient for land-use planning. Public involvement and values, as well as uncertain institutional support, can work at cross purposes to the actions suggested by an HRV assessment, regardless of its accuracy or the soundness of the science behind it. Fostering collaboration and good communication are therefore critical elements in successful implementation of an HRV approach.

REFERENCES

Agee, J.K. (1993). *Fire Ecology of Pacific Northwest Forests.* Island Press, Washington, DC, USA.

Allen, C.D. (2002). Lots of lightning and plenty of people: an ecological history of fire in the upland Southwest. In *Fire, Native Peoples, and the Natural Landscape* (ed. T.R. Vale), pp. 143–193. Island Press, Washington, DC, USA.

Barrett, S.W., DeMeo, T., Jones, J.L., Zeiler, J.D., & Hutter, L.C. (2006). Assessing ecological departure from reference conditions with the rire regime condition class (FRCC) mapping tool. In *Fuels Management – How to Measure Success* (ed. P.L.

Andrews and B.W. Butler), pp. 575–585. Conference Proceedings RMRS-P-41. USDA Forest Service, Ft. Collins, CO, USA.

Barrett, S.W., Havlina, D., & Hann, W.J. (2011). Fire regime condition class: concepts, methods, and applications. In *Proceedings of the 3rd International Association of Wildland Fire Congress*, Spokane, Washington, October 2010.

Cissel, J.H., Swanson, F.J., & Weisberg, P.J. (1999). Landscape management using historical fire regimes, Blue River, Oregon. *Ecological Applications*, **9**, 1217–1231.

Covington, W.W. & Moore, M.M. (1994). Post settlement changes in natural fire regimes and forest structure: ecological restoration of old-growth ponderosa pine forests. *Journal of Sustainable Forestry*, **2**, 153–181.

Daly, C., Conklin, D.R., & Unsworth, M.H. (2009). Local atmospheric coupling in complex topography alters climate change impacts. *International Journal of Climatology*, **30**, 1857–1864.

Doane, D., O'Laughlin, J., Morgan, P., & Miller, C. (2006). Barriers to wildland fire use: a preliminary problem analysis. *International Journal of Wilderness*, **12**(1), 36–38.

ESSA. (2007). Vegetation Dynamics Development Tool user guide, Version 6.0. Prepared by ESSA Technologies Ltd., Vancouver, BC, Canada.

Franklin, J.F. & Dyrness, C.T. (1988). *Natural Vegetation of Oregon and Washington.* Oregon State University Press, Corvallis, OR, USA.

Franklin, J.F. & Forman, R.T.T. (1987). Creating landscape patterns by cutting: ecological consequences and principles. *Landscape Ecology*, **1**, 5–18.

Franklin, J.F., Bledsoe, C.S., & Callahan, J.T. (1990). Contributions of the long-term ecological research program. *BioScience*, **40**(7), 509–523.

Gray, A. & Miller, C. (2006). Vegetation change in the Blue River Landscape study: 1998–2005. http://www.fsl.orst.edu/ccem.

Hann, W.J. & Bunnell, D.L. (2001). Fire and land management planning and implementation across multiple scales. *International Journal of Wildland Fire*, **10**, 389–403.

Hann, W.J., Jones, J., Karl, M.G., et al. (1997). An assessment of ecosystem components in the Interior Columbia Basin and portions of the Klamath and Great Basins: Vol. II. Landscape dynamics of the Basin. General Technical Report PNW-GTR-405. USDA Forest Service, Portland, OR, USA.

Hann, W.J., Shlisky, A., Havlina, D., et al. (2004). Interagency fire regime condition class guidebook, version 3.0, September 2001. http://www.frcc.gov.

Hessburg, P.F., Agee, J.K., & Franklin, J. (2005). Dry forests and wildland fires of the inland Northwest, USA: contrasting the landscape ecology of the pre-settlement and modern eras. *Forest Ecology and Management*, **211**, 117–139.

Johnson, J. & Tirmenstein, D. (2007). Fire regime condition class mapping tool user's guide: version 2.2.0. http://www.niftt.gov.

Keane, R.E., Hessburg, P.F., Landres, P.B., & Swanson, F.J. (2009). The use of historical range and variability (HRV) in landscape management. *Forest Ecology and Management*, **258**, 1025–1037.

Kuenzi, A., Fulé, P., & Sieg, C.H. (2008). Effects of fire severity and pre-fire stand treatment on plant community recovery after a large wildfire. *Forest Ecology and Management*, **225**, 855–865.

LANDFIRE. (2010). LANDFIRE project, USDA Forest Service and USDI. http://www.landfire.gov/index.php.

Macdonald, C., Buttrick, S.C., & Schindel, M. (2006). The condition of Oregon's forests and woodlands: implications for the effective conservation of biodiversity. White paper. The Nature Conservancy, Portland, OR, USA.

Miller, C. & Landres, P.B. (2004). Exploring information needs for wildland fire and fuels management. General Technical Report RMRS-GTR-127. USDA Forest Sercive, Fort Collins, CO, USA.

Moeur, M., Spies, T., Hemstrom, M., et al. (2005). Northwest Forest Plan – the first 10 years (1994–2003): status and trend of later-successional and old-growth forest. General Technical Report PNW-GTR-646. USDA Forest Service, Portland, OR, USA.

Peterson, D.L., Johnson, M.C., Agee, J.K., Jain, T.B., Mckenzie, D., & Reinhardt, E.D. (2005). Forest structure and fire hazard in dry forests of the western United States. General Technical Report PNW-GTR-628. USDA Forest Service, Portland, OR, USA.

Provencher, L., Campbell, J., & Nachlinger, J. (2008). Implementation of mid-scale fire regime condition class mapping. *International Journal of Wildland Fire*, **17**, 390–406.

Rapp, V. (2008). Northwest Forest Plan – the first 10 years (1994–2003): first-decade results of the Northwest Forest Plan. General Technical Report PNW-GTR-720. USDA Forest Service, Portland, OR, USA.

Reinhardt, E.D., Keane, R.E., Calkin, D.E., & Cohen, J.D. (2008). Objectives and considerations for wildland fuel treatment in forested ecosystems of the interior western United States. *Forest Ecology and Management*, **256**, 1997–2006.

Savage, M. & Mast, J.N. (2005). How resilient are southwestern ponderosa pine forests after crown fires? *Canadian Journal of Forestry Research*, **35**, 967–977.

Shindler, B. & Mallon, A.L. (2009). *Public Acceptance of Disturbance-Based Forest Management: A Study of the Blue River Landscape Strategy in the Central Cascades Adaptive Management Area*. Research Paper PNW-RP-581, USDA Forest Service, Portland, OR, USA.

Stainforth, D.A., Aina, T., Christensen, C., et al. (2005). Uncertainty in predictions of the climate response to rising levels of greenhouse gases. *Nature*, **433**, 403–406.

Stephenson, J.R. & Calcarone, G.M. (1999). Southern California mountains and foothills assessment: habitat and species conservation issues. General Technical Report GTR-PSW-172. USDA Forest Service, Albany, CA, USA.

Swanson, F.J., Cissel, J.H., & Reger, A. (2003). Landscape management: diversity of approaches and points of comparison. In *Compatible Forest Management* (ed. R.A. Monserud, R.W. Haynes, and A.C. Johnson), pp. 237–266. Kluwer Academic Publishers, Dordrecht, The Netherlands.

Swetnam, T.L. & Brown, P.M. (2010). Comparing selected fire regime condition class (FRCC) and LANDFIRE vegetation model results to tree-ring data. *International Journal of Wildland Fire*, **19**, 1–13.

Tepley, A.J. (2010). *Age Structure, Developmental Pathways, and Fire Regime Characterization of Douglas-fir-western Hemlock Forests in the Central Western Cascades of Oregon*. PhD dissertation, Oregon State University, Corvallis, OR, USA.

TNC. (2011). Various reports. The Nature Conservancy. http://azconservation.org/projects/southwest_forest_assessment/.

USDA and USDI. (1994). *Pacific Northwest Forest Plan*. USDA and USDI, Portland, OR, USA. http://www.reo.gov/general/aboutNWFP.htm.

Weisberg, P.J. (1998). *Fire History, Fire Regimes, and Development of Forest Structure in the Central Western Oregon Cascades*. PhD dissertation, Oregon State University, Corvallis, OR, USA.

Weisberg, P.J. & Swanson, F.J. (2003). Regional synchroneity in fire regimes of western Oregon and Washington, USA. *Forest Ecology and Management*, **172**, 17–28.

Williams, D.W. & Liebhold, A.W. (2002). Climate change and the outbreak ranges of two North American bark beetles. *Agricultural and Forest Entomology*, **4**, 87–99.

Wong, C.M. & Iverson, K. (2004). Range of natural variability: applying the concept to forest management in central British Columbia. *British Columbia Journal of Ecosystems and Management Extension Note*, **4**, 1–56.

Chapter 14

USING HISTORICAL ECOLOGY TO INFORM WILDLIFE CONSERVATION, RESTORATION, AND MANAGEMENT

Beth A. Hahn[1] *and John L. Curnutt*[2]

[1]USDA Forest Service, Northern Region, Missoula, MT, USA
[2]USDA Forest Service, Eastern Region, Milwaukee, WI, USA

Historical Environmental Variation in Conservation and Natural Resource Management, First Edition. Edited by John A. Wiens, Gregory D. Hayward, Hugh D. Safford, and Catherine M. Giffen.
© 2012 John Wiley & Sons, Ltd. Published 2012 by John Wiley & Sons, Ltd.

14.1 INTRODUCTION

Wildlife ecologists strive to understand how anthropogenic and natural disturbances shape ecosystem composition, structure, and function, and how wildlife responds to these changes at multiple spatial and temporal scales. Wildlife responses to ecosystem dynamics reflect interactions between life-history traits and environmental conditions, as framed by the legacy of history, often referred to as the historical range of variation (HRV). Although ongoing, rapid climate change may lessen the utility of HRV assessments for defining management targets, knowledge of historical ecology remains invaluable in guiding wildlife restoration and conservation efforts (Landres et al. 1999; Jackson & Hobbs 2009; Keane et al. 2009).

In this chapter, we demonstrate the power of historical ecology in conservation and land management through several North American case studies focused on vertebrates. We begin by presenting some background on the use and limitations of wildlife HRV analyses, and then present several examples to illustrate the continuing relevance of historical ecology.

The concept of the ecological niche, developed initially by Grinnell (1917) and subsequently elaborated by Elton (1927), Hutchinson (1957), and others, provides a framework for understanding the factors determining species occurrence and abundance in time and space. One important application of niche theory has been the use of historical ecology to inform wildlife management (Wiens & Donoghue 2004; Willis et al. 2007). Wildlife HRV analyses can provide insight into environmental conditions – including climatic and disturbance regimes – that sustained wildlife in the past and may help to identify limiting factors or key thresholds (Landres et al. 1999, Chapter 1, this book). HRV assessments thus represent one approach to defining the climatic and environmental envelopes that define the spatial limits to species distributions.

Compared to past vegetation, historical wildlife data are scarce and often indirect. Furthermore, inference is typically limited to species presence (Stahl 1996; Tingley & Beissinger 2009). Key sources of wildlife occurrence data include written descriptions from early explorers, trappers, missionaries, and settlers (e.g. Jacques Marquette, seventeenth century [Marquette 1917]; Lewis & Clark, nineteenth century [Lewis et al. 2002]; the Hudson Bay Company, seventeenth to nineteenth centuries); natural-history surveys (e.g. Grinnell & Storer 1924); and historical photographs, maps,

and paintings. In North America, these documentary sources typically date back a few hundred years. These opportunistic inventories usually undersample rare and cryptic species and overreport large, charismatic species. Thus, although historical wildlife-occurrence data can yield insights about broad-scale distribution and change (e.g. Zielinski et al. 2005; Aubry et al. 2007; Tingley et al. 2009), uncertainties remain due to sampling variability (Tingley & Beissinger 2009), erroneous site descriptions, and questionable species identifications (e.g. Shelton & Weckerly 2007). Similarly, species-distribution models derived from historical data may lack sufficient robustness to identify conservation-priority areas across a diverse suite of species (Loiselle et al. 2003). Over longer temporal scales, historical wildlife occurrence data can be derived from the paleoecological record (e.g. Mayfield 1972). With these data, evidence of occurrence is more direct and issues of taxa identification, cryptic taxa, and site descriptions become less of a problem, but other concerns surface. For instance, inference from fossils is strongly affected by specimen durability and burial site conditions (Stahl 1996). Alternatively, DNA from wildlife specimens may be analyzed to determine location and size of population bottlenecks (Avise 1995).

The relevance of HRV in guiding restoration and management objectives is increasingly questioned in the context of climate change (see Chapters 4 and 7, this book). Restoring and managing wildlife populations, however, has always required consideration of ecosystem stressors and disturbances, including climate, at multiple spatial and temporal scales. For instance, the reshuffling of species assemblages into non-analogue communities is a common prediction from models of future climate (Root & Schneider 2002; Parmesan 2006; Botkin et al. 2007). Looking backward, paleoecological evidence from the Pleistocene through the Holocene demonstrates that plant and animal species composition fluctuated, often in association with changing climates, and that past communities rarely reflect contemporary species assemblages (Graham & Grimm 1990; Roy et al. 1996; Stahl 1996). Historical ecology, by documenting the dynamism of ecosystems, provides a wider context for managing and restoring ecosystems than could be derived from current conditions alone (Landres et al. 1999; Jackson & Hobbs 2009; Keane et al. 2009). Although HRV analyses are always based on incomplete data, these analyses allow us to better predict ecosystem resilience, including wildlife species components. HRV, in combination with ecological principles, improves wildlife conservation planning by providing a longer temporal perspective for developing restoration objectives and methods than is possible by limiting considerations to existing ecological conditions.

The following case studies demonstrate that management crafted without insights from the past will suffer from the narrow perspective defined by current conditions. These examples illustrate that even an incomplete understanding of past species distribution, abundance, and biology can result in useful management actions. In fact, when an understanding of historical ecology is available, the management response may appear so clear that, in retrospect, the insights provided through the HRV analysis may appear trivial. Without the basic ecological understanding developed by looking backward, however, management decisions might have been quite different.

14.2 CASE STUDIES

Mountain yellow-legged frog

Surveys conducted by field ecologists in the late 1800s described mountain yellow-legged frogs (the complex of *Rana muscosa* and *Rana sierrae*) as the most abundant vertebrate in montane aquatic habitats at elevations above 1500 m in the Sierra Nevada and Transverse Ranges of Nevada and California (Fig. 14.1; Grinnell & Storer 1924; Storer 1925; Stebbins & Cohen 1995). Over a span of less than a century, populations plunged. Mountain yellow-legged frogs are now absent from 92% of historical sites in the Sierra Nevada and 98% of historical sites in the Transverse Ranges (Vredenburg et al. 2007). Currently, the southern California population is listed as endangered under the United States Endangered Species Act, while the Sierra Nevada population is considered warranted but precluded. What factors led to this dramatic change?

Examination of ecosystem dynamics and frog-population trends over the last 150 years helps to explain the current status of mountain yellow-legged frogs. Prior to the mid-1800s, virtually all Sierran lakes, ponds, and streams above 2100 m were fishless. Waterfalls served as effective barriers to the upward movement of fish (Knapp 1996). Nonnative fish introductions began in the 1850s, and fish stocking was formally initiated by California Department of Fish and Game beginning in the 1920s, continuing through

Fig. 14.1 Range (grey area) of *Rana muscosa* and *Rana sierrae* based on all museum specimens at the California Academy of Sciences and University of California Museum of Vertebrate Zoology. Species boundary is shown by the line. Circles show 225 historic collection sites (1899–1994) that were resurveyed between 1995 and 2005 (Black circles represent extinct populations, n = 211; white circles represent extant populations, n = 14) (reprinted from Vredenburg, V.T., Bingham, R., Knapp, R., Morgan, J.A.T., Moritz, C., and Wake, D. 2007. Concordant molecular and phenotypic data delineate new taxonomy and conservation priorities for the endangered mountain yellow-legged frog. Journal of Zoology 271:361–3744, with permission of John Wiley & Sons, Inc.).

much of the century (Rahel 1997; Patton et al. 1998). In the 1990s, fish stocking ended in National Parks within the Sierra Nevada, but stocking continues in lakes within National Forests, including wilderness areas. As a result, most historically fishless lakes now support nonnative trout, primarily rainbow trout (*Oncorhynchus mykiss*), golden trout (*Oncorhynchus mykiss aguabonita*), and brook trout (*Salvelinus fontinalis*) (Knapp 1996).

In addition to the direct effects of predation, trout competitively exclude mountain yellow-legged frogs from the deep-water habitats that facilitate survival during the extremes of summer and winter weather (Knapp & Matthews 2000). Observations and experiments demonstrate that introduced trout have a strong impact on mountain yellow-legged frog populations. Lakes and ponds with nonnative trout contain fewer mountain yellow-legged frogs than fishless lakes (Bradford 1989; Bradford et al. 1998; Knapp & Matthews 2000; Knapp 2005) and fish-removal experiments lead to rapid increases in frog and tadpole numbers (Knapp & Matthews 1998; Vredenburg 2004; Knapp et al. 2007). Competition and predation result in the complete elimination of frogs from ponds inhabited by persistent trout populations. Consequently, mountain yellow-legged frogs are increasingly restricted to smaller ponds (Bradford et al. 1993; Knapp & Matthews 2000). Thus, the pattern of mountain yellow-legged frog occupancy, both in pond size and geographic distribution, has contracted, leaving the frogs a restricted habitat range with reduced metapopulation connectivity. As a result of reduced snowpack from climate change, the shallower sites, which currently represent the majority of occupied, trout-free habitat, are predicted to experience more frequent drying episodes, potentially lowering frog reproduction and survival (Lacan et al. 2008).

In addition to competitive interactions, the population response to disease outbreaks can also be interpreted through the lens of history. As with other historical information, data on wildlife-disease outbreaks are often scarce. In this case, the fungal disease chytridiomycosis, first detected in 1975 (Ouellet et al. 2005), is now common throughout the frog's range (Knapp & Morgan 2006; Rachowicz et al. 2006; Fellers et al. 2007). Although outbreaks in mountain yellow-legged frog populations typically kill adults and juveniles and lead to population extirpations (Rachowicz et al. 2006), some populations persist despite infection (Briggs et al. 2005). The mechanisms explaining population persistence despite chytridiomycosis are the subject of much research, including the role of between-population differences in frog susceptibility and virulence as well as environmental factors (e.g. water temperature) that could affect disease dynamics.

Historical survey data for mountain yellow-legged frogs from the last centuries, in combination with an understanding of life-history traits and biophysical preferences, suggest that interspecific interactions with nonnative fishes and an emergent disease are the primary population stressors. Trout-removal programs result in larger mountain yellow-legged frog populations, thereby increasing dispersal to and colonization of additional sites and improving population connectivity (Knapp et al. 2007). Improved population connectivity may also facilitate the spread of resistance to chytridiomycosis. Ultimately, restoring entire watersheds to a fishless condition will facilitate the establishment of frog populations at sites characterized by a variety of elevations, water temperatures, and habitat conditions. These same management strategies are relevant when considering mountain yellow-legged frog population responses to climate change. With careful planning, balancing human desires for continued trout fisheries with the need for targeted fish-removal programs is possible (Armstrong & Knapp 2004), as demonstrated by National Park Service and California Department of Fish and Game amphibian restoration plans (Knapp et al. 2007). In this example, understanding the historical patterns of population abundance, connectivity, and spatial dynamics contributed critical information to improve restoration of mountain yellow-legged frogs.

Wood stork

The story of the wood stork (*Mycteria americana*) in the Florida Everglades exemplifies the link between life history, distribution, and the historical variation of the system that it inhabits. The wood stork is a large (85–115 cm tall) wading bird with the majority of its global range in Mexico and northern South America. In the United States, wood storks historically bred in peninsular Florida and, more recently, along the southern Atlantic coast (Rodgers et al. 1996).

Wood stork reproductive success is highly correlated with the timing and spatial distribution of foraging opportunities, which are dependent on environmental conditions. The stork's large size and diet of small fish result in a lengthy breeding season (4 months from egg laying to fledging), during which more than 150 kg of food must be procured to successfully fledge one or two young (Coulter et al. 1999). In the Everglades, the reproductive cue is the concentration of fish that occurs as the landscape dries at the end of the wet summer and fish distributions become increasingly restricted to shrinking pools (Fleming et al. 1994). Food demand increases as the nestlings grow, requiring increased feeding efficiency by adults. Increasing fish densities thus ensure sufficient food for growing nestlings. With the onset of summer rains, rising water levels disperse fish populations and nesting activity concludes, resulting in the abandonment of any remaining nestlings (Rodgers et al. 1996). Therefore, even minor changes in productivity between years appear to mirror shifts in the hydrology of local foraging habitat (Griffin et al. 2008).

The annual link between hydrology and wood stork productivity is critically important in explaining the historical population dynamics of wood storks inhabiting the Everglades. Before major drainage projects, the Greater Everglades Ecosystem consisted of a 45-km-wide expanse of shallow water moving 150 km southward through a low-gradient, wetland landscape originally consisting of vast sawgrass (*Cladium jamaicense*) plains, a series of ridges and sloughs, and short-hydroperiod marl marshes. Vagaries in the annual timing and amount of rain resulted in spatial and temporal variation of flooding and drying. Add fire, hurricanes, and subtle differences in elevation to this large-scale process, and the resulting system is a fine-scale patchwork of soils, plant communities, and hydrologic conditions, changing at different rates over millennia, that once sustained an enormous number of wading birds and many other species of wildlife (Kushlan 1981). The interaction of disturbance processes in the Everglades could destroy or create wood stork habitat at a particular site, but historically, the vast spatial extent of the system assured that suitable habitat was available someplace each year.

Although human habitation of the Everglades dates back about 2000 years, for most of that time, anthropogenic influence was limited to altering the fire regime (Sklar & van der Valk 2002). Intensive manipulation of the Everglades ecosystem began in the late 1800s with the first direct manipulation of the hydrologic regime: flood control and "reclamation" projects. Large-scale drainage, canalization, and damming began in the 1940s (Solecki et al. 1999). Water-control projects decreased the amount of water entering the southern Everglades by 65%, and the amplitude and timing of flood events changed from a seasonal, rain-driven pulse across a large and diverse landscape (i.e. the HRV) to human-mediated releases timed for non-ecological reasons (e.g. flood control) and concentrated in a few spillways (Fennema et al. 1994). These changes resulted in a disproportionate loss of sites with drying pools that concentrated fish (DeAngelis et al. 1998). As a result of the dramatic changes in hydrology and sub-

sequent opportunities for urban and agricultural development, about half of the Everglades ecosystem has been irreversibly transformed to agriculture, housing developments, and roads.

Prior to extensive habitat manipulations, wood storks had seemingly countless sites with high prey densities to support breeding. Based on a review of early twentieth-century observations, the Florida population of wood storks was estimated at around 100 000 individuals, with the greatest concentration of breeding birds occurring in the Everglades (USDI 1999). By the 1980s, the Florida population was reduced to 10% of pre-drainage numbers. As a result of the loss of short-hydroperiod wetlands, entire breeding seasons may be unproductive (Frederick et al. 2009). The original system was large and variable enough that under almost any hydrological condition, the necessary concentration of fish, signaling the beginning of wood stork breeding, would occur at various locations in November or December. With the loss of marginal wetlands, storks do not commence breeding until deeper pools dry sufficiently to concentrate fish. Delays of 1–2 months in nest initiation leave insufficient time for successful fledging of young before the summer rains begin and fish disperse from pools (Frederick et al. 2009).

Although wood stork populations have not fared well in Florida in recent years, knowledge of historical breeding dynamics in the Everglades has been applied to management of the species elsewhere in the United States. Since the 1970s, increasing numbers of wood storks were observed feeding in flooded agricultural fields, drainage ditches, and other artificial habitats during the nonbreeding season as far north as North Carolina (Rodgers et al. 1996). Since 1980, wood storks have expanded their breeding range north into Georgia, South Carolina, and North Carolina (USDI 2007). With proper management of hydrology, the United States breeding population of wood storks has increased from a low of approximately 5000 pairs in the late 1970s to over 8000 in 2006 (USDI 2007). These restoration actions have facilitated range expansion: 70% of the population now breeds north of Lake Okeechobee and the Everglades, suggesting that the viability of the United States breeding population may no longer be dependent upon their historical range (Fig. 14.2; USDI 2007). Although there are no data on the potential response of wood storks to climate change, the recorded range shifts reflect responses predicted for a similarly sized wading bird, the great egret

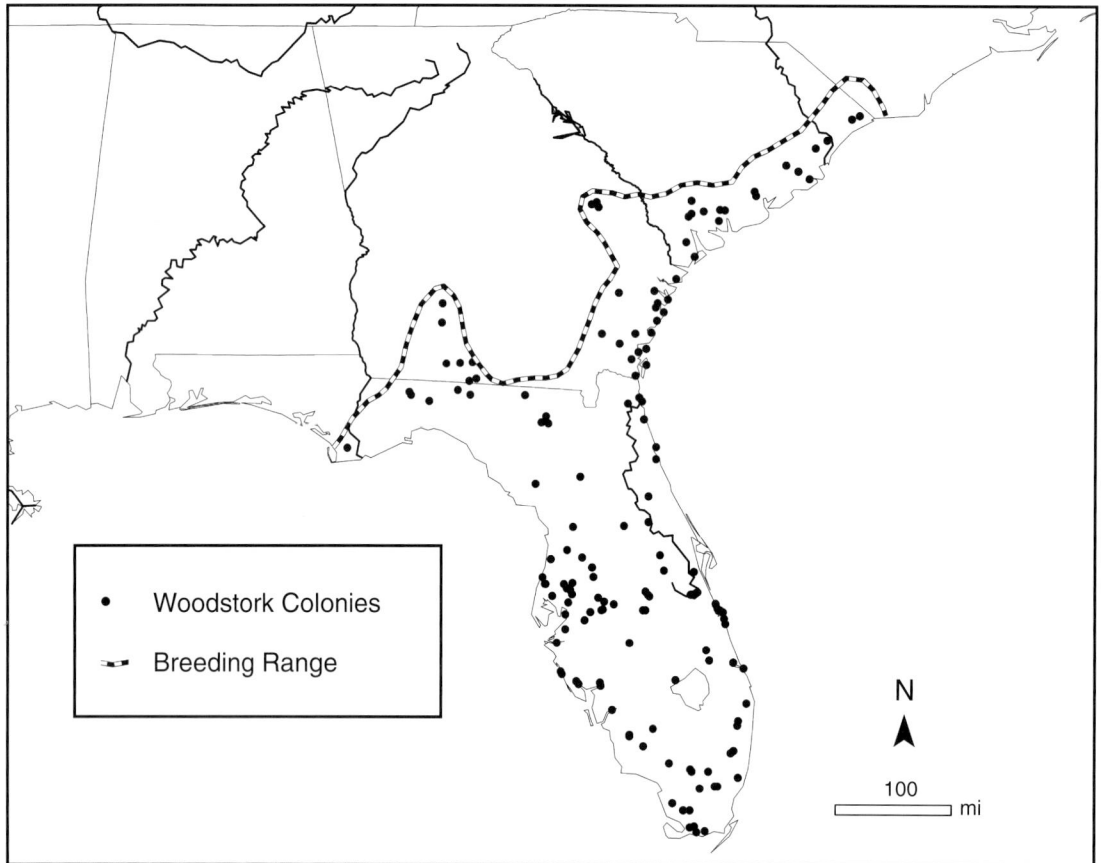

Fig. 14.2 Current wood stork nesting-colony locations, generalized breeding range, and generalized distribution of the United States breeding population of wood storks (reproduced from Brooks, W.B. and Dean, T. 2008. Measuring the biological status of the U.S. breeding population of wood storks. Waterbirds 31 (Special Publication 1):50–59 by permission of Pacific Wildlife Center).

(*Ardea albus*; Matthews et al. 2007). As the wood stork range shifts, historical understanding will contribute to successful management of wetland dynamics to support breeding populations.

Fisher

The fisher (*Martes pennanti*) is a large carnivorous weasel endemic to North America, occurring as three subspecies: the Pacific fisher (*Martes pennanti pacifica*), the eastern fisher (*Martes pennanti pennanti*), and the central fisher (*Martes pennanti columbiana*). Differences in the responses of the eastern and Pacific subspecies to recent environmental changes from historical ecosystem dynamics have important implications for

conservation and management. The Pacific fisher population has been considered for listing under the Endangered Species Act (USDI 2004), whereas the eastern fisher population is expanding and colonizing new habitats.

Fossil evidence first records the presence of fishers in the eastern United States approximately 30000 years ago, whereas evidence for fishers in western Northern America first appeared in the late Holocene, approximately 5000 years ago (Anderson 1994; Graham & Graham 1994). Prior to European settlement of North America, fishers were distributed throughout Canada, New England, and the Great Lakes south to Kentucky and North Carolina. In western North America, fishers were patchily distributed from southern British Columbia into the Sierra Nevada of California and eastward

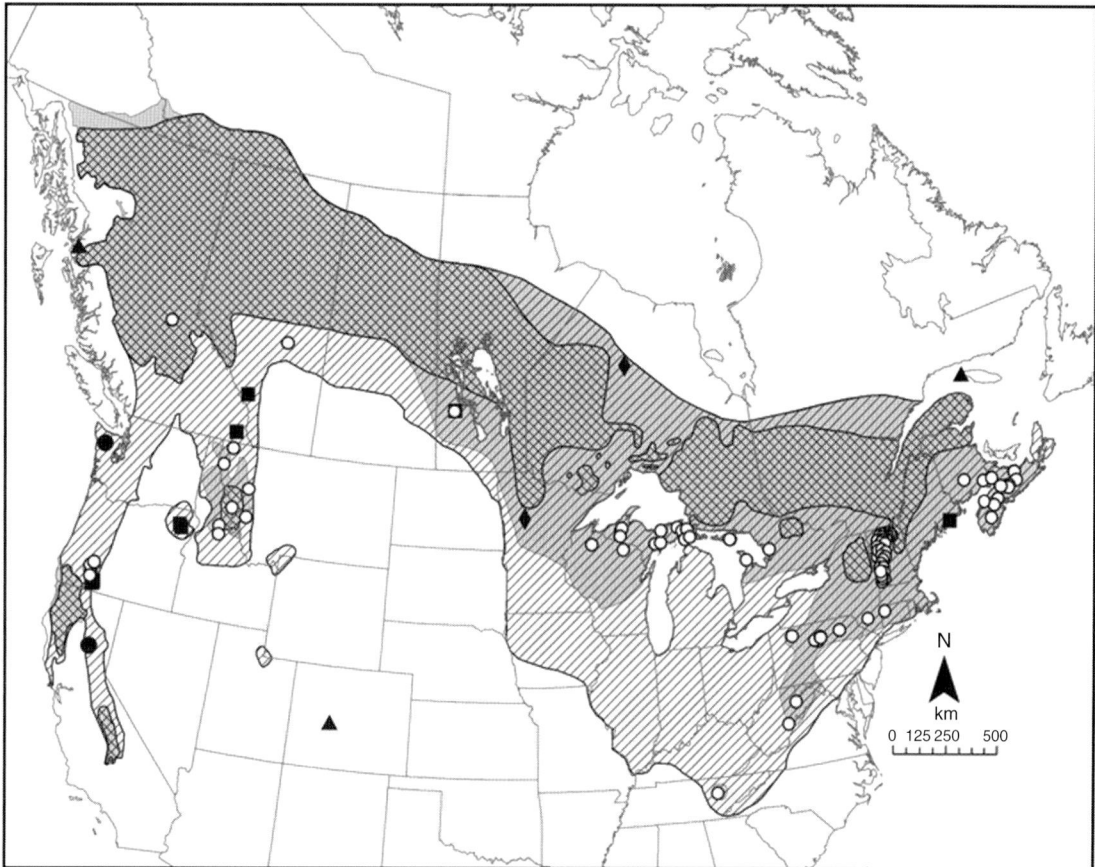

Fig. 14.3 Locations of translocation release sites in relation to the historical fisher range (prior to European settlement; diagonal hatching), the range at its most contracted state (cross hatching), and the current range (shaded). White circles represent successful reintroductions or augmentations, black squares represent failed reintroductions, black diamonds represent reintroductions with unknown outcomes, black circles represent ongoing reintroductions and black triangles represent introductions. (Used with permission of the publisher from Powell, R.A., Lewis, J.C., Slough, B.G. et al. In press. Evaluating translocations of martens, sables, and fishers: testing model predictions with field data. Chapter 6 in Biology and Conservation of Martens, Sables, and Fishers: A New Synthesis (eds K.B. Aubry, W.J. Zielinski, M.G. Raphael, G. Proulx, and S.W. Buskirk). Cornell University Press, Ithaca, NY, USA).

into the mountains of Utah, Idaho, Montana, and Wyoming (Fig. 14.3; Hagmeier 1956; Gibilisco 1994; Powell et al. in press).

Fishers generally inhabit forests at low to moderate elevations with relatively high canopy cover and complex forest structure, including the boreal forests of Canada, mixed deciduous-evergreen forests of eastern North America, and coniferous forests of western North America. Fishers use a variety of successional stages and plant communities, as long as sufficient security cover, prey, coarse downed wood, and snags are present. With the onset of European settlement in North America approximately 400 years ago, however, anthropogenic factors began to exert the greatest influence on fisher populations, primarily through trapping in the eighteenth and nineteenth centuries followed by habitat loss and fragmentation from timber harvest and land-use conversion in the twentieth century (Rand 1944; Hagmeier 1956; Gibilisco 1994; Powell & Zielinski 1994).

The eastern fisher provides an example of successful adaptation to conditions outside of historical ecological conditions. The eastern fisher maintained a wide distribution throughout periods of glacial advance and retreat and for another nine millennia after the arrival of humans (ca. 10 000 BP). Human use of fire would have affected eastern fisher habitat by reducing forests, but fisher habitat remained where the combined effects of humidity and topography preserved tracts of continuous forest, including wooded portions along the prairie-forest interface in Minnesota and Wisconsin (Berg & Kuehn 1994) and in river floodplains and along the shoreline of Lake Michigan (Graham & Graham 1990). As a result of the rapid habitat loss and indiscriminate trapping that accompanied the arrival of European settlers (i.e. habitat and population pressures occurring at spatial and temporal scales quite unlike historical conditions), the fisher was extirpated from southern Minnesota, Wisconsin, and Michigan, and farther south and from most of the southern Appalachians, Pennsylvania, New York, and parts of New England (De Vos 1964; Berg & Kuehn 1994; Ahn et al. 2002). Unlike the bison (*Bison bison*) and wolverine (*Gulo gulo*), however, small populations of fisher persisted in the eastern United States. By the 1950s, the eastern fisher was restricted to northern Minnesota, the northern Adirondack Mountains of New York, northern Maine, and possibly in the mountains of West Virginia (Hagmeier 1956). Since then, habitat and population pressures have abated: the reversion of agricultural lands to forests through the abandonment of farms and limits on trapping have allowed remnant populations to slowly recover in Minnesota (Berg & Kuehn 1994), Ontario (De Vos 1964), and Maine (Ahn et al. 2002). The recovery potential of these remnant populations was limited, however, by short dispersal distances and the consequent reduced interchange among isolated populations (Arthur et al. 1993). Beginning in the late-1950s, fishers were successfully relocated from the growing Minnesota population to heavily forested areas of Michigan and Wisconsin (De Vos 1964). Because fishers use a variety of forest types and switch prey in response to availability (Arthur et al. 1989; Roy 1991; Zielinski et al. 1999), the potential for relocation of fishers was apparent. Additional eastern fisher relocation sites were identified across their historical range, and successful reintroductions in the eastern United States occurred in Vermont (1959), West Virginia (1968), New York (1977), and Pennsylvania (1994).

Human land-use patterns have shaped the distribution of forest ecosystems that have defined the range of the Pacific fisher. Native Americans have inhabited this region for more than 10 000 years (Ames & Maschner 1999) and commonly burned vegetation to affect resource availability (Tuchmann et al. 1996; Deur & Turner 2005; Litman & Nakamura 2007). European exploration of the Pacific Coast began in the sixteenth century, followed by settlement and land development during the eighteenth and nineteenth centuries (Hayes 1999). With the discovery of gold and the subsequent increase in mining, timber harvest dramatically expanded, and livestock grazing and agriculture became more widespread (Beesley 2004; Litman & Nakamura 2007). Forest management over the last 150 years has imposed enormous environmental changes on Pacific fisher habitat, with the decline of late-successional forests and the simplification of forest structure and species composition (Curtis 1997; Thompson et al. 2003). Like the eastern fisher, the Pacific fisher has been challenged by rapid, broadscale habitat loss and trapping-induced mortality. Unlike the eastern fisher, however, the range of the Pacific fisher was historically patchy (i.e. metapopulations were less connected), forest regeneration has not occurred over the same spatial extent as in the eastern United States, and reintroductions in the western United States have been much less successful (Callas & Figura 2008). Despite management efforts, the Pacific fisher has fared poorly in environmental conditions dissimilar from the past. Currently, the Pacific fisher occurs as three to four populations, all of which are geographically isolated from the other populations by distance and anthropogenic and ecological barriers (Lofroth et al. 2010).

Historical ecology explains the extirpation of the fisher from much of its former range and provides the basis to inform restoration and management efforts for different subspecies. The ecological context of both eastern and Pacific fishers has changed dramatically since European settlement. Relatively low reproductive rate and limited dispersal ability made fisher populations vulnerable to anthropogenic changes in the historical environment. However, eastern and Pacific fishers appear to vary in their ecological plasticity, resulting in different population responses to change. The success of the eastern fisher reflects the return of ecosystem conditions to environments similar to historical settings, while at the same time eastern fishers are using widely available habitats that are quite dissimilar to the past, including suburban residential areas and urban sites. In addition, management agencies have facilitated population connectivity through translocations and have greatly reduced trapping mortality. Conversely, the Pacific fisher has more restrictive habitat associations and more isolated populations. To provide habitat, including late-successional forest components, management actions must incorporate and maintain natural-disturbance processes. Fisher populations may be increased through translocations, although attention to source populations is needed to conserve genetic integrity and region-specific adaptations (Pamilo & Savolainen 1999; Storfer 1999). Translocations over the last 60 years have altered the genetic composition of many fisher populations by including individuals from outside of the subspecific native range (Kyle et al. 2001; Drew et al. 2003; Wisely et al. 2004). Restoring the amount and distribution of presettlement Pacific fisher habitat is impossible, but historical analyses emphasize the importance of a network of connected, interacting fisher populations able to respond to fluctuating environmental conditions, including climate change (Lofroth et al. 2010).

14.3 CONCLUSION

Our understanding of the complex, synergistic relationships between dynamic ecosystems and wildlife species components is unavoidably incomplete, especially in the context of climate change. However, information to guide wildlife restoration and management efforts remains a persistent need. Despite seemingly endless restoration opportunities, available resources limit actions. An informed approach based on ecological theory and principles that incorporate historical ecology, while imperfect, helps to define conservation efforts with the highest likelihood of achieving objectives, especially when considered in their current anthropogenic context.

Despite the difficulties in obtaining historical wildlife data, a historical perspective may help managers avoid unintended consequences. Historical assessments can provide invaluable estimates of basic ecological properties such as distribution and relative abundance, as well as insights regarding threshold conditions based on an understanding of conditions in which species have persisted in time and space. The mountain yellow-legged frog example underscores the significance of metapopulation connectivity and provides a baseline for population abundance, demonstrating the dramatic shift in the ecological role of this once-dominant vertebrate in the subalpine environments of the Sierra.

The wood stork example demonstrates the importance of spatial and temporal variation in providing the narrow range of conditions required for successful foraging and reproduction; this information can be used to calibrate hydrologic changes at managed sites to promote successful reproduction. The eastern fisher example illustrates that some wildlife have the capacity to respond favorably to changing conditions that are dissimilar to historical conditions, providing insights toward setting management priorities and conservation expenditures.

In this chapter, we have examined the challenge of managing three species facing a host of threats, such as interspecific predation by introduced species, altered hydrological processes, and habitat fragmentation, degradation, and loss. Some of these threats (e.g. fish stocking) can be modified to reduce population stressors, while other changes are more permanent (e.g. urban development). In each example, an understanding of historical ecology resulted in clear, and rather simple, management actions to address grave conservation problems. Employing the rubric outlined by Romme et al. (see Chapter 17, this book) to formalize the process of confronting management issues with historical understanding is likely to lead to similar management directions for other species.

In the future, wildlife populations will confront the uncertain effects of climate change. An understanding of the response of species to climate shifts in the past (see Chapter 7, this book) underscores the need to incorporate resiliency and redundancy into conservation measures, while emphasizing population connectivity and dispersal. Ultimately, however, effective wildlife management depends on providing the biophysical and human social conditions that facilitate wildlife responses to changing environmental conditions. Wildlife HRV analyses demonstrate that understanding the past provides insights necessary to manage for the future, even with the challenge of an uncertain climate. Tracking wildlife population responses – both positive and negative – to new conditions outside of the historical norm will provide critical information to management in a changing world.

REFERENCES

Ahn, S., Krohn, W.B., Plantinga, A.J., Dalton, T.J., & Hepinstall, J.A. (2002). *Agricultural Land Changes in Maine: A Compilation and Brief Analysis of Census of Agriculture Data,*

1850–1997. Technical Bulletin 182. The University of Main Agricultural and Forest Experiment Station, Orono, ME, USA.

Ames, K., & Maschner, H.D.G. (1999). *Peoples of the Northwest Coast: Their Archaeology and Prehistory.* Thames and Hudson, London, UK.

Anderson, E. (1994). Evolution, prehistoric distribution, and systematics of *Martes.* In *Martens, Sables, and Fishers: Biology and Conservation* (ed. S.W. Buskirk, A.S. Harestad, M.G. Raphael, and R.A. Powell), pp. 13–25. Cornell University Press, Ithaca, NY, USA.

Armstrong, T.W., & Knapp, R.A. (2004). Response by trout populations in alpine lakes to an experimental halt to stocking. *Canadian Journal of Fisheries and Aquatic Sciences,* **61,** 2025–2037.

Arthur, S.M., Krohn, W.B., & Gilbert, J.R. (1989). Habitat use and diet of fishers. *Journal of Wildlife Management,* **533,** 680–688.

Arthur, S.M., Paragi, T.F., & Krohn, W.B. (1993). Dispersal of juvenile fishers in Maine. *Journal of Wildlife Management,* **574,** 868–874.

Aubry, K.B., McKelvey, K.S., & Copeland, J.P. (2007). Distribution and broadscale habitat relations of the wolverine in the contiguous United States. *Journal of Wildlife Management,* **71,** 2147–2158.

Avise, J.C. (1995). Mitochondrial DNA polymorphism and a connection between genetics and demography of relevance to conservation. *Conservation Biology,* **9,** 686–690.

Beesley, D. (2004). *Crow's Range: An Environmental History of the Sierra Nevada.* University of Nevada Press, Reno, NV, USA.

Berg, W.E., & Kuehn, D.W. (1994). Demography and range of fishers and American martens in a changing Minnesota landscape. In *Martens, Sables, and Fishers: Biology and Conservation* (ed. S.W. Buskirk, A.S. Harestad, M.G. Raphael, and R.A. Powell), pp. 262–271. Cornell University Press, Ithaca, NY, USA.

Botkin, D.B., Saxe, H., Araujo, M.B., et al. (2007). Forecasting the effects of global warming on biodiversity. *BioScience,* **57,** 227–236.

Bradford, D.F. (1989). Allotopic distribution of native frogs and introduced fishes in high Sierra Nevada lakes of California: implication of the negative effect of fish introductions. *Copeia,* **1989,** 775–778.

Bradford, D.F., Tabatabai, F., & Graber, D.M. (1993). Isolation of remaining populations of the native frog, *Rana muscosa,* by introduced fishes in Sequoia and Kings Canyon National Parks, California. *Conservation Biology,* **7,** 882–888.

Bradford, D.F., Cooper, S.D., Jenkins, T.M. Jr., Kratz, K., Sarnelle, O., & Brown, A.D. (1998). Influences of natural acidity and introduced fish on faunal assemblages in California alpine lakes. *Canadian Journal of Fisheries and Aquatic Sciences,* **55,** 2478–2491.

Briggs, C.J., Vredenburg, V.T., Knapp, R.A., & Rachowicz, L.J. (2005). Investigating the population-level effects of chytridiomycosis: an emerging infectious disease of amphibians. *Ecology,* **86,** 3149–3159.

Brooks, W.B., & Dean, T. (2008). Measuring the biological status of the U.S. breeding population of wood storks. *Waterbirds,* **31** (Special Publication 1), 50–59.

Callas, R.L., & Figura, P. (2008). Translocation plan for the reintroduction of fishers *Martes pennanti* to lands owned by Sierra Pacific Industries in the northern Sierra Nevada of California. California Department of Fish and Game, Sacramento, CA, USA.

Coulter, M.C., Rodgers, J.A., & Depkin, F.C. (1999). Wood Stork (*Mycteria americana*). In *The Birds of North America Online* (ed. A. Poole), 24 pp. Cornell Lab of Ornithology, Ithaca, NY, USA. (http://bna.birds.cornell.edu/bna).

Curtis, R.D. (1997). The role of extended rotations. In *Creating a Forest for the 21st Century: The Science of Ecosystem Management* (ed. K.A. Kohm and J.F. Franklin), pp. 165–188. Island Press, Washington, DC, USA.

DeAngelis, D.L., Gross, L.J., Huston, M.A., Wolff, W.F., Fleming, D.M., Comiskey, E.J., & Sylvester, S.M. (1998). Landscape modeling for Everglades ecosystem restoration. *Ecosystems,* **1,** 64–75.

Deur, D., & Turner, N.J. (2005). *Keeping It Living: Traditions of Plant Use and Cultivation on the Northwest Coast of North America.* University of British Columbia Press, Vancouver, BC, Canada.

De Vos, A. (1964). Range changes of mammals in the Great Lakes region. *American Midland Naturalist,* **711,** 210–231.

Drew, R.E., Hallett, J.G., Aubry, K.B., Cullings, K.W., Koepf, S.M., & Zielinski, W.J. (2003). Conservation genetics of the fisher *Martes pennanti* based on mitochondrial DNA sequencing. *Molecular Ecology,* **121,** 51–62.

Elton, C. (1927). *Animal Ecology.* Sedgwick and Jackson, London, UK.

Fellers, G.M., Bradford, D.F., Pratt, D., & Wood, L.L. (2007). Demise of repatriated populations of mountain yellow-legged frogs (*Rana muscosa*) in the Sierra Nevada of California. *Herpetological Conservation and Biology,* **2,** 5–21.

Fennema, R.J., Neidrauer, C.J., Johnson, R.A., MacVicar, T.K., & Perkins, W.A. (1994). A computer model to simulate natural Everglades hydrology. In *Everglades: The Ecosystem and Its Restoration* (ed. S.M. Davis and J.C. Ogden), pp. 249–289. St. Lucie Press, Boca Raton, FL, USA.

Fleming, D., Wolff, W., & DeAngelis, D.L. (1994). Importance of landscape heterogeneity to wood storks in Florida Everglades. *Environmental Management,* **18,** 743–757.

Frederick, P., Gawlik, D.E., Ogrden, J.C., Cook, M.I., & Lusk, M. (2009). The white ibis and wood stork as indicators for restoration of the Everglades ecosystem. *Ecological Indicators,* **9,** S83–S95.

Gibilisco, C.J. (1994). Distributional dynamics of modem *Martes* in North America. In *Martens, Sables, and Fishers: Biology and Conservation* (ed. S.W. Buskirk, A.S. Harestad, M.G. Raphael, and R.A. Powell), pp. 59–71. Cornell University Press, Ithaca, NY, USA.

Graham, M.A., & Graham, R.W. (1990). Holocene Records of *Martes pennanti* and *Martes americana* in Whiteside County, Northwestern Illinois. *American Midland Naturalist*, **1241**, 81–92.

Graham, R.W., & Graham, M.A. (1994). The late quaternary distribution of *Martes* in North America. In *Martens, Sables, and Fishers: Biology and Conservation* (ed. S.W. Buskirk, A.S. Harestad, M.G. Raphael, and R.A. Powell), pp. 26–58. Cornell University Press, Ithaca, NY, USA.

Graham, R.W., & Grimm, E.C. (1990). Effects of global climate change on the patterns of terrestrial biological communities. *Trends in Ecology and Evolution*, **5**, 289–292.

Griffin, G., Morris, J., Rodgers, J.A., & Snyder, B. (2008). Changes in wood stork (*Mycteria americana*) nestling success observed in four Florida bird colonies during the 2004, 2005, and 2006 breeding seasons. *Acta Zoologica Academiae Scientiarum Hungaricae*, **54**(Supplement 1), 123–130.

Grinnell, J. (1917). Field tests of theories concerning distributional control. *American Naturalist*, **51**, 115–128.

Grinnell, J., & Storer, T.I. (1924). *Animal Life in the Yosemite*. University of California Press, Berkeley, CA, USA.

Hagmeier, E.M. (1956). Distribution of martin and fisher in North America. *Canadian Field Naturalist*, **704**, 149–168.

Hayes, D. (1999). *Historical Atlas of the Pacific Northwest: Maps of Exploration and Discovery: British Columbia, Washington, Oregon, Alaska, Yukon*. Sasquatch Books, Seattle, WA, USA.

Hutchinson, G.E. (1957). Concluding remarks. *Cold Spring Harbor Symposia on Quantitative Biology*, **22**, 415–427.

Jackson, S.T., & Hobbs, R.J. (2009). Ecological restoration in the light of ecological history. *Science*, **325**, 56–569.

Keane, R.E., Hessburg, P.F., Landres, P.B., & Swanson, F.J. (2009). The use of historical range and variability (HRV) in landscape management. *Forest Ecology and Management*, **258**, 1025–1037.

Knapp, R.A. (1996). Non-native trout in natural lakes of the Sierra Nevada: an analysis of their distribution and impacts on native aquatic biota. In *Sierra Nevada Ecosystem Project* (ed. D.C. Erman), pp. 363–407. Final report to Congress, vol. III. Commissioned reports and background information. Centers for Water and Wildland Resources, University of California, Davis, CA, USA.

Knapp, R.A. (2005). Effects of nonnative fish and habitat characteristics on lentic herpetofauna in Yosemite National Park, USA. *Biological Conservation*, **121**, 265–279.

Knapp, R.A., & Matthews, K.R. (1998). Eradication of nonnative fish by gill-netting from a small mountain lake in California. *Restoration Ecology*, **6**, 207–213.

Knapp, R.A., & Matthews, K.R. (2000). Nonnative fish introductions and the decline of the mountain yellow-legged frog from within protected areas. *Conservation Biology*, **14**, 428–438.

Knapp, R.A., & Morgan, J.A.T. (2006). Tadpole mouthpart depigmentation as an accurate indicator of chytridiomyco-

sis, an emerging disease of amphibians. *Copeia*, **2006**, 188–197.

Knapp, R.A., Boiano, D.M., & Vredenburg, V.T. (2007). Removal of nonnative fish results in population expansion of a declining amphibian (mountain yellow-legged frog, *Rana muscosa*). *Biological Conservation*, **135**, 11–20.

Kushlan, J.A. (1981). Resource use strategies of wading birds. *Wilson Bulletin*, **93**, 145–163.

Kyle, C.J., Robitaille, J.F., & Strobeck, C. (2001). Genetic variation and structure of fisher (*Martes pennanti*) populations across North America. *Molecular Ecology*, **10**, 2341–2347.

Lacan, I., Matthews, K., & Feldman, K. (2008). Interaction of an introduced predator with future effects of climate change in the recruitment dynamics of the imperiled Sierra Nevada yellow-legged frog (*Rana sierrae*). *Herpetological Conservation and Biology*, **3**, 211–223.

Landres, P.B., Morgan, P., & Swanson, F.J. (1999). Overview of the use of natural variability concepts in managing ecological systems. *Ecological Applications*, **9**, 1179–1188.

Lewis, M., Clark, W., & Members of the Corps of Discovery. (2002). *The Journals of the Lewis and Clark Expedition* (ed. G. Moulton). University of Nebraska Press, Lincoln, NE, USA.

Litman, L., & Nakamura, G. (2007). *Forest History*. Forest Stewardship Series 4. University of California ANR Publication 8234. University of California, Davis, CA, USA.

Lofroth, E.C., Raley, C.M., Higley, J.M., et al. (2010). *Conservation Assessment for Fishers (Martes pennanti) in Southcentral British Columbia, Western Washington, Western Oregon, and California*, vol. I. USDI Bureau of Land Management, Denver, CO, USA.

Loiselle, B.A., Howell, C.A., Graham, C.H., Goerck, J.M., Brooks, T., Smith, K.G., & Williams, P.H. (2003). Avoiding pitfalls of using species distribution models in conservation planning. *Conservation Biology*, **17**, 1591–1600.

Marquette, J. (1917). Last voyage, 1674–1675. In *Early Narratives of the Northwest, 1634–1699* (ed. L.P. Kellogg), pp. 261–280. Charles Scribner's Sons, New York, NY, USA.

Matthews, S.N., Prasad, A.M., Peters, M.P., & Iverson, L.R. (2007). *A Climate Change Atlas for 147 Bird Species of the Eastern United States*. Publication NRS-4151. USDA Forest Service, Delaware, OH, USA.

Mayfield, H.F. (1972). Bird bones identified from Indian sites at western end of Lake Erie. *Condor*, **74**, 344–347.

Ouellet, M., Mikaelian, I., Pauli, B.D., Rodrigue, J., & Green, D.M. (2005). Historical evidence of widespread chytrid infection in North American amphibian populations. *Conservation Biology*, **19**, 1431–1440.

Pamilo, P., & Savolainen, O. (1999). Post-glacial colonization, drift, local selection and conservation value of populations: a northern perspective. *Hereditas*, **130**, 229–238.

Parmesan, C. (2006). Ecological and evolutionary responses to climate change. *Annual Review of Ecology, Evolution, and Systematics*, **37**, 637–669.

Patton, T.M., Rahel, F.J., & Hubert, W.A. (1998). Using historical data to assess changes in Wyoming's fish fauna. *Conservation Biology*, **12**, 1120–1128.

Powell, R.A., & Zielinski, W.J. (1994). Fisher. In *The Scientific Basis for Conserving Forest Carnivores: American Marten, Fisher, Lynx, and Wolverine* (ed. L.F. Ruggiero, K.B. Aubry, S.W. Buskirk, and W.J. Zielinski), pp. 38–73. General Technical Report RM GTR-254. USDA Forest Service, Fort Collins, CO, USA.

Powell, R.A., Lewis, J.C., Slough, B.G., et al. (2012). Evaluating translocations of martens, sables, and fishers: testing model predictions with field data. Chapter 6 In *Biology and Conservation of Martens, Sables, and Fishers: A New Synthesis* (ed. K.B. Aubry, W.J. Zielinski, M.G. Raphael, G. Proulx, and S.W. Buskirk). Cornell University Press, Ithaca, NY, USA. In press.

Rachowicz, L.J., Knapp, R.A., Morgan, J.A.T., Stice, M.J., Vredenburg, V.T., Parker, J.M., & Briggs, C.J. (2006). Emerging infectious disease as a proximate cause of amphibian mass mortality. *Ecology*, **87**, 1671–1683.

Rahel, F.J. (1997). From Johnny Appleseed to Dr. Frankenstein: changing values and the legacy of fisheries management. *Fisheries*, **22**, 8–9.

Rand, A.L. (1944). The status of the fisher, *Martes pennanti* (Erxleben), in Canada. *Canadian Field Naturalist*, **58**, 77–81.

Rodgers, J.A. Jr., Kale, H.W. II, & Smith, H.T. (1996). *Rare and Endangered Biota of Florida, vol. V: Birds*. University Press of Florida, Gainesville, FL, USA.

Root, T.L., & Schneider, S.H. (2002). Climate change: overview and implications for wildlife. In *Wildlife Responses to Climate Change: North American Case Studies* (ed. S.H. Schneider and T.L. Root), pp. 1–56. Island Press, Washington, DC, USA.

Roy, K., Valentine, J.W., Jablonski, D., & Kidwell, S.M. (1996). Scales of climatic variability and time averaging in Pleistocene biotas: implications for ecology and evolution. *Trends in Ecology and Evolution*, **11**, 458–463.

Roy, K.D. (1991). *Ecology of Reintroduced Fishers in the Cabinet Mountains of Northwest Montana*. University of Montana, Missoula, MT, USA.

Shelton, S.L., & Weckerly, F.W. (2007). Inconsistencies in historical geographic range maps: the gray wolf as example. *California Fish and Game*, **93**, 224–227.

Sklar, F.H., & A. van der Valk (eds). (2002). *Tree Islands of the Everglades*. Kluwer Academic Publishers, Boston, MA, USA.

Solecki, W.D., Long, J., Harwell, C.C., et al. (1999). Human–environment interactions in South Florida's Everglades region: systems of ecological degradation and restoration. *Urban Ecosystems*, **3**, 305–343.

Stahl, P.W. (1996). Holocene biodiversity: an archaeological perspective from the Americas. *Annual Review of Anthropology*, **25**, 105–126.

Stebbins, R.C., & Cohen, N.W. (1995). *A Natural History of Amphibians*. Princeton University Press, Princeton, NJ, USA.

Storer, T. (1925). A synopsis of the amphibia of California. *University of California Publications in Zoology*, **27**, 1–342.

Storfer, A. (1999). Gene flow and endangered species translocations: a topic revisited. *Biological Conservation*, **87**, 173–180.

Thompson, I.D., Baker, J.A., & Ter-Mikaelian, M. (2003). A review of the long-term effects of post-harvest silviculture on vertebrate wildlife, and predictive models, with an emphasis on boreal forests in Ontario, Canada. *Forest Ecology and Management*, **177**, 441–469.

Tingley, M.W., & Beissinger, S.R. (2009). Detecting range shifts from historical species occurrences: new perspectives on old data. *Trends In Ecology and Evolution*, **24**, 625–633.

Tingley, M.W., Monahan, W.B., Beissinger, S.R., & Moritz, C. (2009). Birds track their Grinnellian niche through a century of climate change. *Proceedings of the National Academy of Sciences of the United States of America*, **106**(S2), 19637–19643.

Tuchmann, E.T., Connaughton, K.P., Freedman, L.F., & Mariwaki, C.B. (1996). *The Northwest Forest Plan: A Report to the President and Congress*. USDA Forest Service, Portland, OR, USA.

USDI. (1999). *South Florida Multi-species Recovery Plan*. U.S. Fish and Wildlife Service, Atlanta, GA, USA.

USDI. (2004). 12-month finding for a petition to list the west coast distinct population segment of the fisher (*Martes pennanti*). *Federal Register*, **69**, 18770–18792.

USDI. (2007). *Wood Stork (Mycteria americana) Five-year Review: Summary and Evaluation*. U.S. Fish and Wildlife Service, Jacksonville, FL, USA.

Vredenburg, V.T. (2004). Reversing introduced species effects: experimental removal of introduced fish leads to rapid recovery of a declining frog. *Proceedings of the National Academy of Sciences of the United States of America*, **101**, 7646–7650.

Vredenburg, V.T., Bingham, R., Knapp, R., Morgan, J.A.T., Moritz, C., & Wake, D. (2007). Concordant molecular and phenotypic data delineate new taxonomy and conservation priorities for the endangered mountain yellow-legged frog. *Journal of Zoology*, **271**, 361–374.

Wiens, J.J., & Donoghue, M.J. (2004). Historical biogeography, ecology and species richness. *Trends in Ecology and Evolution*, **19**, 639–644.

Willis, K.J., Araújo, M.B., Bennett, K.D., Figueroa-Rangel, B., Froyd, C.A., & Myers, N. (2007). How can a knowledge of the past help to conserve the future? Biodiversity conservation and the relevance of long-term ecological studies. *Philosophical Transactions of the Royal Society of London. Series B, Biological Sciences*, **362**, 175–187.

Wisely, S.M., Buskirk, S.W., Russell, G.A., Aubry, K.B., Zielinski, W.J., & Lessa, E.P. (2004). Genetic diversity and structure of the fisher (*Martes pennanti*) in a peninsular and peripheral metapopulation. *Journal of Mammalogy*, **85**, 640–648.

Zielinski, W.J., Duncan, N.P., Farmer, E.C., Truex, R.L., Clevenger, A.P., & Barrett, R.H. (1999). Diet of fishers (*Martes pennanti*) at the southernmost extent of their range. *Journal of Mammalogy*, **80**, 961–971.

Zielinski, W.J., Truex, R.L., Schlexer, F.V., Campbell, L.A., & Carroll, C. (2005). Historical and contemporary distributions of carnivores in forests of the Sierra Nevada, California, USA. *Journal of Biogeography*, **32**, 1385–1407.

Chapter 15

RIVER FLOODPLAIN RESTORATION EXPERIMENTS OFFER A WINDOW INTO THE PAST

Ramona O. Swenson,[1,2] *Richard J. Reiner,*[3] *Mark Reynolds,*[4] *and Jaymee Marty*[5]

[1]Cardno ENTRIX, Sacramento, CA, USA
[2]The Nature Conservancy of California, Cosumnes River Preserve, Galt, CA, USA
[3]The Nature Conservancy, Chico, CA, USA
[4]The Nature Conservancy, San Francisco, CA, USA
[5]The Nature Conservancy, Sacramento, CA, USA

Historical Environmental Variation in Conservation and Natural Resource Management, First Edition. Edited by John A. Wiens, Gregory D. Hayward, Hugh D. Safford, and Catherine M. Giffen.
© 2012 John Wiley & Sons, Ltd. Published 2012 by John Wiley & Sons, Ltd.

15.1   INTRODUCTION

The restoration of a functional ecosystem depends first on understanding the factors that shape historical and future trajectories of pattern and processes, and subsequently matching that understanding to the restoration objectives for a given site (Swetnam et al. 1999; Buergi & Gimmi 2007). This includes an evaluation of the suite of restoration and management options available to the restoration practitioner, since rarely is it feasible or even possible to restore all historical features. In California, historical reconstructions inform restoration of oak woodlands, riparian forests, wetlands, and rivers (Greco & Plant 2003; Whipple et al. 2011). Unfortunately, detailed historical records are difficult to obtain for many systems. In lieu of this information, restoration experiments can provide a window into the past and can advance how we use historical context to improve restoration.

The Cosumnes River case study illustrates an adaptive approach to incorporating history into restoration of a riparian forest and floodplain system at the Cosumnes River Preserve in California. We describe our first attempts at riparian restoration by hand planting, followed by our review of available historical data and field observations and our subsequent iterations as the focus shifted to reestablishing key ecosystem processes. Multidisciplinary monitoring revealed how the floodplain riparian system functioned, providing a window into history that continues to guide restoration of these dynamic ecosystems (Opperman et al. 2010).

California's riparian forests have been greatly reduced statewide, and less than 4% of the historical area remains in the Central Valley due to agricultural conversion and urban development (Katibah 1984). Furthermore, existing remnant riparian forests may function differently compared to historical conditions. Seasonal flooding and hydrological connectivity are prerequisites for forest regeneration and ecologically functional floodplains (Junk et al. 1989; Tockner et al. 2000; Rood et al. 2005). Yet most Central Valley floodplains are severed from their rivers by levees, channelization, and flow regulation (Mount 1995). Large reservoirs for water supply and flood control constructed during the 1940s–1960s have fundamentally altered hydrologic and geomorphic function by eliminating spring flooding, reducing variability of flows, and altering sediment transport (Williams et al. 2009). This river-floodplain disconnect affects several critical functional attributes of floodplains, including reduced nutrient replenishment and associated food-web development as well as decreased variability of flood-dependent habitats (Jeffres et al. 2008).

15.2   COSUMNES RIVER HABITAT RESTORATION

The Cosumnes River is located in the middle of California's Central Valley, which extends 720 km north-south between the Sierra Nevada and Coast Range mountains. Two major rivers, the Sacramento and San Joaquin, drain the valley and meet in an inland delta northeast of the San Francisco Bay Area. The Cosumnes River flows west from the Sierra Nevada into the Delta (Fig. 15.1). Historically, the lower Cosumnes River was a dynamic, multiple-channel system flowing across a low-gradient floodplain dominated by a mosaic of riparian forests, seasonal wetlands, grasslands, and seasonal floodplain lakes (Florsheim & Mount 2002). Currently, over 688 ha of restored and remnant riparian forest, including significant stands of valley oak (*Quercus lobata*) forest, occur along approximately 22 km of the lower Cosumnes River.

The Cosumnes is one of the few Central Valley rivers without a major dam regulating its flows and thus is one of the last remaining tributaries to the Sacramento-San Joaquin system with a quasi-natural hydrograph. The river still maintains a seasonal flow regime typical of Mediterranean systems, experiencing winter flooding with peak flows of up to 2650 m^3/s (1997 flood) and low to no late summer and fall flows (Fig. 15.2). Levees constructed in the area starting in the late 1800s to protect agriculture still constrain much of the river channel (Florsheim & Mount 2002). Levees reduce small intermittent floods but do not protect the floodplain from larger events. Summer low- to no-flow conditions have become exacerbated by excessive groundwater pumping that dropped groundwater levels up to 24 m below mean sea level (Fleckenstein et al. 2004).

The lower Cosumnes River was highlighted as a conservation priority in the early 1980s because of its large blocks of riparian forest, unplowed grassland and savannah, and relatively natural flow regime. In 1987, The Nature Conservancy (TNC) and several state, federal, and private partners established the Cosumnes River Preserve (hereafter Preserve) to protect and restore remnant bottomland riparian

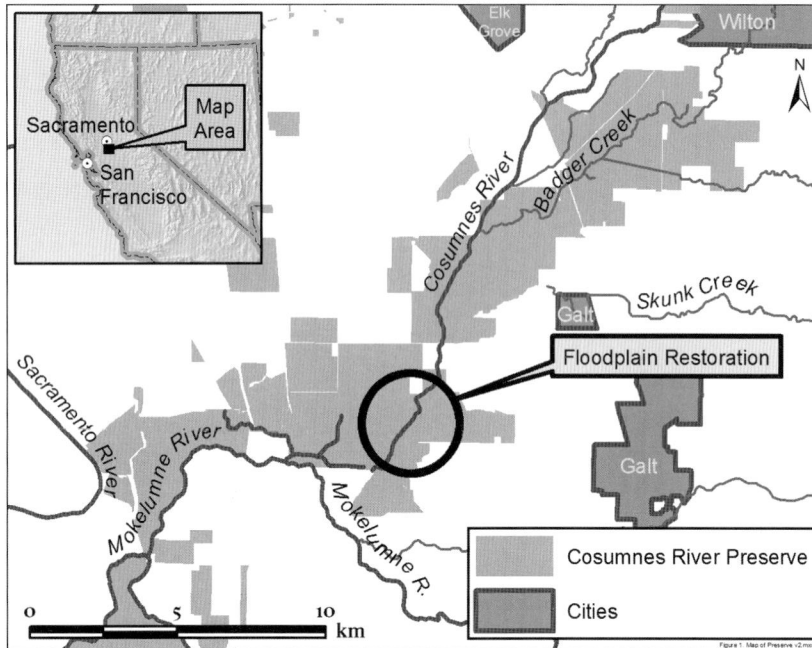

Fig. 15.1 The Cosumnes River Preserve in California's Central Valley.

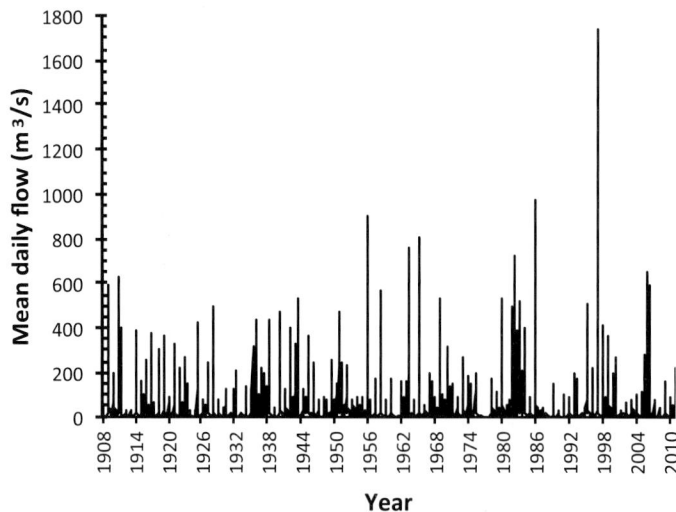

Fig. 15.2 Historical hydrograph for the Cosumnes River (1908–2010) (data source USGS National Water Information System, USGS 11335000 COSUMNES R A MICHIGAN BAR CA; http://waterdata.usgs.gov/nwis/dv/?site_no=11335000& agency_cd=USGS&referred_module=sw).

forests and freshwater wetlands (Swenson et al. 2003). The Preserve currently protects 18 856 ha in the lower Cosumnes watershed and adjacent Delta, including 7080 ha on the lower floodplain (Fig. 15.1).

To increase the amount of valley oak forest, now decimated in the Central Valley, a primary goal for the Preserve was restoration of fallowed agricultural lands to expand and connect existing riparian forest stands. A mature 40-ha valley oak forest remnant, located on flood-prone lands near the restoration sites, was used as a reference site and a model for setting restoration goals. Early efforts focused on hand-planting valley oak acorns and seedlings into former farmland. Hand planting was chosen because observations of nearby fallowed fields showed little recruitment of native trees. Invasive weeds dominated and had reached a steady-state equilibrium (sensu Westoby et al. 1989) that would not transition to forest under existing conditions.

After 5 years of active hand planting it became clear that attempting to recreate the "model" reference forest had serious limitations. Planting by hand proved expensive, some plantings failed or grew slowly, and the understory continued to be exotic species (Reiner 1996). This led TNC staff to reconsider hand planting and to take a closer look at the remnant forest we were using as a model to set restoration goals.

To understand the setting in which the reference forest established, we gathered data from field observations, historical writings, old maps, agency archives, and aerial photos dating from the 1930s (Thompson 1957). Tree coring revealed that most of the forest was composed of 55-year-old, even-aged, valley oak trees established when intensive agriculture had expanded into the area and flood control levees were constructed (Tu 2000). The understory was dominated by jungle-like vines of wild grape (*Vitis californica*) and black-berry (*Rubus ursinus* and non-native *Rubus discolor*). Another smaller area was 33 years old and dominated by Fremont cottonwood (*Populus fremontii*), with young valley oaks in the understory amidst an herbaceous layer of annual and perennial forbs. We also examined sequential historical photos, conducting oral histories from senior neighbors, and uncovered old livestock fences. These investigations revealed a complex history of farming, flooding, and grazing at the time of forest establishment, in conditions that do not exist at present. One forest section was an orchard before 1940. The section had been cleared and then grazed for many

years, which could explain the grass and sedge under-story in this area. It became clear that our reference site had been shaped by a legacy of agriculture, levee construction, a historically higher groundwater table, and fewer non-native plants. The use of this forest as an adequate model for restoration became questionable.

Faced with uncertainty about setting realistic resto-ration goals and using information from the reference site, we looked for a better approach. A critical insight came in 1993 with a newly discovered photo that showed a young 8-ha forest patch where the river tem-porarily breached its levee in 1985. An interview revealed that this flood left substantial sand deposits that the farmer could not remove when he made repairs. Those deposits provided appropriate substrate for natural recruitment of Fremont cottonwood and various willow species (*Salix* spp.) (Fig. 15.3). Further study found seedling valley oaks growing in the understory.

This "accidental" forest inspired TNC to shift focus from horticultural restoration techniques to an eco-logical process-based technique of levee breaching (Fig. 15.4). The main hypothesis was that reintroduc-tion of flooding, a key disturbance regime, would lead to natural reassembly of the historical riparian com-munity (Trowbridge 2007). Site-selection criteria for new levee breaches included ownership, proximity to existing riparian forest stands that were presumed indi-cators of suitable soils, and sites that minimized nega-tive effects on neighboring properties. Assessment of the hydrologic regime included examination of histori-cal streamflow records, first-hand observations of winter flooding, interviews with neighboring land-owners, and limited two-dimensional hydrological modeling to evaluate potential flood risk to adjacent properties.

In October 1995, TNC intentionally breached a levee and restored flooding to a 54-ha agricultural field adja-cent to the "accidental" forest. The first floods in December 1995 deposited extensive sediment (Fig. 15.5). In January 1997, extensive flooding breached more levees upstream and created a second 39-ha res-toration experiment. Construction of a low setback levee protected adjacent farm fields from small late-season floods that inundated this field (Swenson et al. 2003). The two restoration sites are separated by a partial levee, and floodwaters drain from north (upstream field) to south (downstream field) (Fig. 15.3).

Fig. 15.3 Aerial photo of floodplain before (1993) and 15 years after (2010) the levee breach restoration at the Cosumnes River Preserve.

15.3 ECOLOGICAL RESPONSES

The levee breaches functionally reconnected the river to its floodplain for the frequent small floods that historically inundated Central Valley floodplains. The frequency, extent, depth, and duration of inundation were increased, as was sediment deposition. These physical changes in turn created diverse microhabitats. The newly flooded habitat attracted wildlife, including small fish into the flooded forest and flocks of waterfowl on the shallow ponds. What began as a search for more efficient methods of forest restoration, informed by glimpses into the past from reference forest sites, photos, and interviews, now yielded broader unplanned benefits for the floodplain ecosystem.

Restoration projects are rarely monitored and analyzed (Kondolf 1998), and most studies of restoration response have focused only on a few components such as vegetation or birds. TNC collaborated with the University of California Davis (UC Davis), Point Reyes Bird Observatory Conservation Science (PRBO), and others to investigate the physical and biological outcomes.

Hydrology and geomorphology

Flood basins of the Central Valley were historically dry most of the year, but during wet winters became "inland seas" (Florsheim & Mount 2003). On the lower Cosumnes River, winter overbank flows became trapped between alluvial fans and natural levees formed by the Sacramento River and its tributaries. This created floodplain lakes and seasonal marshes that drained through multiple channels. During the late 1800s, sediment from hydraulic mining in the Sierra Nevada raised streambed elevations and water levels and created a backwater effect that increased flood stage and duration on the Cosumnes floodplain. An era of levee construction followed throughout the Central Valley in the late nineteenth and early twentieth century, including banks along the Cosumnes River.

Variable floods on the Cosumnes River have produced a range of geomorphic and ecological outcomes (Florsheim & Mount 2002; Florsheim et al. 2006). This pattern of flood timing and intensity has changed over the past 100 years. The frequency of years with wet

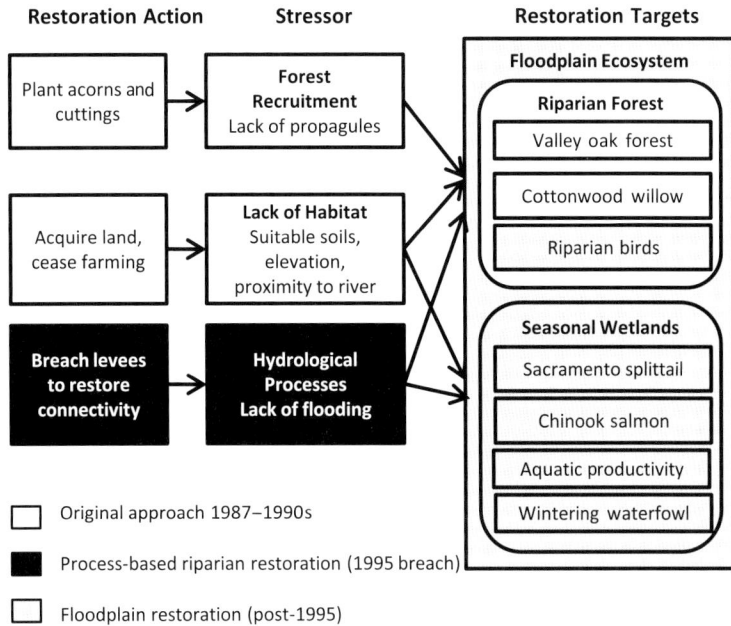

Fig. 15.4 Evolution of the conceptual model of Cosumnes River restoration.

Fig. 15.5 Sand deposited after a 1995 levee breach on the Cosumnes River Preserve (note people for scale) (Photo courtesy of Geoff Fricker).

springs and late floods has decreased since the mid-twentieth century, a trend consistent with a climate-change hypothesis of a rising snow-rainfall transition line and diminishing snowmelt contributions (Booth et al. 2006).

A wide range of hydrologic events is necessary to maintain the integrity of aquatic ecosystems (Poff et al. 1997). Three types of floods encapsulate different floodplain processes: floodplain activation, floodplain maintenance, and floodplain resetting (Opperman et al. 2010). Floodplain activation floods are frequent, small-magnitude floods that reconnect the river and floodplain, often for long duration and several times in a season (Williams et al. 2009; Opperman et al. 2010). Prior to breaching the levees, the lower Cosumnes River overtopped its banks and established connectivity with the floodplain every 5 years, when river flows exceeded approximately $50 \text{m}^3/\text{s}$. After the 1995 breach, flooding occurred earlier and more frequently (1.5-year recurrence interval) at half the flow ($25 \text{m}^3/\text{s}$) (Florsheim & Mount 2003; Florsheim et al. 2006) (Fig. 15.6). Early floods (November–February) had flashy, larger peak flows fed by winter rainstorms, whereas later floods (March–May) had smaller peak flows from snowmelt (Booth et al. 2006). The annual duration of connectivity was 63–158 days during wetter years

(1998–2000, 2005), but only 6–37 days in drier years (2001–2004), and the areal extent of floodplain inundation ranged from 5% in 2001 to 100% in 1998 (Moyle et al. 2007).

Floodplain maintenance floods are higher magnitude and are capable of bank erosion and sediment deposition on the floodplain (Opperman et al. 2010). In the first winter after breaching, flows exceeded $100 \text{m}^3/\text{s}$ and brought a pulse of sediment onto the floodplain. Sands settled out near the levee breach in finger-like deposits up to 5 m deep and a few hundred meters long (Fig. 15.5). Finer silts remained in suspension longer and deposited in thin layers across the floodplain (Florsheim et al. 2006). Subsequent floods reworked floodplain sediments and scoured out channels nearly 4 m below the original elevation of the once-level agricultural field (Florsheim & Mount 2002). During 2000–2003, however, flow events did not exceed $100 \text{m}^3/\text{s}$, and deposition of sand dwindled to minimal levels (Florsheim et al. 2006).

Finally, floodplain resetting floods are rare, very high-magnitude events that produce extensive geomorphic change, such as scouring of floodplain surfaces and channel avulsion (Opperman et al. 2010). In 1997, the flood of record ($2650 \text{m}^3/\text{s}$, 150-yr recurrence interval) breached levees all along the Cosumnes

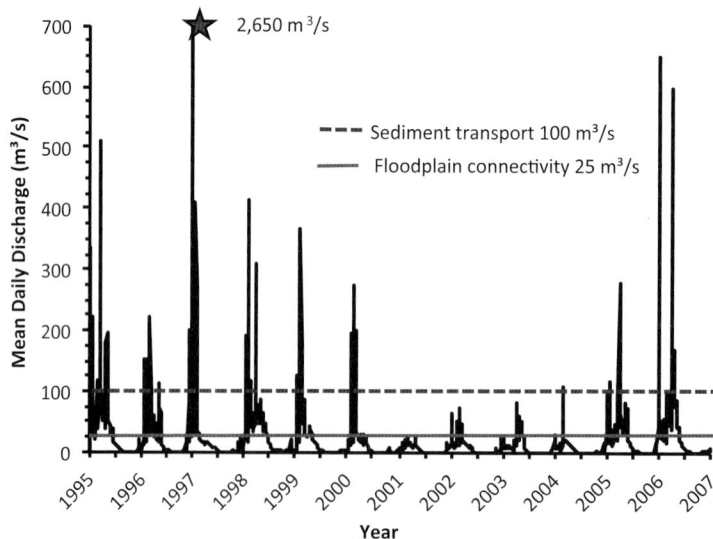

Fig. 15.6 Flood pulses on restored floodplain showing thresholds flows for floodplain activation ($25 \text{m}^3/\text{s}$) and floodplain maintenance ($100 \text{m}^3/\text{s}$) (mean daily discharge, 1995–2006).

River. A pair of breaches opened just upstream of the 1995 breach and were incorporated into a second phase of floodplain restoration (Fig. 15.3).

Frequent, prolonged inundation is essential for activating key processes of an ecologically functional floodplain, in both tropical (Junk et al. 1989) and temperate systems (Williams et al. 2009; Opperman et al. 2010). The restoration of floodplain activation and floodplain maintenance floods at the Cosumnes created a fresh canvas for biotic colonization as new pathways for flooding and draining formed. The natural flow variability and topographic heterogeneity created by the breach affected the temporal and spatial patterns of inundation, which in turn produced diverse microhabitats.

Plant succession and community structure

Riparian plant communities are shaped by inundation dynamics (Mahoney & Rood 1998) and height above the water table (Stromberg et al. 1991), which are both influenced by floodplain topography (Florsheim & Mount 2002). However, species colonization may not follow predictable patterns. At Cosumnes, millions of cottonwood (*P. fremontii*) seedlings quickly colonized the fresh sand deposited on the leveled agricultural field by May 1996. Yet seedling survival was low, with one out of every 2000 seedlings surviving to 1998, presumably due to desiccation during the hot, dry summer (Tu 2000). The sprouting of twigs and branches buried in the sand, however, was the sole source of recruitment for willows (*Salix* spp.). Similarly, vegetative cottonwood recruitment outnumbered seedlings six to one after 2 years (Tu 2000). Recruitment and survival of cottonwood and willow trees was greatest on the highest sandbar areas that drained earliest (Trowbridge 2007) (Fig. 15.2). On lower, less-drained elevations, tree seedlings (cottonwood, willow, Oregon ash [*Fraxinus latifolia*]) recruited and grew less vigorously. Native herbaceous plants dominated this area. Scoured areas that held water into the summer had only herbaceous vegetation, and a few long-lasting ponds had emergent vegetation (*Scirpus acutus, Scirpus fluviatilis, Typha angustifolia*) and other wetland plants.

As of 2010, 106 species (64 native, 42 non-native) had colonized the floodplain (TNC, unpublished data). A mixture of annual and perennial forbs established immediately following the levee breaches (Trowbridge 2007). Since 2000, the floodplain habitat has trended toward dominance of mostly native perennial species, with most annual species becoming undetectable in monitoring plots by 2010 (TNC, unpublished data). Away from the sand deposits, tree recruitment has been slow and patchy. The flagship species, valley oak, has not recruited quickly in the floodplain, with seedlings found in one monitoring plot in 2000 and in only three monitoring plots by 2010 (TNC, unpublished data). This could be due to dispersal limitation and lack of propagules, or prolonged inundation of the floodplain for several years following the breach. Valley oak acorns exhibit poor germination and seedling survival under long-term flooded conditions (Trowbridge 2007). The frequency of other woody species like Oregon ash, willow, and cottonwood has also increased slowly in that same decade (TNC, unpublished data). If rapid establishment of woody riparian tree species had been the sole metric of success, then process-based restoration might have been deemed a disappointment in the early phase of restoration. Unlike many planted sites on leveled fields, however, the restored floodplain has greater complexity and variety in its plant communities due to the topographic heterogeneity. Fourteen years after the levee breach, the restored floodplain is a diverse mosaic of early-successional forest, herbaceous vegetation, and seasonal marsh (Fig. 15.3).

The restoration of a more natural flood regime, coupled with increased topographic heterogeneity, more closely resembles historical conditions and favors native plant species. Many of these native plants are adapted to the seasonal inundation, and the improved topography creates a variety of habitat niches for a number of these native species to coexist. That the restored condition tends to favor native plants is a big advantage for restoration practitioners because much of the early maintenance on a restoration site focuses on weed control while native vegetation becomes established. Recently, however, aggressive weedy species like yellow water primrose (*Ludwigia* spp.), perennial pepperweed (*Lepidium latifollum*), and even annual grasses such as annual ryegrass (*Lolium multiflorum*) have begun to invade specific niche habitats and are the focus of weed-control efforts.

Birds

The mosaic of habitats at the restored floodplain and adjoining remnant forest support a diverse and abundant avifauna, including many species of neotropical migrant songbirds, waterfowl, and waterbirds that have declined throughout the west (Humple & Geupel 2002). This variety of birds depends on diverse physical structure and plant community composition. Riparian bird species composition, abundance, and nesting-success observations were collected between 1995 and 2006 in mature riparian forest and restoration sites between 1 and 22 years of age (Wood et al. 2006). Consistent with other studies along the Sacramento River (Gardali et al. 2006), bird-species composition varied by restoration site age, plant composition, and structure. Bird species diversity increased with maturity of restoration sites. Species-specific abundance was positively correlated with measures of age, composition of tree species (cottonwood and oak), and presence of shrubs and dense herbaceous understory such as mugwort (*Artemisia douglasiana*) and blackberry (*Rubus* spp.) (Nur et al. 2008).

Trends in abundance of birds also varied between restored areas. Some species declined in the mature forest, while other species increased or were stable in younger restoration sites. In contrast, Sacramento River sites showed increases in bird abundance in mature forest as nearby restoration sites matured (Gardali et al. 2006). The difference might be explained by observations in mature forest at the Cosumnes Preserve of nest predation by black rats (*Rattus rattus*) and nest parasitism by Brown-headed Cowbirds (*Molothrus ater*) (Hammond 2008). Preservation of mature, remnant riparian habitat alone may not be sufficient to maintain populations of some bird species (Wood et al. 2006).

Flood timing and depth also affected some species. When flooding persisted into the spring breeding season, song sparrows (*Melospiza melodia*) suffered reduced nesting success in that year, but greater survival the following year. This could be due to changes in vegetative cover, insect abundance, and/or predator activity or predator populations associated with flooding.

The bird studies added a new facet to our understanding of how the historical floodplain functioned. The fidelity of bird species to specific riparian composition and structure highlights the importance of encouraging a mosaic of vegetation, as seen in dynamic riparian systems with floodplain resetting events that remove mature forest and create patches of early-successional habitat (Golet et al. 2008). Early-successional stands make up less than 20% of total riparian forest habitat on the Cosumnes Preserve,

which is likely a far lower percentage than what would have existed historically. This information has led Cosumnes Preserve management to emphasize the need to restore a larger proportion of the forests with species representing this early-successional state.

Aquatic biota

Perhaps the greatest surprise of the Cosumnes restoration lay beneath the surface of the floodwaters. Our original objective was to restore riparian forest based on an overstory species composition of a relict forest. A closer examination of this reference site during flood events revealed an ephemeral yet highly productive aquatic ecosystem, including unusually large juvenile Chinook salmon (*Oncorhynchus tshawytscha*). This led to studies examining relationships among flooding patterns, primary and secondary productivity, and fish use of the floodplain.

Biological productivity and recruitment are more dependent on the duration and timing of inundation than the magnitude of flows (Poff et al. 1997; Booth et al. 2006; Opperman et al. 2010). With prolonged inundation, the Cosumnes floodplain progressed from a physically driven system when connected to the river floods to a biologically driven, pond-like system with increasing temperature and productivity once inflow ceased. Periodic small floods boosted aquatic productivity of phytoplankton by delivering new pulses of nutrients, mixing waters, and exchanging organic materials with the river. Aquatic productivity was greater in floodplain ponds than in river sites, and zooplankton biomass increased rapidly following each flood event (Ahearn et al. 2006; Grosholz & Gallo 2006).

Floodplain inundation creates habitat for spawning and foraging fish that can access the floodplain to take advantage of the abundant invertebrates and vegetated shallows (Sommer et al. 2001; Moyle et al. 2007). The Sacramento splittail (*Pogonichthys macrolepidotus*) is a large native minnow, with adults spawning on inundated vegetation in February to March, and juveniles rearing on the floodplain then departing when it drains in April to May (Ribeiro et al. 2004; Moyle et al. 2007). Juvenile Chinook salmon also access the floodplain in February to March to forage prior to emigrating to the sea (Moyle et al. 2007). Growth rates of juvenile Chinook were faster on ephemeral floodplain habitats than in the river (Jeffres et al. 2008).

Table 15.1 Common fish species on the Cosumnes floodplain (Moyle et al. 2007).

Species		Native or exotic	User group	Spawning months on floodplain
Catostomus occidentalis	Sacramento sucker	N	River spawner	
Cyprinus carpio	Common carp	E	Floodplain spawner	March–June
Gambusia affinis	Western mosquitofish	E	Perennial pond	April–July
Lepomis macrochirus	Bluegill	E	Forager	May–July
Lepomis microlophus	Redear sunfish	E	Forager	May–July
Menidia beryllina	Inland silverside	E	Perennial pond	April–July
Micropterus salmoides	Largemouth bass	E	Forager	–
Notemigonus crysoleucas	Golden shiner	E	Forager	–
Oncorhynchus tshawytscha	Chinook salmon	N	River spawner	–
Pogonichthys lepidotus	Sacramento splittail	N	Floodplain spawner	March–May
Pomoxis nigromaculatus	Black crappie	E	Forager	April–July
Ptychocheilus grandis	Sacramento pikeminnow	N	River spawner	–

Restoring the historical pattern of seasonal inundation created habitat uniquely suited for floodplain-dependent native fishes and less hospitable for non-native fish (Moyle et al. 2007). The lower Cosumnes River harbors 32 species of fishes, about half of them native, and 22 species spawn or forage on the floodplain (Table 15.1). Native fishes that evolved with California's seasonal precipitation use the floodplain early in the year. In contrast, non-native species that evolved in temperate regions with year-round precipitation arrive later and remain longer on the floodplain (April–July), spawn under warmer conditions, and are stranded more often when the floodplain drains and ponds dry out. Non-native fishes dominate perennial habitat such as ditches and floodplain ponds (Feyrer et al. 2004; Moyle et al. 2007). An optimal flood regime for native fishes should include early-season, cold-water events that persist long enough for bursts in algal and invertebrate productivity, followed by spring draining of the floodplain before it warms seasonally and favors non-native species (Crain et al. 2004). However, this hydrological pattern now occurs less often than historically (Booth et al. 2006), which may constrain restoration opportunities for fish.

15.4 DISCUSSION

Ecological restoration projects seek to restore not only species composition and habitat structure, but also dynamic processes and ecological function, often to some historical reference condition. Effective land management and restoration must be based on an understanding of historical pattern and processes, including anthropogenic effects, although obtaining adequate information can be challenging (Buergi & Gimmi 2007). The study of appropriate reference sites and conditions can provide insights into past ecological conditions, historical ranges of variation, and restoration feasibility (Hughes et al. 2005). Using reference sites as a window into the past can guide sustainable restoration of ecological function into the future. However, this approach must be balanced against realities of irrevocably altered systems that do not function as before, multiple restoration trajectories, and future change that could render a return to historical conditions impossible or irrelevant (Dufour & Piégay 2009; Seavy et al. 2009).

Our adaptive approach to restoration at the Cosumnes River Preserve has tracked advances in the theory and practice of floodplain and riparian restoration over the past 25 years. Early efforts sought to replicate a template or reference site by planting desired plant species, such as valley oak, to rapidly achieve a climax successional forest. In the early 1990s, theories such as the Flood Pulse Concept (Junk et al. 1989) highlighted the importance of river-floodplain connectivity and heralded an era of process-based restoration, as exemplified by the levee breaches. By the early 2000s, insights from geomorphology and landscape ecology shed light on the role of variable floods in shaping topography and creating microhabitats for vegetation recruitment (Mahoney & Rood 1998; Trowbridge 2007). In recent years, threshold floods that trigger different ecological processes have received attention (Williams et al. 2009; Opperman et al. 2010), as illustrated by pulses of productivity and fish growth. In retrospect, rather than simply meeting the early goal of reestablishing oak forest, restoration of floodplain processes changed the trajectory of the floodplain ecosystem and facilitated multiple conservation objectives (Hughes et al. 2005; Dufour & Piégay 2009).

We believe our experiences contain several important lessons for conservation practitioners:

• *Use historical information and reference sites judiciously.* Restoration projects often use historical conditions and reference sites as templates for defining objectives and evaluating success. However, this approach, if narrowly adhered to, has limitations and potential pitfalls (Hughes et al. 2005). Too much focus on past conditions as a guide in restoration of highly altered systems can obscure more feasible opportunities to restore processes and positively affect the overall ecological trajectory (Dufour & Piégay 2009). In California's Central Valley, riparian habitats were degraded so extensively by the early twentieth century that historical information on ranges of variation and comparable riverine reference sites was unavailable (Katibah 1984). To compound this problem, most forest remnants occur on dammed river systems that now experience altered flow regimes (Williams et al. 2009). Because rivers are dynamic and variable, it may be more appropriate to use restoration trajectories rather than a fixed snapshot in time (Hughes et al. 2005). Another option is to develop a "guiding image" using a combination of historical data, available reference sites, and process-based empirical models to specify the restoration potential for a healthy river system (Jansson et al. 2005; Palmer et al. 2005).

• *Physical and ecological processes drive the system.* Reestablishment of physical processes is necessary to restore ecosystem function (Poff et al. 1997; Wohl et al. 2005). Case studies such as the Cosumnes River show the many benefits of restoring driving processes such as flood dynamics that reinforce and amplify ecological processes. To achieve ecosystem function, resilience, and biodiversity goals in highly altered ecosystems, restoration of abiotic process may need to be coupled with active ecological management (Suding et al. 2004).

• *You cannot recreate the past – and you may not want to.* The two breach events (1995 and 1997) initially produced different plant communities with different successional trajectories, despite similar location and hydrological regime. Recent theory suggests that degraded systems may respond unpredictably to restoration, resulting in multiple equilibrium states or even unstable states (Suding et al. 2004). Unique conditions at the time of initial colonization events can have a legacy effect on successional pathways and restoration outcomes (Poff et al. 1997). As Wohl et al. (2005) put it, "a river has a history that continues to influence its present and future." Further complications, such as climate change, may make a return to historical reference conditions impossible (Seavy et al. 2009). In the future under climate change, the Cosumnes River will likely have fewer spring floods, with serious consequences for the amount and productivity of seasonally inundated floodplain habitat.

• *You can get more than you expected – if you are paying attention and manage adaptively.* Without monitoring and evaluation, opportunities to learn and improve restoration are often lost (Kondolf 1998). The levee breaches provided a glimpse into a functioning floodplain ecosystem and a better understanding of natural-process restoration. We maximized learning through a coordinated monitoring program. We discovered whole components of the ecosystem previously unknown, including thresholds of flooding necessary to achieve functions such as sediment deposition, algal and zooplankton productivity, and native fish habitat. Monitoring also revealed a mosaic of microhabitats and vegetation structure that help maintain bird diversity. With these extensive studies, we broadened our restoration goals and strategies to encompass many components of a functional floodplain ecosystem and refined future restoration in a cycle of adaptive management.

The Cosumnes restoration efforts evolved from a simple, single-species-focused planting model to mimic a forest reference site to a more analytical, process-based model of ecological function, which provides a deductive framework to guide future restoration (Jansson et al. 2005; Palmer et al. 2005). In the wake of the levee-breach studies, we have systematically assessed the restoration potential at other sites along the Cosumnes River based on the feasibility of restoring flooding. Those sites where frequent long-duration flooding is possible will be designated for seasonal wetland and riparian habitat using levee breaching; other sites with higher elevations that preclude frequent flooding will be restored with more traditional horticultural planting techniques. The lessons of the Cosumnes River permeate water policy and resource management in the Central Valley (Williams et al. 2009; Opperman et al. 2010), where hundreds of millions of dollars are spent to restore riparian, wetland, and riverine habitats for declining fish species (Sommer et al. 2007).

Fish ecologist John Magnusun lamented that, because our natural frame of reference is limited to our lifetime's experience, we live in the "invisible present," effectively unable to see the long-term processes of cause and effect (Magnuson 1990). Use of historical information, reference sites, and long-term research attempts to surmount this dilemma for restoration ecology. By expanding our frame beyond the invisible present, restoration is more likely to succeed.

Our approach to restoration of the Cosumnes River floodplain ecosystem brings a kind of pragmatism to the dilemma of the invisible present. Although this restored system lacks many of the hydrological and ecological features of presettlement Central Valley rivers, maintaining the assisted-floodplain ecosystem is a feasible and sustainable approach to support diverse ecological processes, communities, and species. As such, the "accidental forest" created by an unplanned levee breech has served not only as a window into the past, but one into the future as well.

ACKNOWLEDGMENTS

We thank Jeff Mount and Peter Moyle for spearheading the UC Davis Cosumnes Research Group. Jeff Opperman, Carson Jeffres, Nadav Nur, and Chrissy Howell contributed ideas and data, Geoff Fricker and Seth Paine provided graphics.

REFERENCES

Ahearn, D.S., Viers, J.H., Mount, J.F., & Dahlgren, R.A. (2006). Priming the productivity pump: flood pulse driven trends in suspended algal biomass distribution across a restored floodplain. *Freshwater Biology*, **51**, 1417–1433.

Booth, E., Mount, J., & Viers, J. (2006). Hydrologic variability of the Cosumnes River floodplain. *San Francisco Estuary and Watershed Science*, **4**(2), article 2. http://repositories.cdlib.org/jmie/sfews/vol4/iss2/art2.

Buergi, M. & Gimmi, U. (2007). Three objectives of historical ecology: the case of litter collecting in Central European forests. *Landscape Ecology*, **22**, 77–87.

Crain, P.K., Whitener, K., & Moyle, P.B. (2004). Use of a restored Central California floodplain by larvae of native and alien fishes. In *Early Life History of Fishes in the San Francisco Estuary and Watershed* (ed. F. Feyrer, L.R. Brown, R.L. Brown, and J.J. Orsi), pp. 125–140. American Fisheries Society, Bethesda, MD, USA.

Dufour, S. & Piégay, H. (2009). From the myth of a lost paradise to targeted river restoration: forget natural references and focus on human benefits. *River Research and Applications*, **25**, 568–581.

Feyrer, F., Sommer, T.R., Zeug, S.C., O'Leary, G., & Harrell, W. (2004). Fish assemblages of perennial floodplain ponds of the Sacramento River, California, USA, with implications for the conservation of native fishes. *Fisheries Management and Ecology*, **11**, 335–344.

Fleckenstein, J., Anderson, M., Fogg, G., & Mount, J. (2004). Managing surface water-groundwater to restore fall flows in the Cosumnes River. *Journal of Water Resources Planning and Management*, **130**, 301–310.

Florsheim, J.L. & Mount, J.F. (2002). Restoration of floodplain topography by sand-splay complex formation in response to intentional levee breaches, lower Cosumnes River, California. *Geomorphology*, **44**, 67–94.

Florsheim, J.L. & Mount, J.F. (2003). Changes in lowland floodplain sedimentation processes: pre-disturbance to post-rehabilitation, Cosumnes River, California. *Geomorphology*, **56**, 305–323.

Florsheim, J.L., Mount, J.F., & Constantine, C.R. (2006). A geomorphic monitoring and adaptive assessment framework to assess the effect of lowland floodplain river restoration on channel–floodplain sediment continuity. *River Research and Applications*, **22**, 353–375.

Gardali, T., Holmes, A.L., Small, S.L., Nur, N., Geupel, G.R., & Golet, G.H. (2006). Abundance patterns of landbirds in restored and remnant riparian forests of the Sacramento River, California, USA. *Restoration Ecology*, **14**, 391–403.

Golet, G.H., Gardali, T., Howell, C.A., et al. (2008). Wildlife response to riparian restoration on the Sacramento River. *San Francisco Estuary and Watershed Science*, **6**(2), article 1. http://repositories.cdlib.org/jmie/sfews/vol6/iss2/art1/.

Greco, S.E. & Plant, R.E. (2003). Temporal mapping of riparian landscape change on the Sacramento River, Miles 196–218, California, USA. *Landscape Research*, **28**, 405–426.

Grosholz, E. & Gallo, E. (2006). The influence of flood cycle and fish predation on invertebrate production on a restored California floodplain. *Hydrobiologia*, **568**, 91–109.

Hammond, J.L. (2008). *Identification of Nest Predators and Reproductive Response of the Modesto Song Sparrow, Melospiza melodia Mailliardi, to Experimental Predator Removal*. Master's thesis, Humboldt State University, Arcata, CA, USA.

Hughes, F.M.R., Colston, A., & Mountford, J.O. (2005). Restoring riparian ecosystems: the challenge of accommodating variability and designing restoration trajectories. *Ecology and Society*, **10**(1), 12. http://www.ecologyandsociety.org/vol10/iss1/art12.

Humple, D.L. & Geupel, G.R. (2002). Autumn populations of birds in riparian habitat of California's Central Valley. *Western Birds*, **33**, 34–50.

Jansson, R., Backx, H., Boulton, A.J., et al. (2005). Stating mechanisms and refining criteria for ecologically successful river restoration: a comment on Palmer et al. (2005). *Journal of Applied Ecology*, **42**, 218–222.

Jeffres, C.A., Opperman, J., & Moyle, P.B. (2008). Ephemeral floodplain habitats provide best growth conditions for juvenile Chinook salmon in a California river. *Environmental Biology of Fishes*, **83**, 449–458.

Junk, W.J., Bayley, P.F., & Sparks, R.E. (1989). The flood pulse concept in river-floodplain systems. *Canadian Journal of Fisheries and Aquatic Sciences*, **106**, 110–127.

Katibah, E. (1984). A brief history of riparian forests in the Central Valley of California. In *California Riparian Systems: Ecology, Conservation and Productive Management* (ed. K. Hendrix and R. Warner), pp. 23–29. University of California Press, Berkeley, CA, USA.

Kondolf, G.M. (1998). Lessons learned from river restoration projects in California. *Aquatic Conservation*, **8**, 39–52.

Magnuson, J.J. (1990). Long-term ecological research and the invisible present. *BioScience*, **40**(7), 495–501.

Mahoney, J.M. & Rood, S.B. (1998). Streamflow requirements for cottonwood seedling recruitment: an integrative model. *Wetlands*, **18**, 634–645.

Mount, J.F. (1995). *Rivers and Streams: The Conflict between Fluvial Process and Land Use.* University of California Press, Berkeley, CA, USA.

Moyle, P.B., Crain, P.K., & Whitener, K. (2007). Patterns in the use of a restored California floodplain by native and alien fishes. *San Francisco Estuary and Watershed Science*, **5**(3), 1–27. http://baydelta.ucdavis.edu/files/crg/reports/Moyle FloodplainfishMS-26nov.pdf.

Nur, N., Ballard, G., & Geupel, G.R. (2008). Regional analysis of riparian bird species response to vegetation and local habitat features. *Wilson Journal of Ornithology*, **120**(4), 840–855.

Opperman, J.J., Luster, R., McKenney, B.A., Roberts, M., & Meadows, A.W. (2010). Ecologically functional floodplains: connectivity, flow regime, and scale. *Journal of the American Water Resources Association*, **46**, 211–226.

Palmer, M.A., Bernhardt, E.S., Allan, J.D., et al. (2005). Standards for ecologically successful river restoration. *Journal of Applied Ecology*, **42**, 208–217.

Poff, N.L., Allan, J.D., Bain, M.B., et al. (1997). The natural flow regime: a new paradigm for riverine conservation and restoration. *BioScience*, **47**, 769–784.

Reiner, R.J. (1996). The Cosumnes River Preserve: 1987–95 fertile ground for new conservation ideas. *Fremontia*, **24**, 16–19.

Ribeiro, F., Crain, P.K., & Moyle, P.B. (2004). Variation in condition factor and growth in young-of-the-year fishes in floodplain and riverine habitats of the Cosumnes River, California. *Hydrobiologia*, **527**, 77–84.

Rood, S.B., Samuelson, G.M., Braatne, J.H., Gourley, C.R., Hughes, F.M.R., & Mahoney, J.M. (2005). Managing river flows to restore floodplain forests. *Frontiers in Ecology and the Environment*, **3**(4), 193–201.

Seavy, N.E., Gardali, T., Golet, G.H., et al. (2009). Why climate change makes riparian restoration more important than ever: recommendations for practice and research. *Ecological Restoration*, **27**(3), 330–338.

Sommer, T.R., Nobriga, M.L., Harrell, W.C., Batham, W., & Kimmerer, W.J. (2001). Floodplain rearing of juvenile Chinook salmon: evidence of enhanced growth and survival. *Canadian Journal of Fisheries and Aquatic Sciences*, **58**(2), 325–333.

Sommer, T.R., Armor, C., Baxter, R., et al. (2007). The collapse of pelagic fishes in the upper San Francisco Estuary. *Fisheries*, **32**(6), 270–277.

Stromberg, J.C., Patten, D.T., & Richter, B.D. (1991). Flood flows and dynamics of Sonoran riparian forests. *Rivers*, **2**(3), 221–235.

Suding, K.N., Gross, K.L., & Housman, D.R. (2004). Alternative states and positive feedbacks in restoration ecology. *Trends in Ecology and Evolution*, **19**, 46–53.

Swenson, R.O., Whitener, K., & Eaton, M. (2003). Restoring floods to floodplains: riparian and floodplain restoration at the Cosumnes River Preserve. In *California riparian Systems: Processes and Floodplain Management, Ecology and Restoration* (ed. P.M. Faber), pp. 224–229. 2001 Riparian Habitat and Floodplains Conference Proceedings. Riparian Habitat Joint Venture, Sacramento, CA, USA.

Swetnam, T.W., Allen, C.D., & Betancourt, J.L. (1999). Applied historical ecology: using the past to manage for the future. *Ecological Applications*, **9**(4), 1189–1206.

Thompson, J. (1957). *The Settlement Geography of the Sacramento-San Joaquin Delta, California: Palo Alto, California.* Doctoral dissertation, Stanford University, Palo Alto, CA, USA.

Tockner, K., Malard, F., & Ward, J.V. (2000). An extension of the flood pulse concept. *Hydrological Processes*, **14**, 2861–2883.

Trowbridge, W.B. (2007). The role of stochasticity and priority effects in floodplain restoration. *Ecological Applications*, **17**, 1312–1324.

Tu, M. (2000). *Vegetation Patterns and Processes of Natural Regeneration in Periodically Flooded Riparian Forests in the Central Valley of California.* Doctoral dissertation, University of California, Davis, CA, USA.

Westoby, M., Walker, B., & Noy-Meir, I. (1989). Opportunistic management for rangelands not at equilibrium. *Journal of Range Management*, **42**, 266–274.

Whipple, A.A., Grossinger, R.M., & Davis, F.W. (2011). Shifting baselines in a California oak savanna: nineteenth century data to inform restoration scenarios. *Restoration Ecology*, **19**, 88–101.

Williams, P.B., Andrews, E., Opperman, J.J., Bozkurt, S., & Moyle, P.B. (2009). Quantifying activated floodplains on a lowland regulated river: its application to floodplain restoration in the Sacramento Valley. San Francisco Estuary and Watershed Science. http://repositories.cdlib.org/jmie/sfews/vol7/iss1/art4.

Wohl, E., Angermeier, P.L., Bledsoe, B., et al. (2005). River restoration. *Water Resources Research*, **41**, W10301. doi:10.1029/2005WR003985.

Wood, J.K., Nur, N., Howell, C.A., & Guepel, G.R. (2006). Overview of Cosumnes riparian bird study and recommendations for monitoring and management. Report to the California Bay-Delta Authority Ecosystem Restoration Program from PRBO Conservation Science. http://baydelta.ucdavis.edu/files/crg/reports/PRBO_Overview_Woodetal2006.pdf.

Chapter 16

STREAMS PAST AND FUTURE: FLUVIAL RESPONSES TO RAPID ENVIRONMENTAL CHANGE IN THE CONTEXT OF HISTORICAL VARIATION

Daniel A. Auerbach,[1] *N. LeRoy Poff,*[1]
Ryan R. McShane,[1] *David M. Merritt,*[1,2]
Matthew I. Pyne,[1] *and Thomas K. Wilding*[1]

[1]Colorado State University, Fort Collins, CO, USA
[2]USDA Forest Service, Natural Resource Research Center, Fort Collins, CO, USA

16.1 INTRODUCTION

Changes at the global scale jeopardize the sustainable management of biodiversity and ecosystem services in streams and rivers (Baron et al. 2002; Poff 2009). Climatic shifts, land-use changes and the spread of invasive species complicate the goal of balancing ecosystem protection with water-resource development. As the environmental drivers of riverine systems move outside of their historical range of variation, researchers and managers must confront the possibility of novel ecological states and their consequences for currently valued resources.

Unprecedented ecological conditions in rivers demand management strategies that balance knowledge of historical patterns with the rapid assimilation of new information. For example, stakeholders might ask whether a valued population of cold-water game fish will persist within its current range when winters are shorter and summers are warmer. Addressing this concern requires synthesizing historical information, such as evolved thermal tolerances within a natural flow context, with expectations regarding the population's physiological or behavioral adaptability in the face of contemporary factors that include interaction with introduced species, harvesting pressure, habitat loss or augmentation, and trends in nutrient or sediment inputs due to altered land-use practices. The task of integrating current observations with historical understanding poses challenges to the formulation of clear policy regarding the management of rivers and their watersheds. Fortunately, a general, process-based framework that builds from the study of many unique rivers provides the necessary conceptual foundation for addressing this task.

Riverine ecosystems are characterized by ongoing physical fluxes and numerous feedbacks that link organisms and their shared habitat (Allan & Castillo 2007; Box 16.1). Ecological states observed in rivers, such as population levels or nutrient concentrations,

---

Box 16.1 Five principles of riverine ecosystem function.

River and stream ecosystems consist of strongly coupled biological and physical elements that interact in the context of specific climatic, physiographic, and biogeographic settings. Defining "ecological states" as population abundances, species assemblages, or biogeochemical concentrations, we can identify several critical characteristics of these systems.

1. Conditions in streams and rivers naturally vary between regions and within watersheds from headwaters to outlet. River networks also transition among different ecological states through time such as cycling between periods of high and low flows. Differences in these patterns of variation through time permit distinguishing rivers in terms of historical flow conditions and provide the basis for appropriate management targets.

2. Hydrologic, geomorphic, and biological processes combine to determine ecological states. As low points on the landscape, streams receive water, sediment, wood, and nutrients from the surrounding terrestrial landscape, and thereby integrate processes occurring throughout watersheds. Streams display feedbacks between these domains, such as the effect of riparian vegetation on sediment movement, but the physical attributes of a system often help determine the temporal scale of relevance to historically informed management.

3. Discharge, or streamflow, is often a particularly important determinant of ecological states in rivers and acts as a key control on sediment flux, temperature fluctuations, wood inputs, and nutrient concentrations. Flow regimes mediate between environmental drivers and ecosystem responses and may be manipulated to achieve desired ecological states. As a record of changes through time, adjusting a river's flow regime represents an explicitly historical approach to management.

4. Population, community, and biogeochemical dynamics in rivers and streams are innately linked. Management that alters the relative abundance of one or several species may alter nutrient or solute concentrations and vice versa. The pathways by which these interactions occur are likely to be complex, context-specific, and sometimes counterintuitive. These interactions warrant an experimental approach to manipulating rivers and streams.

5. River and stream systems can recover (sometimes rapidly) from degradation, if recovery efforts address the causes as well as the symptoms of detrimental forces. In addition to fostering appropriate habitat conditions, management that accounts for species' natural histories, in terms of demographic processes and biotic interactions, is more likely to achieve lasting success.

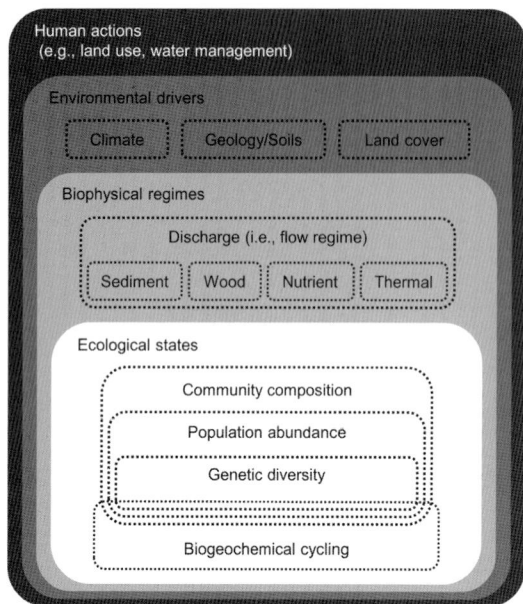

Fig. 16.1 A conceptual representation of the hierarchical organization of stream and river ecosystems. Ecological states, such as community composition or genetic diversity, are proximately driven by biophysical regimes in flow, sediment, wood, nutrients, and temperature, with the latter also strongly influenced by the flow regime (the characteristic temporal sequence of measured discharge). In turn, these environmental regimes are controlled by atmospheric and terrestrial drivers. Ultimately, human actions can force each of the underlying environmental drivers and biophysical regimes.

are proximately driven by environmental regimes in flow, sediment, wood, and temperature (Box 16.1, Fig. 16.1). These environmental regimes are in turn controlled by broader atmospheric, terrestrial, and anthropogenic drivers such as precipitation, land use, and water management. This dynamic interplay creates tremendous spatial and temporal variation in river networks. Understanding and influencing these mediating regimes, even as they continue to shift, is critical to achieving desired ecological states. This perspective generates a focused set of mechanistic questions and research priorities concerning changes in historical relationships, for instance between fish populations and flow or temperature regimes.

Analyzing river ecosystems in terms of their characteristic biophysical variation, particularly their flow regimes,

provides a sound basis for anticipating responses to rapid environmental change and can thereby suggest meaningful management actions. Accordingly, we focus on the relationship between drivers outside of their historical range of variation, altered flow regimes, and consequent ecological states. We caution, however, that management to manipulate flow regimes must be undertaken within the context of the many other complex factors that influence ecological integrity in rivers. For example, food web dynamics (e.g. Power et al. 2008), terrestrial subsidies (Baxter et al. 2005), or landscape influences (Allan 2004) may interact with changing flow regimes to influence the outcomes of a particular flow-based management action. In addition, we note that despite the necessity of historical knowledge, its insufficiency as a complete guide to the future poses a basic challenge to the formulation of effective management strategies in an uncertain future.

Historical variation as a double-edged sword

The difficulty of accurately defining "novel" conditions in rivers illustrates the importance of historical information. Along with our increasingly detailed knowledge of river ecosystems has come an appreciation of the extent to which humans have modified the banks, channels, and biota we encounter. Reduced spawning runs of salmon have disrupted the flow of marine nutrient subsidies to riparian forests (Naiman et al. 2002). Widespread damming and land-use change have altered patterns of discharge and reduced connectivity throughout channel networks, with detrimental impacts for many organisms and ecosystem functions (Nilsson et al. 2005; Poff et al. 2007). Agricultural development, urbanization, and airborne deposition of contaminants have altered nutrient fluxes (Allan 2004), and non-native species introductions have disrupted community interactions (Rahel & Olden 2008). The magnitude and history of these changes can undermine the perception of natural baseline conditions.

Contemporary watersheds may also carry a legacy of human actions prior to the usual period of record captured by instrumented monitoring. Examples in North America include the continental-scale adjustment of flow and sediment regimes following widespread beaver trapping during the eighteenth and nineteenth centuries, the removal of large woody debris from channels to facilitate transportation, the catastrophic channel modifications from hydraulic mining and tie-drives, and the massive input and storage of sediment after widespread deforestation (Wohl 2005). Seen in the light of these changes, some current ecosystem states could be viewed as transient recovery phases rather than stable endpoints. Thus, knowledge of a river's past may reveal that the reference conditions that guide certain decisions are the product of human impacts as well as natural evolutionary and geomorphic forces (Wohl 2005; Humphries & Winemiller 2009; Fig.16.1).

Complicating the role of historical information is the recognition that "stationarity is dead" (Milly et al. 2008). Many current water-resource-management practices are grounded in the assumption that climatic drivers and associated fluvial processes have a stable and well-characterized statistical range of variation. Time series of discharge data extending back as far as the late nineteenth century are often used in defining present-day habitat distributions (e.g. floodplain boundaries, depth profiles) or in designing reservoir operation plans. Yet environmental drivers that have shifted outside of their historical range of variation weaken the premise that historical information supports predictions concerning future ecological states. In addition, many riverine ecosystems already have no apparent analog in the historical record and require treatment as such (e.g. cold, clear-water reaches below dams on rivers that were historically warm and turbid; highly salinized and acidified streams in urban areas). Failure to acknowledge these changes risks decisions that are encumbered by false expectations concerning relationships that have not persisted. Researchers and managers share a responsibility to consider how a non-stationary climate and further modification of channel structure, hydrology, and biological assemblages will combine to undermine current assumptions regarding riverine ecosystem dynamics.

Understanding the record of history is critical to effective management, but knowledge of past ecosystem forms and processes cannot eliminate uncertainty from the decision-making process. Stewardship of rivers requires that we consider the direction, magnitude, and variability of expected changes in the context of previous observations, but with a license to experiment. Grounding management experiments in a process-based framework that emphasizes the relevance of flow regimes to riverine ecosystem function (Box 16.1) will improve efforts to anticipate and adapt to projected climatic and socioeconomic changes.

16.2 FLOW REGIMES STRUCTURE ECOLOGICAL COMPLEXITY

Aquatic and riparian ecosystems are controlled by patterns of variation in key environmental regimes, particularly streamflow, sediment, nutrient concentrations, wood inputs, and temperature (Fig. 16.1). Streamflow varies in terms of its magnitude (how much flow?); duration (how long is a particular flow level sustained?); frequency (how often does a particular flow level occur?); rate of change (how quickly does the flow change?); and timing and predictability (when do particular flow levels occur?) (Poff et al. 1997). Differences in these dimensions of the flow regime, both within and between drainage networks, generate a physical habitat template (Southwood 1988) that selects for organisms with certain functional characteristics (Poff & Ward 1990; Townsend & Hildrew 1994; Poff et al. 2006b). For instance, a turbulent stretch of river is more likely to contain organisms with a preference for faster current velocity and coarse substrates; a calmer, forested section may be dominated by invertebrates with the ability to shred falling leaves; and a system dominated by frequent flooding will tend to contain mobile species with short generation times.

Flow dynamics "filter" species from a regional candidate pool according to traits for life history, physiology, morphology, and behavior. Regional or watershed-scale filters (e.g. annual minimum temperature, flood frequency) interact with localized, site-scale filters (e.g. velocity, depth, presence of predator/prey) to influence the distribution and abundance of organisms such as invertebrates and fish (Poff 1997; Fausch et al. 2002; Wiens 2002). In addition to these niche constraints, dispersal limitation and source-sink dynamics likely play a role in determining the composition of some stream communities and may also be subject to flow regimes (Lowe et al. 2006; Hitt & Angermeier 2008; Winemiller et al. 2010).

Streamflow directly filters ecological responses by constraining habitat dimensions and by setting the sequence of mortality-inducing disturbances for organisms of varying vulnerability. Flow also operates indirectly by influencing other physical regimes: temperature, sediment, wood inputs, and nutrient concentrations (Doyle et al. 2005; Fig. 16.1). Tracking changes through time in this "master variable" provides a perspective on observed ecological states in rivers that emphasizes the biological significance of a mixture of high and low flows (and their frequency, duration and timing), rather than a single minimum or average condition (Poff et al. 1997; Bunn & Arthington 2002). The plot of measured discharge against a given duration is known as a hydrograph, and the set of daily, weekly, yearly, or interannual hydrographs provides fundamental insight into a river's ecological function (Figs. 16.2 and 16.3). The "natural flow regime" is the distribution of flows that would occur in the absence of major development and which is associated with a characteristic set of ecological states in a river network (Poff et al. 1997; Fig. 16.2). The dyna-

mism of rivers dictates that efforts to understand and manage them incorporate the patterns revealed in this key record of historical processes.

Species can adapt to natural flow regimes, especially when extreme events limit growth or reproductive success, so that critical behaviors, such as foraging, mating, dispersal, and establishment, come to follow closely a characteristic sequence of flows (Lytle & Poff 2004; Lytle et al. 2008). The tight relationship between the structure and function of river ecosystems and flow-regime characteristics, both natural and altered, is well documented (Bunn & Arthington 2002; Poff & Zimmerman 2010). For example, flashy streams (with rapid transitions between high and low flows) favor habitat and diet generalists (Poff & Allan 1995; Roy

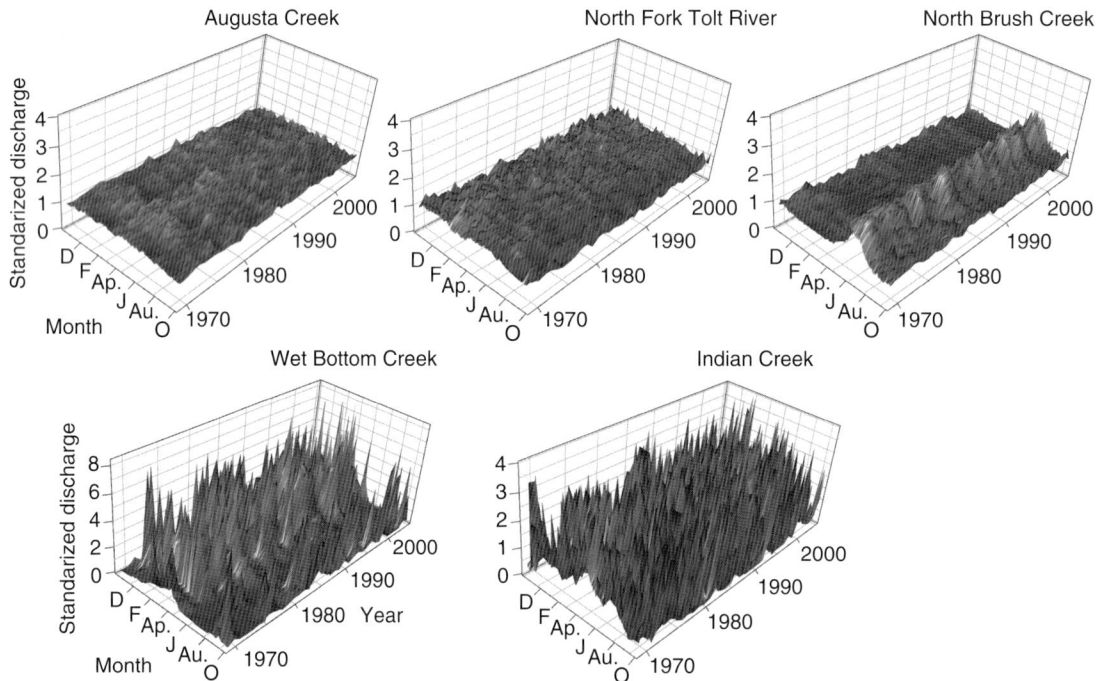

Fig. 16.2 Hydrographs illustrating regional differences in flow regime. Daily flow is shown for the 1969–2005 water years. All five streams drain catchments of similar size (90–125 km²), and flow was normalized to enable comparison between sites (natural-log-transformed and divided by mean daily flow). The three upper hydrographs represent perennial streams, and the lower two represent intermittent streams. Following the stream classification terminology of Poff (1996), Wet Bottom Creek in Arizona is a "harsh intermittent" stream that experiences extended periods of zero flow punctuated by large flow events caused by sporadic, intense rainstorms. Indian Creek is a "flashy intermittent" stream in Illinois subject to short periods of zero flow. Augusta Creek is a "superstable groundwater" stream in Michigan buffered from short-term variation in precipitation by a large groundwater input. North Fork Tolt River in Washington is a "winter rain" system. North Brush Creek in Wyoming is a classic "snowmelt" stream.

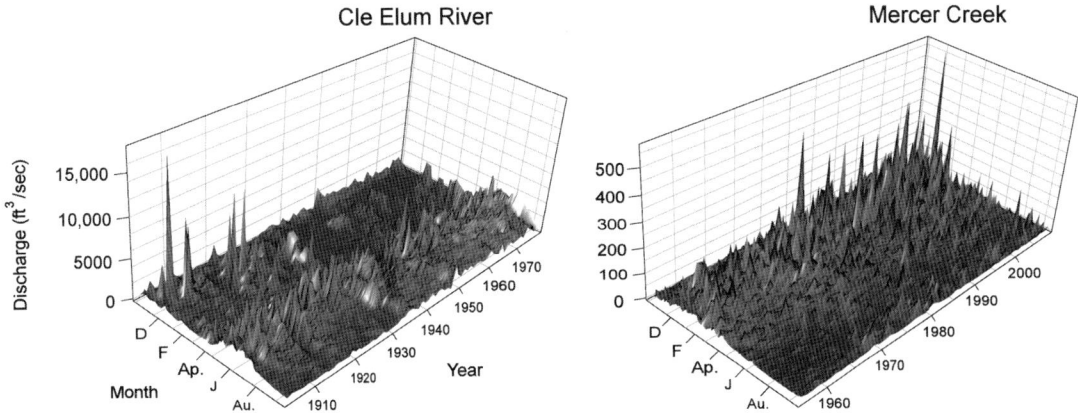

Fig. 16.3 Human influences on flow regimes. Hydrograph of Cle Elum River (left panel) on the east slope of the Cascade Range in Washington, showing the influence of a storage dam on natural variation in flow. This hydrograph encompasses the water years 1904–1978. Cle Elum River is naturally a "rain-and-snow" stream that shows peak flows from both winter rains and spring snowmelt. In 1933, an irrigation dam constructed upstream of the gage reduced winter peak flows and raised late summer low flows. Hydrograph of Mercer Creek (right panel) in Bellvue, Washington, showing a subtle effect of urbanization on a small stream. Over the period of record (1956–2009), the population of Bellevue grew from 7658 (1950) to 109 569 (2000), which is reflected in the increasing frequency and intensity of high flows seen in the winter since the 1970s. Annual flood events (flows exceeding the 75th percentile flow) increased from 31 to 49, on average, between 1955 and 2000.

et al. 2005), peak flows sustain the recruitment of native riparian vegetation (Merritt & Poff 2010), and timing of high flows relative to species' life histories can prevent invasion of non-native fish (Fausch et al. 2001) or modify entire food webs (Power et al. 2008).

Consequently, the natural flow paradigm argues that alteration of these sequences will likely affect the biological communities and overall integrity in rivers and riparian zones (Poff et al. 1997; 2010; Arthington et al. 2006; Fig. 16.3). Patterns of flows clearly regulate the basic dynamics of riverine ecosystems – the challenge is to transform a historically informed understanding of flow–ecology relationships into mechanistic models that project how fluvial systems will respond to novel conditions associated with rapid climate change.

16.3 FLOW REGIMES AS A MANAGEMENT TARGET

Changing regional precipitation and temperature patterns will necessitate action to maintain desired ecological states and ecosystem goods and services. In contrast to climate, geology, or physiography, stream-

flow within a drainage network can be manipulated to mitigate global-scale changes. Management actions, such as revegetation and dam reoperation, can influence watershed hydrology, delaying or diminishing undesirable effects associated with climate change. For instance, releases of stored winter rains could compensate to some degree for reduced snowmelt runoff peaks in a warming mountainous region. Careful adjustment of flow regimes begins with measurement of the existing and potential flows in a river. This information permits formulating flow targets and assessing progress toward them (Richter et al. 2006). Given the foundation provided by these data, several general issues are relevant to the design of site-specific plans.

Political and economic constraints are often as relevant to the design of environmental regime adjustments as the limits of ecological knowledge. Whenever possible, we encourage identifying the actions most likely to lead to desirable ecological states and then reconciling these alternatives with societal needs, perhaps iteratively (Palmer et al. 2009; Poff 2009). Such a process involves translating policy objectives (e.g. "clean water," "biodiversity") into implementation methods (e.g. intercept storm runoff with adequate riparian buffers; maintain connectivity via flows

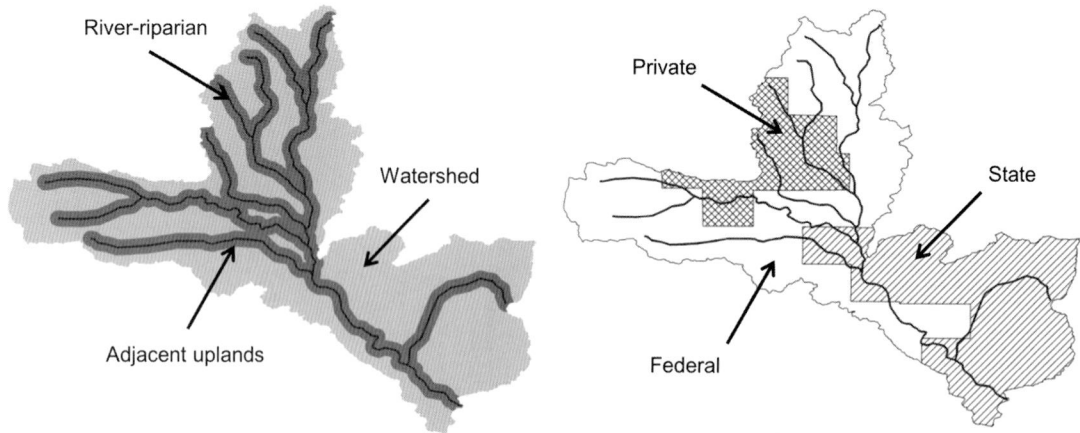

Fig. 16.4 Hypothetical watershed illustrating scale and boundary issues. The left panel indicates three scales of relevance to management, and the right panel indicates how these scales might intersect with ownership divisions. Changing environmental regimes will manifest as altered ecological states at several scales within a river network, and sensitivity to the scale hierarchy will improve mitigation and adaptation efforts. When political boundaries do not align with ecological boundaries at various scales, coordination and cooperation among landholders will be required to ensure habitat connectivity and to prevent contradictory management activities.

that allow access to floodplains and side channels). Including acceptable ranges of variation around targeted average levels of a desired ecosystem state is critical to designing approaches that are robust to both climatic and socioeconomic change (Landres et al. 1999). Management plans that accommodate natural variability will establish the importance of flexibility and may stand a greater chance of meeting their criteria for success over time.

Key considerations for strategic management

Three additional concerns factor into planning appropriate environmental regimes: mismatches between institutional jurisdictions and ecologically meaningful boundaries (Cumming et al. 2006), hierarchical spatial scales of influence (Poff 1997; Poole 2002), and the persistence and connectivity of habitat (Fausch et al. 2002; Pringle 2003).

Eliminating boundary mismatches will be impossible in many circumstances, but they nonetheless inform flow-regime interventions (and most other manipulations) because most global changes are

unlikely to follow existing political boundaries or organizational divisions. Successfully mitigating the effects of environmental drivers that have departed from their historical range is likely to involve new partnerships between public and private authorities, with greater coordination and cooperation between agencies, stakeholders, and researchers (Cash et al. 2006; Fig. 16.4). For example, irrigators and wildlife authorities might seek ways to incentivize water delivery that preserves a semblance of critical spawning or seeding cues even as other changes diminish these flow-regime elements.

The branching structure and habitat orientation imposed by unidirectional flow distinguish the effects of scale in river systems, producing ecological states that are the result of local factors nested within the network context (Poole 2002; Grant et al. 2007; Rodriguez-Iturbe et al. 2009). For example, species assemblages and nutrient cycling in a particular reach reflect the interaction of flows with the channel substrate and hydraulic geometry (i.e. pool vs riffle), with the surrounding riparian area (i.e. extensive floodplains vs confined canyons), with neighboring up- and downstream reaches, and with more distant terrestrial portions of the watershed (Ward 1989; Fausch et al.

2002; Wiens 2002). Sensitivity to the effects of hierarchical scale can improve management interventions by informing the feasibility of counteracting large-scale mismatches with relatively small-scale interventions (Fig. 16.4). For instance, constructing pools for a fish species that requires them may incur significant, lasting costs if the species does not fit within the regional thermal profile or if watershed-wide land-use change produces an unsuitable sediment regime (Pretty et al. 2003). Selective timber harvesting throughout a watershed might diminish soil erosion relative to clear cutting, but a simultaneously intensive riparian grazing program could counteract benefits to aquatic organisms by promoting bank destabilization. Similarly, reshaping channel profiles with heavy machinery may not yield the desired biodiversity outcomes if the surrounding species pool has been significantly depleted, is blocked by movement barriers, or includes aggressive invasive species. These examples are certainly not intended to discourage local efforts, but rather to emphasize the need to consider the consequences of possible actions – flow manipulations or otherwise – across multiple spatial scales (Bond & Lake 2003; Brown et al. 2011; Fig. 16.4).

Attention to the habitat patch networks that correspond to spatial variation in flows can also enhance the design of robust management strategies. Stream-dwelling organisms depend on heterogeneous conditions that support foraging and reproduction and that act as refugia during disturbance events (Schlosser 1991; Palmer et al. 1995; Magoulick & Kobza 2003). Refugia exist at multiple scales, from interstitial spaces between cobbles and localized thermal habitats to landscape features such as tributary and side channels. Connectivity between these patches is necessary to permit adequate gene flow, tracking of spatiotemporally varying resources, and the possibility of regional "rescue" following local extirpation (i.e. metapopulation dynamics) (Fagan 2002; Falke & Fausch 2010; Winemiller et al. 2010). Flow regimes can mediate how in-channel infrastructure and watershed land use influence both the movement of organisms and the extent or accessibility of habitat (Stanford et al. 1996; Pringle 2003). Manipulating flows or fluvial forms to protect refugia and their connectivity may aid some species in their struggle to adapt (Magoulick & Kobza 2003; Matthews & Wickel 2009; Palmer et al. 2009). For example, the importance of pools and hyporheic zones for stream organisms in existing arid areas might justify efforts to promote such habitats in streams

expected to shift from continuous to intermittent flow (Humphries & Baldwin 2003).

Niches lost, niches gained

Complete information is never available when confronting riverine management issues both because of logistical limitations and because of the ongoing evolution of ecological systems. Nonetheless, an adaptive, experimental approach that accounts for the concerns outlined above positions managers to handle novel trends and fluctuations in environmental drivers. We propose that shifting climate and land-use drivers will affect riverine ecosystems in three conceptually distinct ways (Fig. 16.5). First, currently occurring niches – the endpoint of the multiscaled, multidimensional

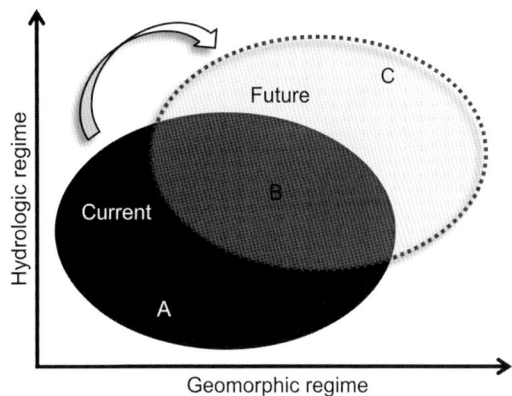

Fig. 16.5 Shifts in the range of variation of environmental regimes will alter available niches in rivers. Hydrologic and geomorphic regimes in river networks may transition from a current range of variation (black) to a future one (white). Multiple interacting regimes form niches in streams and rivers, but are reduced here to these two dimension for simplicity. The three zones (A–C) represent ecological states (i.e. "niches") within the current and future biophysical regimes: zone A (black) denotes entirely lost ecological states under future hydro-geomorphic regimes; zone B (gray) depicts states that persist either locally or regionally; and zone C (white) corresponds to "novel" states that have never existed within the current range of variation. The types of changes on a particular river will help determine management strategies. Respectively, these might involve intensive flow manipulation to preserve lost hydrologic cues, ongoing monitoring for further shifts, or a program of experimentation to determine new best practices.

filtering processes described above – may be entirely lost from a region of interest (zone A in Fig. 16.5). Second, such niches may be relatively unchanged in a current location or regionally retained but geographically displaced (zone B in Fig. 16.5). Third, entirely new or regionally novel niches may emerge (zone C in Fig. 16.5).

Real river ecosystems will almost certainly mix these simplified scenarios (Fig. 16.6), especially if flow regimes buffer countervailing effects from altered large-scale drivers. Yet, they constitute useful cases for thinking about changes in flow-regime components. For instance, large or discontinuous shifts in the mean or variance of maximum daily flows might correspond to niche loss or gain. Figure 16.6 illustrates this as the loss of black and gain of white segments in a hypothetical river network. By contrast, smaller shifts in maximum daily flows might result in a relatively minor spatial niche displacement, as illustrated by the transition of some segments from black to gray in Fig. 16.6.

The persistence of a flow regime associated with a valued species or desirable ecosystem state presumably constitutes the best-case scenario. Under these circumstances, management can strive to "do no harm" by focusing on monitoring or maintenance to ensure that

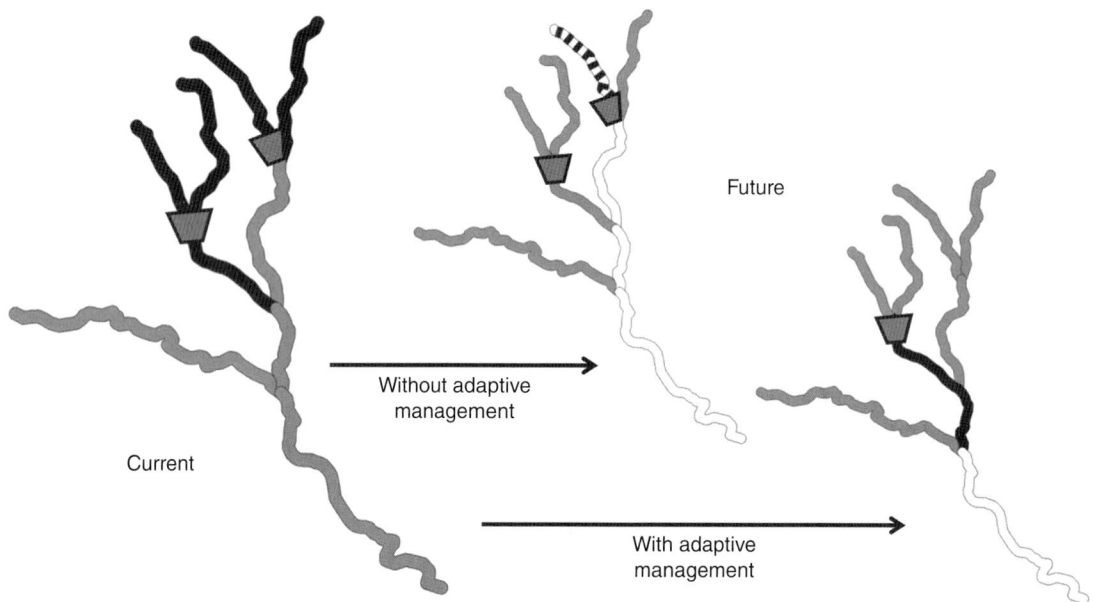

Fig. 16.6 Altered ecological states in the river network. Colors correspond to Fig. 16.5: black segments represent lost conditions, gray denotes persistence, and white the emergence of novel states. Gray trapezoids represent water infrastructure such as dams and diversions. Current environmental regimes (left panel) support distinct species assemblages adapted to a particular range of variation. Shifts in the range of variation may occur in the absence of management adjustments (upper right panel) or within an adaptively managed watershed (lower right panel). Even without adaptive management (upper right), certain segments could retain desired ecological states (gray stays gray), but some segments would likely change states (black becomes gray) while others are entirely lost ("extinction" of black). Dam operation could further disconnect the network, obstructing colonization from other segments (black becomes dashed). In contrast, adaptive management (lower right) that includes re-operation of an existing dam might permit the local preservation of an otherwise vulnerable ecological state (black stays black), or its regional persistence in other segments farther downstream (gray becomes black). Additionally, dam removal could allow colonization from other portions of the network (black becomes gray rather than dashed). The adaptive management scenario broadens the potential number of ecological states within the network to include those that might otherwise be lost and those that match stakeholder values.

further global change does not drive the system past a flow threshold at which the target niche is lost. In a similarly optimistic vein, flow-regime changes could also deliver management benefits if an undesirable ecosystem state is replaced by a desirable one (e.g. niche changes leading to reduction of the detrimental effects caused by an invasive species).

Unfortunately, in many cases, the niche consequences of flow-regime changes will be considered detrimental. The difficulty of trying to "go against the flow" by artificially sustaining desired conditions will generate tough choices between complex and costly logistics and the need to communicate to stakeholders that it is unreasonable to preserve existing states. As an example, consider a shift from a spring-snowmelt to winter-rain flow regime, resulting in an earlier and lower magnitude annual peak flow (e.g. North Brush Creek to North Fork Tolt River in Fig. 16.2). The persistence of a species with reproductive timing that is adapted to predictable, yearly peak flows will then depend on its ability to adjust to new cues phenotypically or genetically or on managers' ability to deliver flows that sufficiently mimic the current sequence (Schlaepfer et al. 2002; Lytle & Poff 2004). If the altered peak flow timing exceeds the species' adaptive capacity, and over-allocated water resources in the basin preclude adequate flow manipulation, then regional extirpation is plausible. Species already facing unfavorable flow regimes, such as Plains cottonwood (*Populus deltoides*) in western US rivers, are at potentially greater risk under this scenario (Stella et al. 2006; Merritt & Poff 2010).

Entirely new niches present yet another set of challenges. The scope of continental or global flow regimes may provide precedents for locally novel circumstances (e.g. present day arid-region streams serving as models for systems undergoing desertification), but when no such precedent is available, managers will need to incorporate predictive modeling and a diversified program of trial and error. No single method will constitute the best response to the loss, creation, or displacement of niches within river and stream ecosystems. Legal constraints or early indications of irreversible damage may call for intensive manipulations. Careful monitoring of a system that adjusts on its own may be most appropriate in other circumstances. Successful resource management and conservation will creatively pursue a combination of approaches with clear evaluation standards and will avoid entrench-

ment by shifting resources to those that deliver the greatest ecological return.

Classification to enhance dialogue and decisions

A flow-regime paradigm emphasizes the importance of variability, but the complexity of patterns in rivers can become overwhelming, particularly when social or political mandates presume managers act within straightforward, tractable systems. Classification of flow types (e.g. snowmelt vs rainfall, perennial vs intermittent) affords a means of dealing with some of this complexity while retaining essential relationships between flow pattern and ecological outcome. Differences in precipitation, vegetation, and topography support the differentiation of regimes in terms of biologically relevant hydrograph components (e.g. frequency of peak and low flows, seasonality; Fig. 16.2). This facilitates the association of certain regime types with targeted ecological outcomes across similar reaches, sub-watersheds, or basins, and can expedite efforts to monitor progress toward these goals (Arthington et al. 2006; Poff et al. 2010). Flow-regime classification can also play a role in the reconstruction of appropriate hydrographs when large dams and extensive land-use modification mask the underlying natural variation (Fig. 16.3). Representative typologies have been developed for the United States (e.g. Poff 1996), New Zealand (Biggs et al. 1990), Australia (Kennard et al. 2010), and France (Snelder et al. 2009) as well as various other nations.

Classifications are defined by recent historical climatic conditions, and this raises the issue of their applicability to future conditions. Climate change is projected to modify historical flow regimes significantly (Doll & Zhang 2010) and may induce entirely novel hydro-climatological conditions (Williams et al. 2007). In some instances, this will necessitate the revision of flow-regime types, and this possibility constitutes a strong argument for the maintenance and expansion of the gauge networks that provide critical data for any flow-based approach. Many watersheds, however, are likely to experience shifts in temperature and precipitation patterns that are within the total range of all systems considered for classification (e.g. among all types shown in Fig. 16.2). That is, novel conditions for a particular river may or may not be novel at some broader scale (e.g. continental). In some cases, a river or stream may transition between currently recognizable flow-regime classes, such that management of the newly altered system can take advantage of knowledge regarding existing systems with similar hydrologic properties. For example, anticipating a transition from perennial to ephemeral flows (Seager et al. 2007), managers in a drying region might consult with those who have already experienced the challenges of sustaining desired ecological states under periodic drought conditions.

Certain forms of water-resource infrastructure have distinct signatures in the hydrograph, and typologies may include patterns such as the regular discharge spikes associated with "hydropeaking" (releases from hydropower installations that follow from the intensity of electrical demand). Classification may also help prioritize resources by indicating if and how a flow regime is already subject to significant human alteration (Bunn & Arthington 2002; Graf 2006; Poff et al. 2006a; 2007; Fig. 16.3). In some instances, climatic shifts may be of less relevance to desired ecological states, and climate-related mitigation programs may or may not adequately address impacts related to other types of change. For example, consistent, intermediate-magnitude flows below an irrigation dam might harm a native species accustomed to sporadic or seasonal fluctuations in discharge or allow non-native species preferring stabilized flows to flourish (Marchetti & Moyle 2001). In this case, an assessment of the ecological consequences of altered precipitation in the basin would need to begin with an account of the dam's impacts and the degree of flexibility in its operation.

Efforts to apply a flow classification approach to pressing issues in river management have progressed rapidly and continue to be refined with the availability of new data and analysis tools. The ecological limits of hydrologic alteration (ELOHA) framework seeks to integrate scientific knowledge regarding the importance of heterogeneity in the hydrologic, geomorphic, and biological attributes of fluvial environments with social priorities for the use and protection of those systems (Poff et al. 2010). Incorporating the needs of multiple stakeholders, the ELOHA process provides a means of quantitatively relating flows to ecological states at a regional scale, thereby generating a context for changes in a particular watershed. Poff et al. (2010) describe the steps of the ELOHA framework in detail. Briefly, they include building a hydrologic foundation, classifying segments appropriately, constructing defensible flow-response models, evaluating uncertainty, and building toward consensus among stakeholders. Our emphasis here has been on flow regimes, but because of the importance of the interaction between discharge and the geomorphic context, ELOHA involves local geomorphic subclassification to better stratify ecological responses. Although the ELOHA methodology will not be suited to all issues in riverine ecosystem management, its recent implementations demonstrate the value of addressing novel environmental conditions in terms of flow (http://conserveonline.org/workspaces/eloha). The approach enables managers facing climatic or land-use uncertainty to refine their expectations for ecological responses by conducting a scenario analysis using flow-response models developed for a region. ELOHA also embodies the more general goal of framing resource management and conservation decisions within the understanding of basic ecological processes gained through research and historical analysis.

16.4   SUMMARY

In this chapter, we have argued that a process-based understanding of river ecology (Box 16.1, Fig. 16.1) offers a means of effectively evaluating resource-management options in the face of rapid global change. Contemporary ecological states in rivers reflect a hierarchical set of drivers (Fig. 16.1) with characteristic ranges of variation over a recent historical period. These relationships provide the basis for anticipating rivers' responses as their drivers depart from these historical ranges. Monitoring and adjusting the properties of key environmental regimes, particularly the flow regime, will provide an opportunity for management to mitigate or adapt to large-scale shifts. Recognizing the importance of variable flows and their interaction with multiscaled habitat networks is critical to developing management strategies that sustain the heterogeneity required for diverse, functioning river ecosystems. Despite the political and economic costs of an experimental approach to management, operational flexibility is vital to successfully integrating knowledge of rivers' past with their ongoing alteration. We remain hopeful that the dedication and ingenuity of the many managers committed to ecological and socioeconomic health will overcome the difficulties posed by a rapidly changing world.

REFERENCES

Allan, J.D. (2004). Landscapes and riverscapes: the influence of land use on stream ecosystems. *Annual Review of Ecology, Evolution, and Systematics*, **35**, 257–284.

Allan, J.D. & Castillo, M.M. (2007). *Stream Ecology Structure and Function of Running Waters*. Springer, Dordrecht, The Netherlands.

Arthington, A.H., Bunn, S.E., Poff, N.L., & Naiman, R.J. (2006). The challenge of providing environmental flow rules to sustain river ecosystems. *Ecological Applications*, **16**, 1311–1318.

Baron, J.S., Poff, N.L., Angermeier, P.L., et al. (2002). Meeting ecological and societal needs for freshwater. *Ecological Applications*, **12**, 1247–1260.

Baxter, C.V., Fausch, K.D., & Saunders, W.C. (2005). Tangled webs: reciprocal flows of invertebrate prey link streams and riparian zones. *Freshwater Biology*, **50**, 201–220.

Biggs, B.J.F., Duncan, M.J., Jowett, I.G., Quinn, J.M., Hickey, C.W., Daviescolley, R.J., & Close, M.E. (1990). Ecological characterization, classification, and modeling of New-Zealand Rivers – an introduction and synthesis. *New Zealand Journal of Marine and Freshwater Research*, **24**, 277–304.

Bond, N.R. & Lake, P.S. (2003). Local habitat restoration in streams: constraints on the effectiveness of restoration for stream biota. *Ecological Management & Restoration*, **4**, 193–198.

Brown, B.L., Swan, C.M., Auerbach, D.A., Campbell-Grant, E.H., Hitt, N.P., Maloney, K.O., & Patrick, C. (2011). Metacommunity theory as a multi-species, multi-scale framework for studying the influence of river network structure on riverine communities and ecosystems. *Journal of the North American Benthological Society*, **30**, 310–327.

Bunn, S.E. & Arthington, A.H. (2002). Basic principles and ecological consequences of altered flow regimes for aquatic biodiversity. *Environmental Management*, **30**, 492–507.

Cash, D.W., Adger, W.N., Berkes, F., et al. (2006). Scale and cross-scale dynamics: governance and information in a multilevel world. *Ecology and Society*, **11**(2), article 8. http://www.ecologyandsociety.org/vol11/iss2/art8/.

Cumming, G.S., Cumming, D.H.M., & Redman, C.L. (2006). Scale mismatches in social-ecological systems: causes, consequences, and solutions. *Ecology and Society*, **11**(1), article 14. http://www.ecologyandsociety.org/vol11/iss1/art14.

Doll, P. & Zhang, J. (2010). Impact of climate change on freshwater ecosystems: a global-scale analysis of ecologically relevant river flow alterations. *Hydrology and Earth System Sciences*, **14**, 783–799.

Doyle, M.W., Stanley, E.H., Strayer, D.L., Jacobson, R.B., & Schmidt, J.C. (2005). Effective discharge analysis of ecological processes in streams. *Water Resources Research*, **41**, W11411. doi:10.1029/2005WR004222.

Fagan, W.F. (2002). Connectivity, fragmentation, and extinction risk in dendritic metapopulations. *Ecology*, **83**, 3243–3249.

Falke, J.A. & Fausch, K.D. (2010). From metapopulations to metacommunities: linking theory with empirical observations of the spatial population dynamics of stream fishes. In *Community Ecology of Stream Fishes: Concepts, Approaches and Techniques* (ed. D.A. Jackson and K.B. Gido), pp. 207–233. American Fisheries Society Symposium 73. American Fisheries Society, Bethesda, MD, USA.

Fausch, K.D., Taniguchi, Y., Nakano, S., Grossman, G.D., & Townsend, C.R. (2001). Flood disturbance regimes influence rainbow trout invasion success among five holarctic regions. *Ecological Applications*, **11**, 1438–1455.

Fausch, K.D., Torgersen, C.E., Baxter, C.V., & Li, H.W. (2002). Landscapes to riverscapes: bridging the gap between research and conservation of stream fishes. *Bioscience*, **52**, 483–498.

Graf, W.L. (2006). Downstream hydrologic and geomorphic effects of large dams on American rivers. *Geomorphology*, **79**, 336–360.

Grant, E.H.C., Lowe, W.H., & Fagan, W.F. (2007). Living in the branches: population dynamics and ecological processes in dendritic networks. *Ecology Letters*, **10**, 165–175.

Hitt, N.P. & Angermeier, P.L. (2008). Evidence for fish dispersal from spatial analysis of stream network topology. *Journal of the North American Benthological Society*, **27**, 304–320.

Humphries, P. & Baldwin, D.S. (2003). Drought and aquatic ecosystems: an introduction. *Freshwater Biology*, **48**, 1141–1146.

Humphries, P. & Winemiller, K.O. (2009). Historical impacts on river fauna, shifting baselines, and challenges for restoration. *Bioscience*, **59**, 673–684.

Kennard, M.J., Pusey, B.J., Olden, J.D., MacKay, S.J., Stein, J.L., & Marsh, N. (2010). Classification of natural flow regimes in Australia to support environmental flow management. *Freshwater Biology*, **55**, 171–193.

Landres, P.B., Morgan, P., & Swanson, F.J. (1999). Overview of the use of natural variability concepts in managing ecological systems. *Ecological Applications*, **9**, 1179–1188.

Lowe, W.H., Likens, G.E., McPeek, M.A., & Buso, D.C. (2006). Linking direct and indirect data on dispersal: isolation by slope in a headwater stream salamander. *Ecology*, **87**, 334–339.

Lytle, D.A. & Poff, N.L. (2004). Adaptation to natural flow regimes. *Trends in Ecology & Evolution*, **19**, 94–100.

Lytle, D.A., Bogan, M.T., & Finn, D.S. (2008). Evolution of aquatic insect behaviours across a gradient of disturbance predictability. *Proceedings of the Royal Society of London. Series B. Biological Sciences*, **275**, 453–462.

Magoulick, D.D. & Kobza, R.M. (2003). The role of refugia for fishes during drought: a review and synthesis. *Freshwater Biology*, **48**, 1186–1198.

Marchetti, M.P. & Moyle, P.B. (2001). Effects of flow regime on fish assemblages in a regulated California stream. *Ecological Applications*, **11**, 530–539.

Matthews, J.H. & Wickel, A.J. (2009). Embracing uncertainty in freshwater climate change adaptation: a natural history approach. *Climate and Development*, **1**, 269–279.

Merritt, D.M. & Poff, N.L. (2010). Shifting dominance of riparian *Populus* and *Tamarix* along gradients of flow alteration in western North American rivers. *Ecological Applications*, **20**, 135–152.

Milly, P.C.D., Betancourt, J., Falkenmark, M., Hirsch, R.M., Kundzewicz, Z.W., Lettenmaier, D.P., & Stouffer, R.J. (2008). Stationarity is dead: whither water management? *Science*, **319**, 573–574.

Naiman, R.J., Bilby, R.E., Schindler, D.E., & Helfield, J.M. (2002). Pacific salmon, nutrients, and the dynamics of freshwater and riparian ecosystems. *Ecosystems*, **5**, 399–417.

Nilsson, C., Reidy, C.A., Dynesius, M., & Revenga, C. (2005). Fragmentation and flow regulation of the world's large river systems. *Science*, **308**, 405–408.

Palmer, M.A., Arensburger, P., Botts, P.S., Hakenkamp, C.C., & Reid, J.W. (1995). Disturbance and the community structure of stream invertebrates: patch-specific effects and the role of refugia. *Freshwater Biology*, **34**, 343–356.

Palmer, M.A., Lettenmaier, D., Poff, N.L., Postel, S.L., Richter, B., & Warner, R. (2009). Climate change and river ecosystems: protection and adaptation options. *Environmental Management*, **44**, 1053–1068.

Poff, N.L. (1996). A hydrogeography of unregulated streams in the United States and an examination of scale-dependence in some hydrological descriptors. *Freshwater Biology*, **36**, 71–91.

Poff, N.L. (1997). Landscape filters and species traits: towards mechanistic understanding and prediction in stream ecology. *Journal of the North American Benthological Society*, **16**, 391–409.

Poff, N.L. (2009). Managing for variability to sustain freshwater ecosystems. *ASCE Journal of Water Resources Planning and Management*, **135**, 1–4.

Poff, N.L. & Allan, J.D. (1995). Functional-organization of stream fish assemblages in relation to hydrological variability. *Ecology*, **76**, 606–627.

Poff, N.L. & Ward, J.V. (1990). Physical habitat template of lotic systems: recovery in the context of historical pattern of spatiotemporal heterogeneity. *Environmental Management*, **14**, 629–645.

Poff, N.L. & Zimmerman, J.K.H. (2010). Ecological responses to altered flow regimes: a literature review to inform the science and management of environmental flows. *Freshwater Biology*, **55**, 194–205.

Poff, N.L., Allan, J.D., Bain, M.B., et al. (1997). The natural flow regime. *Bioscience*, **47**, 769–784.

Poff, N.L., Bledsoe, B.P., & Cuhaciyan, C.O. (2006a). Hydrologic variation with land use across the contiguous United States: geomorphic and ecological consequences for stream ecosystems. *Geomorphology*, **79**, 264–285.

Poff, N.L., Olden, J.D., Vieira, N.K.M., Finn, D.S., Simmons, M.P., & Kondratieff, B.C. (2006b). Functional trait niches of North American lotic insects: traits-based ecological applications in light of phylogenetic relationships. *Journal of the North American Benthological Society*, **25**, 730–755.

Poff, N.L., Olden, J.D., Merritt, D.M., & Pepin, D.M. (2007). Homogenization of regional river dynamics by dams and global biodiversity implications. *Proceedings of the National Academy of Sciences of the United States of America*, **104**, 5732–5737.

Poff, N.L., Richter, B.D., Arthington, A.H., et al. (2010). The ecological limits of hydrologic alteration (ELOHA): a new framework for developing regional environmental flow standards. *Freshwater Biology*, **55**, 147–170.

Poole, G.C. (2002). Fluvial landscape ecology: addressing uniqueness within the river discontinuum. *Freshwater Biology*, **47**, 641–660.

Power, M.E., Parker, M.S., & Dietrich, W.E. (2008). Seasonal reassembly of a river food web: floods, droughts, and impacts of fish. *Ecological Monographs*, **78**, 263–282.

Pretty, J.L., Harrison, S.S.C., Shepherd, D.J., Smith, C., Hildrew, A.G., & Hey, R.D. (2003). River rehabilitation and fish populations: assessing the benefit of instream structures. *Journal of Applied Ecology*, **40**, 251–265.

Pringle, C. (2003). What is hydrologic connectivity and why is it ecologically important? *Hydrological Processes*, **17**, 2685–2689.

Rahel, F.J. & Olden, J.D. (2008). Assessing the effects of climate change on aquatic invasive species. *Conservation Biology*, **22**, 521–533.

Richter, B.D., Warner, A.T., Meyer, J.L., & Lutz, K. (2006). A collaborative and adaptive process for developing environmental flow recommendations. *River Research and Applications*, **22**, 297–318.

Rodriguez-Iturbe, I., Muneepeerakul, R., Bertuzzo, E., Levin, S.A., & Rinaldo, A. (2009). River networks as ecological corridors: a complex systems perspective for integrating hydrologic, geomorphologic, and ecologic dynamics. *Water Resources Research*, **45**, W01413. doi:10.1029/2008 WR007124.

Roy, A.H., Freeman, M.C., Freeman, B.J., Wenger, S.J., Ensign, W.E., & Meyer, J.L. (2005). Investigating hydrologic alteration as a mechanism of fish assemblage shifts in urbanizing streams. *Journal of the North American Benthological Society*, **24**, 656–678.

Schlaepfer, M.A., Runge, M.C., & Sherman, P.W. (2002). Ecological and evolutionary traps. *Trends in Ecology & Evolution*, **17**, 474–480.

Schlosser, I.J. (1991). Stream fish ecology: a landscape perspective. *Bioscience*, **41**, 704–712.

Seager, R., Ting, M.F., Held, I., et al. (2007). Model projections of an imminent transition to a more arid climate in southwestern North America. *Science*, **316**, 1181–1184.

Snelder, T.H., Lamouroux, N., Leathwick, J.R., Pella, H., Sauquet, E., & Shankar, U. (2009). Predictive mapping of the natural flow regimes of France. *Journal of Hydrology*, **373**, 57–67.

Southwood, T.R.E. (1988). Tactics, strategies and templets. *Oikos*, **52**, 3–18.

Stanford, J.A., Ward, J.V., Liss, W.J., Frissell, C.A., Williams, R.N., Lichatowich, J.A., & Coutant, C.C. (1996). A general protocol for restoration of regulated rivers. *Regulated Rivers: Research & Management*, **12**, 391–413.

Stella, J.C., Battles, J.J., Orr, B.K., & McBride, J.R. (2006). Synchrony of seed dispersal, hydrology and local climate in a semi-arid river reach in California. *Ecosystems*, **9**, 1200–1214.

Townsend, C.R. & Hildrew, A.G. (1994). Species traits in relation to a habitat templet for River Systems. *Freshwater Biology*, **31**, 265–275.

Ward, J.V. (1989). The 4-dimensional nature of lotic ecosystems. *Journal of the North American Benthological Society*, **8**, 2–8.

Wiens, J.A. (2002). Riverine landscapes: taking landscape ecology into the water. *Freshwater Biology*, **47**, 501–515.

Williams, J.W., Jackson, S.T., & Kutzbacht, J.E. (2007). Projected distributions of novel and disappearing climates by 2100 AD. *Proceedings of the National Academy of Sciences of the United States of America*, **104**, 5738–5742.

Winemiller, K.O., Flecker, A.S., & Hoeinghaus, D.J. (2010). Patch dynamics and environmental heterogeneity in lotic ecosystems. *Journal of the North American Benthological Society*, **29**, 84–99.

Wohl, E. (2005). Compromised rivers: understanding historical human impacts on rivers in the context of restoration. *Ecology and Society*, **10**(2), article 2. http://www.ecology andsociety.org/vol10/iss2/art2/ES-2005-1339.pdf.

Chapter 17

A FRAMEWORK FOR APPLYING THE HISTORICAL RANGE OF VARIATION CONCEPT TO ECOSYSTEM MANAGEMENT

William H. Romme,[1] *Gregory D. Hayward,*[2,3] *and Claudia Regan*[3]

[1]Colorado State University, Fort Collins, CO, USA
[2]USDA Forest Service, Alaska Region, Anchorage, AK, USA
[3]USDA Forest Service, Rocky Mountain Region, Lakewood, CO, USA

Historical Environmental Variation in Conservation and Natural Resource Management, First Edition. Edited by John A. Wiens, Gregory D. Hayward, Hugh D. Safford, and Catherine M. Giffen.

17.1 INTRODUCTION

Ecosystems are changing at an ever-increasing pace. Seemingly intractable issues of climate change, invasive species, and accelerating human demands on natural resources increasingly dominate land-management and conservation concerns (Millar et al. 2007). Faced with broadscale ecological change and increased social demands, how can managers and stakeholders – the people who use and care about the future of public and private lands – assess the current state of the ecosystems with which they are involved, identify the most significant and urgent threats to ecological integrity, and prioritize management actions in the face of potentially paralyzing uncertainty? We suggest that an understanding of the local history of an ecosystem, along with the broad understanding of system dynamics that comes from historical ecology, can provide a useful foundation for evaluating possible future scenarios and developing appropriate management responses. More specifically, the historical range of variation (HRV) concept can be placed within an analytical framework to help guide decisions on how to manage natural resources in the face of uncertainty about the future. Consistent with its use elsewhere in this book, we define HRV as *the variation of ecological characteristics and processes over scales of time and space that are appropriate for a given management application.*

Early uses of the HRV concept aimed to determine whether an ecological system was within or outside of HRV. The HRV condition often was tacitly assumed to be the management target, without explicitly considering whether a return to that historical state was technically feasible or socially desirable. Early HRV assessments also were plagued by incomplete historical data, questions about how to incorporate the impacts of indigenous people during the historical reference period, and complications related to the spatial and temporal scale of analysis (Chapters 1, 5, and 6 develop these issues in greater detail). Together with a growing awareness of the rate and magnitude of current and future environmental changes, these difficulties have led some to conclude that the HRV concept is of little or no practical use in land management or conservation.

A central theme of this book is that HRV is indeed of practical use, but it needs to be applied within a richer conceptual framework than a simple dichotomous question of "inside" or "outside" of some HRV. This framework should incorporate both ecological and social aspects of land management to facilitate informed discussion and decision making by all parties involved (Duncan et al. 2010; see Chapter 1, this book).

In this chapter we outline a framework for incorporating HRV-derived insights into an evaluation of current ecological conditions and options for management over the next 20–30 years. During this time, we can expect significant climatic change – with progressively warmer temperatures, greater drought stress, and more frequent disturbance by fires, storms, and insect outbreaks (IPCC 2007; Spracklen et al. 2009; Gray & McCabe 2010) – as well as continuing pressures of human population growth, resource consumption, and non-native species invasions. By mid-century, climatic changes and human pressures are projected to be of such magnitude that entirely new ecosystems could replace current ecosystems over large areas of the Earth (Malhi et al. 2009; Coops & Waring 2011). Such enormous ecological changes call for new ways of thinking about and managing landscapes, ways that we cannot yet envision. A new approach is needed.

17.2 A FRAMEWORK FOR APPLYING HRV TO ECOSYSTEM MANAGEMENT

We apply a sequence of six steps for evaluating the current state of an ecosystem (relative to the historical condition), for assessing how the system is likely to respond to climate change and other environmental stressors during the next 20–30 years, and for developing a plan of action (see Table 17.1 and Fig. 17.1). The first step (#1 in Table 17.1 and Fig. 17.1) is to clearly identify the ecosystem or ecosystem components being evaluated. The system can be defined in any way that is relevant to specific management concerns.

The second step (#2) is to assess the current status of major structural, compositional, and functional elements of the ecosystem. For each ecosystem component of interest (e.g. species composition, amount of course wood, extent of bark beetle infestation), this assessment should clearly describe the range of conditions that commonly occurred during the relevant historical time frame. This time frame (the reference period) will differ among ecosystems and among ecosystem components because ecological rhythms are inherently variable. For example, where disturbances tend to be frequent and low-intensity and are followed

Table 17.1 A framework for ecological evaluation and identification of management objectives as informed by the historical range of variation (HRV). Also see Fig. 17.1.

1. Identify/define the ecosystem of interest.
    a. identify the type of ecosystem, often described by vegetation type
    b. identify spatial extent of assessment
    c. identify temporal extent of assessment
2. Within the focal ecosystem, identify the major elements or components of HRV (both structural and functional) whose current condition
    a. is similar to common conditions during the reference period
    b. was likely uncommon during the reference period
    c. probably never occurred during the reference period
3. Of the elements whose current condition is similar to common conditions during the reference period, determine whether that condition is socially acceptable.
    a. If yes, then no further management action is required except for maintenance of this element.
    b. If no, then evaluate whether management actions that would modify the element to be more socially acceptable would also be ecologically viable (since the management action could move the element to an uncommon condition that may not be sustainable without significant input of effort and resources).
4. Of the elements whose current condition was likely uncommon or probably never occurred during the reference period, identify those that are most critical to maintaining ecosystem integrity, that is, those that in their current condition threaten long-term ecological integrity.
5. Of those elements identified in #4 above, evaluate whether HRV provides clues about potential intervention points and management options (including restoration of select elements of HRV as a potential but not an exclusive management option).
6. For elements identified in #5 above for which restoration of historical condition is both socially desirable and ecologically feasible, develop specific programs to restore these elements.

by rapid recovery, a relatively short time period may reveal the essential system dynamics. In contrast, many ecosystems are characterized by large, infrequent disturbances (Turner & Dale 1998); these events may appear rare or even unprecedented within a short time frame but may have profoundly shaped many ecosystems when viewed at a longer and more relevant temporal scale (Chapter 5, this book). Based on an understanding of the historical range of conditions, the current condition can be placed in context as *commonly occurring, rarely occurring*, or *probably never occurred historically*. This categorization differs from the simple "inside" versus "outside" distinction that has sometimes been used and sidesteps some of the attendant problems. For example, it often is impossible to determine whether a particular condition was just barely "inside" or just barely "outside" of HRV; often, such a fine distinction does not matter for management purposes. Table 17.2 lists some elements that managers may wish to consider during an HRV assessment, although the list is not exhaustive.

The process now bifurcates as ecological elements whose current state is similar to what was common in the past are evaluated separately from those elements that are currently very different from reference-period conditions. If an element today is similar to conditions common in the past (#3 in Table 17.1 and Fig. 17.1), then managers may have no *ecological* concerns about its continuing integrity. If society also finds this condition acceptable, then the only management activity needed may be simply maintenance. However, characteristics that were ecologically common in the past may be *socially* unacceptable, that is, they may lie outside the social range of variability (SRV; Duncan et al. 2010). An example would be a forest in which the historical fire regime was characterized by stand-replacing crown fires but the specific location is at the wildland–urban interface. Even though high-severity fire would be a commonly occurring, natural event in the HRV for this system, such fire would not be socially acceptable because of the threats to human life and property. In this situation, the SRV would call for moving the system away from the historically common ecological condition, to try to eliminate high-severity fires by altering fuel conditions and aggressively suppressing all ignitions. Even if the final management

(1) Description of the system, including spatial and temporal scale and reference period

(2a) Elements or components whose current condition is similar to common conditions during the reference period

(2b) Elements or components whose current condition was uncommon during the reference period

(2c) Elements or components whose current condition probably never occurred during the reference period

(3) Is the condition is socially acceptable?

NO

YES

(4) Does the current condition threaten ecological integrity?

NO

YES

(3b) Identify potential actions to change the condition, possibly including a historically uncommon or unprecedented condition, in which case evaluate the proposed action starting from 2b or 2c

(3a) Maintain current condition

Evaluate whether the condition is socially acceptable (step 3)

(5) Does HRV suggest suitable mitigation or restoration trea tments?

YES

NO

(6) Develop a mitigation or restoration plan based on HRV

Develop a mitigation plan from other sources

Is the proposed plan socially acceptable?

YES

NO

Implement

Develop an alternative plan

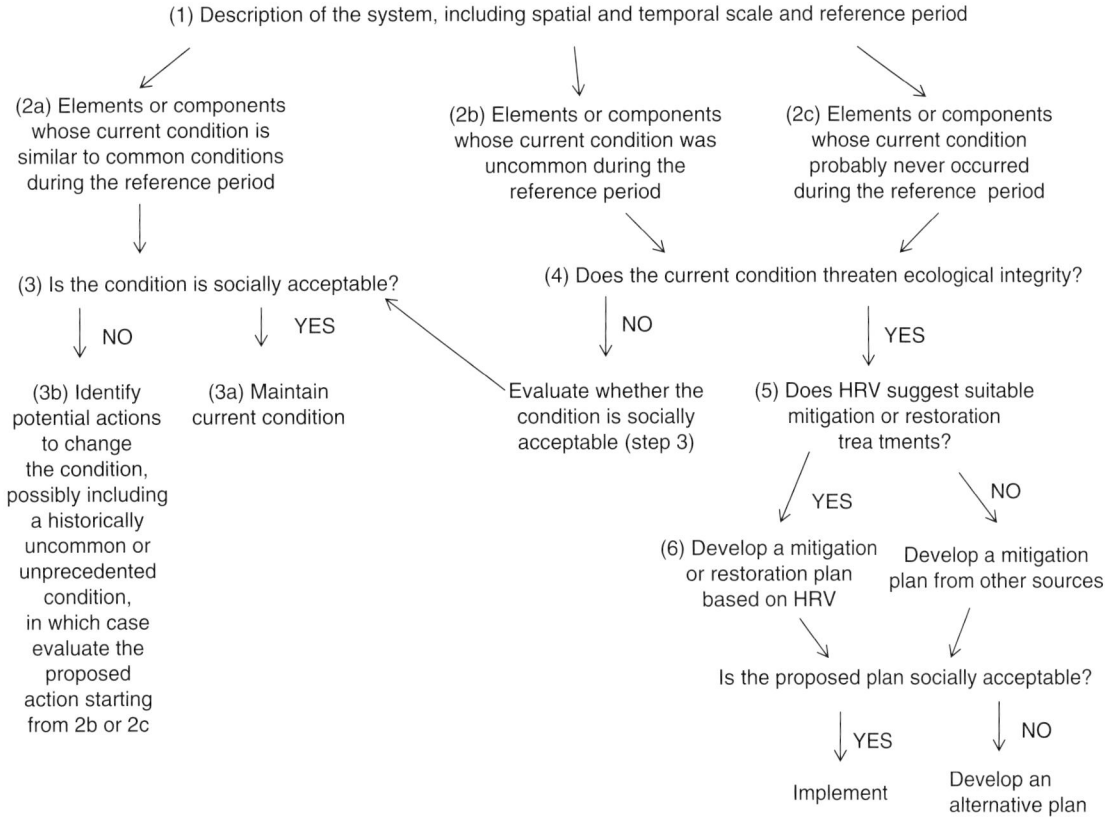

Fig. 17.1 Framework for applying historical ecology concepts to land management. See also Table 17.1.

decision is to attempt to move the ecosystem into a rare or nonoccurring historical state (in this case by eliminating all high-severity fire), the HRV concept nevertheless plays two critical roles. First, it informs the public of the technical difficulties that will be faced and high levels of management effort that will be needed to eliminate high-severity fire from an ecosystem where such disturbances have a natural tendency to occur. Second, the HRV assessment identifies other elements of the system that will be strongly affected by altering this key ecological process (e.g. regeneration and persistence of fire-dependent plant species); these elements would then be evaluated separately in the context of the framework shown in Table 17.1 and Fig. 17.1.

Step #4 (Table 17.1, Fig. 17.1) evaluates those elements of the ecosystem that currently differ signifi-

cantly from the range of common conditions in the past, conditions that are near the limits of HRV, far beyond HRV, or entirely unprecedented. The key point is that not all departures from HRV have the same consequences for ecological integrity and resilience. The presence of recently arrived non-native species, for example, is by definition historically unprecedented. Some of these species profoundly alter the structure and function of ecosystems. For instance, cheatgrass (*Bromus tectorum*) altered the fire regimes of many western shrublands, resulting in systems changing from shrubland to annual grassland; cheatgrass, therefore, would be considered a serious ecological threat in a still-intact western shrubland (Billings 1990). Other non-native species, however, have caused very little change. For example, if dandelion (*Taraxacum officinale*) is present at low density in a mountain

Table 17.2 Examples of ecological elements that may be important foci for HRV assessments.

Structural
   Physiognomy (e.g. forest, shrubland, grassland, wetland)
   Biomass distribution (above and belowground)
   Age- or size-class distribution of dominant species.
   Frequency distribution of patch sizes
   Frequency distribution of stand structural stages and stand ages
   Amounts and distribution of coarse wood
   Numbers and distribution of legacy elements such as snags
   Structure and composition of soil crusts
   Etc. . . .
Compositional
   Plant species present and their usual relative abundance
   Animal species present (including mammals, birds, other vertebrates, and invertebrates) and their usual relative abundance
   Biota of soil communities, including bacteria, fungi, algae, and invertebrates
   Functional
   Primary production
   Biogeochemical cycling
   Disturbance regimes
   Etc. . . .
Ecosystem services
   Clean water
   Aesthetics
   Sustainable commodity production (timber, forage, etc.)
   Recreation
   Etc.

grassland but all of the historically important grasses and forbs still predominate, the dandelion would not be considered a major threat even though its presence represents a departure from HRV.

After identifying the structural and functional elements of an ecosystem that represent both (1) a departure from conditions that were common in the HRV, and (2) a threat to continued ecological integrity and resilience, the next task is to develop management strategies for ameliorating the threat (#5 in Table 17.1 and Fig. 17.1). In many cases, HRV provides options or ideas for successful treatment. For example, in some southwestern ponderosa pine (*Pinus ponderosa*) forests, low-severity surface fires no longer occur at their his-

torical frequency, and high-severity crown fires now occur with increasing frequency (Chapter 1, this book). We know that the historical low-severity fire regime was maintained in part by a particular forest structure. Appropriate management actions can restore this structure (#6 in Table 17.1 and Fig. 17.1, and see Chapter 1, this book). Once again, however, SRV must be a part of this process. If the public understands the ecological rationale and approves of the action, then management can move forward (in this case by an active forest-restoration program); if the public does not understand or does not agree, then more work is needed. Sometimes, a knowledge of historical variation will not tell one much about the most effective strategy for dealing with a particular problem: if invasive plants are the issue, then managers and the public will need to turn to botanists, integrated pest managers, and weed science for potential solutions.

In the following sections, we develop the application of this framework by examining three case studies (illustrated in Fig. 17.2 and summarized in Table 17.3). Rather than present these case studies in full detail, we highlight selected features of each to illustrate the application of historical information to land-management dilemmas. Although the three examples involve issues of fire management in western conifer forest ecosystems, they illustrate very different historical conditions that lead to different conclusions about current conditions and appropriate management responses, especially when juxtaposed with social desires. Each case study is presented in the same way: as a series of numbered steps that reference the steps outlined in Table 17.1 and Fig. 17.1.

17.3   CASE STUDY I: YELLOWSTONE LODGEPOLE PINE LANDSCAPE

1 *Identify/define the ecosystem of interest.* Coniferous forests dominated by lodgepole pine (*Pinus contorta* var. *latifolia*) cover some 80% of Yellowstone National Park as well as extensive portions of surrounding national forests and other high-elevation lands in the Greater Yellowstone Ecosystem of northwestern Wyoming and adjacent portions of Montana and Idaho (Fig. 17.2a, Table 17.3). Most of this area is wilderness managed by federal land agencies. The most appropriate historical reference period encompasses the past 10 000 years. Although this is a longer time period than is commonly used in HRV assessments, good information

Fig. 17.2 Three ecosystems developed as case studies in a framework for applying the historical range of variation concept to ecosystem management. (a) Lodgepole pine forests in Yellowstone National Park, showing rapid forest regeneration in areas burned in the large 1988 fires; (b) mixed-conifer-aspen forests in the San Juan Mountains of southwestern Colorado; (c) piñon-juniper woodlands in Mesa Verde National Park, Colorado, showing an area on the ridgetop where a very old woodland burned in 2000, and an area on the adjacent slope that was missed by the 2000 fire and has not burned in many centuries. (Photographs by W.H. Romme.) Please refer to Colour Plate 7.

is available about ecological conditions and dynamics during this period. Such a long look backward provides important context for a system that has weathered substantial climatic fluctuations in the past and is shaped in part by infrequent but large disturbances.

2a *Identify the major elements/components whose current condition is similar to common conditions during the reference period.* The composition and structure of the Yellowstone lodgepole pine forest ecosystem today are very similar to conditions that were common throughout the reference period. Compositionally, the Greater Yellowstone Ecosystem is a unique area in the continental United States in that all of its pre-Columbian fauna and flora is still present, including the large

mammalian carnivores (gray wolf, *Canis lupis*, and grizzly bear, *Ursus arctos horribilis*) (YNP 2010). The basic coniferous-forest physiognomy of the system has been the same for the past 10 000 years, and lodgepole pine has been the dominant tree species throughout this time (Whitlock 1993; Millspaugh et al. 2000).

2b *Identify the major elements/components whose current condition was likely uncommon during the reference period.* The number and size of severe fires that have occurred in the Yellowstone ecosystem since the mid-1980s were unprecedented during the previous 100 years. In particular, fires in 1988 burned >400 000 ha (Christensen et al. 1989) and may have been larger than any fires within the previous 300

Table 17.3 Summary of three western United States case studies illustrating the application of HRV concepts to assessment and management of ecosystems in anticipation of climate change and other environmental stressors during the coming 20–30 years. See text for details.

	Ecosystem		
	Yellowstone lodgepole pine forest	**Southern Rockies mixed-conifer-aspen forest**	**Colorado Plateau piñon-juniper woodland**
1. Identify system of interest	High-elevation forests in and around Yellowstone Park	Mid-elevation forests in southwestern Colorado	Persistent woodlands in southwestern Colorado
Reference period	past 10 000 years	400 years prior to late 1800s	700 years prior to late 1900s
2a. Elements whose current condition is similar to common conditions during the reference period	Intact native flora, fauna, and soils	Mostly intact native flora, fauna, and soils	Mostly intact native flora, fauna, and soils
2b. Elements whose current condition was likely uncommon during the reference period	Fire regime (apparently more area burned in past 25 years than in previous 300 years)	Fire regime and landscape structure (longer fire intervals and greater proportion of high-density stands today)	Insect disturbance regime (recent severe tree mortality)
2c. Elements whose current condition probably never occurred during the reference period	• Structures and activities of an industrial society • Non-native plants	• Structures and activities of an industrial society • Non-native plants	• Structures and activities of an industrial society • Shortened fire rotation and domination of burned areas by non-native plants
3. HRV condition socially acceptable?	Yes for intact native flora, fauna, soils, and ecosystem services	Yes for intact native flora, fauna, soils, and ecosystem services	Yes for intact native flora, fauna, soils, and ecosystem services
4. Elements whose current condition was likely uncommon or never occurred during the reference period that threaten long-term ecological integrity and resilience	None . . . the system has persisted through a wide range of climatic conditions and fire frequencies, and current postfire trajectories appear normal	None in the *short term* . . . but in the *long term* certain fire-maintained landscape elements eventually may be lost, and socially unacceptable large fires may occur	Shorter fire rotations and non-native invasion of burned areas may cause a state transition from native woodland to unprecedented assemblage of non-native herbs
5. HRV provides clues to potential management options?	N/A	Naturally wide variation in historical disturbance frequency and severity allows flexibility and patience in returning fire and fire-maintained elements	HRV indicates that prescribed burning would not be "restoration," but new ideas from other fields are needed to deal with the ecological degradation now occurring
6. Example of implementation of management options based on insights from HRV	Ongoing natural fire policy	Collaborative ecosystem management by Uncompahgre Partnership	Not using prescribed fire for restoration purposes

years (as suggested by Fig. 2 in Romme & Despain 1989), although it is possible that the fires in the early 1700s or earlier might have burned as large or larger areas. Within this framework for evaluating ecosystem conditions, however, it does not matter whether the size of the 1988 fires was truly unprecedented. The important point is that the 1988 fires were near or beyond the typical maximum size of historical fires, and therefore a potential cause for concern.

2c *Identify the major elements/components whose current condition probably never occurred during the reference period.* The major elements of the Yellowstone lodgepole pine forest landscape that are clearly unprecedented historically are (1) roads, buildings, and other human developments, which tend to be clustered in patches or along corridors and are absent over large expanses of wild country, and (2) a few non-native plant species, such as Canada thistle (*Cirsium arvense*).

3 *Of the elements whose current condition is similar to common conditions during the reference period, determine whether that current condition is socially acceptable.* The persistence of Yellowstone's full complement of pre-Columbian fauna and flora is not only socially acceptable but is one of the most highly valued aspects of the Park. There are concerns that bison (*Bison bison*) and wolves may pose threats to domestic livestock outside the Park by spreading disease or preying on animals (YNP 2010). Management agencies actively manage to mitigate these threats, and few people would advocate eliminating bison and (perhaps arguably) wolves. Maintaining this element of historical ecological conditions (all native fauna and flora) is, in fact, a primary management goal in Yellowstone Park. The forests also provide important ecosystem services, such as clean water and opportunities for backcountry recreation.

4 *Of the elements whose current condition was likely uncommon or probably never occurred during the reference period, identify those that are most critical to maintaining ecosystem integrity, that is, those that in their current condition threaten long-term ecological integrity.* The non-native plant species established in Yellowstone's high-elevation forests generally are not aggressive invaders and are largely restricted to areas of local soil disturbance such as trailheads and roadsides (Turner et al. 1997). Notably, the non-native species do not dominate burned areas, as has occurred in other ecosystems (see the third case study below); in fact, Canada thistle has declined or disappeared from many of the burned areas where it was first documented after 1988 (Wright and Tinker 2012). Despite

the non-natives being an unprecedented element of the system, they appear to pose little threat to long-term ecological integrity in this situation, and no special management actions appear necessary to prevent their further spread. It probably makes sense to spend limited resources on more serious ecological issues, either within the lodgepole pine system or in other ecosystems.

What about fire? Although recent fires have been large – perhaps even unusually so – other aspects of the current and historical fire regimes in Yellowstone suggest that the recent upsurge in fire size and frequency does not pose a serious threat to the ecological integrity or resilience of Yellowstone's lodgepole pine landscape (Romme & Turner 2004). Because fires are naturally infrequent in this ecosystem, we need a longer time frame than just a few recent centuries to evaluate whether recent and projected future changes in fire frequency pose a serious threat to the ecological integrity of Yellowstone's lodgepole pine ecosystem. Studies of pollen and charcoal preserved in lake sediments provide this information. Millspaugh et al. (2000) documented patterns in vegetation and fire frequency on the Yellowstone Central Plateau since the postglacial arrival of lodgepole pine ca. 11 000 years ago. Climate has fluctuated, with warmer, drier summers during the early Holocene than has characterized the last 1000 years. The number of large fires per millennium also has varied, from 10–15 in the early Holocene to 2–4 during the last 1000 years. Yet coniferous forest persisted throughout this period with lodgepole pine as either the dominant or codominant species in the area. This historical information tells us that the lodgepole pine forest ecosystem can tolerate a wide range of climatic conditions and fire frequencies without losing its essential characteristics.

Post-1988 studies also have revealed that burned lodgepole pine forests recovered quickly, initially via sprouting from surviving below-ground plant structures and then by new seedling recruitment, with little change in species composition (Turner et al. 1997). In fact, no native species was imperiled by the 1988 fires, and some native plant species (e.g. *Ceanothus velutinus*) actually require fire to complete their life cycles by stimulating germination of a dormant soil seed bank. Moreover, surprisingly little soil or nutrient loss occurred after the 1988 fires, in part because of uptake by the rapidly recovering plant community and also via nutrient immobilization by soil microbes (Turner et al. 2007; Turner 2010). Local density and growth rate of young lodgepole pine vary in response to prefire stand structure and local fire behavior, but regeneration is adequate to ensure full restocking almost everywhere that lodgepole pine was present before the fires. Net primary production in some stands already is approaching or equaling levels seen in mature forests (Turner et al. 2004). In sum, the forests appear to be recovering very effectively and in a manner similar to what happened after fires in previous centuries.

It follows from the persistence of lodgepole pine forests despite substantial environmental variability during the Holocene, and from the rapid recovery of Yellowstone's lodgepole pine forest ecosystem after the large and severe 1988 fires, that this ecosystem should be relatively resilient in the face of changing climate and fire regime. The more frequent future droughts and fires that are projected for the northern Rockies (Spracklen et al. 2009; Gray & McCabe 2010) may lie within the range of environmental conditions that Yellowstone's lodgepole pine ecosystems have already endured; at least this may be so during the next 20–30 years. With more frequent future fire, the distribution of lodgepole pine age classes will shift toward a greater preponderance of younger stands, but coniferous-forest ecosystems dominated by lodgepole pine likely will still blanket extensive landscapes.

In view of the historical and recent resilience of Yellowstone's lodgepole pine forests, it appears that this vegetation type in Yellowstone, and perhaps in much of the Rocky Mountain region, may require no special management attention or restoration program to ensure its persistence for at least the next few decades. Instead, resources probably could be directed at more immediately imperiled ecosystems, such as the ones described below. A formal SRV assessment has not been conducted for this ecosystem, but it is likely that the current management emphasis on maintaining historical composition, structure, and dynamics using minimal management intervention would fit within the SRV. An exception would be in specific locations where vulnerable human structures lie within lodgepole pine forest. Romme et al. (2004) explore some ways in which prescribed burning and fuel reduction might be used to reduce this fire threat while maintaining the integrity and natural character of the landscape as a whole. Because we identified no major threats to the current or near-future integrity of Yellowstone's lodgepole pine ecosystems and no major social concerns with the current condition or management emphasis, the analysis stops here.

Therefore, we do not address steps #5 and #6 in this case study.

We must close this case study by acknowledging that, by mid-century, climatic conditions could deteriorate so severely that the Yellowstone Plateau is no longer a favorable habitat for lodgepole pine (Coops & Waring 2011). This could mean that lodgepole begins to be replaced by more drought-tolerant species such as Douglas fir and ponderosa pine (Bartlein et al. 1997). However, it is not known how well the generally infertile soils of the Yellowstone Plateau (Despain 1990) would support these other tree species; an alternative future scenario is that today's coniferous forests would be replaced by grasslands or other non-forest vegetation. An ecological change of this magnitude would far exceed any Holocene precedent, and such a transformation of the landscape would call for fundamentally new approaches to management – a challenge that goes beyond the scope of this analysis.

17.4   CASE STUDY II: MIXED-CONIFER-ASPEN FORESTS IN THE SOUTHERN ROCKY MOUNTAINS

1 *Identify/define the ecosystem of interest.* For this case study, we consider mixed-conifer-aspen forests in the San Juan Mountains and Uncompahgre Plateau of southwestern Colorado (Fig. 17.2b, Table 17.3). Forests at middle elevations (2300–3000 m) are among the most diverse ecosystems in this region, supporting variable mixtures of ponderosa pine, southwestern white pine (*Pinus strobiformis*), Douglas fir, white fir (*Abies concolor*), subalpine fir (*Abies lasiocarpa*), blue spruce (*Picea pungens*), Engelmann spruce (*Picea engelmannii*), and quaking aspen (*Populus tremuloides*) (Romme et al. 2009a; Vankat 2011). The natural-disturbance regime, dominated by fire and insect outbreaks, is equally variable, as is the range of potential trajectories of postdisturbance vegetation development (Jones 1974). The mixed-severity fire regime that characterized the historical period included large and small fires, high- and low-severity fires, and highly variable intervals between fires (Bigio et al. 2010). Consequently, a single site can support different stand structures and species compositions depending on local environmental conditions and the nature of the last disturbance (Romme et al. 2009a). For example, a low-severity fire or bark-beetle outbreak on a south-facing slope at 2600 m may result in an open stand of ponderosa pine, Douglas fir, and aspen, whereas a severe fire on the same site may produce a nearly pure stand of aspen (Romme et al. 2009a).

The historical reference period for this analysis includes the several centuries just prior to widespread Euro-American settlement of the region in the late 1800s. This spatial and temporal scope provides sufficient information and context to understand the behavior of dominant ecological processes.

2a *Identify the major elements/components whose current condition is similar to common conditions during the reference period.* Apparently all of the native plant species are still present in mixed-conifer-aspen forests of southwestern Colorado, although the cover and biomass of palatable species are lower in some areas than was common before the introduction of livestock in the late 1800s (Paulson & Baker 2006). Soils also are still intact and functioning normally in most areas. The native fauna is still largely intact, except that wolves and grizzlies have been lost, and the relative abundance of some species probably has changed in response to the changes in forest structure described below.

2b *Identify the major elements/components whose current condition was likely uncommon during the reference period.* Some of the biggest changes in mixed-conifer-aspen forests since the reference period are seen in fire intervals and stand structures. Before 1900, composite fire intervals within an area of ca. 1 km^2 in the San Juan Mountains ranged from 25 to 40 years on drier sites (Grissino-Mayer et al. 2004) to 100+ years on more mesic sites (Aoki 2010). However, there were no extensive fires in mixed-conifer-aspen forests of southwestern Colorado from the late 1800s until 2002, when the large Missionary Ridge fire occurred. Although it was not historically unusual for individual stands to escape fire for such a long interval, this >100-year lack of fire throughout the mixed-conifer-aspen landscape was unprecedented in the past several centuries (Anderson et al. 2008; Bigio et al. 2010). The lack of low-severity surface fires, which periodically thinned the forests, may be especially significant (Bigio et al. 2010). Although neither the structure nor the composition of any individual stand today appears unprecedented in the context of HRV, at the scale of the entire landscape, mixed-conifer-aspen stands having high tree densities and basal areas probably are more extensive today than they were at most times before 1900. Concurrently, stand types created or enhanced by fire, such as young aspen and low-density pine stands, have become less abundant during the twenti-

eth century (Romme et al. 2009a). This shift in landscape structure toward a preponderance of high-density stands and shade-tolerant species probably is due primarily to the paucity of recent fires, but also was favored by the generally moist climatic conditions of the past century, which were especially conducive to tree establishment and survival (Swetnam & Betancourt 1998).

Without the exclusion of fire over the past century, up to four large fires might have occurred. As a result, when these forests eventually do burn, the fire interval at a stand level will be close to or beyond the upper limit of historical fire intervals and the fire may be near or beyond the maximum historical fire intensity (heat release) and severity (impacts on organisms and the environment). Similarly, the densely connected forest landscape may now be vulnerable to fires near or beyond the maximum size and severity of historical fires at the landscape level (Bigio et al. 2010). Increased burning over the next few decades is likely, given the projected deepening of recent southwestern drought conditions (Seager et al. 2007). Even with more fire, however, we anticipate that the physiognomy of this vegetation type will continue to be forest, in part because of the diversity of tree species that respond differently to fire. Nevertheless, shifts in species composition are likely. Aspen, which sprouts readily after fire, may increase, but Douglas fir and white fir, which reproduce by seed and may be unable to disperse adequate seed into large, severely burned openings, may decrease. Moreover, some or many of the major species may not be able to tolerate the extreme environmental changes projected by most climate models for late in the twentieth century.

2c *Identify the major elements/components whose current condition probably never occurred during the reference period.* The most striking departures from HRV are the presence of roads, houses, and other human developments associated with an industrial society, as well as the impacts of the economic and recreational activities of those people. Other major elements that are clearly unprecedented in HRV are the non-native plants, particularly Canada thistle and cheatgrass, which have invaded some disturbed mixed-conifer-aspen stands.

3 *Of the elements whose current condition is similar to common conditions during the reference period, determine whether HRV is socially acceptable.* People value the intact soils, flora, and fauna as well as the scenic vistas resulting from the intact forests of the region. Sustainable timber and forage production are additional ecosystem services provided by well-functioning southwestern mixed-conifer-aspen forest ecosystems.

4 *Of the elements whose current condition was likely uncommon or probably never occurred during the reference period, identify those that are most critical to maintaining ecosystem integrity, that is, those that in their current condition threaten long-term ecological integrity.* Neither the long time since fire nor the preponderance of stands having high basal area appear to pose any *immediate* threat to the ecological integrity of mixed-conifer-aspen forests in southwestern Colorado. The recent large fires have burned at mixed severity, similar to historical fires although perhaps toward the extreme end of historical variation (Bigio et al. 2010). Vegetation responses to recent fires also appear to be similar to what occurred after historical fires, even in high-severity burned patches, and no unusual loss of soil or nutrients has been observed (Romme et al. 2009a). If fires continue to be rare over the next several decades, however, the fire regime will move progressively farther beyond the common conditions that characterized the HRV. Some fire-maintained vegetation components, such as low-density pine forests and young aspen stands, will become increasingly scarce. High-severity fires may begin to burn over larger areas than the public is comfortable with, and it will become increasingly difficult to protect powerlines, communication facilities, and other vulnerable infrastructure from fire damage. Thus, although we identify no serious threats to ecological integrity in the short-term (i.e. within the next few years), we do believe that fire or other forest-disturbance processes need to be returned to this system within the next decade or two to prevent eventual loss of ecological integrity due to uncharacteristically large and severe fires.

5 *Of those elements identified in #4 above, evaluate whether HRV provides clues about potential intervention points and management options.* An important feature of the historical disturbance regime in this ecosystem was the inherent variability in frequency, severity, and spatial scale of disturbance (Romme et al. 2009a). We are beginning to see the high-severity fire component again after the long twentieth-century hiatus, but we still do not have anything like the historical frequency and extent of low-severity and moderate-severity surface fires that formerly thinned stands and probably reduced the potential for high-severity burning except under extreme weather conditions (Bigio et al. 2010). Restoration of a full range of fire behaviors, sizes, and severities, or a program that emulates the effects of

historical fires via mechanical thinning or harvest, could be an effective strategy for enhancing the resilience and ecological integrity of mixed-conifer-aspen forest ecosystems in the face of an uncertain future. The great variability of HRV disturbance regimes means that historical fire intervals and severities need not be mimicked exactly. In fact, managers probably should not even attempt to reproduce past fire patterns because the climate and general environmental context are different today than in previous centuries.

6 *For elements identified in #5 above for which restoration of historical condition is both socially desirable and ecologically feasible, develop specific programs to restore these elements.* A fortunate aspect of the mixed-conifer-aspen forest system is that, although we face significant challenges, we also have time to develop ecologically sound management plans and to build public understanding and support for the treatments that eventually will be applied. For example, a program designed to allow more frequent natural fire and to increase the representation of rare landscape elements in mixed-conifer-aspen forests is being developed and implemented by a collaborative community partnership focused on the Uncompahgre Plateau in southwestern Colorado. The Uncompahgre Partnership (http://www.upproject.org/) brings together federal land-management agencies, university researchers, local environmental groups, loggers, ranchers, and other interested members of the community. Funded by grants from the National Forest Foundation and other sources, the group is implementing a program of mechanical thinning and timber harvest in strategic locations to restore lower-density pine stands, to stimulate regeneration of young aspen forests, and to modify landscape-level fuel conditions such that future mixed-severity fires can be allowed to burn over large areas without unacceptable risk to life, property, or resources. The group is not attempting to recreate the stand or landscape configuration that existed at any specific time in the past, but is seeking to increase the landscape diversity and resilience of the ecosystem to the current and projected conditions of the twenty-first century. Similar collaborative efforts are underway in other parts of the Southwest.

17.5   CASE STUDY III: PIÑON-JUNIPER WOODLANDS OF MESA VERDE

1 *Identify/define the ecosystem of interest.* One of the major vegetation types in Mesa Verde National Park in southwestern Colorado is persistent piñon-juniper woodland (Romme et al. 2009b), characterized by centuries-old piñon (*Pinus edulis*) and junipers (*Juniperus osteosperma* and *Juniperus scopulorum*), high tree density and canopy cover, and a sparse herbaceous understory (Fig. 17.2c, Table 17.3). Despite low herbaceous cover, the flora is surprisingly diverse, and includes several statewide and national endemic species (Floyd & Colyer 2003). The historical disturbance regime included periodic canopy thinning by insects, disease, and drought, plus infrequent but stand-replacing wildfire (Floyd et al. 2004). Park management emphasizes protection and interpretation of the rich array of cultural resources, including 700 to 800-year-old multistory stone dwellings nestled in cliff alcoves. The historical reference period for this analysis extends from the Ancestral Puebloan abandonment of Mesa Verde (ca. 1300) until the time of widespread Euro-American settlement of the region in the late 1800s. The temporal and spatial scope of this evaluation is somewhat narrower than we would prefer for a system having an infrequent, spatially extensive disturbance regime. However, the ecosystem prior to abandonment was not comparable to what developed from that time to the present, because people apparently removed much of the woodland for agricultural purposes prior to 1300 (Flint-Lacey 2003). Nevertheless, the scope of historical analysis that is possible in this system is sufficient to provide important insights for management.

2a *Identify the major elements/components whose current condition is similar to common conditions during the reference period.* As with mixed-conifer-aspen forests, most of the pre-Columbian flora and fauna remain intact, as do most soils.

2b *Identify the major elements/components whose current condition was likely uncommon during the reference period.* Extensive mortality of piñon, and to a lesser extent juniper, occurred in 2002–2004, driven by drought, unusually warm temperatures, and an outbreak of piñon Ips bark beetles (*Ips confusus*) (Breshears et al. 2005). Although somewhat similar mortality events occurred previously and Ips is native to the region, the magnitude of piñon mortality probably was near or beyond the historical maximum, with ca. 30% of piñon trees in the park dying over a 5-year period, with proportionally greater mortality in larger size classes (Floyd et al. 2009).

2c *Identify the major elements/components whose current condition probably never occurred during the reference*

period. As with the above case studies, the accoutrements and activities of industrial society now are present, although direct human impacts are concentrated within a small portion of the Park area. The most dramatic broadscale change in the Park's piñon-juniper woodlands has been the increased frequency and extent of large severe wildfires since 1995. The fire rotation (time required to burn a cumulative area equal in extent to the entire Park) decreased from ca. 400 years before 1995 to only ca. 20 years after 1995; in 1995, most of the woodland had not burned in at least 700 years, but since 1995, more than half of the Park has burned (Floyd et al. 2004). Increasingly frequent spring droughts are a primary reason for the increase in burning since 1995. Recent large fires have been mostly high-severity, stand-replacing burns that consumed almost all of the soil O-horizon. Where sprouting shrubs (e.g. *Quercus gambelii* and *Amelanchier utahensis*) were abundant prior to the fires, vigorous resprouting began in the fall of the year of the fire and resulted in rapid reestablishment of plant cover. This resprouting process represents the normal historical postfire response in mountain shrublands of the region (Floyd et al. 2000), and non-native plant invasion was relatively minor in these areas (Floyd et al. 2006). In contrast, where non-sprouting shrubs such as *Purshia tridentata* dominated the understory, the bare soils exposed by the fire were invaded by non-native plant species, notably muskthistle (*Carduus nutans*) and cheatgrass (Floyd et al. 2006). These non-native species now constitute most of the plant cover in many burned areas.

3 *Of the elements whose current condition is similar to common conditions during the reference period, determine whether HRV is socially acceptable.* The native flora and fauna are well accepted and valued by local residents and visitors; in addition to appreciation of native biodiversity, visitors enjoy seeing the archaeological features within the context of the native vegetation and wildlife.

4 *Of the elements whose current condition was likely uncommon or probably never occurred during the reference period, identify those that are most critical to maintaining ecosystem integrity, that is, those that in their current condition threaten long-term ecological integrity.* Despite the severity of the recent tree-mortality event caused by drought and insects, the woodlands appear to be recovering from this disturbance; surviving small trees are growing larger, and new seedlings are being observed in almost all areas that were not also subjected to fire

(Floyd et al. 2009). Where recent fires have been followed by abundant establishment of non-native species, however, natural recovery either is not occurring or is occurring so slowly that dominance by native species cannot be expected for many decades. The woodland flora appears unable to compete effectively with the aggressive non-native herbs that are taking over burned areas. High-intensity fires also can damage the Park's archaeological resources, and the loss of ancient piñon and juniper trees diminishes the aesthetic experience for many visitors. The piñon-juniper woodland in Mesa Verde has little history of fire and the dominant tree species lack adaptations to fire, but the woodland now is being subjected to frequent, high-severity fire. Thus, the current disturbance regime is far beyond anything this vegetation type has experienced in the past 700 years. One of the greatest concerns is that expanding cheatgrass populations will fuel even shorter fire intervals, thereby accelerating the degradation of the remaining patches of ancient woodland (Turner et al. 2008).

5 *Of those elements identified in #4 above, evaluate whether HRV provides clues about potential intervention points and management options.* Unfortunately, this is a case in which historical data do not provide much help in developing a specific strategy for maintaining ecological integrity in the face of climate change and invasive species. Rather than finding solutions in the past, we need creative new approaches for reining in the expansion of non-native plants and restoring a more natural trajectory of woodland recovery in burned areas. One promising idea is to plant seeds of native grasses immediately after fires. Preliminary results indicate that this method can substantially reduce the densities of non-native species, although it does not eliminate the non-natives altogether (Floyd et al. 2006).

Our historical understanding does indicate some things that we should *not* do. Unlike southwestern ponderosa pine forests, where restoration of formerly frequent fire is an urgent priority (Chapter 1, this book), we should not conduct prescribed burning to restore this ecosystem because frequent fire is not a key process in persistent piñon-juniper woodland. This striking difference between piñon-juniper and ponderosa-pine vegetation in the use of fire for restoration illustrates the importance of tailoring management to local ecological conditions and avoiding "one-size-fits-all" treatments.

17.6 SUMMING UP: USING HRV TO HELP COPE WITH UPCOMING ENVIRONMENTAL CHANGE

An understanding of the HRV of an ecosystem aids in planning and managing for what are likely to be dramatic future environmental changes in at least three important ways. First, HRV informs us of the range of conditions, climatic and otherwise, in which the biota and ecological processes that we value today have persisted in the past and therefore will likely have the capacity to persist in the future. For example, Yellowstone's lodgepole pine forest ecosystem has been shaped in part by high-severity fires occurring at highly variable frequencies throughout the past 10 000 years. This is a nonequilibrium system with no single most-common state. The biota have numerous adaptations to high-severity fire, and ecosystem processes like energy flow and biogeochemistry are highly conservative in the aftermath of fire. In contrast, southwestern ponderosa-pine forest ecosystems (Chapter 1, this book) developed within a relatively narrow range of fire frequencies and severities during previous centuries, resulting in a more-or-less steady-state ecosystem structure and function. This formerly stable condition in ponderosa-pine forests now is threatened by a profound change in the disturbance regime. The biota are well adapted to fire, but only to a regime of frequent, low-severity fires. Different yet is the persistent piñon-juniper woodland ecosystem, which lacks a history of or effective adaptations to almost any kind of fire.

Second, our understanding of HRV enables us to evaluate how resilient a particular ecosystem may be in the face of impending climate change and other environmental threats. This information permits managers to prioritize treatments and expenditures of limited resources in a quasi-triage mode. For example, Yellowstone's lodgepole pine forest ecosystem has not been damaged by recent high-severity fires and probably is not threatened by increases in fire frequency and severity – at least not as projected for the next 20–30 years. This is because the native flora and fauna have effective fire adaptations and because a similar ecosystem composition and structure has persisted through similar environmental changes in the past. Stand age classes may shift toward a greater preponderance of younger stands, but the ecosystem will continue to function largely as it has historically. In contrast, southwestern ponderosa-pine forests and piñon-juniper woodlands on the Colorado Plateau have little history of high-severity fires and do not recover readily when such fires occur today. In a triage mode, therefore, Yellowstone's lodgepole pine forests probably are a relatively low priority for treatment or restoration; scarce resources should be devoted to other systems, notably southwestern ponderosa-pine forests and piñon-juniper woodlands on the Colorado Plateau, that are less resilient and at greater risk of severe ecological degradation even within the next 20–30 years.

Third, once management and restoration priorities have been identified and endorsed by the public, HRV information can help us identify effective treatments for increasing the ecological integrity and resilience of ecosystem types that are most vulnerable to damage from climate change or other stressors. For example, we know that southwestern ponderosa-pine forest ecosystems are beautifully adapted to the stand structures, fire frequencies, and fire behaviors of the eighteenth and nineteenth centuries, and we have documented the eighteenth to nineteenth century fire regime and associated stand structures in great detail (Chapter 1, this book). In contrast, we know that it is not necessary or even desirable to return Yellowstone's lodgepole pine forests or Southern Rocky Mountain mixed-conifer-aspen forests to any particular historical state; stand and landscape structure in these ecosystems naturally fluctuates over time. All that may be necessary to maintain ecological integrity in these ecosystems over the next 20–30 years is to allow fires and insects to behave approximately as they did historically.

The three case studies developed here are all drawn from Rocky Mountain and southwestern forest and woodland ecosystems, because these are the ecosystems with which we are most familiar. However, we suggest that the framework for analysis presented here may be broadly applicable to other forests and ecosystems across (and beyond) North America. Our hope is that the structured application of the HRV concept illustrated in Table 17.1 and Fig. 17.1 and described in the three case studies will contribute another useful conceptual tool to managers' toolbox for coping with climate and environmental change (Millar et al. 2007).

REFERENCES

Anderson, R.S., Allen, C.D., Toney, J.L., Jass, R.B., & Bair, A.N. (2008). Holocene vegetation and fire regimes in subalpine and mixed conifer forests, southern Rocky Mountains, USA. *International Journal of Wildland Fire*, **17**, 96–114.

Aoki, C. (2010). *Fire History and Serotiny in the Rocky Mountains of Colorado*. Thesis, Colorado State University, Fort Collins CO, USA.

Bartlein, P.J., Whitlock, C., & Shafer, S.L. (1997). Future climate in the Yellowstone National Park region and its potential impact on vegetation. *Conservation Biology*, **11**, 782–792.

Bigio, E., Swetnam, T.W., & Baisan, C.H. (2010). A comparison and integration of tree-ring and alluvial records of fire history at the Missionary Ridge Fire, Durango, Colorado, USA. *The Holocene*, **20**, 1047–1061.

Billings, D.W. (1990). *Bromus tectorum*: a biotic cause of ecosystem impoverishment in the Great Basin. In *The Earth in Transition: Patterns and Processes of Biotic Impoverishment* (ed. G.M. Woodwell), pp. 301–322. Cambridge University Press, Cambridge, UK.

Breshears, D.D., Cobb, N.S., Rich, P.M., et al. (2005). Regional vegetation die-off in response to global-change-type drought. *Proceedings of the National Academy of Sciences of the United States of America*, **102**(42), 15144–15148.

Christensen, N.L., Agee, J.K., Brussard, P.F., et al. (1989). Interpreting the Yellowstone fires of 1988. *BioScience*, **39**, 678–685.

Coops, N.C. & Waring, R.H. (2011). A process-based approach to estimate lodgepole pine (*Pinus contorta* Dougl.) distribution in the Pacific Northwest under climate change. *Climatic Change*, **105**, 313–328.

Despain, D.G. (1990). *Yellowstone Vegetation*. Roberts Rinehart, Boulder, CO, USA.

Duncan, S.L., McComb, B.C., & Johnson, K.N. (2010). Integrating ecological and social ranges of variability in conservation of biodiversity: past, present, and future. *Ecology and Society*, **15**(1). Article 5. http://www.ecologyandsociety.org/vol15/iss1/art5/.

Flint-Lacey, P.R. (2003). The ancestral Puebloans and their piñon-juniper woodlands. In *Ancient Piñon-juniper Woodlands: A Natural History of Mesa Verde Country* (ed. M.L. Floyd), pp. 309–319. University Press of Colorado, Boulder, CO, USA.

Floyd, M.L. & Colyer, M. (2003). Beneath the trees: shrubs, herbs, and some surprising rarities. In *Ancient Piñon-juniper Woodlands: A Natural History of Mesa Verde Country* (ed. M.L. Floyd), pp. 31–60. University Press of Colorado, Boulder, CO, USA.

Floyd, M.L., Romme, W.H., & Hanna, D. (2000). Fire history and vegetation pattern in Mesa Verde National Park. *Ecological Applications*, **10**, 1666–1680.

Floyd, M.L., Hanna, D.D., & Romme, W.H. (2004). Historical and recent fire regimes in piñon-juniper woodlands on Mesa Verde, Colorado, USA. *Forest Ecology and Management*, **198**, 269–289.

Floyd, M.L., Hanna, D., Romme, W.H., & Crews, T. (2006). Predicting and mitigating weed invasions to restore natural post-fire succession in Mesa Verde National Park, Colorado, USA. *International Journal of Wildland Fire*, **15**, 247–259.

Floyd, M.L., Clifford, M., Cobb, N.S., Hanna, D., Delph, R., Ford, P., & Turner, D. (2009). Relationship of stand characteristics to drought-induced mortality in three Southwestern piñon-juniper woodlands. *Ecological Applications*, **19**, 1223–1230.

Gray, S.T. & McCabe, G.J. (2010). A combined water balance and tree ring approach to understanding the potential hydrologic effects of climate change in the central Rocky Mountain region. *Water Resources Research*, **46**, W05513. doi:10.1029/2008WR007650.

Grissino-Mayer, H.D., Romme, W.H., Floyd, M.L., & Hanna, D.D. (2004). Climatic and human influences on fire regimes of the southern San Juan Mountains, Colorado, USA. *Ecology*, **85**, 1708–1724.

IPCC. (2007). *Climate Change 2007: The Physical Science Basis*. Contribution of Working Group I to the fourth assessment report of the Intergovernmental Panel on Climate Change. Cambridge University Press, Cambridge, UK and New York, USA.

Jones, J.R. (1974). Silviculture of southwestern mixed conifers and aspen: the status of our knowledge. Research Paper RM-122. USDA Forest Service, Fort Collins, CO, USA.

Malhi, Y., Aragao, L.E.O.C., Galbraith, D., et al. (2009). Exploring the likelihood and mechanism of a climate-change-induced dieback of the Amazon rainforest. *Proceedings of the National Academy of Sciences of the United States of America*, **106**(49), 20610–20615.

Millar, C.I., Stephenson, N.L., & Stephens, S.L. (2007). Climate change and forests of the future: managing in the face of uncertainty. *Ecological Applications*, **17**, 2145–2151.

Millspaugh, S.H., Whitlock, C., & Bartlein, P.J. (2000). Variations in fire frequency and climate over the past 17 000 years in central Yellowstone National Park. *Geology*, **28**, 211–214.

Paulson, D.D. & Baker, W.L. (2006). *The Nature of Southwestern Colorado: Recognizing Human Legacies and Restoring Natural Places*. University Press of Colorado, Boulder, CO, USA.

Romme, W.H. & Despain, D.G. (1989). Historical perspective on the Yellowstone fires of 1988. *BioScience*, **39**, 695–699.

Romme, W.H. & Turner, M.G. (2004). Ten years after the 1988 Yellowstone fires: is restoration needed? In *After the Fires: The Ecology of Change in Yellowstone National Park* (ed. L.L. Wallace), pp. 318–361. Yale University Press, New Haven, CT, USA and London, UK.

Romme, W.H., Turner, M.G., Tinker, D.B., & Knight, D.H. (2004). Emulating natural forest disturbance in the wildland-urban interface of the Greater Yellowstone Ecosystem of the United States. In *Emulating Natural Forest Landscape Disturbances: Concepts and Applications* (ed. A.H.

Perera, L.J. Buse, and M.G. Weber), pp. 243–250. Columbia University Press, New York, USA.

Romme, W.H., Floyd, L., & Hanna, D. (2009a). Historical range of variability and current landscape condition analysis: South Central Highlands Section, southwestern Colorado and northwestern New Mexico. Colorado Forest Restoration Institute, Colorado State University, Fort Collins, CO, USA. http://warnercnr.colostate.edu/docs/frs/cfri/RommeSECentralHRV.pdf.

Romme, W., Allen, C., Bailey, J., et al. (2009b). Historical and modern disturbance regimes of piñon-juniper vegetation in the western U.S. *Rangeland Ecology and Management*, **62**, 203–222.

Seager, R., Ting, M., Held, I., et al. (2007). Model projections of an imminent transition to a more arid climate in southwestern North America. *Science*, **316**, 1181–1184.

Spracklen, D.V., Mickley, L.J., Logan, J.A., Hudman, R.C., Yevich, R., Flannigan, M.D., & Westerling, A.L. (2009). Impacts of climate change from 2000 to 2050 on wildfire activity and carbonaceous aerosol concentrations in the western United States. *Journal of Geophysical Research-Atmospheres*, **114**, D20301. doi:10.1029/2008JD 010966.

Swetnam, T.W. & Betancourt, J.L. (1998). Mesoscale disturbance and ecological response to decadal climatic variability in the American Southwest. *Journal of Climate*, **11**, 3128–3147.

Turner, C.E., Romme, W., Chew, J., et al. (2008). The Frame project – a collaborative modeling approach to natural resource management at Mesa Verde National Park, Colorado. In *The Colorado Plateau III: Integrating Research and Resources Management for Effective Conservation* (ed. C. van Riper III and M.K. Sogge), pp. 23–41. University of Arizona Press, Tucson, AZ, USA.

Turner, M.G. (2010). Disturbance and landscape dynamics in a changing world. *Ecology*, **91**, 2833–2849.

Turner, M.G. & Dale, V.H. (1998). Comparing large, infrequent disturbances: what have we learned? *Ecosystems*, **1**, 493–496.

Turner, M.G., Romme, W.H., Gardner, R.H., & Hargrove, W.W. (1997). Effects of fire size and pattern on early succession in Yellowstone National Park. *Ecological Monographs*, **67**, 411–433.

Turner, M.G., Tinker, D.B., Romme, W.H., Kashian, D.M., & Litton, C.M. (2004). Landscape patterns of sapling density, leaf area, and aboveground net primary production in post-fire lodgepole pine forests, Yellowstone National Park (USA). *Ecosystems*, **7**, 751–775.

Turner, M.G., Smithwick, E.A.H., Metzger, K.L., Tinker, D.B., & Romme, W.H. (2007). Inorganic nitrogen availability following severe stand-replacing fire in the greater Yellowstone ecosystem. *Proceedings of the National Academy of Sciences of the United States of America*, **104**, 4782–4789.

Vankat, J.L. (2011). Post-1935 changes in forest vegetation of Grand Canyon National Park, Arizona, USA: Part 2 – Mixed conifer, spruce-fir, and quaking aspen forests. *Forest Ecology and Management*, **261**, 326–341.

Whitlock, C. (1993). Postglacial vegetation and climate of Grand Teton and southern Yellowstone National Park. *Ecological Monographs*, **63**, 173–198.

Wright, B.R. & Tinker, D.B. (2012). Canada thistle (*Cirsium arvense* (L.) Scop.) dynamics in young, postfire forests in Yellowstone National Park, Northwestern Wyoming, USA. *Plant Ecology*, **213**, 613–624.

YNP. (2010). Yellowstone resources and issues. Yellowstone National Park Division of Interpretation, WY, USA.

Section 5

Global Perspectives

John A. Wiens[1,2]

[1]PRBO Conservation Science, Petaluma, CA, USA
[2]University of Western Australia, Crawley, WA, Australia

The previous chapters, and especially the case studies in Section 4, convey a distinctly North American perspective on historical ecology and historical variation and how they have been used in resource management and conservation. But 'history' and 'variation', and the ways in which they affect ecological systems, are by no means the solitary purview of North American ecologists or managers, nor are North American ecosystems the only ones affected by history and exhibiting variation. History and variation are everywhere.

The five chapters in this section provide a sampling of different perspectives on history, variation, and the ecology of systems in various parts of the world. Tony Sinclair draws on his decades of experience in the African Serengeti, David Lindenmayer provides case studies from Australia, Keith Kirby uses a map on his wall to launch into a discourse on English woods, Brandon Bestelmeyer asks whether all the talk of historical variation in forested ecosystems is relevant to arid and semiarid rangelands (in North America, to be sure, but certainly different from forests), and Yrjö Haila extends the forest perspective to consider the cultural history of forests and forestry in Fennoscandia. Collectively, these chapters (extended essays, really) illustrate that, no matter where one is in the world or what type of ecological system one studies, history has left its mark. More to the point of this book, they also show how a knowledge of history can contribute ecological insights or help to frame management and conservation practices.

Historical Environmental Variation in Conservation and Natural Resource Management, First Edition. Edited by John A. Wiens, Gregory D. Hayward, Hugh D. Safford, and Catherine M. Giffen.
© 2012 John Wiley & Sons, Ltd. Published 2012 by John Wiley & Sons, Ltd.

Chapter 18

ECOLOGICAL HISTORY GUIDES THE FUTURE OF CONSERVATION: LESSONS FROM AFRICA

A.R.E. Sinclair

University of British Columbia, Vancouver, BC, Canada

Historical Environmental Variation in Conservation and Natural Resource Management, First Edition. Edited by John A. Wiens, Gregory D. Hayward, Hugh D. Safford, and Catherine M. Giffen.
© 2012 John Wiley & Sons, Ltd. Published 2012 by John Wiley & Sons, Ltd.

18.1   INTRODUCTION

There has been no greater controversy in the management of wildlife of savannah Africa than the culling of African elephants (*Loxodonta africana*) in protected areas. In the 1960s, culling was proposed on the grounds that elephants were destroying the vegetation, being out of balance with their environment. During this time, culling was implemented in Kruger Park, South Africa (and subsequently continued for 40 years), Zimbabwe, Uganda and Kenya among other countries (Laws et al. 1975; Whyte et al. 2003). Having killed 300 elephants in Tsavo Park, Kenya, scientists proposed that a total of 3000 be culled from the population of 20 000. Public furore reached a crescendo in 1968, the scientists were asked to leave, and the Kenya National Park conservationists decreed on principle that no culling within protected areas be allowed forthwith, a situation that continues today. However, the controversy left unanswered the question: Why were elephant populations increasing in the first place? The simplistic argument that they were being pushed into protected areas by expanding human populations was unconvincing because not only were elephant birth rates very high inside these parks (contrary to expectations from the displacement hypothesis), but also birth rates were high outside the protected areas.

A similar question arose with the discovery of increasing numbers of wildebeest (*Connochaetes taurinus*) and African buffalo (*Syncerus caffer*) in Serengeti Park, Tanzania, in 1965 (Sinclair 1973). The reaction of park managers was concern that the system was out of balance and management culling was necessary. Indeed, a project was initiated in the early 1970s to assess the feasibility of commercial cropping for financial profit.

Some 50 years of subsequent ecological research, together with historical information at different timescales, now provides a context that allows both answers to these questions and an assessment of future global conservation.

18.2   THE GREAT RINDERPEST

Rinderpest, a viral disease of cattle that occurs naturally in Asia, was introduced into Ethiopia in 1887 by cattle brought from India by Italian invaders. There is no evidence that the disease was ever in Africa prior to this event. Rinderpest spread to West Africa, then south through East Africa, reaching the Cape by 1896. The resulting panzootic killed over 95% of cattle throughout Africa; in many cases, complete herds died off, and famines decimated the human population of Africa. This ranks as one of the greatest socioecological disasters in human history, on a par with the plague in Europe and smallpox in the New World. By 1892, widespread famine occurred in Ethiopia, Somalia, southern Sudan and eastern Africa (Pankhurst 1966). Explorers report starving pastoralists on the Serengeti plains in 1891 (Baumann 1894). Buffalo and many species of antelope, particularly wildebeest, were also decimated.

The repercussions from this panzootic have had a profound influence on the ecology of the Serengeti over the last century. The Serengeti ecosystem, as defined by the movements of the migratory wildebeest in the present period (1950–2010), had very few human inhabitants in the nineteenth century. Early travellers in the 1860s reported only nomadic elephant hunters in the central woodland areas and no farmers or pastoralists (Wakefield 1870). Baumann's (1894) map of human settlements in 1891 showed none within the present ecosystem, and human boundaries were similar to or even outside those of the present. As at present, there were agriculturalists in the west and pastoralists in the east. Both groups suffered severe declines from the rinderpest, famine and secondary outbreaks of smallpox.

These events in the 1890–1900 period, together with a wildebeest population that we have calculated would have been less than 50 000 animals, would have resulted in a broad change in ecological processes. Over vast areas, the savannah experienced a severe reduction of burning – fires are all caused by humans – and a major regeneration of young trees. The outbreak of trees, which has been documented empirically, occurred both inside and outside the current protected area (Sinclair et al. 2008). The dense vegetation in turn resulted in another secondary epidemic, namely typanosomiasis (sleeping sickness), which is spread by tsetse flies (*Glossina* spp.), because tsetses thrive in dense vegetation. Once tsetses invade an area, cattle cannot live there, so humans were unable to return. Indeed, humans and cattle did not return until the vegetation surrounding the Serengeti had been cleared by mechanical means and by deliberate burning in the 1930–1950 period (Ford 1971).

Meanwhile, rinderpest epizootics occurred every 10–20 years over the period 1900–1963. Hence, for over half a century, cattle and wild bovines remained at low density throughout the Serengeti. The virus then disappeared from wildlife populations as a result of a cattle vaccination campaign. The evidence for this disappearance came from antibody titres in blood sera collected from animals of known age – some 80–100% of wildebeest and buffalo born in the 1950s had rinderpest antibodies, but only 50% of those born during 1960–1962 had antibodies and none born after 1963 suffered from the disease. By eliminating the disease from the domestic reservoir, the vaccination programme effectively protected wildlife from infectious yearling cattle, and consequently the disease died out rapidly in wild populations (Sinclair 1977; Plowright 1982).

Only ruminants are affected by rinderpest, the greater morbidity being in species more closely related to cattle. Thus, buffalo were most affected, followed by wildebeest. Infections were also reported in giraffe (*Giraffa camelopardalis*) and warthog (*Phacochoerus aethiopicus*), but other ruminants appear to have been less influenced by the disease.

Following the disappearance of rinderpest in 1963, juvenile survival in wildebeest and buffalo doubled. Both populations increased exponentially, with buffalo increasing from an estimated low of 15 000 and levelling out after 1973 near 75 000 and wildebeest growing from a low of 200 000 to 1.3 million after 1977. Most significantly, zebra (*Equus burchelli*), which as non-ruminants are not affected by rinderpest, have remained at constant numbers, around 200 000, for 45 years from 1958 to 2003. This species provided the circumstantial evidence that rinderpest reduced the numbers of susceptible ruminants, and the disappearance of the disease allowed their increase.

The removal of this disease provided the conditions for detecting the regulating mechanism operating in the two populations. Both were regulated through density-dependent adult mortality. This mortality took place in the dry season and was caused by a decline in per-capita food supply (Sinclair 1977; Mduma et al. 1999). The severe drought of 1993 reduced the food supply and caused a 30% drop in both populations. The wildebeest have subsequently rebounded to their previous level. These observations have shown that, contrary to earlier predictions, the populations could be self-regulated through the availability of their

resources, and interference from park managers was unnecessary.

The most important lesson that comes out of these studies is that the cause of changes in ungulate populations and their predators, and in the vegetation, as we first recorded them in the 1960s were not understood until they were placed in the context of the long-term events of the rinderpest panzootics. Both the historical account of the 1890s and the blood antibody data collected in the 1960s were integral to explaining these events. In summary, it was the historical evidence from the 1890s, combined with the antibody data over the period when rinderpest disappeared, that explained the increase in populations.

One other aspect of the great rinderpest involved the change in tree densities. The tree regeneration did not have any impact on the grazing ungulates such as wildebeest, but it was relevant to elephant populations, which were rebounding through causes that I relate below. Rinderpest resulted in the exodus of humans from around Serengeti, the consequent reduction of burning, and a pulse of tree regeneration in 1890. This pulse in young trees resulted some 70 years later in the mature trees that provided the food for the increasing numbers of elephants in Serengeti the 1960s (Sinclair et al. 2008).

18.3 THE IVORY TRADE

Insights from history extend beyond the critical understanding developed after evaluating the rinderpest and typanosomiasis epizootics in light of cattle management and woodland development. In the nineteenth century, Zanzibar Island was an Arab state under the Sultan of Oman. It was the base from which Arab caravans set out for the interior of Africa. The central staging post was Tabora in the centre of what is now Tanzania. From Tabora, these caravans went west into Congo and north along both the western and eastern sides of Lake Victoria. One such route went across the Serengeti plains, through the central woodlands and then west to Lake Victoria (Wakefield 1870; Farler 1882). Other routes started from the Kenya and Mozambique coasts, and again from Khartoum travelling west and south. These routes effectively covered east, central and southern Africa.

These caravans developed to serve the ivory trade, moving ivory from throughout Africa to India and Arabia. The system of hunters, traders and transport

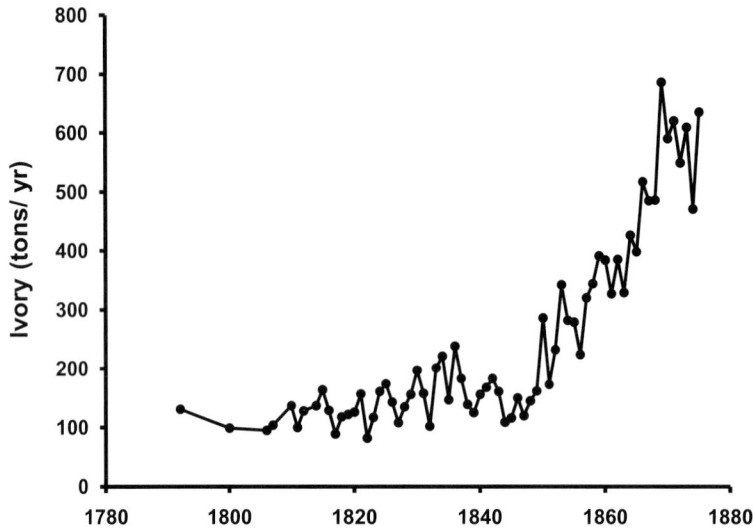

Fig. 18.1 Imports of elephant ivory from Africa to England in the mid-nineteenth century, which caused the collapse of elephant populations in Africa. The record stops in 1875 (data from Sheriff 1987).

operated on a small scale for a thousand years or more, but starting around 1840, the trade expanded rapidly as the demand for ivory in Europe and North America suddenly increased. The fundamental motivation for the increase in demand was the Industrial Revolution lead by Britain in the 1780s. By 1840, the increase in wealth in Britain created a demand for luxuries such as pianos and snooker tables – ivory was used for piano keys, billiard balls and even knife handles. Imports of ivory to Britain were steady before 1840 but thereafter increased linearly until about 1880 (Fig. 18.1) (Sheriff 1987). Although the import record stops at this point, export records from Zanzibar and Khartoum show a rapid decline to near zero after 1890 (Spinage 1973).

The Arab caravans purchased ivory from the African tribes that hunted elephants within their territories. The hunting was so heavy that elephants were eradicated. This extirpation started near the east coast in the 1850s and then extended westward as the caravans were obliged to travel farther to find ivory. By the 1890s, elephants were so scarce throughout Africa that exports collapsed (Sheriff 1987). Tsavo Park, Kenya, which is now famous as an elephant park and in the early nineteenth century the domain of the Wakamba elephant hunters, had no elephants in the 1890s (Patterson 1907). The central Serengeti woodlands had no resident tribes in the 1860s but sup-

ported nomadic Wandorobo elephant hunters (Wakefield 1870), implying elephants were present at that time, but they were gone in the period 1890–1938 (Baumann 1894; Johnson 1929; Moore 1938). A similar situation occurred in southern Africa – Chobe Park, Botswana, had large numbers of elephants in the mid-1800s, but all had been hunted by 1900 (Moe et al. 2009).

The absence of elephants in their preferred high-density areas such as Tsavo allowed the vegetation to grow from an open savannah during the period of elephant use into dense thickets. Indeed, the cover in Tsavo was so dense that Patterson (1907) had to walk while crouched through tunnels made by rhino. Such a situation would be inconceivable today. Recovery of the elephant population in savannah Africa took the first half of the twentieth century, starting from the few remaining groups in remote areas. Surviving elephants in Lorian swamp of northeast Kenya, and the Nile in Uganda expanded with the collapse of the ivory trade. It was not until the 1950s that elephants were seen commonly both in and out of protected areas. Elders of the Waikoma tribe remembered seeing two elephants in the west of the Serengeti in the 1930s, the first they had ever seen (J. Hando, pers. comm.). In 1958, there were some 800 elephants in Serengeti, in 1961, over 1000, and by 1965, numbers had increased to 2000 and a decade later, they had reached 3000. The same increases occurred in most areas where elephants lived. Although some of this increase must have been due to compression from expanding human activity outside, it was certainly not the only cause for increase or even the main cause; the populations were growing back from the low levels of a century earlier.

The decline and subsequent growth of elephant populations had significant consequences for savannah vegetation throughout Africa. The dense vegetation that was a feature of savannah Africa in the first half of the twentieth century, following the collapse of elephant populations in the previous 50 years, provided abundant food and allowed elephant numbers to increase. As elephant populations grew, the biomass of trees and shrubs declined, as would be expected. In Tsavo, a drought in 1971 expedited this process and some 8000 elephants died. Subsequently, despite predictions that Tsavo would become a desert (Laws 1969), the vegetation has recovered and a savannah of trees and shrubs dominates the park in the 2000s.

Several lessons emerge from the review of the ivory trade in the 1840–1890 period. Firstly, when biologists first looked at elephant populations and their habitats in the 1960s, they had little context to interpret the changes taking place. Consequently, they interpreted the decline of savannah tree populations as both unstable and aberrant ecology. Now that we appreciate that the habitats had previously changed to dense thickets of trees and shrubs due to the absence of elephants for some 60 years, an alternative explanation of the population trends in the 1960s emerged. Secondly, the trends in elephant numbers and vegetation biomass observed in the 1960s can be seen as standard biological adjustments following the major disturbance to elephant populations a hundred years before. Therefore, such readjustments should not be thought of as aberrant, implying that populations are somehow 'out of control'. Indeed, there is evidence that normal regulating mechanisms operate in elephant populations (Sinclair & Metzger 2009). Birth rates, recruitment of juveniles and adult mortality rates change as elephant populations grow and as they respond to social and environmental pressure. Thirdly, it follows that management intervention through the culling of elephants may not have been needed then and may not be necessary in future if the protected areas are sufficiently large to allow population fluctuations. Small areas will always be vulnerable to disturbance from megaherbivores because the scale is smaller than the

normal movements of the animals. Thus, understanding ecological history over the past two centuries allows for a better interpretation of population dynamics and a more informed conservation practice for the future.

18.4 PALEOECOLOGY OF THE SERENGETI PLAINS

Insights into the ecological dynamics of the Serengeti come not only from recent history but also from long-term paleo-history records of vegetation. The stable carbon isotopic composition of soils can be used to elucidate the vegetation of an area. Woody dicot plants have C3 photosynthesis, whereas tropical grasses have C4 processes. Soil carbonate formed from pure C4 plants has a delta ^{13}C value of +2 per mil, whereas that formed from C3 plants has a value of −10 per mil (Cerling 1992). Modern soils from the Serengeti short grass plains have a delta ^{13}C value of +0.5 per mil. Thus, the values show that earlier records were from woodland and savannah, and this vegetation gradually changed to grassland as conditions became more arid.

Palaeosols from the past 4 million years have been analysed from two locations on the Serengeti plains, Olduvai Gorge and Laetoli (Cerling 1992; Peters et al. 2008). The range of delta ^{13}C values at different times shows a consistent trend at both sites (Fig. 18.2). Peters et al. (2008) document these trends. Some 4 million years ago on the present Serengeti plains, the vegetation was woodland with some riverine forest similar to northern Serengeti today. This interpretation is supported by ungulate fossils that include duikers and bushbuck species similar to those found in forest habitats today. By 3 million years BP, conditions had become drier and a bushland mixture of shrubs and small trees similar to that found in Tsavo today predominated. These habitats persisted until 1 million years ago, when more open savannah developed, again becoming drier. Grazing ungulates predominated, and there was an absence of species such as reedbuck (*Redunca sp.*) and hippo (*Hippopotamus amphibious*), indicating that lakes and rivers were not present. Finally, it was not until some 500000 or even as recently as 100000 years ago that the present treeless plains developed (Andrews & Bamford 2008).

The important conclusion that emerges from these paleontological studies is that habitats have been continually changing due to persistent shifts in climatic

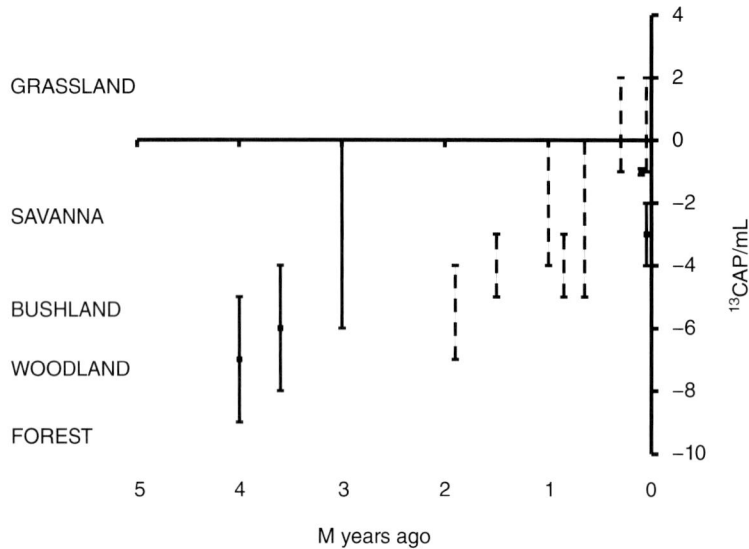

Fig. 18.2 The range of delta ^{13}C values in fossil soils at two sites on the Serengeti plains, Laetoli (solid bars) and Olduvai (broken bars). The types of vegetation which produce these values are shown on the left. (Reconstructed from Peters et al. 2008.) The vegetation has changed consistently over the past 4 million years.

conditions. As habitats change, so do the animals that depend on them. We see a gradual shift of the carnivore types as well as their diversity due to the change in herbivore prey species. This conclusion challenges the implicit premise of protected areas – that once a legal boundary is in place, it should remain ecologically stationary forever and management should act to keep it thus. This is the present Conservation Paradigm.

18.5 THE NEW CONSERVATION PARADIGM: CONTINGENCY RESERVES

The most important point following from the palae-oecological evidence that environments, habitats and fauna are constantly changing, even in the absence of anthropogenically induced climate shifts, is that they will continue to change in the future. It could be argued, however, that trends over a period of 4 million years, as we have seen at Olduvai Gorge in Serengeti, are too slow to be of any practical use for conservation at the present time – we might reasonably consider one to two hundred years ahead. To address this concern, I look at two other sites.

The pollen record in Tsavo (Gillson 2004) shows repeated switches between savannah and open grassland. Before 800 AD there was grassland, from 800 to 1530 AD there was savannah, then grassland again until 1820, and since then savannah. Thus, particular habitats existed from 300 to 700 years. More importantly, a change occurred only 190 years ago and took place quite suddenly. Similarly, in eastern Brazil, the endangered *Araucaria* forests that are fast disappearing – and causing concern for conservationists – have, from pollen analysis, only been in their present sites for some 200 years, there being a grassland prior to that (Behling & Pillar 2007). Clearly, these forests moved in from somewhere else, and they are now moving elsewhere as local conditions become less suitable.

The important conclusion coming from the studies of climate and habitats is that not only may ecosystems change suddenly – only a few decades are necessary – but they change at a timescale relevant to future conservation planning. The long-term research in Serengeti (Sinclair et al. 2008) has elucidated the mechanisms for such change. Climate shifts due to global warming allow predictions on which direction the change will take place (Thuiller et al. 2006), but the type and mechanism of change must be ascertained from the past.

The current conservation paradigm for protected areas is to demarcate a legal boundary that is then kept inviolate, as far as possible. However, the evidence from the historical record provides two conclusions that support a new conservation paradigm. The first is that all ecosystems will change, perhaps to a quite different state, possibly quite suddenly, and possibly quite soon. The Serengeti as we know it today is unlikely to be the same, or even similar, a century from now.

Secondly, if we value the current ecosystems, then we will have to make provisions for them to move into new locations. Thus, the historical evidence shows that climate is continually changing on a scale of centuries, or even faster, which means the conditions for plant communities will alter. Habitats we value now will move outside the present boundaries, as seen for those in Tsavo and the *Araucaria* forests of Brazil, and other habitats will take over. Animal communities will follow in lockstep with their habitats. The Serengeti migration of ungulates will move to new areas to follow new weather patterns and food supply. Carnivores will go with them. At present, such moves will take them directly into areas of agriculture and even villages and towns. A human-nature conflict is inevitable.

To avoid these human-nature conflicts, conservation planning is faced with having to set aside what amounts to 'contingency reserves', areas to be protected that currently may not have conservation value but may be needed in the future. Such a policy, the new conservation paradigm, will require a sea-change in ecological understanding by policy makers.

REFERENCES

Andrews, P. & Bamford, M. (2008). Past and present vegetation ecology of Laetoli, Tanzania. *Journal of Human Evolution*, **54**, 78–98.

Baumann, O. (1894). *Durch Massailand zur Nilquelle*. Berlin. Reprinted 1968. Johnson Reprint Corporation, New York, USA.

Behling, H. & Pillar, V.D. (2007). Late Quaternary vegetation, biodiversity and fire dynamics on the southern Brazilian highland and their implication for conservation and management of modern Araucaria forest and grassland ecosystems. *Philosophical Transactions of the Royal Society B: Biological Sciences*, **362**, 243–251.

Cerling, T.E. (1992). Development of grasslands and savannas in East Africa during the Neogene. *Palaeogeography, Palaeoclimatology, and Palaeoecology (Global and Planetary Change Section)*, **97**, 241–247.

Farler, J.P. (1882). Native routes in East Africa from Pangani to the Masai country and the Victoria Nyanza. *Proceedings of the Royal Geographical Society*, **4** (new series), 730–742.

Ford, J. (1971). *The Role of Trypanosomiases in African Ecology*. Clarendon Press, Oxford, UK.

Gillson, L. (2004). Testing non-equilibrium theories in savannas: 1400 years of vegetation change in Tsavo National Park, Kenya. *Ecological Complexity*, **1**, 281–298.

Johnson, M. (1929). *Lion: African Adventure with the King of Beasts*. G.P. Putnam's Sons, New York, USA.

Laws, R.M. (1969). The Tsavo research project. *Journal of Reproduction & Fertility. Supplement*, **6**, 495–531.

Laws, R.M., Parker, I.S.C. & Johnstone, R.C.B. (1975). *Elephants and Their Habitats*. Oxford University Press, Oxford, UK.

Mduma, S.A.R., Sinclair, A.R.E. & Hilborn, R. (1999). Food regulates the Serengeti wildebeest population: a 40-year record. *Journal of Animal Ecology*, **68**, 1101–1122.

Moe, S.R., Rutina, L.P., Hyttenborn, H. & du Toit, J.T. (2009). What controls woodland regeneration after elephants have killed big trees? *Journal of Applied Ecology*, **46**, 223–230.

Moore, A. (1938). *Serengeti*. Country Life, London, UK.

Pankhurst, R. (1966). The Great Ethiopian famine of 1888–1892: a new assessment. Part 2. *Journal of the History of Medicine and Allied Sciences*, **21**, 271–294.

Patterson, J.H. (1907). *The Maneaters of Tsavo*. Reprinted 1996. Pocket Books, New York, USA.

Peters, C.R., Blumenschine, R.J., Hay, R.L., et al. (2008). Paleoecology of the Serengeti-Mara ecosystem. In *Serengeti III: Human Impacts on Ecosystem Dynamics* (ed. A.R.E. Sinclair, C. Packer, S.A.R. Mduma and J.M. Fryxell), pp. 47–94. University of Chicago Press, Chicago, IL, USA.

Plowright, W. (1982). The effects of rinderpest and rinderpest control on wildlife in Africa. *Symposium of the Zoological Society of London*, **50**, 1–28.

Sheriff, A. (1987). *Slaves, Spices and Ivory in Zanzibar*. James Currey, Oxford, UK.

Sinclair, A.R.E. (1973). Population increases of buffalo and wildebeest in the Serengeti. *East African Wildlife Journal*, **11**, 93–107.

Sinclair, A.R.E. (1977). *The African Buffalo*. University of Chicago Press, Chicago, IL, USA.

Sinclair, A.R.E. & Metzger, K. (2009). Advances in wildlife ecology and the influence of Graeme Caughley. *Wildlife Research*, **36**, 8–15.

Sinclair, A.R.E., Hopcraft, J.G.C., Olff, H., Mduma, S.A.R., Galvin, K.A. & Sharam, G.J. (2008). Historical and future changes to the Serengeti ecosystem. In *Serengeti III: Human Impacts on Ecosystem Dynamics* (ed. A.R.E. Sinclair, C. Packer, S.A.R. Mduma and J.M. Fryxell), pp. 7–46. University of Chicago Press, Chicago, IL, USA.

Spinage, C.A. (1973). A review of ivory exploitation and elephant population trends in Africa. *East African Wildlife Journal*, **11**, 281–289.

Thuiller, W., Broennimannw, O., Hughes, G., Alkemade, J.R.M., Midgley, J.G.F. & Corsi, F. (2006). Vulnerability of African mammals to anthropogenic climate change under conservative land transformation assumptions. *Global Change Biology*, **12**, 424–440.

Wakefield, T. (1870). Routes of native caravans from the coast to the interior of East Africa. *Journal of the Royal Geographical Society*, **11**, 303–338.

Whyte, I.J., van Aarde, R. & Pimm, S.L. (2003). Kruger's elephant population: its size and consequences for ecosystem heterogeneity. In *The Kruger Experience: Ecology and Management of Savanna Heterogeneity* (ed. J.T. du Toit, K.H. Rogers and H.C. Biggs), pp. 332–348. Island Press, Washington, DC, USA.

Chapter 19

ECOLOGICAL HISTORY HAS PRESENT AND FUTURE ECOLOGICAL CONSEQUENCES – CASE STUDIES FROM AUSTRALIA

David Lindenmayer

The Australian National University, Canberra, ACT, Australia

19.1   INTRODUCTION

Ecology is often regarded as a case-study discipline (Shrader-Frechette & McCoy 1993), and using case studies to develop general messages can be a critically important approach to inform resource-management practices and conservation actions (Fazey et al. 2006). Here, I present a personal perspective on the intersection between the ecological history of ecosystems and the management and conservation of those systems based on two case studies. The first study is drawn from a major research program that has continued for more than 27 years in the wet ash forests of the Central Highlands of Victoria, southeastern Australia (Lindenmayer 2009). The other is from Booderee National Park, Jervis Bay Territory (also in southeastern Australia), which has been running for a shorter period (9 years) (Lindenmayer et al. 2008b). Although the bulk of my comments focus on these case studies, there are some more general lessons that emerge that may have much broader applicability, and I touch briefly on them in the concluding section.

19.2   CASE STUDY #1 – THE WET ASH FORESTS OF VICTORIA, SOUTHEASTERN AUSTRALIA

The first case study is from the montane ash forests of the Central Highlands of Victoria. Mountain Ash (*Eucalyptus regnans*), Alpine Ash (*Eucalyptus delegatensis*), and Shining Gum (*Eucalyptus nitens*) dominate these forests. This region is close to Melbourne, Australia's second largest city, and it is notable because some of the forests it supports contain the tallest flowering plants on earth (with some trees exceeding 100 m in top height). The montane ash forests of the Central Highlands of Victoria also (1) are the source of most of the water for Melbourne, (2) provide habitat for a range of high-profile species such as the nationally endangered Leadbeater's Possum (*Gymnobelidues leadbeateri*), and (3) support a major timber and pulpwood industry, with almost all logging currently taking place by clear-cutting methods, although postfire salvage logging (which also involves clear-cutting) often occurs following high-severity conflagrations such as those which occurred in 1939, 1983 (Lindenmayer et al. 2008a), and most recently, in 2009.

Wildfire, logging, and biodiversity are intimately interlinked through ecological history and contemporary management in montane ash forests, sometimes in unexpected and/or subtle ways. Many of these interrelationships have shaped montane ash forests in ways that have a major influence on current biotic assemblages, current forest management practices, and current wildlife management strategies. They also will have a significant effect on future fire and biodiversity management (Lindenmayer 2009). To understand these interrelationships, it is important first to provide some background on natural disturbance regimes in montane ash forests. The typical natural disturbance pathway in these forests is a high-severity, stand-replacing, or partial stand-replacing fire. Such fires kill most overstory eucalypt trees in montane ash forests. Unlike most Australian eucalypts, these trees rarely resprout from epicormic shoots. New cohorts of trees germinate *en masse* from seeds shed from the burned crowns of trees during a wildfire (Ashton 1975). These new young trees grow at truly phenomenal rates for the first 70 years – sometimes exceeding 1 m per year. However, they do not produce viable seed in the crown until they are 20 years of age, so if a second fire occurs before trees reach sexual maturity, stands of montane ash forest are eliminated.

Although wildfires can be stand-replacing or partial stand-replacing events in montane ash forests, they nevertheless leave significant biological legacies (sensu Franklin et al. 2000). Biological legacies are parts of the previous forest stand carried through the fire to the recovering forest. These can be large dead trees, living fire-scarred trees, large pieces of coarse woody debris, plants that have survived (even as seeds), rhizomes or rootstocks, and residual populations of animals. A suite of major studies over the past two decades has shown that biological legacies remaining from past fires play many key ecological roles in montane ash forests. For example, they speed post-disturbance recovery of many populations of plants and animals, and they have a strong influence on the successional trajectory of postfire forests. As an example, burnt forests with some living fire-scarred trees become multiaged stands; that is, forests with two or more age cohorts of overstory eucalypt trees. These kinds of stands are important because they support the highest species richness of arboreal marsupials, and they provide the high-quality habitat for individual species such as Leadbeater's Possum (Lindenmayer 2009). Multiaged stands are also where the highest known density for aboveground carbon biomass occurs anywhere in the world, with measured values exceeding 2000 tonnes per hectare (Keith et al. 2009).

Ecological history can affect many aspects of the interplay between disturbance regimes, biological legacies, and biodiversity in montane ash forests. First, the time elapsed since a previous fire influences the age of a stand of montane ash forest. This, in turn, affects the number, type, and spatial pattern of biological legacies that can occur after a wildfire. For example, more large living and dead trees with hollows will persist or be created after a fire in an old-growth forest than if a young regrowth stand is burnt. This is because the nature of cavity ontogeny in montane ash trees means that small-diameter regrowth trees that are burnt do not develop hollows suitable for use by vertebrates (Gibbons & Lindenmayer 2002). The occurrence and abundance of these biological legacies matters for a wide range of species of cavity-dependent vertebrates which can either (1) persist in burnt stands if they have access to shelter and den sites in trees with hollows or (2) recolonize burnt stands by occupying regenerating stands within 10–20 years after a fire. The majority of these species are absent from burnt but then regenerating stands if these kinds of cavity-tree resources are rare or absent (Lindenmayer 2009). Therefore, the sequence of fires in an area can have important impacts on stand age and stand structure, which in turn influences habitat suitability for a range of animal species. Notably, the process of cavity development in montane ash forests is slow and mediated by termites and fungi (lacking the accelerating effects of excavators like woodpeckers) (Lindenmayer 2009).

A second impact of ecological history relates to post-disturbance forest management. Post-disturbance salvage logging was widespread after wildfires in 1939 and 1983 and is now occurring after the recent 2009 wildfires. These operations remove extensive numbers of biological legacies that result from these fires. For example, the extent of multiaged montane ash forest was reduced from an estimated 30% of stands to currently less than 10% (McCarthy & Lindenmayer 1998). Similarly, past postfire salvage logging removes large numbers of dead trees with hollows. Indeed, recent work suggests that because past salvage logging programs (especially after the 1939 fires) were so pervasive, it may take 150–200 years for forests to recover key structural attributes such as large trees with hollows that are a critical habitat component for many cavity-dependent species in montane ash forests (Lindenmayer & Ough 2006). That work also suggests

that the colonization of salvage-logged forests by some species of cavity-dependent wildlife will be greatly accelerated by the retention of existing dead and living large trees with hollows in harvested areas.

A third impact of ecological history relates to traditional (green tree) clear-cut logging of montane ash forests. Clear-cutting leads to long-term stand structural simplification, converting old-growth stands to the start of the successional process, but lacking a large proportion of the biological legacies that characterize young regenerating forests after natural disturbance (Lindenmayer 2009). Field studies over the past 27 years indicate that the paucity of biological legacies in young clear-cut and regenerated stands makes them unsuitable as habitat for many species, particularly cavity-dependent ones.

There may be other effects of past logging history in montane ash forests. Recent work tentatively suggests that the history of clear-cut logging (including salvage logging) might have altered critical aspects of fire regimes such as increased fire severity (Lindenmayer et al. 2009). The mechanisms for this change remain unclear, but they may include:

1 *Changing microclimates and fuel characteristics.* These changes can lead to increased drying of understory vegetation and the forest floor through reductions in water-holding mosses and the humus layer. This can alter the amount, type, and moisture content of fuels.
2 *Altering the structure and plant species composition of stands.* Such changes not only alter microclimatic conditions as described above, but also can change stocking densities and patterns of trees, inter-crown spacing of trees, and other forest attributes such as plant-species composition. This, in turn, can influence fire behavior. For example, clear-cutting leads to the development of dense stands of regrowth saplings that can create more available fuel than if the forest is not clear-cut.
3 *Altering patterns of landscape cover.* Logging operations change natural patterns of spatial juxtaposition of different kinds of forest stands, resulting in homogenization of the age-class structure. This, in turn, may change the way that fire spreads through landscapes. For example, some areas traditionally characterized by an absence of fire may become more susceptible to fires that spread from adjacent, more flammable, logged areas. Notably, places such as closed-water catchments where logging operations have never occurred contain stands of old-growth forest with evidence of past fires, but those disturbances have been

repeated, low-severity fires that have not been stand-replacing and instead have produced multiaged forests (Lindenmayer et al. 2000). Importantly, a broader view of natural-disturbance history regimes in montane ash forests suggests the existence of a range of disturbance pathways beyond the single simple one of complete stand-replacing wildfires.

Interrelationships between clear-cutting and wild-fire may create the potential for negative feedbacks between human disturbance, natural disturbance, and declining stand age across montane ash landscapes. That is, young regrowth forests resulting from clear-cutting have a higher risk of burning and then being salvage logged. They may then be "fire-trapped" (Fig. 19.1), making it difficult for them to avoid being re-burnt and precluding them from reaching an old-growth stage (typically 200+ years). An additional major risk is recurrent fire at intervals of less than 20 years. In such cases, montane ash forest would be lost entirely and replaced by other vegetation types, such as stands dominated by understorey *Acacia* spp. trees (Lindenmayer 2009). This would have substantial con-sequences for plant and animal species associated with montane ash forest.

In summary, the history of clear-cutting and postfire salvage logging has influenced the spatial distribution

and abundance of key biological legacies such as large trees with hollows throughout montane ash forests (Lindenmayer et al. 1991). It also has affected the com-position of stand age classes in montane ash forest landscapes. This, in turn, has influenced the spatial distribution and abundance of a suite of cavity-tree associated animals in these forests (Lindenmayer et al. 1994). Hence, the present distribution patterns of large tree and animal species are a reflection of ecologi-cal history, particularly past natural and human dis-turbance events, combinations of the two (i.e. salvage logging), and the cumulative effects of both (i.e. logging-induced changes in fire regimes).

Ecological history, forest management, fire management, and biodiversity conservation

The findings outlined above have several important implications for forest management, postfire manage-ment practices, and biodiversity conservation in montane ash forests, and likely other forest types as well. First, existing stands of old growth and multiaged forest have unique values that would be lost through either clear-cut logging or salvage logging. This is because they are places where (1) significant numbers

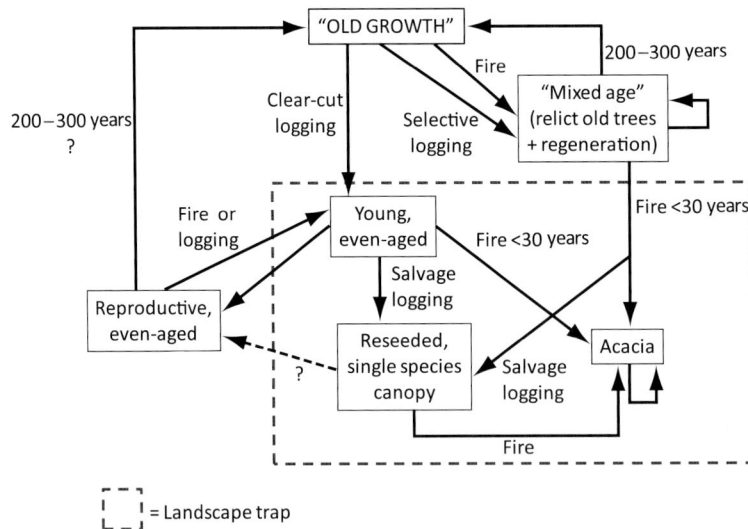

Fig. 19.1 Conceptual model of the interplay between forest age, wildfire, clear-cut logging, and salvage logging in montane ash forests. The hatched line encompasses a landscape fire trap in which there are negative feedbacks between natural and human disturbances that preclude the development of older, less fire-prone stands. Please refer to Colour Plate 8.

of biological legacies occur – both in the absence of fire and after major wildfires; (2) a range of species of conservation significance is most likely to occur; and (3) ecological recovery is likely to be fastest if a wildfire occurs. Moreover, remaining old-growth stands will assume increasing importance as more of the remaining old forest is logged, burnt, or both. The ecological values of old-growth stands are well recognized by government authorities responsible for the management of montane ash forests, which are now exempt from harvesting, including postfire salvage logging. As a consequence, a zoning system has been adopted in forests broadly designated for wood production. These zones delineate areas that are and are not available for timber and pulpwood production. For example, there are Special Protection Zones where the primary management action is the conservation of species such as Leadbeater's Possum (Macfarlane et al. 1998). In essence, this means that ecological history has underpinned the location of different parts of the forest where different kinds of management actions have priority. A second important implication of the ecological history of fire, salvage logging, and clear-cutting is the relative paucity of large trees with hollows across substantial areas of regrowth montane ash forest. The problems for biodiversity conservation created by such kinds of stand simplification are now well recognized. In response, alternative silvicultural systems to traditional clear-cutting practices are being tested in an attempt to better retain and/or create new cohorts of large trees in logged areas, as well as to develop more stands with multiple-age cohorts of trees. These new systems are modeled on the Variable Retention Harvest System developed in northwestern North America (Franklin et al. 1997) and are based on retaining islands of protected unlogged forest within harvest units. Importantly, studies of natural-disturbance regimes and landscape history indicate that flat and undulating areas with southerly aspects are those where multiaged stands are most likely to occur naturally (Mackey et al. 2002), and this has been used to guide where alternative island-retention logging systems would be best applied (Lindenmayer 2009).

A third important implication of the intersection of ecological history of fire, salvage logging, and clear-cutting is how it can change people's perceptions of what is "natural" in many ecosystems. This change in perception is akin to the shifting baseline concept widely used in marine ecosystems (Baum & Myers 2004) but less well known in forest environments. In

Victorian montane ash forests, forest managers failed to recognize that multiage stands were once far more widespread because evidence of them had largely been lost from wood-production areas. This stemmed from recurrent logging, high-intensity wildfires, and postfire salvage logging leading to widespread even-aged regrowth stands. In contrast, work in unlogged water catchments has highlighted that the prevalence of multiaged stands is much greater than previously recognized (McCarthy & Lindenmayer 1998).

19.3 CASE STUDY #2 – BOODEREE NATIONAL PARK, JERVIS BAY TERRITORY, SOUTHEASTERN AUSTRALIA

My second (and shorter) case study is also from southeastern Australia and also has key underlying themes of fire, vegetation cover, and biodiversity conservation. It is based on 6600-ha Booderee National Park, in the Jervis Bay Territory. The area is notable for its extraordinary diversity of vegetation types in a relatively small area – ranging from warm temperate rainforest to open woodland and coastal heathland. It also supports highly diverse floral and faunal assemblages, including a suite of endangered species such as the Eastern Bristlebird (*Dasyornis brachypterus*) and the Barkley's Grevillea (*Grevillea barkleyana*), of which the park supports some of the most significant remaining populations.

Fire is a key ecological process in Booderee National Park, and the area has been subject to many past wildfires and extensive prescribed burning (Baker 2000). Indeed, Booderee National Park has some of the best-kept fire-history records of any location in southeastern Australia. Two key areas of work have particular relevance to the theme of ecological history and contemporary management and conservation. The first was a detailed longitudinal investigation of postfire recovery of various vertebrate groups following a major conflagration in 2003, in which approximately half of Booderee National Park was burnt. The second was an empirical study of relationships between fire history and site-level vertebrate species richness (see Lindenmayer et al. 2008b for both studies).

A key (and somewhat surprising) finding of our first study quantifying postfire ecological recovery was that, after the fire in 2003, the majority of individual bird species and the bird assemblage as a whole in most vegetation types recovered within 2 years to what it had been prior to the fire. There also were few, if any, major differences in populations of individual bird species and the bird assemblage between areas that had been burnt in 2003 and those that had not. Post-fire conditions, particularly high levels of rainfall, appear to have promoted vegetation growth and structural regeneration, aiding bird species recovery in most vegetation types. In addition, the presence of biological legacies remaining after a fire either facilitated the persistence of a given species on a site or promoted the speed or recovery of some species to pre-disturbance levels. For example, the amount of unburned vegetation within a burnt site and in the area immediately surrounding a site was a significant explanatory variable in logistic regression models for observed species richness in 2004 (immediately following the 2003 wildfire) and also in statistical models developed for a number of individual species.

In the second study, historical data on fire were linked with detailed empirical data gathered since 2002 on the occurrence and abundance of birds, reptiles, mammals, and amphibians at 110 permanent field sites. An example of the many results derived from this work was that although most of the bird assemblage recovered quickly following the 2003 fire, longer term fire history was found to have had a highly significant effect on observed bird species richness. Richness was reduced by 9.1% per fire over the past 35 years, irrespective of the vegetation type that was examined; it was lowest on sites burnt five times since 1974.

Ecological history, fire management, and biodiversity conservation

The results outlined above have some important implications for fire-management practices and their intersection with biodiversity conservation in Booderee National Park. First, it is clear that recovery after a single fire did not reflect the long-term effects of multiple fires on overall bird species richness at a site. Hence, there is a need for the careful application of subsequent planned fires such as hazard reduction or prescribed burns to any given site. Because of the relationships between increasing numbers of fires and reduced observed species richness at a site, fire history should guide decisions about where to apply such planned burns. For example, places where there is a history of many past fires should be exempt from additional near- and medium-term planned fires such as those used for

prescribed burning. In addition, there is considerable value in identifying places that have been fire-free for a prolonged period, quantifying the characteristics of the vegetation structure and plant species composition in those areas, and then attempting to maintain those features as part of management interventions in other parts of Booderee National Park.

A second important implication arising from the empirical results of work at Booderee National Park is that biological legacies matter – an outcome reinforcing one of the key themes in the first case study. Biological legacies either facilitate the persistence of species on burnt sites or speed their postfire recovery. Hence, the presence of some species on sites will be influenced, in part, by the legacies that remain on a site following a previous disturbance (Franklin et al. 2000). Informed fire-management practices need to ensure that the ecological integrity of biological legacies is not impaired. For example, in some jurisdictions, fire-suppression activities involve back burning that attempts to halt a fire front or "black-out" unburnt areas not "consumed" by a fire. This may lead to a loss of key biological legacies like unburnt patches or patches initially subject to low-severity fire (Backer et al. 2004). Data from Booderee National Park on the importance of the amount of unburnt vegetation within and surrounding a site suggest that these practices have the potential to negatively influence the postfire response of some elements of the biota. The lesson from ecological history in this case is that contemporary management practices should avoid practices like black-out burning wherever possible because of their effects on biodiversity, particularly postfire ecological recovery. Such practices can reduce landscape heterogeneity (such as the mosaic of burned and unburned patches), and this also can have negative impacts on biodiversity (Lindenmayer et al. 2008b) as well as likely postfire ecological recovery.

Issues associated with the integration of fire management and biodiversity conservation are critical ones for Booderee National Park and many other similar places, both in southeastern Australia and elsewhere in the world. This is because an array of studies has predicted that fire size, fire frequency, and/or fire severity will increase in the near future in many parts of the world as a consequence of climate change (Cary 2002; Lenihan et al. 2003). The potential for increasing fire frequency and severity has led to widespread calls for more prescribed burning and more intensive fire-suppression efforts once a major fire event has started. The impacts of these actions on biodiversity conservation need to be carefully considered, particularly in the context of fire history and the biological legacies that result from natural, spatially patchy fires but are missing from the uniformly burnt areas that can result from fire-suppression activities like black-out burning.

19.4 CONCLUDING COMMENTS

It is clear from work in many parts of the world that ecological history can have a significant effect on current ecological conditions, contemporary management, and on future management. The two case studies I have described illustrate this key point. However, there are other important lessons from these case studies. In particular, some historical effects can last longer and be more pervasive than many people previously believed (cf Jones & Schmitz 2009). The loss of large cavity trees in montane ash forests and the prolonged period for recruitment (spanning one to two centuries) is a classic example. A second lesson relates to debates about the effects of climate change. Both case studies in this chapter indicate that any effects of future climate change would not take place on a blank canvas. Rather, the impacts of climate change will be influenced by the ecological history of past disturbances as well as past and current management actions. For example, the regeneration niches of tree species in montane ash forests are predicted to be severely challenged by potential future changes in temperature and rainfall; hence, there may be interrelationships with the amount of forest regenerating after logging, after fire, or both. Indeed, climate change is also likely to drive substantial changes in fire regimes in southeastern Australia (Cary 2002; Steffen et al. 2009). It is therefore critical to identify and then manage the factors that may exacerbate problems associated with the alterations in fire regimes driven by climate change. If preexisting factors such as logging change fire proneness, then interactions between logging and climate change could lead to cumulative negative impacts, including those on biodiversity. Conversely, recent work in Amazonia suggests that some kinds of forest may have an inherent "resilience" to climate change through maintaining mesic microclimate conditions if they remain undisturbed by other factors like logging (Mahli et al. 2009). This finding suggests that careful consideration must be given to past disturbance regimes and present management practices to better prepare natural environments for future changes in climate, with the overarching aim of reducing risks of potential interaction effects.

ACKNOWLEDGMENTS

The content of this chapter has been informed by numerous collaborative partnerships with many colleagues over the past 25 years. In particular, I thank Ross Cunningham and Jeff Wood for outstanding statistical insights that have led to some of the most important discoveries in the Victorian forests and at Jervis Bay.

REFERENCES

Ashton, D.H. (1975). The root and shoot development of *Eucalyptus regnans* F. Muell. *Australian Journal of Botany*, **23**, 867–887.

Backer, D.M., Jensen, S.E., & McPherson, G.R. (2004). Impacts of fire suppression activities on natural communities. *Conservation Biology*, **18**, 937–944.

Baker, J.R. (2000). The Eastern Bristlebird: cover-dependent and fire-sensitive. *Emu*, **100**, 286–298.

Baum, J.K. & Myers, R.A. (2004). Shifting baselines and the decline of pelagic sharks in the Gulf of Mexico. *Ecology Letters*, **7**, 135–145.

Cary, G. (2002). Importance of a changing climate for fire regimes in Australia. In *Flammable Australia* (ed. R.A. Bradstock, J.E. Williams, and A.M. Gill), pp. 26–48. Cambridge University Press, Melbourne, Australia.

Fazey, I., Fazey, J.A., Salisbury, J.G., & Lindenmayer, D.B. (2006). The nature and role of experiential knowledge for environmental conservation. *Environmental Conservation*, **33**, 1–10.

Franklin, J.F., Berg, D.E., Thornburgh, D.A., & Tappeiner, J.C. (1997). Alternative silvicultural approaches to timber harvest: variable retention harvest systems. In *Creating A Forestry for the 21st Century* (ed. K.A. Kohm and J.F. Franklin), pp. 111–139. Island Press, Covelo, CA, USA.

Franklin, J.F., Lindenmayer, D.B., MacMahon, J.A., et al. (2000). Threads of continuity: ecosystem disturbances, biological legacies and ecosystem recovery. *Conservation Biology in Practice*, **1**, 8–16.

Gibbons, P. & Lindenmayer, D.B. (2002). *Tree Hollows and Wildlife Conservation in Australia*. CSIRO Publishing, Melbourne, Australia.

Jones, H.P. & Schmitz, O.J. (2009). Rapid recovery of damaged ecosystems. *PLoS ONE*, **4**(5), e5653.

Keith, H., Mackey, B.G., & Lindenmayer, D.B. (2009). Re-evaluation of forest biomass carbon stocks and lessons from the world's most carbon-dense forests. *Proceedings of the National Academy of Sciences of the United States of America*, **106**, 11635–11640.

Lenihan, J.M., Drapek, R., Bachelet, D., & Neilson, R.P. (2003). Climate change effect on vegetation distribution, carbon, and fire in California. *Ecological Applications*, **13**, 1667–1681.

Lindenmayer, D.B. (2009). *Forest Pattern and Ecological Process: A Synthesis of 25 Years of Research*. CSIRO Publishing, Melbourne, Australia.

Lindenmayer, D.B. & Ough, K. (2006). Salvage logging in the montane ash eucalypt forests of the Central Highlands of Victoria and its potential impacts on biodiversity. *Conservation Biology*, **20**, 1005–1015.

Lindenmayer, D.B., Cunningham, R.B., Nix, H.A., Tanton, M.T., & Smith, A.P. (1991). Predicting the abundance of hollow-bearing trees in montane ash forests of south-eastern Australia. *Australian Journal of Ecology*, **16**, 91–98.

Lindenmayer, D.B., Cunningham, R.B., & Donnelly, C.F. (1994). The conservation of arboreal marsupials in the montane ash forests of the Central Highlands of Victoria, south-eastern Australia: VI. The performance of statistical models of the nest tree and habitat requirements of arboreal marsupials applied to new survey data. *Biological Conservation*, **70**, 143–147.

Lindenmayer, D.B., Cunningham, R.B., Donnelly, C.F., & Franklin, J.F. (2000). Structural features of old growth Australian montane ash forests. *Forest Ecology and Management*, **134**, 189–204.

Lindenmayer, D.B., Burton, P.J., & Franklin, J.F. (2008a). *Salvage Logging and Its Ecological Consequences*. Island Press, Washington, DC, USA.

Lindenmayer, D.B., Wood, J.T., Cunningham, R.B., et al. (2008b). Testing hypotheses associated with bird responses to wildfire. *Ecological Applications*, **18**, 1967–1983.

Lindenmayer, D.B., Hunter, M.L., Burton, P.J., & Gibbons, P. (2009). Effects of logging on fire regimes in moist forests. *Conservation Letters*, **2**, 271–277.

Macfarlane, M.A., Smith, J., & Lowe, K. (1998). Leadbeater's Possum (*Gymnobelideus leadbeateri*) recovery plan, 1998–2002. Department of Natural Resources and Environment, Government of Victoria, Melbourne, Australia.

Mackey, B., Lindenmayer, D.B., Gill, A.M., McCarthy, M.A., & Lindesay, J.A. (eds.). (2002). *Wildlife, Fire and Future Climate: A Forest Ecosystem Analysis*. CSIRO Publishing, Melbourne, Australia.

Mahli, Y., Aragao, L.E., Galbraith, D., et al. (2009). Exploring the likelihood and mechanism of a climate-change-induced dieback of the Amazon rainforest. *Proceedings of the National Academy of Sciences of the United States of America*, **106**, 20610–20615.

McCarthy, M.A. & Lindenmayer, D.B. (1998). Multi-aged Mountain Ash forest, wildlife conservation and timber harvesting. *Forest Ecology and Management*, **104**, 43–56.

Shrader-Frechette, K.S. & McCoy, E.D. (1993). *Method in Ecology: Strategies for Conservation*. Cambridge University Press, Cambridge, UK.

Steffen, W., Burbidge, A., Hughes, L., et al. (2009). *Australia's Biodiversity and Climate Change*. CSIRO Publishing, Melbourne, Australia.

Chapter 20

A VIEW FROM THE PAST TO THE FUTURE

Keith J. Kirby

Natural England, Peterborough, UK

Historical Environmental Variation in Conservation and Natural Resource Management, First Edition. Edited by John A. Wiens, Gregory D. Hayward, Hugh D. Safford, and Catherine M. Giffen.

20.1   INTRODUCTION

Above my fireplace is a map of part of Essex, a county in southern England just to the east of London, showing a scatter of small woods and parks, a small meandering river, towns and villages. I was born in one of those villages and fished in the river; most of the woods are still there. The broad pattern of this landscape was set long before the Messrs Chapman and Andre published their map in 1777 and has remained surprisingly constant since. The detail has changed dramatically, however, even in the 50 years that I have known the area: pastures have been ploughed up, hedges removed, and broadleaved coppices cleared and replanted with faster-growing conifer trees.

Another map, by Rocque from 1761, shows the land north-west of Oxford, around Wytham Woods (Savill et al. 2010). Large areas shown as open common grazing are now either enclosed fields or have become woodland. The Woods today are a diverse mix of tree species, some planted, some not; the trees range from massive oaks (*Quercus robur*) a metre or more in diameter to dense thickets of 40-year-old ash poles (*Fraxinus excelsior*). Under the trees there are places awash with bluebells (*Hyacinthoides non-scripta*) in the spring, but elsewhere they are rare.

As an ecologist, I want to understand how these landscapes function, why certain species are where they are; what causes the differences in the structure and composition of Wytham Woods. Without that knowledge I am hampered in my attempts to conserve those elements we value, or to advise on the sustainability of future land use more generally. So, I need an historical perspective, because the processes, sites and landscapes are influenced by events that happened decades, centuries, even millennia ago.

Developing and maintaining a long view of an area has in some respects become easier in the last few decades, but more difficult in others. The Internet makes it easier to access old maps (see for example http://www.old-maps.co.uk/index.html for nineteenth-century maps for England) and historical documents and to search for papers across disciplines by keywords. Geographic Information Systems and online aerial photographic records make it easier to pick out changes in landscapes over time. However, people tend to have less time to devote to one site or landscape, to really get to know them. In the United Kingdom, PhD studies typically last 3–4 years, management plans for nature reserves tend to be revised every 5 years, and staff in research institutes, universities, conservation agencies change jobs every 10 years or so in order to progress in their careers.

In this essay I explore the value of an historical perspective, focusing on changes in woodland cover in England and on one particular site, Wytham Woods, over three broad time periods. The first deals with the reshaping of the natural landscape by the spread of farming, which in Britain started with the Neolithic cultures about 6000 years ago. The second considers the changes from the end of the Mediaeval era (roughly 600 years ago) onward, which led to the landscape patterns of the late nineteenth/early twentieth century that are often used as the template for nature conservation in England. Finally, I consider the dramatic land-use changes of the last 60 years. All three 'views from the past' have lessons that may help us prepare for the future, not because the future will be the same as the past, but because it will be different.

20.2   A LANDSCAPE TRANSFORMED (C. 6000–600 YEARS BP)

What would a natural landscape look like?

Much of Europe has a long history of active land use; there are not large areas of 'untouched wilderness', but only areas where the influence of humans is less rather than more obvious. This is particularly true for England. The potential 'natural' vegetation cover is predominantly broadleaved deciduous forest (Peterken 1996), but currently only about 5% of the land surface is broadleaved woodland. About another 4% is forested, but largely comprised of non-native coniferous plantations created during the last century (Forestry Commission 2001). Agriculture employs about 70% of the contemporary landscape.

Six thousand years ago, however, farming was just getting underway. Evidence suggests that the landscape was largely tree-covered; one model points toward extensive deep, dark forests, such as occur in the old-growth areas of the Bialowieza Forest in Poland (Falinski 1986) or in the Ent-haunted forests of Tolkein's *Lord of the Rings*. Nothing like this survives in England today in terms of extent, composition, or structure.

Is Bialowieza Forest an appropriate template for what a natural landscape might look like? Vera (2000) proposed that much of north-west Europe, including much of Britain, would naturally have been a mix of open and forested landscapes, maintained by large herds of wild cattle (aurochsen, *Bos primigenius*), European bison (*Bison bonasus*) and wild horse (*Equus ferus*). Such a model could explain the occurrence in the fossil plant and insect record of species that now occur mainly in more open countryside. The modern natural analogue for this type of model would be landscapes of parkland and wooded pastures.

Arguments against the 'Vera hypothesis' as a general model of pre-Neolithic landscapes have been summarized by Hodder et al. (2009), but the debate has, as discussed later, influenced approaches to modern conservation.

Fragmented woodland, but species survival

Whatever the structure of the pre-Neolithic landscape, it is clear that after about 6000 years BP, tree pollen and sub-fossil remains of beetles associated with trees declined. The Roman invasion about 2000 years ago resulted in an extensively farmed countryside and, hence, not a densely wooded one (Rackham 2003). After the Roman occupation ended (c410 AD), there would have been changes in land use, but not perhaps the great resurgence of woodland cover that was once suggested for the Anglo-Saxon-Viking period (c500–1000 AD) (Pryor 2006). Just north of where I was born are a small church built in part from oaks felled in this period and a deer park referred to in 1045 AD (Rackham 2003); the landscape was thus already well-organised and tended.

From the woodland accounts in the Domesday Book, Rackham (2003) constructed a distribution map at roughly 10km^2 scale of woodland distribution for much of England in 1086. He estimated the total woodland cover at only about 15%. Smout et al. (2005) similarly argue that much of Scotland had already been deforested by this time. [As an aside, I might note that this does not put Britain in a very good position to lecture the developing world about curbing forest loss!]

If woodland cover declined from perhaps 80% cover 6000 years ago to less than 20% around 1000 years ago (and to about 5% 100 years ago), we might expect a corresponding loss of species as the woodland habitats became more fragmented and isolated amongst farmland. Certainly various large mammals (wolf *Canis lupus*, bear *Ursus arctos*, beaver *Castor fiber*) were much reduced and did eventually go extinct (Yalden

1999); there is also some evidence for loss of some of the specialised dead-wood fauna (Buckland 2005). On the whole, however, there does not seem to have been the catastrophic decline in species richness across woodland species groups that might have been predicted from ecological or island biogeography theory. Perhaps what prevented this were the length of time over which the changes took place – species had time to move to new areas – and the heterogeneous nature of the matrix between the surviving woodland patches. Scattered trees persisted amongst the grasslands, heaths, and so on, that were created out of the forest, and many species appear to have been able to adapt to these new, more open conditions.

The chalk grassland, acid heathland and other open habitats created by the early forest clearances have become highly valued for their distinctive plant and animal communities (Ratcliffe 1977; Rackham 1986), but also as part of our cultural heritage. Amongst writings from around the nineteenth century, for example, heaths form the backdrop to the novels of Thomas Hardy set in south-west England; Wordsworth celebrates the grandeur of the open fells of the Lake District; another poet, John Clare, concentrates on the meadows and commons of Northamptonshire in middle England.

Box 20.1 explores how the above general pattern of landscape development played out in one particular place, Wytham Woods.

20.3   THE FORMATION OF THE 'MODERN LANDSCAPE' (FROM ABOUT 600–60 YEARS BP)

Maps of England become more common and more accurate in their depiction of the landscape from about 1500 AD. The spread of printing increased the number of books, documents and accounts that were produced and have survived. Woodland ecologists are fortunate that forests and woods tend to be featured on maps and in written records because they are of economic and military significance – the organisation in Britain responsible for mapping the country is still called the *Ordnance* Survey. Woods and forest are also relatively easy to distinguish and therefore depict accurately within accounts and maps. In rare instances, individual woods can be traced back hundreds of years (Rackham 2003) through a combination of such map and documentary evidence.

Box 20.1 Development of the landscape at Wytham Woods (Savill et al. 2010).

Wytham Woods is relatively unusual in lowland England in having within it a well-preserved pollen record covering the whole of the last 10000 years. The initial spread of pine (*Pinus sylvestris*) woodland about 10000 years ago was followed by establishment of mixed forest in which lime (linden) (*Tilia* spp.) and elm (*Ulmus* spp.) were conspicuous (5000–6000 BP). These species appear to have been common over much of lowland England. Lime and elm both declined over the next few thousand years, and neither is a significant component of the Woods today. Increases in the pollen of grasses and plants of disturbed ground from about 4000 years BP suggest forest removal within the local landscape. Evidence clearly indicates that people were living on Wytham Hill in Roman times. The gales of the late 1980s in an 'ancient' (Peterken 1977) part of Wytham Great Wood blew down trees and up with the roots came Romano-British hearths, pottery, and other artefacts. Elsewhere in Britain, and on the continent, evidence of former occupation within ancient woodland sites is becoming common; and in some cases the effects on the woodland soils can be very long-lasting.

The Domesday Book makes no significant mention of the woods at Wytham, but subsequently there are references to them being used as source of 'bundles of thornes' – probably for firewood for the nuns of nearby Godstow Nunnery – and King Henry I (1069–1135) had rights to take red (*Cervus elaphus*) and roe deer (*Capreolus capreolus*) from the Woods. Much of the area that is now wooded was in the Mediaeval period open common grazing, albeit with scattered trees. Pollarded survivors from this common grazing period are probably the oldest trees in the modern woods. Originally they grew in much more open conditions and their branches would have been regularly cropped.

In the 1970s, Rackham (2003) and Peterken (1977) proposed that woods that had a long history (ancient woods), although perhaps not primary in the sense of never having been cleared, were the highest priority for modern-day conservation. These ecosystems are more likely to contain rare and threatened species as well as features and evidence of past land use that are a valuable part of our cultural heritage. Historians and conservationists elsewhere in Europe have similarly focused their attention on ancient forests to understand the relationship between history and woodland characteristics; for example, Verheyen et al. (1999).

Historical maps and records for other habitats and features in the countryside are less common than for woodland, but landscape historians and historical ecologists are able to identify specific regional or local patterns. For example, Rackham (1986) argued that some field and boundary patterns in eastern England may be over a thousand years old. By contrast, many of the hedges in the midlands derive from the Enclosure Acts of the eighteenth and nineteenth centuries, which allowed landowners to divide open common land into enclosed fields. Woodland fragmentation happened two or more millennia ago, but much of the break-up of Hardy's Dorset heathland or the great wetlands of the Fens in eastern England occurred only in the last few hundred years.

Documents and maps can also indicate how land was managed, or at least how it was meant to be managed (even now what is written in a management plan may not be what happens on the ground!). So eighteenth-century accounts detail how old the trees were when stands were cut or what the material was used for, which may give an indirect guide to age: firewood harvests were likely to be from young coppice poles, timbers for large buildings from older trees. There may be records of the efforts made to protect forest regrowth from grazing, or of fines on peasants who allowed their cattle to graze in the woods when they should not. These inferences may be supported by what survives on the ground in terms of woodland archaeological features or the structure of the trees themselves (Rackham 2003).

From these records we can build a picture of how particular landscapes have evolved (e.g. Box 20.2) and hence which species may have increased or decreased as a consequence.

20.4 A QUICKENING PACE OF LAND-USE CHANGE (FROM ABOUT 60 YEARS BP ONWARD)

During the Second World War, the shipping blockade around Britain focused attention on increasing production from existing farmland and forests. Large areas

Box 20.2 Wytham Woods – from mediaeval to modern (Savill et al. 2010).

The land around Wytham Hill passed to the Earls of Abingdon following the dissolution of the monasteries after 1540, but the main reorganisation of the landscape took place in the second part of the eighteenth and first part of the nineteenth centuries.

The open fields around the Woods were enclosed and allocated to individual farms with the implied goal of 'improving the land'. Within what are now the Woods, common grazing rights were abolished, and some of the open grassland started to develop into woodland. However, grazing by rabbits (*Oryctolagus cuniculus*), an introduction to Britain in the eleventh century, kept other areas open. Reports suggest 1000–2000 rabbits were killed annually on the estate until the 1950s, when they were virtually eliminated by the disease myxomatosis.

The separate coppices around the hill were being actively worked in the nineteenth century. There are many records for sales of wood and timber from the Estate. Gradually the coppices were linked up, partly from secondary succession on the former commons but also by deliberate woodland creation. The boundaries between old and new woodland are often still visible on the ground as low earth banks and ditches; old coppice stools (stumps) in the ancient woodland areas on one side of the bank contrast with the old pollards and open-grown trees on the other side in the former common grazing areas.

New tree species were introduced to the Woods in the nineteenth century: sycamore (*Acer pseudoplatanus*), hybrid lime (*Tilia vulgaris*), larch (*Larix* spp.), and Scot's pine. Extensive planting of beech (*Fagus sylvatica*) took place, and some of these trees now provide valuable dead-wood habitat. These old beech established before the introduction from North America of the grey squirrel (*Sciurus carolinensis)*, which now destroys many young beech crops. Grey squirrels did not occur in Wytham before the 1920s.

of permanent grassland were ploughed and between a third and a half of all woods were felled, at least in part. The abundance of young growth in many woods and of *Betula* spp – early successional trees – reflects this widespread disturbance.

The strong focus on production continued for several decades. At the same time, interest in conservation was increasing. These opposing desires led to inevitable conflicts, which came to a head in the 1970s and 1980s. Major losses of habitats and species were documented in this period (Natural England 2008), including declines in range and abundance of woodland butterflies such as the pearl-bordered fritillary (*Boloria euphrosyne*), breeding birds such as the nightingale (*Luscina megarhynchos*) and mammals such as the dormouse (*Muscardinus avellanarius*).

The social and political tension stimulated debate regarding public support from grants, tax relief and other instruments for a range of land-use practices, including forestry. As a consequence, public funding is increasingly directed towards specific public goods rather than support for production (Kirby 2003). Another outcome, since the 'Earth Summit' in 1992, has been an increasing effort to create and restore habitats and to improve the condition of areas that have become neglected in recent years.

20.5 CONTRIBUTIONS OF HISTORICAL ECOLOGY TO MODERN CONSERVATION

An understanding of English forest history is crucial to informing current conservation. Given the extent of past deforestation, there is a good case for increasing forest cover. Do we wish to try to restore some sort of natural forest? If so, how to do it, how much and where? Surviving ancient woods are likely to be the most valuable for wildlife, but where are they and how do we protect them? If we wish to restore wildlife to areas of intensive farmland, should we aim for large blocks of native trees and woods, or is there value in distributing this as scattered trees and small copses?

There are no large natural areas left in England, but there is interest in, for example, allowing some woodland nature reserves such as Lady Park Wood to develop with only minimum human intervention (Peterken 1996). Recently there has been increased interest in applying this 'hands-off' approach to larger areas, inspired in part by the example of the Oostvaardersplassen Reserve in the Netherlands (Vera 2000). We cannot recreate the pre-Neolithic landscape in England, but the discussions as to what it might have been like or the role of large herbivores in determining its structure, for example, have informed the debate about future 'rewilding'/wilderness projects and the benefits that they might bring (Taylor 2005; Kirby 2009).

The work of Rackham (2003) and Peterken (1977) established that ancient woods, defined as those believed to have supported woodland cover at least back to 1600 AD, were a priority for conservation. In 1981, a project was initiated to identify all such sites over 2 ha in size – there proved to be about 22 000 in England. This provisional ancient woodland inventory (it is continuing to be revised as new data become available) has been influential in woodland conservation through providing a basis for policies and regulations to conserve ancient woods across the landscape rather than in just a few special protected sites (Goldberg et al. 2007).

Looking to the future, ancient woods are increasingly seen as potential species-rich nodes for conservation networks and land mosaics with the capacity to rebuild biodiversity under conditions of expected climate change. Historical studies suggest a wide range of species survived the transition from pre-Neolithic forests through to mediaeval and post-mediaeval times because of the mix of woods, trees, meadows, heaths, and so on, across the whole countryside, not just in isolated sites. This reinforces suggestions that future conservation should aim at creating large-scale but heterogeneous landscapes, with more trees, but not necessarily always in large blocks.

Historical ecology can help to offset the short 'institutional memory' of many organisations. For example, of the 2500 staff in Natural England (the Government body concerned with landscape and biodiversity conservation in England), 40% are less than 40 years old, so will have only limited direct memory of how sites and landscapes looked even 30 years ago. The long-term impact of short-lived events (major storms or disease outbreaks, for example) on the structure and composition of woods may not be recognised as such. Box 20.3 illustrates for Wytham Woods some of the key events of just the last 60 years that are crucial to an understanding of why it looks as it does.

A related issue is that because an individual's historical perspective is necessarily limited, there is a risk that people will accept current environmental conditions as the norm – the shifting baseline syndrome. Long-term trends may not immediately be recognised because the mean annual change is smaller than the year-to-year fluctuations. It may be difficult to separate, say, the effects of climate change, the cumulative

---

Box 20.3 Wytham Woods – some key events affecting its current state from the last 60 years (Savill et al. 2010).

1955 – Virtual elimination of rabbits in the Woods leads to cohort of ash and sycamore establishment over large areas of the Woods.

1950s – Management plan guided by University Forestry Department results in large areas planted (or replanted) with fast growing (often non-native) tree species.

1960s – Decision to give priority to research over forest production led to reduced management (little thinning or felling) in the woods until the late 1980s.

1970s – Dutch Elm Disease begins to eliminate elms as tall canopy trees.

1980s – Increasing deer numbers limit further tree regeneration, remove low cover such as bramble and the low branches of trees and shrubs and encourage a shift to a predominantly grassy ground flora.

1990s – Efforts increased to clear scrub succession to maintain/restore rich grassland communities leading to more open habitats in the woods

2000s – Efforts at deer management succeed in reducing deer numbers; vegetation starts to show signs of recovery.

---

impact of grazing in woodland, or the impact of the shift from coppice management to high-forest treatments (Kirby et al. 2005; Hopkins & Kirby 2007).

Historical ecology can help people appreciate that the English landscapes, habitats and species assemblages we now value have, over a longer time span, been shaped by past farming and forestry practices. Landscapes are dynamic; as land management changes, so do the habitats and species assemblages that survive. We have to choose what we want to conserve and how we wish to do it.

For example, in Wytham the lack of native lime dates back to a decline in prehistoric times; should we reintroduce it? The coppice boundaries are of mediaeval origin, but the woods now spread beyond them: how much of the more recent woodland do we want to clear to restore the limestone grassland that once was widespread? Sycamore was planted into the woods in the nineteenth century, but has naturalised over much of the north slope: do we accept this introduction as part of the future composition of the woods or seek to restrict its spread? Should we keep some small conifer plantations as an example of the impact of mid-twentieth century forestry policies? The answers to these types of question depend on the particular circumstances of the site – its social, ecological and geographic context.

20.6   CONCLUSIONS

As demonstrated by a long look backwards across the countryside of England, we cannot recreate the past. Today's starting point is unique; historical ecology demonstrates that trajectories are conditional on legacies from the past and drivers in the future.

More generally we can say that:
• human management changed the natural vegetation cover of England long ago;
• despite enormous loss of forest across the island, the pace and pattern of change resulted in fewer species extinctions than might have been expected by the scale of the loss of forests;
• the current forest cover reflects direct and indirect human impacts at a variety of scales, from the deliberate planting of non-native coniferous trees to allowing populations of deer to spread, with serious consequences for the structure and composition of native woodland; therefore,
• conservation decisions for both forested and unforested lands across England must take account of short- and long-term historical patterns by considering what we want to retain from the past and create for the future.

The future pressures on habitats and species will be different in character and magnitude, whether from land-use change, climate change, new pests and diseases, or unpredictable factors such as societal priorities or technological shifts. In the nineteenth century, demand for oak bark for tanning leather made oak coppice systems economic; development of alternative chemical processes eliminated that market; what will be the most valuable wood product for the twenty-first century? Over the last 30 years there has been a considerable increase in awareness and support for conservation, but will this survive under conditions of increasing economic austerity? We can expect to

observe unique new patterns as history unfolds. Past conditions represent poor targets for future management, but by understanding how past changes have influenced the current state, we are in a better position to explore the impacts of future changes.

We can now return to the two eighteenth-century maps with which I started this essay. The woods shown on them are likely to be ancient; where they survive, bluebells and other rare woodland species are most likely to occur. These woods should consequently be a focus for conservation effort. They likely existed as scattered, much modified fragments of the former natural forests, probably for at least a thousand years. However, in between the forest patches would have been a rich mosaic of meadows, heathland and scattered trees, which allowed other species to thrive and also facilitated movement of forest species between the fragments of ancient woodland. More recently, much of this mosaic has been grossly simplified to a handful of intensively managed crops. The challenge for twenty-first-century conservation is not just how to maintain and expand the woods that are left, but what to encourage to fill the gaps in between as well.

REFERENCES

Buckland, P.C. (2005). Palaeoecological evidence for the Vera hypothesis. In Large herbivores in the wildwood and in modern naturalistic grazing system. (ed. K.H. Hodder, J.M. Bullock, P.C. Buckland and K.J. Kirby), pp. 62–116. English Nature Research Report 648, Peterborough, UK.

Falinski, J.B. (1986). *Vegetation Dynamics in Temperate Lowland Primeval Forests*. Junk, Dordrecht, The Netherlands.

Forestry Commission. (2001). *National Inventory of Woodland and Trees*. Forestry Commission, Edinburgh, UK.

Goldberg, E.A., Kirby, K.J., Hall, J.E. & Latham, J. (2007). The ancient woodland concept as a practical conservation tool in Great Britain. *Journal for Nature Conservation*, **15**, 109–119.

Hodder, K.H., Buckland, P.C., Kirby, K.J. & Bullock, J.M. (2009). Can the mid-Holocene provide models for re-wilding the landscape in Britain? *British Wildlife*, **20**(Suppl.), 4–15.

Hopkins, J. & Kirby, K.J. (2007). Ecological change in British broadleaved woodland since 1947. *The Ibis*, **149**(Suppl.), 29–40.

Kirby, K.J. (2003). Woodland conservation in privately-owned cultural landscapes: the English experience. *Environmental Science and Policy*, **6**, 253–259.

Kirby, K.J. (2009). Policy in or for the wilderness. *British Wildlife*, **20**(Suppl.), 59–62.

Kirby, K.J., Smart, S.M., Black, H.I.J., Bunce, R.G.H., Corney, P.M. & Smithers, R.J. (2005). Long term ecological change in British woodland (1971–2001). English Nature Research Report 653, Peterborough, UK.

Natural England. (2008). *State of the Natural Environment 2008*. Natural England, Sheffield, UK.

Peterken, G.F. (1977). Habitat conservation priorities in British and European woodlands. *Biological Conservation*, **11**, 223–236.

Peterken, G.F. (1996). *Natural Woodland*. Cambridge University Press, Cambridge, UK.

Pryor, F. (2006). *Britain in the Middle Ages*. Harper Collins, London.

Rackham, O. (1986). *The History of the Countryside*. Dent, London.

Rackham, O. (2003). *Ancient Woodland (Revised Edition)*. Castlepoint Press, Dalbeattie, UK.

Ratcliffe, D.A. (1977). *A Nature Conservation Review*. Cambridge University Press, Cambridge, UK.

Savill, P.S.S., Perrins, C., Kirby, K.J. & Fisher, N. (2010). *Wytham Woods, Oxford's Ecological Laboratory*. Oxford University Press, Oxford, UK.

Smout, T.C., MacDonald, A.R. & Watson, F. (2005). *A History of the Native Woodlands of Scotland, 1500–1920*. Edinburgh University Press, Edinburgh, UK.

Taylor, P. (2005). *Beyond Conservation: A Wildland Strategy*. Earthscan and BANC, London, UK.

Vera, F.W.M. (2000). *Grazing Ecology and Forest History*. CABI, Wallingford, UK.

Verheyen, K., Bossuyt, B., Hermy, M. & Tack, G. (1999). The land use history (1278–1990) of a mixed hardwood forest in western Belgium and its relationship with chemical soil characteristics. *Journal of Biogeography*, **26**, 1115–1128.

Yalden, D.W. (1999). *The History of British Mammals*. Poyser, London.

Chapter 21

IS THE HISTORICAL RANGE OF VARIATION RELEVANT TO RANGELAND MANAGEMENT?

Brandon T. Bestelmeyer

New Mexico State University, Las Cruces, NM, USA

Historical Environmental Variation in Conservation and Natural Resource Management, First Edition. Edited by John A. Wiens, Gregory D. Hayward, Hugh D. Safford, and Catherine M. Giffen.

21.1 INTRODUCTION

Rangelands occupy about 40% of the Earth's terrestrial surface, comprising grasslands, shrublands, and deserts (Havstad et al. 2007). Over a billion people live in rangelands, engaged primarily in pastoralism and limited crop agriculture for subsistence. Rangelands can be strongly coupled human-natural systems. They can harbor spectacular biodiversity and beauty, while at the same time providing food and soul-enriching connections between humans and the natural world. Resources can be exploited via livestock production such that many historical or natural elements are sustained (Brown & McDonald 1995). The sustainable partnership of people and rangelands, however, has not been easy in many places. Recent estimates suggest that many of the world's rangelands are degraded. Degradation effects range from transient structural changes to persistent shifts in species composition or loss of soil and biological productivity. Globally, rangeland management will become even more difficult. Degradation is expected to accelerate with increases in human population and regional aridification (Millennium Ecosystem Assessment 2005). In this chapter, I offer some thoughts on the use of historical range of variation concepts (hereafter HRV) to address the core problem of managing change in rangelands. I argue that analysis of HRV can provide evidence of ecological potential and the attributes comprising rangeland health, but that HRV is not by itself a firm basis for rangeland management. This view echoes themes developed elsewhere in this book (Hayward et al., Chapter 3, this book). I further suggest that "ecological sites" and state-and-transition models are useful tools with which to organize information about HRV and how HRV varies across landscapes and regions.

21.2 HRV IN PRACTICE

The role of HRV in rangelands, as in other ecosystems, has been to establish reference conditions or benchmarks to understand the ecological processes causing system change (Landres et al. 1999). Change in rangelands can happen very gradually, so that it becomes difficult to recognize over long time frames (Pauly 1995). In other cases, change can happen so rapidly and completely that it is not clear change happened at all. Recent gap analysis vegetation maps in the US

portion of the northern Chihuahuan Desert provide a good example. Areas mapped as desert shrubland provide no indication that they were grassland only a century ago. The uninitiated might think that shrublands have always dominated these landscapes. Similar problems with the perception of change in the Chihuahuan Desert were evident as early as 1927, shortly after degradation took place. A report from the US Forest Service Jornada Experiment Station (now the USDA-ARS Jornada Experimental Range) stated, "During the past years, the public domain has been so badly depleted that the loss of grass density . . . has not been recognized" (Campbell & Canfield 1927). Stockmen of the time could not understand why carrying capacity had declined.

Conversely, we can imagine incorrectly that the past condition of an area was different than it is now. Narratives about the encroachment of juniper trees often overlook data suggesting that dense juniper stands occupied parts of the landscape for hundreds of years (Romme et al. 2009). In such areas, appeals to restore historical conditions to these landscapes by removing juniper would be misguided (Belsky 1996).

The specification of appropriate benchmarks and models of change for specific land areas is important to rangeland managers for two reasons. First, benchmarks and models provide clues to otherwise unrecognized potential for recovery of desired states. In areas where historical degradation has occurred, the geophysical setting and remnant vegetation can be used to recognize ecological potential (i.e. the vegetation that an area could support) and the interventions needed to achieve it. Second, benchmarks and past changes provide a warning of what might occur in areas not yet degraded – a ghost of degradation future. With some appreciation of the limits of ecosystem resiliency and the consequences of degradation, managers can better evaluate the costs and benefits of their decisions.

With these functions in mind, the concept of HRV has been woven into two tools that are now a primary focus of rangeland science. "Ecological sites" comprise units of a land-classification system that distinguishes areas according to ecological potential; that is, by the soils, landforms, and climate that interactively determine the availability of limiting resources and environmental stresses for plants (USDA 2003). In the United States, ecological sites are made spatially explicit by their relationship to soil mapping units of the National Cooperative Soil Survey, but they can be recognized anywhere by considering the landscape position and soil profile of a point. A collection of ecological sites indirectly specifies how the properties of HRV vary across landscapes and regions. They thus help to distinguish inherent differences in ecological potential (e.g. soils of varying fertility) from the effects of historical disturbances or management impacts.

State-and-transition models describe the dynamic properties of ecological states that are observed on an ecological site. State-and-transition models specify the nature of HRV for a site via the dynamics of plant communities within a reference (often historical) state (Bestelmeyer et al. 2009). They also describe the causes of shifts to other states featuring distinct community dynamics. The models provide two important sources of information for managers. First, they describe changes in patterns leading to state changes that are difficult to reverse (i.e. degradation thresholds). These early-warning indicators can be used to recognize management strategies that promote (or diminish) resilience and to prioritize prophylactic interventions. Second, the models describe the indicators of restoration thresholds that slow or preclude a return of the system to the characteristics of the HRV or other desired conditions. These indicators are used to choose appropriate restoration interventions and to decide when intervention is not worth the effort.

Ecological sites and state-and-transition models evolved from the need to recognize and manage for desirable ecological states in heterogeneous and dynamic environments. Both tools, or something like them, have been developed in many parts of the world to apply to rangelands, forests, and even aquatic ecosystems (Hobbs & Suding 2009; Lester & Fairweather 2009). It is important to recognize that ecosystem classifications and models take different forms depending upon how we perceive the functioning of ecosystems. In riparian ecosystems, for example, "proper functioning condition" is conceived as an intimate coevolution of geomorphology, hydrology, and vegetation across scales (Prichard 1998). In forests of the western United States, HRV or the "natural range of variability" is modeled as a shifting mosaic of coexisting successional stages (Keane et al. 2002). These are coarse-scaled models in which the states are tied to disturbance regimes integrating entire landscapes. Rangeland models, on the other hand, divide landscapes into components differentiated by soils, topographic position, and hydrology (ecological sites, or in common managerial usage, simply "sites"). Normal or natural temporal variation within these units is defined by the successional or nonequilibrial dynamics of species through periods of patchy and varying rainfall or in response to localized natural disturbances. This view reflects how rangelands are used by pastoralists the world over, including the value of different forage species at different times and places and how those (and other) species respond to management (Illius & O'Connor 1999). HRV is used to understand how management uses and climatic variability affect rangeland conditions, usually tailored to specific sites or sets of spatially associated sites. HRV concepts reflect the scales at which particular types of ecosystems are viewed and managed.

The conditions of management interest, and therefore how to interpret HRV, are not globally uniform within rangelands. In the United States, the use of HRV as a management benchmark is codified or implied in a number of procedures (Laycock 2003; Schussman & Smith 2006). These procedures emerge from a preoccupation with pre-European history as the ideal condition. This interpretation is weakening somewhat with increasing evidence that early peoples dominated many North American ecosystems and then succumbed to disease to produce the benchmarks (Mann 2005). In other places, the long-term histories of human effects on the landscape are more fully recognized. For example, instead of HRV, Europeans often focus explicitly on "naturalness," based on the current function of ecosystems without human interference (Carrión & Fernandez 2009). In contrast, Mongolian, African, and Middle Eastern pastoralists embrace human connections and focus largely on the sustainability of net primary production and desirable forage species, reflecting a pervasive human reliance on rangeland products (e.g. Warren 2002). However HRV is used, the common thread across cultures appears to be the search for sustainability; the capacity to recover desired species during favorable periods. In arid rangelands, this capacity is based on feedbacks between soil processes and vegetation growth that, if disrupted, lead to reduced production, species loss, and other universally undesirable effects. Thus, HRV may not constitute a management goal per se (although it often does), but rather is evidence for a set of processes that represent a management goal. Those processes are most directly described by "rangeland health" (National Research Council 1994).

21.3 RANGELAND HEALTH

Rangeland health and the closely related "landscape function" use suites of indicators to provide evidence of the relationships between desirable ecosystem traits and feedback processes that support those traits, usually involving surface-soil properties. The term "health" has created controversy and numerous critics (e.g. Lackey 2001). The health metaphor has numerous problems (see Sundt 2010), but most importantly, it evokes a moral imperative for a particular set of environmental conditions, raising the thorny question of who gets to define them and how to define them (Warren 2002). This problem, however, afflicts any attempt to define land-management targets. As alluded to above, the specification of any value-laden benchmark, be it "health," "integrity," "resilience," "sustainability," or "degradation," is culturally relative and should be collaboratively developed (e.g. Reed et al. 2008).

While rangeland health as a general ecological principle is difficult to support empirically, rangeland health considered as a methodological approach offers several advantages to rangeland managers, especially when compared to the use of idealized historical conditions to specify management targets. Rangeland health seeks to provide information on the ecological processes governing change rather than the reconstruction of often poorly understood historical dynamics (Ludwig et al. 1997; Pyke et al. 2002). Benchmark conditions can then be defined, if desired, as those environmental conditions that stabilize soil resources, intercept and retain water resources, and convert those resources into biomass – usually, but not always, of native species. Indications that the desirable feedbacks are weakening or undesirable ones are taking hold foretell a loss of those conditions. Thus, rangeland-health approaches can be a source of useful indicators to guide current management. Information about HRV can provide complementary evidence about the ecological processes that support desired conditions.

Rangeland health is not a panacea. It can be difficult to distinguish variations in rangelands that are due to normal, climate-driven, nonequilibrial behavior versus those that signal the breakdown of feedbacks due to anthropogenic effects (Fig. 21.1a). Drought clearly displaces rangelands from a highly productive condition. In fact, drought is considered a cause of degradation in the international aid arena (Reynolds et al. 2007). Yet drought also represents a natural process to which

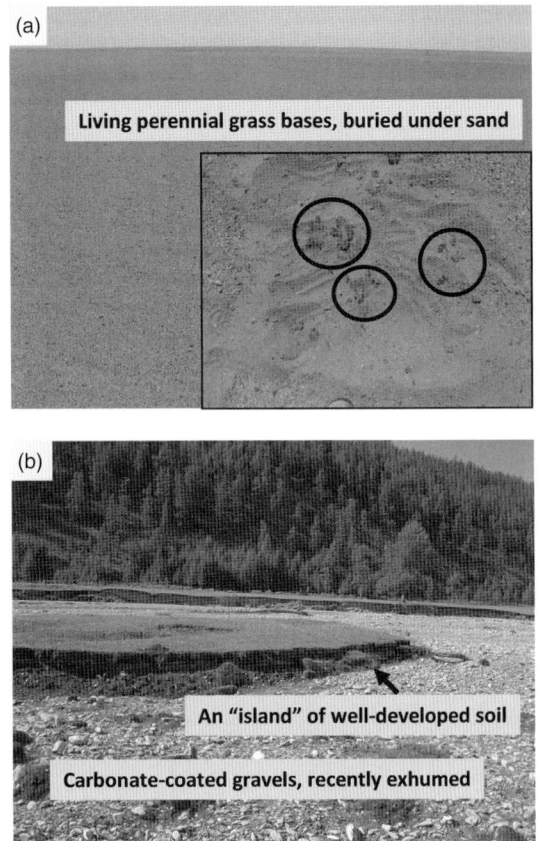

Fig. 21.1 Two cases where the clues to rangeland health are subtle, both from Mongolian rangelands. (a) A rangeland area in Omnogobi aimag (desert steppe) that we thought was completely desertified until we looked under the thin veneer of eolian sand deposits to find perennial grass crowns (inset photo, in circles). The site is clearly affected by drought and probably by grazing use, but was more resilient than we thought. (b) An area in Hovsgol aimag (forest steppe), featuring a channel and a grassy island in its middle. The sides of the island and adjacent bank reveal a similar, deep mollic soil surface horizon, indicating that the channel is recent and was derived from a formerly grassy valley bottom or meadow. Eroded clumps of the soil and grass occur across the channel. The cobbles lining the bottom of the channel are covered in powdery calcium carbonate, further indicating recent erosion from formerly stable subsoil. Casual observers might assume that they were looking at a natural, intermittent stream bed.

rangelands have been exposed for thousands of years (Buck & Monger 1999) and therefore a part of the HRV. Drought followed by high rainfall leads to a periodic decoupling between grazer populations and forage resources. The decoupling can allow vegetation recovery, even as it causes human suffering in the short term (Hobbs et al. 2008).

Although climatic variability can obscure long-term, directional changes in rangeland conditions, certain indicators can reveal the signature of altered feedbacks and an approach to thresholds. For example, soil organic-matter-rich A-horizons take centuries to millennia to develop in grasslands and change very slowly. The thinning or loss of these horizons clearly indicates an abrupt change in the system relative to the processes described by HRV (Fig. 21.1b). Thus, carefully selected process indicators can allow rangeland ecologists to detect deviations from HRV or "healthy" conditions through the haze of nonequilibrium, even when historical references are lacking.

21.4 CLIMATE CHANGE AND ALTERNATIVE REGIME MODELS

Directional climate change represents a critical problem for defining management targets via either rangeland health or the use of historical benchmarks. Climate change alters the geophysical environment directly and so alters ecological potential. Ecological potential limits the management targets that are realistically available. Thus, fundamental changes in ecological potential with climate change may limit the utility of HRV to directly inform management benchmarks. The HRV of the twentieth century may be as unattainable as the Pleistocene conditions desired by some conservation biologists (Donlan et al. 2005).

Ecological sites and state-and-transition models provide a partial solution to this problem. Ecological sites are based on geography, soil profiles, and landform, so these relatively static properties will continue to filter the effects of climatic drivers. Ecological sites will continue to be relevant for stratifying the landscape even though the species occurring within them may change. State-and-transition models were expressly developed to describe these changes and to recognize the evolution of barriers to ecological potential. Irrecoverable degraded states are effectively treated as the new ecological potential based on an evaluation of ecological mechanisms, while maintaining the his-

torical connection to an ancestral state described by HRV.

In the same way, we could also recognize another hierarchical level above the alternative states of the state-and-transition model – alternative regimes (see Walker et al. 2009 for precedent). Although the term regime is often conflated with "state" (Scheffer & Carpenter 2003), I define regime here as a suite of co-occurring drivers and conditions, and their feedbacks to management, that characterize a discrete period of time. Thus, a "regime shift" is caused by a change of sufficient magnitude in one or more drivers to alter relationships with other drivers and management. Each regime is characterized by a different state-and-transition model. Different regimes define different rules by which local management interacts with other contemporary processes to cause state transitions. Directional climate change, and other long-term directional changes, can cause a regime shift such that the rules embedded in a state-and-transition model are changed. We use the regime shift to formally recognize that a new suite of potential management targets are available.

For example, the current management benchmarks and state-and-transition models of the northern Chihuahuan Desert near Las Cruces, New Mexico, may already characterize a previous regime. Furthermore, the conventional interpretation of HRV may be based on an even older regime (Fig. 21.2). The shift from the "historical" to the "recent" regime occurred in the late 1800s as a consequence of the dramatic increase in domestic herbivore populations, the extirpation of Native American populations with the attendant loss of fire and reduced use of shrubs for food and fuel (Fredrickson et al. 2006), and the end of the Little Ice Age that is postulated to have led to widespread black grama (*Bouteloua eriopoda*) grassland establishment (Neilson 1986). Rangeland managers developed the state-and-transition models for the recent regime. These models describe how intensive grass defoliation and soil disturbance can lead to grass mortality, coupled with reduced capacity of grass to recover alongside the ability of shrubs to colonize readily. A shrub-free landscape is generally regarded as impossible, so managers seek instead to maintain desert grassland states where they still exist.

It can be argued that the northern Chihuahuan Desert has undergone a shift to an "incumbent" regime. The incumbent regime features increased awareness of the fragility of desert grassland and

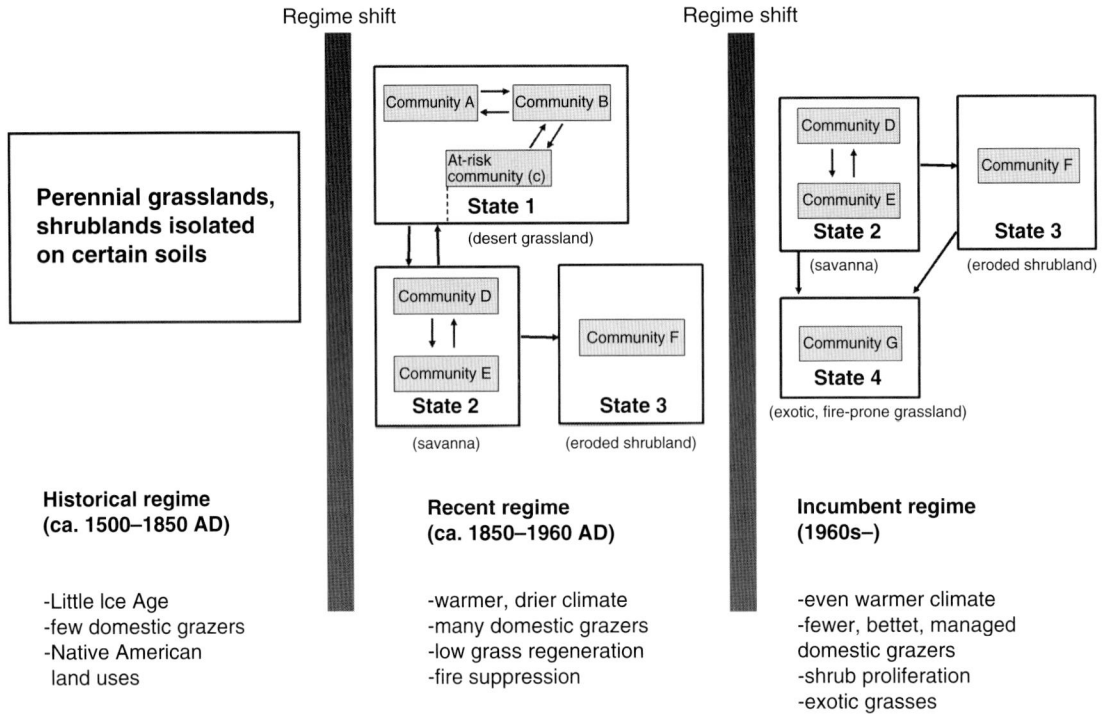

Fig. 21.2 An example of an alternative regime model with embedded state-and-transition models for an area of sandy soils in the northern Chihuahuan Desert, styled after Walker et al. (2009). The distinguishing features of each regime are described. The state-and-transition model for the historical regime is undefined because there are no data on variation. The state-and-transition models for the other regimes feature states within which plant community variation (e.g. nonequilibrial dynamics) is described. Alternative states denote persistent changes in ecosystem function and the plant communities that can be observed as a consequence. Recovery of previous states sometimes occurs via restoration actions (see Briske et al. 2008).

reduced cattle stocking rates, a shift to exurban amenity-based values, the proliferation of invasive species such as Lehmann lovegrass (*Eragrostis lehmanniana*), and the encroachment of shrubs into nearly all areas. Climate change will likely accelerate these coupled changes. In the new state-and-transition model, managers may be looking to maintain a dwindling native grass component in shrub savannas. Fire, a useful tool to control shrubs in the previous regime, may lead to expansion of the exotic grass in the incumbent regime. Within this new regime, managers might also consider the positive attributes of Lehmann lovegrass; it is a robust perennial grass, it is palatable to livestock, and it stabilizes soil and water resources.

Extensive stands of the invasive Lehmann lovegrass may be preferable to eroding shrublands in some areas.

21.5 CONCLUSION

HRV analysis is clearly useful for rangeland management, but it should not be used exclusively to establish management targets. HRV helps us to understand the functioning of ecosystems at a point in time when they were minimally affected by human activities (the United States) or at least less affected than now (other places). It serves as a temporal baseline to explain how we got here, if not where we can go from here (Swetnam et al. 1999). Rangeland health, on the other hand, appears to have utility as an organizing concept for rangeland management that spans a range of cultural value systems and geographic histories. In many places, HRV can provide evidence of rangeland health attributes. This is true even in places like Mongolia and Africa that have been grazed for thousands of years. But HRV should only provide evidence for a broader range of ecosystem behaviors than is presently observed, not define broad societal and management goals (Jackson & Hobbs 2009). To do the latter, we will need to understand the states that are available and the mechanisms by which management can maintain or attain them, considering both the regime and the ecological sites in question. In summary, HRV is relevant to rangeland management, but primarily in the context of state changes, regime shifts, and societal expectations about what rangeland "health" should entail. In a changing world, rangeland management must exploit an ecological toolbox that is much expanded from the traditional reliance on HRV.

REFERENCES

Belsky, A.J. (1996). Viewpoint: western juniper expansion: is it a threat to arid northwestern ecosystems? *Journal of Range Management*, **49**, 53–59.

Bestelmeyer, B.T., Tugel, A.J., Peacock, G.L., et al. (2009). State-and-transition models for heterogeneous landscapes: a strategy for development and application. *Rangeland Ecology and Management*, **62**, 1–15.

Briske, D.D., Bestelmeyer, B.T., Stringham, T.K., & Shaver, P.L. (2008). Recommendations for development of resilience-based state-and-transition models. *Rangeland Ecology and Management*, **61**, 359–367.

Brown, J.H. & McDonald, W. (1995). Livestock grazing and conservation of southwestern rangelands. *Conservation Biology*, **6**, 1644–1647.

Buck, B.J. & Monger, H.C. (1999). Stable isotopes and soil-geomorphology as indicators of Holocene climate change, northern Chihuahuan Desert. *Journal of Arid Environments*, **43**, 357–373.

Campbell, R.S. & Canfield, R.H. (1927). Annual report, Jornada Range Experiment Station. USDA Forest Service.

Carrión, J.S. & Fernandez, S. (2009). The survival of the "natural potential vegetation" concept (or the power of tradition). *Journal of Biogeography*, **36**, 2202–2203.

Donlan, C.J., Greene, H.W., Berger, J., et al. (2005). Re-wilding North America. *Nature*, **436**, 913–914.

Fredrickson, E.L., Estell, R.E., Laliberte, A., & Anderson, D.M. (2006). Mesquite recruitment in the Chihuahuan desert: historic and prehistoric patterns with long-term impacts. *Journal of Arid Environments*, **65**, 285–295.

Havstad, K.M., Peters, D.P.C., Skaggs, R., et al. (2007). Ecosystem services to and from rangelands of the western United States. *Ecological Economics*, **64**, 261–268.

Hobbs, N.T., Galvin, K.A., Stokes, C.J., et al. (2008). Fragmentation of rangelands: implications for humans, animals, and landscapes. *Global Environmental Change*, **18**, 776–785.

Hobbs, R.J. & Suding, K.N. (2009). *New Models for Ecosystem Dynamics and Restoration*. Island Press, Washington, DC, USA.

Illius, A.W. & O'Connor, T.G. (1999). On the relevance of non-equilibrium concepts to arid and semiarid grazing systems. *Ecological Applications*, **9**, 798–813.

Jackson, S.T. & Hobbs, R.J. (2009). Ecological restoration in the light of ecological history. *Science*, **325**, 567–568.

Keane, R.E., Parsons, R.A., & Hessburg, P.F. (2002). Estimating historical range and variation of landscape patch dynamics: limitations of the simulation approach. *Ecological Modelling*, **151**, 29–49.

Lackey, R.T. (2001). Values, policy and ecosystem health. *BioScience*, **51**, 437–443.

Landres, P.B., Morgan, P., & Swanson, F.J. (1999). Overview of the use of natural variability concepts in managing ecological systems. *Ecological Applications*, **9**, 1179–1188.

Laycock, W.A. (2003). Lessons from the past: Have we learned from our mistakes? *Arid Land Research and Management*, **17**, 359–367.

Lester, R.E. & Fairweather, P.G. (2009). Modelling future conditions in the degraded semi-arid estuary of Australia's largest river using ecosystem states. *Estuarine, Coastal and Shelf Science*, **85**, 1–11.

Ludwig, J.A., Tongway, D.J., Freudenberger, D.O., Noble, J.C., & Hodgkinson, K.C. (eds.). (1997). *Landscape Ecology, Function and Management: Principles from Australia's Rangelands*. CSIRO Publishing, Melbourne, Australia.

Mann, C.C. (2005). *1491: New Revelations of the Americas Before Columbus*. Knopf Press, New York, USA.

Millennium Ecosystem Assessment (2005). *Ecosystems and Human Well-Being: Desertification Synthesis*. World Resources Institute, Washington, DC, USA.

National Research Council. (1994). *Rangeland Health: New Methods to Classify, Inventory, and Monitor Rangelands*. National Academy Press, Washington, DC, USA.

Neilson, R.P. (1986). High resolution climatic analysis and Southwest biogeography. *Science*, **232**, 27–34.

Pauly, D. (1995). Anecdotes and the shifting baseline syndrome of fisheries. *Trends in Ecology and Evolution*, **10**, 430.

Prichard, D. (1998). *Riparian Area Management: A User Guide to Assessing Proper Functioning Condition Under the Supporting Sciences for Lotic Areas*. Publication TR 1737-15. USDI Bureau of Land Management, Denver, CO, USA.

Pyke, D.A., Herrick, J.E., Shaver, P., & Pellant, M. (2002). Rangeland health attributes and indicators for qualitative assessment. *Journal of Range Management*, **55**, 584–597.

Reed, M., Dougill, A., & Baker, T. (2008). Participatory indicator development: what can ecologists and local communities learn from each other? *Ecological Applications*, **18**, 1253–1269.

Reynolds, J.F., Stafford-Smith, D.M., Lambin, E.F., et al. (2007). Global desertification: building a science for dryland development. *Science*, **316**, 847–851.

Romme, W.H., Allen, C.D., Bailey, J.D., et al. (2009). Historical and modern disturbance regimes, stand structures, and landscape dynamics in piñon-juniper vegetation of the western U.S. *Rangeland Ecology and Management*, **62**, 203–222.

Scheffer, M. & Carpenter, S.R. (2003). Catastrophic regime shifts in ecosystems: linking theory to observation. *Trends in Ecology and Evolution*, **18**, 648–656.

Schussman, H. & Smith, E. (2006). Historical range of variation for potential natural vegetation types of the Southwest. Prepared for the USDA Forest Service by The Nature Conservancy, Tucson, AZ, USA. http://azconservation.org/downloads/multi/category/sw_forest_assessment/ [12/16/2010].

Sundt, P. (2010). Conceptual pitfalls and rangeland resilience. *Rangelands*, **32**, 30–33.

Swetnam, T.W., Allen, C.D., & Betancourt, J.L. (1999). Applied historical ecology: using the past to manage the future. *Ecological Applications*, **9**, 1189–1206.

USDA. (2003). *National Range and Pasture Handbook*. USDA National Resource Conservation Service, Washington, DC, USA.

Walker, S., Cieraad, E., Monks, A., et al. (2009). Long-term dynamics and rehabilitation of woody ecosystems in dryland South Island, New Zealand. In *New Models of Ecosystem Dynamics and Restoration* (ed. R.J. Hobbs and K.N. Suding), pp. 78–95. Island Press, Covelo, CA, USA.

Warren, A. (2002). Land degradation is contextual. *Land Degradation and Development*, **13**, 449–459.

Chapter 22

KNOWING THE FENNOSCANDIAN TAIGA: ECOHISTORICAL LESSONS

Yrjö Haila

University of Tampere, Tampere, Finland

Historical Environmental Variation in Conservation and Natural Resource Management, First Edition. Edited by John A. Wiens, Gregory D. Hayward, Hugh D. Safford, and Catherine M. Giffen.
© 2012 John Wiley & Sons, Ltd. Published 2012 by John Wiley & Sons, Ltd.

It is possible to love a small acreage in Kansas as much as John Muir loved the entire Sierra Nevada. That is fortunate, for the wilderness of the Sierra will disappear unless little pieces of nonwilderness become intensely loved by lots of people.

Wes Jackson (1991)

22.1 INTRODUCTION: HEURISTICS

A major part of the forests of northwestern Europe belong to the vast boreal forest or taiga, the largest uniform biome on the Earth, stretching across both northern continents. The general biogeography of the boreal zone is well charted (Shugart et al. 1992). It is spread between the northern summer limit of temperate airmasses and the southern winter limit of arctic airmasses; in other words, it is characterized by relatively warm summers and cold or very cold winters.

Recent climatic history has left a particularly distinct mark on the biogeography of the taiga. It is, by and large, a product of the Pleistocene Ice Age, the last two and a half million years or so. Most of the current taiga has been covered several times with ice sheets 2–3 km thick. The spread of the most recent continental glaciers peaked 21 000 years ago. In the perspective of the shaping of the northern biota, the time since the latest glaciation is merely a blink of an eye. Such dramatic and very recent variations in ecological conditions mean that the distant past does not offer any model for pondering seriously upon what a "natural" state of the taiga might be like.

But could we perhaps construct a model of a natural stationary state of the taiga, characteristic of the few millennia that have lapsed since the end of the latest glaciation, and use that as a standard for assessing its present ecological state? This question has no simple answer. In the first place, conditions in the taiga have been changing continuously, due mainly to climate variations and plant invasions. It is questionable whether a stationary state has been reached at all, or at what spatial and temporal scale this might have happened. But another and ultimately more important complication is that humans have been essential actors in the dynamics of the taiga ever since the melting of the most recent ice sheets.

That ecological formations are dynamic is no big news any more. A dynamic view has been integrated into conservation since the 1980s, recently in the shape of a "natural variability paradigm" (for a review concerning the Fennoscandian taiga, see Kuuluvainen 2002). However, while emulating natural dynamics is a good way to let biological diversity take care of itself (Haila 2004), the notion of "naturalness" deserves another thought. All organisms change the environment in which they live. So do humans. There is nothing "unnatural" in this principle. Contrary to what conservationists sometimes say, people have been aware of their environmental dependencies since time immemorial (see Haila 1999). We need to take the whole kaleidoscope of human-modified landscapes seriously, from cities and gardens to old-growth forests and everything in between.

Following the lead articulated by Wes Jackson in the epigram, we must broaden our concept of what environments are important and why. My aim in this chapter is to explore the history of forest use, using the Fennoscandian taiga as a case study. The ecological features of the Fennoscandian taiga are known probably better than those of any other forested region on the Earth, thanks to strong research traditions in forestry and forest ecology going back to the late nineteenth century, and the history of human intrusions into the forests of this region is also known reasonably well.

I adopt an ecosocial perspective, leaning on the heuristics that ecology is present in human sustenance in an analogous fashion as physiology is present in the human body (Haila & Dyke 2006). In an important sense, the body is physiology. Similarly, human sustenance is ecology. As heuristics, the ecosocial perspective suggests that the material basis of human sustenance is constructed through circular causality, driven by the shaping and reshaping of local ecologies by human livelihood practices, and of human practices shaped by ecological feedbacks. Human sustenance depends on resourceful modification of the environment; in modern parlance, on recognizing and learning to use particular ecosystem services.

The primary question is: "How should cultural resources be harnessed toward fitting together human goals, interests, and endeavors and ecological necessities in the taiga?" I am particularly interested in what past human exploitation of ecosystem services in the taiga may teach us about the coupling of human-induced and natural processes. Human material success on the Earth has been spectacular, but this success has brought about the sustainability crisis we are facing.

The contextual specificity of conservation issues has received attention lately (e.g. the articles in Lindenmayer & Hobbs 2007). My premise is that contextual specificity is also a good lead for tracing ecosocial dynamics. In other words, to expose the cultural heritage of views and uses of nature, we need to chart as thoroughly as possible what ecosystem services people have utilized and the kinds of relationships people have developed with the ecological formations from which the services were derived. Knowledge of nature has been a necessary intermediate step in such relationships since the dawn of history. The ecology of human sustenance may well be the first field of human practice in which systematic knowledge began to accumulate in the past. People had to collect and catch edible stuff from their surroundings long before they started to learn geometry by building temples and measuring plots of land.

To begin, a brief review of the ecosocial history of the Fennoscandian taiga examines how the demand for forest-based products produced extensive ecological changes in the northern European taiga.

22.2   NETWORKS: FROM ECOLOGY TO ECONOMY TO ECOLOGY

Compared with other parts of the boreal zone, Fennoscandia has been under a particularly intensive human influence. Denmark and the coastal plains south of the Baltic Sea developed into strongholds of early northern European agriculture by 4000–3000 years BC. Permanent agricultural settlements spread to southern Sweden and other fertile lands bordering upon the Baltic during the subsequent millennia, reaching southern Finland in the last millennium BC. Another group of tribes, subsisting on hunting and fishing, arrived from the east and colonized the northern parts of Fennoscandia at around the same time. The Sami people are their modern descendants. None of these communities was isolated from other human communities. The colonists brought along their networks of exchange on which they depended for getting necessities such as salt and special materials such as flint for preparing tools.

Early permanent settlements brought about changes in their immediate environments, but it was demand for forest-based products from the outside that triggered more extensive ecological changes in the northern European taiga. Before industrialization, fur and grain were the two main products traded to the outside. In Finland, swidden (slash-and-burn) cultivation was a livelihood practice that allowed for colonizing inner parts of the country. It remained a major form of forest use in eastern parts of Finland and in Karelia (the region on both sides of the eastern border of Finland) until the late nineteenth century. Grain was grown typically on recently burned land only for a couple of years; thereafter, the land was used as pasture for cattle. Agriculture and wide-range hunting were combined into a major form of livelihood over the centuries of the early iron age (Taavitsainen 1987).

Large-scale demand for wood originated in Fennoscandia with the onset of mining and metallurgy in south-central Sweden in the late Middle Ages. Iron and copper provided an economic basis for the Swedish imperium that dominated the Baltic world from the sixteenth to the eighteenth century (Kirby 1990). Mining companies were given special privileges for exploiting forests for firewood and charcoal.

Tar and timber were the main wood-based products that integrated northern European lands into the developing European economy in the eary modern era (Åström 1988; Kirby 1990; Kuisma 1993; Tasanen 2004). Swedish merchants dominated the trade in the northern Baltic, and England came to be the most important export destination. Tar was originally a more important export than timber. Tar production commenced at the northern fringes of the Baltic in the mid-seventeenth century and peaked in the eighteenth century. Some 80% of the product came from the Finnish side of the Gulf of Bothnia. Tar production spread to inner parts of Finland along rivers flowing down to the Gulf of Bothnia. St. Petersburg, established at the eastern tip of the Gulf of Finland in 1703, drew forest-based products from southeastern parts of the country.

Norway was at first the leading producer of timber for the English market. In the seventeenth century, the volume of Norwegian timber export was 10 times larger than that of Sweden. However, Norwegian timber export declined during the eighteenth century when accessible old-growth forests along the coast had been cut down. Sweden passed Norway in terms of volume of timber export, the main source areas being forested lands along the rivers flowing from Scandinavian mountains to the Gulf of Bothnia.

This brief summary brings forth the important role of transport for shaping the availability of forest-based products. It is no wonder that Norway was the first mass-scale supplier of timber from Fennoscandia. The Finnish lands east of the Gulf of Bothnia were shaped by differential transport costs into three zones according to the dominant product (summarized by Åström 1978). Timber was transported from areas close to the coast, tar from more distant parts of the catchments of the river systems, and backwoods were exploited for grain by slash-and-burn cultivation. The development of such distinct zonality was driven by a kind of path dependency. For instance, the long duration of slash-and-burn cultivation in the vicinity of villages in the inland of eastern Finland and Karelia eliminated the chances of using forest for any other purpose: there simply were no mature woods around.

Finally, the industrialization of timber production in the mid-nineteenth century and the origin of paper and pulp industries in the 1860s changed the perception of forest exploitation. In earlier times, acquisition of forest products was purely extractive, but shortages changed the situation. Intensive exploitation aims at active channeling of ecological potential into valuable products following an agricultural model, or "silviculture" in North-American vocabulary. Intensive forestry was gradually adopted from the mid-nineteenth century in all parts of Fennoscandia, albeit with differences that have to do with structures of forest ownership and government institutions (Tasanen 2004). Iron and copper works acted as pathbreakers by adopting a stand rotation system.

Following the expansion of forest industries, intensive modern forestry with clear-felling, soil preparation, seeding or planting, and fertilizing was adopted after World War II both in Sweden and in Finland with active governmental support. In Finland, an extensive project to drain peatlands was also put into force. The shift to intensive silviculture resulted in a reorganization of the entire production apparatus for wood-based products. Under ideal circumstances, silviculture aims at achieving a high degree of control, both of the growth of forests and the behavior of forest owners (Jokinen 2006 analyzes the Finnish case).

Haila and Levins (1992) adopted the term "ecohistorical period" to characterize historically distinct formations produced by regimes of human resource use and the ensuing ecological changes. Ecohistorical periods are shaped by changes that occur over large areas in a uniform fashion. The creation of zones of forest use east of the Gulf of Bothnia desribed above provides an example, and dominant forms of land use such as slash-and-burn cultivation provide others.

Ecohistorical periods have a social dimension, brought about by networks of cooperation between different human actors and the integration of particular elements of nature into such networks. An interesting example is provided by Sven-Erik Åström (1988), who noted the importance of one-blade peasant sawmills and hand-sawing of high-quality timber in expanding the extent of Finnish timber export in the nineteenth century. In this case, a particular technology, combined with evolving skills of the practioners and a trading network, created a long-lasting way of using the forest. The story is symptomatic of a more general phenomenon. Economic historians have noted that the trade in bulk commodities such as grain and timber was one of the factors that provided an impetus for the early economies of Europe by drawing an increasing proportion of the population into market exchange (Jones 1988). The procurement of tar and timber served a similar function in the northern Baltic lands. Thus, the improvement of transport connections through waterway clearance and construction tied rural areas into economic networks. Forest-based industrialization produced a more even distribution of income in rural areas than, say, localized mining compounds.

The aim of constructing ecosocial histories is not to create historical classifications, however. Rather, the aim is to identify dynamically important aspects of the use of nature and how these have molded human cultures. We need to take another look at the ecological features of the taiga that contributed to the human material success in this region.

22.3 FLEXIBILITIES: WHAT THE TAIGA AFFORDS

The ice age glaciations exerted a strong unifying influence on the biota of Fennoscandia. The resulting set of stand-forming forest trees includes only one pine, one spruce, two birches, and one aspen. New species are added on the list farther to the east, toward the Ural Mountains: Siberian larch, sembra pine, and fir. In the course of the glaciations, the number of tree species has declined because of extinctions. A likely heritage of the Pleistocene history is robustness of the biota. Organisms inhabiting the taiga have had to cope with drastic fluctuations in environmental conditions.

The ground vegetation of Fennoscandian forests is dominated by a relatively small group of plants, with local variation according to topography, moisture, and soil type (Esseen et al. 1992; Tonteri 1994). Typical stands are uniform – these stands formed the basis of the forest site-type classification constructed by the Finnish geobotanist A.K. Cajander in the early twentieth century. Relatively speaking, the gradients of variation in upland forests are much less steep than in regions with larger differences in elevation. The plant cover on the slopes of Scandinavian mountains is more variable, also thanks to mild maritime winters and high precipitation.

There is no doubt that the exploitation of forest resources has brought about major changes in forest nature all over Fennoscandia, but coming up with a precise assessment of ecological consequences of the changes is very difficult. Humans have modified the taiga to a remarkable degree. It is, in fact, next to impossible to destroy the ecological potential of a boreal forest. Effects of toxic pollution from mining and industrial compounds get diluted after the pollution discontinues, as demonstrated by examples from Canada and northern Russia; the productive potential of a forest stand can be channeled to a new set of tree species; a stand can be transformed to a field, so diverting the potential to cultivars instead of trees; and so on. The successes of human modification to the taiga imply that qualitatively specific criteria and metrics are needed for ecological assessments.

I think reasonable criteria must be congruent with the type of resource use that is of concern, and the criteria must be applied in a scale-specific manner. How else to assess the consequences of clearing a forest for a field than within an appropriate landscape perspective, for instance? A qualitatively specific approach allows constructing "phase space" representations that describe the possible ways in which a particular type of forest can change under a particular regime of exploitation (e.g. Dyke 1988; Haila & Dyke 2006).

For example, the earliest exploitation regime comprised direct extraction of specific products – for instance, hunting fur animals and thus influencing directly their population dynamics. Sable marten and beaver were hunted into extinction in Fennoscandia and northern Russia, the former before the modern era, the latter surviving in a preserve in Norway but vanishing from other parts of Fennoscandia by the mid-nineteenth century. The use of timber was originally based on selective picking of suitable trees for particular purposes. Extensive land-use practices such as slash-and-burn cultivation made use of land in a similar extractive manner.

Carrying capacity in the conventional sense offers a preliminary measuring stick for the success versus failure of extractive resource use. Species extinctions speak for themselves. There also were wood shortages close to villages in southern parts of Finland and in central Sweden already in the eighteenth century. The demand of charcoal for mining and metallurgy, and the harvesting of pine stands for tar production created local shortages at least in central Sweden and in eastern Finland, respectively, although estimates in this regard are unsatisfactory. Wasteful methods such as slash-and-burn cultivation and pasturage also lowered the productivity of forest soil, that is, the carrying capacity for timber. In the mid-nineteenth century, Finnish foresters made estimates of area relations of forests that were good for different purposes. The ratio of mature pine or spruce forest to swidden and home-use forest in several central and eastern Finnish districts was on the order of 1:1 to 1:3, with the huge extreme of 1:341 in one particular region (Heikinheimo 1987).

The modification of ecological conditions depends not only on what happens on a site, but also relies directly on contact networks created by transportation, transfer of technologies and skills, and markets. This is what makes regional differentiation in conditions important, and what also strengthens differentiation through positive ecohistorical feedback loops. One can find examples of both declining productivity as in landscapes of slash-and-burn cultivation, and of increasing productivity in landscapes with successful drainage.

To sum up, various characteristics of the forest ecosystem can be modified to provide services. What services are available depends on what has been done in the past to make them appear. From a historical perspective, human material practices and ways of life have been co-constructed with ecological processes, involving both local and regional scale feedback loops. Unveiling the constraints and enablements of such processes is a formidable analytic challenge in its own right. A consistent characteristic of the northern Fennoscadian case has been the integration of forest production into rural livelihoods on a long-term basis. Farming and forestry have been closely tied together, particularly in Finland. Such integration has shaped perceptions of forests and molded a cultural heritage that cannot be ignored when pursuing viable conservation strategies.

22.4 PERCEPTIONS: WHAT THE TAIGA WAS, IS, AND WILL BE

There are two sides to human attitudes about forests. On the one hand, the forest is, in Robert Pogue Harrison's (1992) fine phrase, a "shadow of civilization": a wild standard against which the cultural worlds created by humans have been envisaged. Harrison builds up a normative argument that areas bounding human culture have to be preserved as a counterpoint to the community established in a "clearing" inside, as it were. "Without such outside domains, there is no inside in which to dwell" (Harrison 1992). It would be difficult to disagree with this argument.

The contrast painted by Harrison has a material history. Archaeologist Ian Hodder (1990) has persuasively argued that a contrast between human-marked abode and wildness outside became inscribed into spaces surrounding the earliest permanent settlements in European prehistory some 6000–7000 years ago. Hodder describes the contrast using the terms "domus versus agrios" – tamed versus wild. He deciphers this interpretation from symbolic marks in archaeological sites of that era that repeat in different ways a domesticated versus wild borderline. The emphasis on the symbolic dimension is particularly important. People have made symbols true by living according to their dictates, as numerous anthropologists have pointed out.

On the other hand, lots and lots of people have always lived in and been dependent on the forest. This is, of course, the case in the tropics, but to a certain extent in the taiga as well. In this regard, the historical background of Finns deviates from that of Swedes or Norwegians. A phrase often repeated, that Finns have a special relationship with forests, may have some historical truth in it. Even in the modern era, life in northern and eastern parts of Finland depends more on local forest environments than in other Nordic lands, with the possible exception of northern Sweden. Furthermore, Finns were sent to colonize and tame southern Swedish backwoods in the sixteenth and seventeenth centuries as well as the backwoods of Ingria, the county to the south and west of St. Petersburg, in the seventeenth century.

But the conditions of traditional rural life in the Finnish forest country were very harsh indeed. A well-known Finnish ethnologist is reputed for having remarked glibly that "The Finnish people has always hated their forests." History books reproduce heartbreaking photographs of people toiling on burned land in slash-and-burn cultivation. The harshness of traditional rural life triggered in the nineteenth century famous controversies about the feasibility of a modern, forest-based economy. Some prominent economic thinkers regarded such a thought as an oxymoron.

In other words, there has probably been an inherent love–hate relationship in the Nordic perceptions of forests. But whatever the case, the forest has been a participant in the subsistence networks and practices of people living in northern and eastern Fennoscandia. Dependence stimulates protection. The Finnish epic poetry as well as stories and mythologies of other long-term inhabitants of the taiga underline the necessity of staying on good terms with gods and goddesses of the forest. In later times, extractive uses of forests gave rise to new perspectives on forest protection. The central European tradition of reserving hunting grounds for the royal court extended as far to the north as the Åland Islands in the Baltic Sea. Systematic views of forest protection grew up in tandem with scientific forestry. The early ideals were about sustainable yield of timber, but the failures of this perspective and reasons for them are familiar by now. The integration of nature conservation in the modern sense into forest management is also a story often told.

22.5 CONCLUSIONS

The future of the taiga does not depend on nature conservation alone. Conservation science has learned to know invaluable aspects of forest ecology, but public perception of forest, including a symbolic perspective, is decisive as well. The forest is not a preserve, to be considered independently of human presence. The forest is an inherent part of our abode. We need to learn to know the forest, in the same sense as we know our home.

Broad participation in conservation efforts, coupled with an increased understanding of forests, is therefore necessary. Pathways toward this goal have been opened up in Finland as a response to serious setbacks caused by a politically top-down conservation procedure adopted in the implementation of the Natura 2000 programme of the European Union in the late 1990s (see Hiedanpää 2002). The half dozen years of experimenting with voluntary conservation have produced positive results, when assessed using ecological criteria (Mönkkönen et al. 2008). However, I think that formal scientific criteria defined from today's perspective are not that important. Areas in which resource use is regulated attain new ecological values with time. Not only forest destruction is dynamic; forest preservation is dynamic, too. A critical first step in forest conservation is to get people to appreciate and cherish particular features of their forests. Positive attitudes and practices speak for themselves and turn infectuous.

Optimization is in vogue nowadays in conservation thinking, but the setting is multidimensionally dynamic and straighforward optimization is an unachievable goal. There are two simple rules of thumb to follow: first, species richness correlates with biomass (Wilkinson 2006); second, native habitats need protection wherever they are seriously threatened (Lindenmayer & Hobbs 2007).

I have, mainly implicitly, pursued a historical understanding of ecosystem services in this chapter. Ecosystem services are services only if and when they are actually used. Using services modifies the services. Successful opening up of specific ecosystem services by humans has made the material success of our species possible. But success of one type tends to close off potentials for success of other types. For this reason, it is worth trying tools of thinking such as ecohistorical periods and phase space representations of environmental change. Our main challenge is to assess the viability of alternative types of success and draw conclusions.

Success can be quite rapid, too, particularly regarding species protection. When I got hooked on birds in my secondary school years, whooper swan and crane were considered seriously threatened in Finland; they were thought of as emblematic residents of virgin bogs and wetlands. Now, both of these species have a strong presence around the summer home by a southern Finnish lake in a thoroughly agricultural landscape at which I write this chapter. No one in their wildest dreams could have thought half a century ago that this might be possible.

REFERENCES

Åström, S.-E. (1978). *Natur och byte: ekologiska synpunkter på Finlands ekonomiska historia.* [Nature and exchange: ecological perspectives on Finnish economic history; in Swedish.] Söderströms, Ekenäs, Finland.

Åström, S.-E. (1988). *From Tar to Timber: Studies in Northeast European Forest Exploitation and Foreign Trade 1660–1860.*

Commentationes Humanarum Litterarum 85, The Finnish Society of Sciences and Letters, Helsinki, Finland.

Dyke, C. (1988). *The Evolutionary Dynamics of Complex Systems: A Study in Biosocial Complexity.* Oxford University Press, New York, USA.

Esseen, P.A., Ehnström, B., Ericson, L., & Sjöberg, K. (1992). Boreal forests: the focal habitats of Fennoscandia. In *Ecological principles of nature conservation* (ed. L. Hansson), pp. 252–325. Elsevier, London, UK.

Haila, Y. (1999). Socioecologies. *Ecography*, **22**, 337–348.

Haila, Y. (2004). Making sense of the biodiversity crisis: a process perspective. In *Philosophy of Biodiversity* (ed. M. Oksanen and J. Pietarinen), pp. 54–88. Cambridge University Press, Cambridge, UK.

Haila, Y. & Dyke, C. (eds). (2006). *How Nature Speaks: The Dynamics of the Human Ecological Condition.* Duke University Press, Durham, NC.

Haila, Y. & Levins, R. (1992). *Humanity and Nature: Ecology, Science and Society.* Pluto Press, London, UK.

Harrison, R.P. (1992). *Forests: The Shadow of Civilization.* The University of Chicago Press, Chicago, IL, USA.

Heikinheimo, O. (1987). The impact of swidden cultivation on forests in Finland. [Extracts from a Finnish original, published in 1915.] *Suomen Antropologi*, **12**, 199–206.

Hiedanpää, J. (2002). European-wide conservation versus local well-being: the reception of Natura 2000 Reserve Network in Karvia, SW-Finland. *Landscape and Urban Planning*, **61**, 113–123.

Hodder, I. (1990). *The Domestication of Europe: Structure and Contingency in Neolithic Societies.* Basil Blackwell, London, UK.

Jackson, W. (1991). Nature as the measure for a sustainable agriculture. In *Ecology, Economy, Ethics: The Broken Circle* (ed. F.H. Bormann and S.R. Kellert), pp. 43–58. Yale University Press, New Haven, CT, USA.

Jokinen, A. (2006). Standardization and entrainment in forest management. In *How Nature Speaks: The Dynamics of the Human Ecological Condition* (ed. Y. Haila and C. Dyke), pp. 198–217. Duke University Press, Durham, NC, USA.

Jones, E.L. (1988). *Growth Recurring: Economic Change in World History.* Clarendon Press, Oxford, UK.

Kirby, D. (1990). *Northern Europe in the Early Modern Period: The Baltic World 1492–1772.* Longman, Burnt Mill, Harlow, Essex, UK.

Kuisma, M. (1993). *Metsäteollisuuden maa.* [Green gold and capitalism. Finland, forests and world economy, 1620–1920; in Finnish with extensive English summary.] Societas Historica Finlandiae, Helsinki, Finland.

Kuuluvainen, T. (2002). Natural variability of forests as a reference for restoring and managing biological diversity in boreal Fennoscandia. *Silva Fennica*, **36**(1), 97–125.

Lindenmayer, D.B. & Hobbs, R.J. (eds). (2007). *Managing and Designing Landscapes for Conservation: Moving from Perspectives to Principles.* Blackwell, Malden, MA, USA.

Mönkkönen, M., Ylisirniö, A.-L., & Hämäläinen, T. (2008). Ecological efficiency of voluntary conservation of boreal-forest biodiversity. *Conservation Biology*, **23**, 339–347.

Shugart, H.H., Leemans, R., & Bonan, G.B. (eds.). (1992). *A Systems Analysis of the Global Boreal Forest*. Cambridge University Press, Cambridge, UK.

Taavitsainen, J.-P. (1987). Wide-range hunting and swidden cultivation as prerequisites of iron age colonization in Finland. *Suomen Antropologi*, **12**, 213–233.

Tasanen, T. (2004). *Läksi puut ylenemähän*. [The history of silvicuture in Finland from the medieval to the break-through of forest industry in the 1870s; in Finnish with extensive English summary.] The Finnish Forest Research Institute, Helsinki, Finland.

Tonteri, T. (1994). Species richness of boreal understorey forest vegetation in relation to site type and successional factors. *Annales Zoologici Fennici*, **31**(1), 53–60.

Wilkinson, D.M. (2006). *Fundamental Processes in Ecology: An Earth Systems Approach*. Oxford University Press, Oxford, UK.

Section 6

Challenges for the Future

REFLECTIONS ON THE RELEVANCE OF HISTORY IN A NONSTATIONARY WORLD

Julio L. Betancourt

US Geological Survey National Research Program Water Resources Division, Tucson, AZ, USA

Historical Environmental Variation in Conservation and Natural Resource Management, First Edition. Edited by John A. Wiens, Gregory D. Hayward, Hugh D. Safford, and Catherine M. Giffen.

23.1   INTRODUCTION

Until now, looking back at historical observations has been a serviceable way to estimate future conditions in natural resource management and water planning. But climatic means and variances are changing in a directional way, altering the frequency, intensity, magnitude, and scale of droughts, floods, fires, and other disturbances. Adaptive strategies will depend critically on the prognosis of how future conditions will deviate from the present and the past.

If you manage a national park known and perhaps named for a particular resource (e.g. Saguaro, Joshua Tree, or Sequoia National Park), how will you know if the resource will be stable or unstable under climate change? Is a salt marsh, the hotspot for biodiversity in the coastal wildlife refuge that you manage, going to be freshened by sea-level rise; how far in advance will you need to advocate the purchase of adjacent properties inland? In fire-prone ecosystems, how do you mimic natural fire regimes to protect lives and properties, while conserving ecological goods and services, under accelerated land use, increasingly ubiquitous human ignitions, and a changing climate? And if you are a water planner, how do you balance water supply and demand and minimize risks to life and property when you cannot foresee the future? Will your system have more water or less water, at the right or the wrong time, and for how many more customers at what consumptive rate? Do increased drought risks justify new water-conservation policies and water-supply reservoirs? Do increased flood risks require new flood-control reservoirs and/or flood-plain regulation?

The answers to these questions depend largely on what the future holds and what the past can and cannot tell us in a no-analog world. Adaptation to climate change will require retooling of traditional methods and developing alternative ones that are better suited for managing resources and ecosystems under a nonstationary climate. Is history irrelevant or indispensable in formulating new approaches to forecast and manage an uncertain future?

I will focus here on two topics, plant migration and ecological synchrony, to illustrate emerging issues in global-change ecology that might benefit from historical insights. My examples are mostly from the western United States, where I have worked for nearly three decades. My premise is that the challenges of a nonstationary and continually changing climate in the twenty-first century demand more deliberate and strategic frameworks for acquiring and incorporating historical data and insights that can inform adaptive management. As examples, I discuss two existing frameworks, historical range of variation (HRV) and hydrologic stationarity (HS), and how they are compromised by global change. I conclude with some specific recommendations for historical ecology in a changing world.

23.2   FORECASTING SPECIES MOVEMENTS IN THE NEAR TERM: LESSONS FROM THE HOLOCENE IN THE WESTERN UNITED STATES

In response to future global and regional warming, many species will adjust their distributions. These shifts will have important consequences for biodiversity, conservation biology, and ecosystem processes and services. Currently, forecasts for distributional shifts rely on bioclimatic models (also called envelope models, ecological niche models, habitat models, or species distribution models) based on statistical descriptions of the climate space that species presently occupy. Numerous techniques, including general linear models, general additive models, climate envelope models, classification and regression trees, and genetic algorithms, are used to make these projections. Mechanistic or process-based species distribution models have been developed, but they require life history and physiological data that are available for only a few species or are limited to specific biomes (e.g. PHENOFIT, a model that emphasizes temperate species and areas of the world; Morin et al. 2007).

The shortcomings of bioclimatic models – mainly that the complexities and spatiotemporal dynamics of populations and ecosystems are not considered – generally are acknowledged, although mostly in passing (Jackson et al. 2009). Rates and pathways of imminent migrations will depend on many things, including life history, new barriers (e.g. habitat fragmentation) and conduits (e.g. roadways, ecological corridors), landscape structure, changing disturbance regimes, and ecological interactions among native and non-native organisms. Ecological and evolutionary legacies, and not just changing climates, will shape the trajectory of migrational responses to a changing climate. The pace of migration will also depend on the overprinting of anthropogenic climate changes on a backdrop of natural climate variability at interannual to multidecadal scales (Brown & Wu 2005; Gray et al. 2006). These complexities confound not only forecasts for natural migrations, but also prospects for so-called assisted migrations engineered to conserve species in the face of climate change (McLachlan et al. 2007).

For example, bioclimatic modeling indicates that much of the present southern range of Joshua tree (*Yucca brevifolia*) in the Mojave Desert will become unsuitable by the end of the twenty-first century (Cole et al. 2011). Climatic projections imply that suitable areas will open north and east of its current distribution, but the fossil record suggests that Joshua tree will be slow to migrate and may need to be transplanted to maintain viable populations into the next century. Potentially suitable areas for these transplants, however, are currently experiencing broadscale invasions by winter annual grasses such as red brome (*Bromus madritensis* var. *rubens*) and cheatgrass (*Bromus tectorum*) and frequent and large wildfires. To be successful and not futile, assisted migration of Joshua tree will have to consider both grass invasions and wildfires in the transplant areas, risks that also are projected to change with the climate (Bradley et al. 2010; Abatzoglou & Kolden 2011).

The record of Holocene vegetation change indicates that, at least for long-lived woody plants such as Joshua tree, migrational responses to climate are complex and anything but instantaneous, playing out over centuries and even millennia at regional scales. My colleagues and I have used fossil woodrat (*Neotoma*) middens (Betancourt et al. 1990) and other proxies to develop detailed migrational histories for several dominant plants in western North America, including *Larrea tridentata* (creosote bush) (Hunter et al. 2001), *Juniperus osteosperma* (Utah juniper) (Lyford et al. 2003), *Pinus edulis* (Colorado pinyon) (Betancourt et al. 1991, Lanner and Van Devender 1998, Gray et al. 2006), single-needle pinyon (*Pinus monophylla*) (Nowak et al. 1994, Lanner and Van Devender 1998), and *Pinus ponderosa* var. *scopulorum* (ponderosa pine) (Norris et al., unpublished). The midden record of Holocene plant migrations is hardly perfect. Although it may do a better job than other paleomethods at detecting small, isolated populations that can persist for millennia along cliffsides, determining first colonizations is still much like searching for a needle in a haystack. Perhaps the safest statement is that the midden record does a decent job of documenting the expansion phase of migrations. With that disclaimer,

the midden record provides a few insights relevant to forecasting future plant migrations.

Bioclimatic models mostly target the morphological species, but migrational responses to climate change surely happen at the level of the genotype. The top two branches of a classification tree for the bioclimatic distribution of *P. ponderosa* are its two races, var. *ponderosa* and var. *scopulorum* (Norris et al. 2006). Except in a narrow zone of introgression near their northern limits, the two races occupy mostly mutually exclusive climate space. As might be expected, they have independent and distinct migrational histories (Norris et al., unpublished). Likewise, in the American deserts, the three polyploid races of *Larrea tridentata* apparently reorganized their distributions individualistically in response to postglacial climates (Hunter et al. 2001). Clearly, bioclimatic models that target the morphological species are too broad a brush for assessing potential distributional shifts to a nonstationary climate.

Past and future invasions may proceed neither as a front-like advance nor a random pattern of long-distance colonization, but rather in ways that are partly predictable from the environmental mosaic. Isolated sites that are more suitable than intervening areas can exist hundreds of kilometers from the peripheries of continuous distributions. *Juniperus osteosperma* jumped about 400 km from southern Wyoming to warm and dry limestone sites at its current northern limits near the Wyoming–Montana border in the middle Holocene, and then backfilled intervening areas in the late Holocene (Lyford et al. 2003). Migrations often proceed episodically, with periods of distributional inertia alternating with rapid advances. Invasion rates may be governed not just by internal population dynamics, but by climate variability at multidecadal to millennial timescales as well. For *J. osteosperma*, there was a hiatus of ~2600 years between colonization of northern disjunct populations near the Montana border and expansion throughout central Wyoming. Presumably, a lengthy period of unfavorable climate in the intervening areas caused this delay. Unfortunately, paleoclimate records for the central Rockies (e.g. Shuman et al. 2009) are not sufficiently detailed or in close enough agreement to develop an independent chronology of climatic phenomena that may have paced *J. osteosperma* colonization and expansion. The fossil record offers useful insights, but it is still fragmentary, incomplete, and perhaps not yet detailed enough to inform predictive models of future migrations.

23.3 ECOLOGICAL SYNCHRONY AND ITS IMPORTANCE FOR ECOLOGICAL FORECASTING

Scientists and land managers often state that the most useful predictions of ecological responses to a changing climate are apt to be near term (10–30 years) and local, where most land-management decisions currently are made. If we could cope with short-term climate variability, it would certainly help prepare us to adapt to climate change. Near-term ecological forecasting has not been a principal goal of ecology or natural resource management, in part because climatologists long have argued that the reliable forecasting window is a few days to a season at most. What if fast advances in climatology, combining our understanding of both anthropogenic forcing and natural climatic variability, could give us reliable predictions a decade in advance? What kinds of information and models would we need to forecast the ecological consequences and inform management decisions?

Given biological complexity and the idiosyncracies of local ecosystems, to what extent does climate variability drive ecological variance, for which processes, and at what temporal and spatial scales? For example, in the western United States, do we have enough information and understanding to hindcast large-scale demographic and disturbance patterns, or even migrational responses, to the known climate variability over the past millennium? We know from dendrochronology that the interannual climate signal is robust and regional in tree growth, mainly because cool-season precipitation and soil moisture at the start of the growing season are synchronous across large areas. This is why we can cross-date a ponderosa pine growing in southern Arizona with a Douglas fir (*Pseudotsuga menziesii*) in northern New Mexico. It is also why forest disturbances associated with fires and insects are synchronous across regional scales in lockstep with El Niño Southern Oscillation and other large-scale climatic modes (Swetnam & Betancourt 1998; Kitzberger et al. 2007; Trouet et al. 2010). We know much less, however, about the degree and extent of regional synchrony in demography because there have been no systematic, tree-ring-based surveys of establishment and mortality in western woodlands and forest, even for the past few centuries.

The timing of colonization and expansion phases of migration may be strongly linked to large-scale disturbances synchronized regionally by extreme droughts and secular tends in temperature. In the wake of these synchronized disturbances, succession may favor an incumbent, a poorly represented, or even an extralocal species. *P. edulis* colonized its northernmost isolated outposts in southeastern Colorado between 940 and 430 years BP (Anderson & Feiler 2009), at Owl Canyon north of Ft. Collins, Colorado, between 1290 and 420 years ago (Betancourt et al. 1991), and at Dutch John Mountain (Flaming Gorge National Recreation Area) on the Wyoming–Utah border between 1000 and 800 years BP (Jackson et al. 2005).

The contemporaneity of these long-distant colonizations at three very distant sites along the periphery of the *P. edulis* range suggests that it was driven by a singular climatic event. At Dutch John Mountain, tree-ring reconstructions show that this local expansion was facilitated by Utah juniper mortality during a severe thirteenth-century drought (the so-called Great Drought associated with Anasazi abandonment of the Four Corners area), with replacement by pinyon in the ensuing wet period (Gray et al. 2006). The Great Drought terminated an unusual string of extreme droughts during the Medieval Climatic Anomaly (MCA) between 700 and 1100 years BP (Cook et al. 2004; Meko et al. 2007) attributed to a cold tropical Pacific and warm North Atlantic (Graham et al. 2007; Feng et al. 2011).

Based on charcoal in the alluvial stratigraphic record, the MCA was characterized by stand-replacing fires in ponderosa pine and mixed-conifer forests across distant sites in the West, including the South Fork of the Payette in central Idaho (Pierce et al. 2004; Pierce & Meyer 2008), the northern Yellowstone area in northwestern Wyoming (Meyer et al. 1995), and the Sacramento Mountains in southern New Mexico (New 2007; Frechette & Meyer 2009). Large-scale tree die-offs may have been widespread during the MCA, particularly near the dry, lower limits of pinyons and other western conifers, but direct evidence is lacking. Pinyon extirpation near its lower limits at Chaco Canyon, New Mexico between 1200 and 500 years ago was attributed to overharvesting of fuelwood by the Anasazi (Betancourt and Van Devender 1981). An alternative and untested view is that this extirpation was caused by tree die-offs suffered during the same mid-twelfth-century drought that caused Anasazi abandonment at Chaco Canyon and/or the late-thirteenth-century drought that caused the abandonment at Mesa Verde and elsewhere in the Four Corners region.

Likewise, large-scale, drought-induced die-offs punctuate pinyon demography in the last century, with at least two episodes of widespread mortality (the 1950s and 2002–2003 droughts; Swetnam & Betancourt 1998; Breshears et al. 2005). Widespread changes in structure and composition of pinyon-juniper woodlands have been studied for the most recent drought (e.g. Mueller et al. 2005), but there is currently no plan to monitor succession in the die-off areas, and perhaps not enough time (only a decade) has passed to judge if less common or extralocal species are invading the impact areas. The notion that synchronous disturbances can lead to invasion from the south or lower elevations, now that it is warmer, is germane to documented episodes of widespread mortality across other vegetation types in the United States and elsewhere (Allen et al. 2010).

Climate change also may impact the spatial and temporal scales of synchrony in ecological processes. Post and Forchhammer (2004) documented increases in the spatial autocorrelation (i.e. synchrony) of both local weather and caribou (*Rangifer tarandus*) populations in Greenland with recent warming. In south-central and southwest Alaska, spruce beetle (*Dendroctonus rufipennis*) eruptions since the 1990s are within the historical geographic range, but they exhibit greater spatiotemporal synchrony – a greater percentage of sites record high-severity infestations – than at other times in the past 250 years (Sherriff et al. 2011). Longer and warmer growing seasons in the western United States probably will increase the synchrony of both fires and insect outbreaks within and across regions, and over the short term may accelerate succession and increase synchrony in tree demography. In 50 years, we may be managing even-aged stands of conifers on an unprecedented scale. Ecological synchrony should be considered an important aspect of historic range of variation that may be heavily impacted by climate change and land use.

Identification of scales over which ecological patterns and processes are synchronous will be necessary to guide and minimize monitoring, maximize the generality of research results, and enhance the predictability of ecological responses to climate. Forecasting ecological responses to climate variability and change will require information about the spatial scaling of their effects on local population dynamics. What kinds of observational and proxy data and networks do we need to identify and understand ecological synchrony in the West? Once understood, what is the predictive

value of this ecological synchrony? How will future climate change and variability affect ecological synchrony across metapopulations, communities, and ecosystems, and how might we manage for it? How will ecosystems shaped by ecological synchrony at a particular timescale of climate variability react to longer term climatic change? What role will synchrony play in propagating ecological responses to climate change through adjacent regions?

Plant migrations and ecological synchrony are but two examples where new frameworks and better historical data are needed to inform predictions of ecological responses to climate change. This will require new paradigms with few precedents. Below I discuss two frameworks, HRV and HS, that rely on historic information to bound the future behavior of ecological and hydrological systems. In the eyes of some, and in the face of climate change, these two frameworks may have outlasted their usefulness. Yet these are precisely the kinds of frameworks (perhaps with different underlying assumptions) that can continue to use historical information to guide adaptive management of natural resources and ecosystems under a changing climate.

23.4   EXAMPLES OF SUCCESSFUL AND PERHAPS OUTDATED HISTORICAL FRAMEWORKS: HRV AND HS

HRV and the more established and broadly applied concept of HS differ in history, approaches, and breadth of application in resource and risk management. Both HRV and HS assume that natural systems fluctuate within an unchanging or well-defined envelope of variability, and it is now widely recognized that this assumption of stationarity is compromised by global change.

In HS, hydrologic variation is regarded as a very noisy, random process in which the statistical mean and moments of sampled distributions do not vary with time. Stationary processes can be characterized by probability distribution functions or their derivatives, probability density functions (pdf's), and these are used to derive design, operational, and regulatory values (e.g. the 100-year flood). HS usually relies on standardized discharge measurements made across a network of fixed gauges along both regulated and unregulated streams. These measured discharges result from the interplay among several hydrologic processes (precipitation, evapotranspiration, infiltration, runoff) over the watershed above the gauge, and as such can represent large areas.

The HS concept evolved out of the Harvard Water Program in the 1950s, at a time when adequate guidance was lacking at the federal level for the design and planning of water systems (Maass et al. 1962). The Harvard Water Program integrated economic theory and engineering practice in a democratic political arena (Reuss 2003). Its central tenet, HS, has been applied routinely and globally in water-system engineering and planning. Despite occasional grumblings about nonstationarity in streamflow time series due to directional changes in watershed conditions (e.g. urbanization) and low-frequency climate variability (decadal to century scale) (Webb & Betancourt 1992), HS is widely accepted by water-system planners and engineers. Under a stable climate, HS has simplified the task of managing water systems in an economically efficient and effective manner. In many cases, HS has been institutionalized, including reliance on conventional flood-frequency analysis for floodplain engineering and management (Interagency Advisory Committee on Water Data 1982), or the reliance on mean annual streamflow to negotiate and sustain the Colorado River Compact (National Research Council 2007). HS depends heavily on stream-gauge networks limited to the last hundred years, and often much less. In the western United States, however, there is an increasing and creative use of streamflow reconstructions from tree rings to identify and consider natural streamflow "ensembles" in water planning (Gangopadhyay et al. 2009). Ancient flood deposits have been used to estimate flood magnitude and frequency, and thus supplement and extend the gauged record, but paleoflood hydrology remains controversial and narrowly applicable in contemporary flood-risk management (Baker 2008).

In HRV, on the other hand, ecological variation is estimated as the range of some condition or process that has occurred in the past – in essence, some bounded variation in behavior over time. The metrics for these conditions and processes are diverse (e.g. burned area, fire interval, stand structure, species diversity, patch size) and require long-term and spatially explicit historical data. More often than not, these data are plagued by poorly understood uncertainties and come from a few sites that may or may not be representative of the overall landscape. HRV metrics are reconstructed and not gauged directly, nor are they standardized or prescribed for universal use. Fire inter-

val, for example, can be variously estimated using different statistical techniques, temporal depth, and sampling densities. In some aspects of HRV, such as parameterization of climate variability in landscape models, stationarity often is assumed (Keane et al. 2009).

In comparison to HS, the actual use of the HRV concept in ecosystem management is a relatively young endeavor, with the first deliberation in the 1990s taking advantage of new spatial-analytical and much improved historical analyses (Landres et al. 1999). In large part, HRV reflected the fusion of disturbance and sustainability theory in forest ecology and management. HRV also has been mostly a provincial affair, applied to some North American forests and aquatic ecosystems, but hardly at all to grasslands and shrublands or to other continents. In the United States, use of HRV in national-forest planning was endorsed by the Healthy Forest Restoration Act and is incorporated in planning directives at the national level (USDA Forest Service 2005).

Recently, an international team of hydrologists argued that, in view of the realities of an uncertain and already changing climate, HS should be declared dead (Milly et al. 2008). Projected changes in runoff during the lifetime of major infrastructure projects that were built to last for decades are now large enough to push hydroclimates beyond the range of historical behavior over much of the globe. This nonstationarity can entail not just directional trends in hydroclimatic means and variances at a site or a region that can be attributed to rising greenhouse-gas emissions (Barnett et al. 2008), but also a breakdown in cross-regional patterns of hydroclimatic variability.

Over the last millennium, a persistent north–south seesaw, pivoted ~40°N, characterized snowpack variability over in the Rocky Mountains, reflecting antiphased regional responses of precipitation to Pacific Ocean variability. In the 1980s, however, this seesaw dissipated as unprecedented springtime warming caused snowpack to decline across the entire length of the Rockies (Pederson et al. 2011). The implications for water planning are clear: snowpack in the Columbia, Missouri, and Colorado River Basins provides 60–80% of the water supply for 70 million people. Much less obvious are the ecological implications of this disruption in subcontinental snowpack patterns.

Ecological responses to climate may themselves be changing and contributing to nonstationarity. Reductions in temperature sensitivity are evident in tree-ring width and density records from northern-latitude forests since the 1950s. This co-called "divergence problem" – the underestimation of actual warming by reconstructed temperatures from tree rings – has consequences for large-scale patterns of forest growth and the global carbon cycle (D'Arrigo et al. 2008). In western forests, earlier springs and snowmelt are advancing and lengthening the fire season (Westerling et al. 2006) and at desert elevations, invasions by non-native winter annual grasses (e.g. *Bromus madritensis* var. *rubens* and *Schismus* spp.) are changing long-standing fire climatologies. Because these annual grasses respond synoptically to wet winters and then burn in the arid foresummer (0.5 million hectares from two fires in 2005), El Niños now are rivaling La Niñas as the worst fire years, inflating the area burned in the southwestern United States. After wet El Niño winters, wide ignition fronts can now spread from the desert valleys into adjacent woodlands and forests in the highlands, where historically fires were more extensive after dry La Niña winters (Swetnam & Betancourt 1998). This change in fire climatology and connectivity from the desert to the mountains could impact the entire landscape mosaic. Global change also contributes to changes in land use, landscape connectivity, and biological invasions that are capable of transforming ecoysystems with or without climate change.

A continuously and rapidly changing climate, along with other accelerated land and resource use to support a growing human population, compromises formal application of HRV and, as some have argued, may greatly diminish the utility of historical analogs and reference conditions as sensible targets for ecosystem management (Millar et al. 2007; Williams & Jackson 2007; Stephenson et al. 2010). Moreover, the basic paradigms of environmental and natural resources law now will have to shift from preservation and restoration (by definition stationary assumptions) to increasing resilience and adaptive capacity under a continuously changing and nonlinear climate. As Craig (2010) has stated, "this will require different kinds of legal amendments, and perhaps even new laws, for different regulatory contexts."

So are HS, HRV, and the historical record still relevant in this brave new world fraught with nonstationarity? How do we adjust or replace HS and HRV? The template for a new water-resources paradigm seems straightforward, at least on the surface. Nonstationary, probabilistic models can describe the temporal evolu-

tion of hydrologic means and variances, with estimates of uncertainty under a changing climate and land surface. This will require a stable institutional platform for climate predictions and climate-information delivery, to include mandatory and extensive training of hydrologists, engineers, and managers in nonstationarity, uncertainty, and climate modeling (Milly et al. 2008). Ironically, stationarity was used in the past to justify discontinuation and relocation of stream gauges, but adjusting for changing hydrology in a nonstationary world will necessitate maintaining the location of existing stream gauges.

For ecologists and natural resource managers, nonstationarity poses a more daunting challenge. Ecological responses to climate and land-use change are inherently complex, interconnected, slow and fast, linear and nonlinear, and most likely full of surprises (Burkett et al. 2005). Although there are exceptions, it could be argued that to date theoretical ecology and ecological models focused little on how climate variability (much less climate change) influence ecosystems at annual to decadal timescales (the scales at which most human adaptation is likely to happen). Moreover, except in a few select cases (e.g. Craine et al. 2007), ecologists fail to invest in the kinds of long-term and continental-scale ecological monitoring necessary to adequately identify temporal and spatial patterns in ecological variation and explore their relationship with climate. Unlike physical systems (climate and water), biological systems have not been monitored uniformly with broadly applicable metrics and agreed-upon standards over a long period of time and across the country, much less the world. Consequently, there is a lack of both fundamental understanding and appropriate baselines. The utility of historical ecology may be compromised in a changing world but, in the absence of direct long-term observational data at regional to continental scales, the paleorecord may be all we have.

23.5 MANAGING ECOSYSTEMS UNDER AN UNPREDICTABLE AND CONTINUOUSLY CHANGING CLIMATE

In the western United States, protracted livestock grazing and fire suppression have allowed fuels to accumulate unnaturally in some vegetation types, while rapid expansion of the wildland–urban interface increasingly puts people and property in harm's way. In the late 1990s, the National Fire Plan (and the related Healthy Forests Initiative and Healthy Forests Restoration Act) was established to undertake "a long-term hazardous fuels reduction program in order to reduce the risks of catastrophic wildland fire to communities and natural resources while restoring forest and rangeland ecosystems to closely match their historical structure, function, diversity, and dynamics." In support of this initiative, Congress established the USDA–DOI Joint Fire Sciences Program to provide credible research on the use of fire surrogates to reduce fuels. In the decade since, the federal government has spent ~$170 million on fire science and ~ $1 billion on fuels treatments implemented over 25 million hectares; the American Recovery and Reinvestment Act (the stimulus package) of 2009 added another $0.5 billion for hazardous fuel reduction. These expenditures are arguably this Nation's single largest investment in federal lands science and management over the last decade.

Despite considerable investment to reduce accumulated fuels, one could argue that large-scale disturbances driven by severe drought and warming already are rendering western woodlands and forests vulnerable to abrupt and irreversible change. During the past 30 years, earlier and warmer growing seasons have increased the frequency, intensity, extent, and synchrony of drought-related disturbances. Warmer springs may also be accelerating postfire succession, promoting even greater synchrony. The result is a variable climate and flashy landscape capable of broadscale, multispecies die-offs followed by unnatural surges in plant recruitment. The challenge no longer may be whether or not to reduce stem densities and thus avoid catastrophic fires, but rather how to forecast, monitor, and engineer the products of succession on a subcontinental scale.

Instead of focusing solely on fuels reduction, which may or may not be sustainable in the long term, land managers may do better to engineer the products of succession. Wise management of western ecosystems might well involve systematic and proactive efforts to throw Nature out of synchrony. We can passively adjust to the consequences of climate change, or we can strive to identify points and scales of intervention where the outcomes of succession can be manipulated under a changing climate. It is cheaper and more effective to intervene during the earliest stages of succession, but this will have to be done with clear purpose (e.g. to maintain biodiversity or conserve ecological goods and services), sound science, and unprecedented

coordination across jurisdictions and regions. Such regional management efforts will require a sustained and well-organized series of post-disturbance treatments to disrupt ecological synchrony (making the managed "natural" systems patchier and thus more resilient to climatic variability and change) and focused "plantings" to ensure that the right mix of genotypes and species are locally available to hedge against predicted changes in climate. The next decade will pose many challenges for both scientists and land managers who, lacking relevant frameworks and information, may not be equipped to anticipate and manage ecosystems undergoing dramatic change.

23.6 REFOCUSING THE ACQUISITION AND APPLICATION OF HISTORICAL EVIDENCE TO SERVE AN UNCERTAIN AND COMPLICATED FUTURE

So how do historical ecologists, in particular, contribute significantly to forecasting and managing ecological change under a nonstationary climate? Traditionally, paleoecologists focused on likely times and places of ecosystem change, concentrating their studies on ecotones, range limits, the last Ice Age (within radiocarbon time, or the last 50 000 years), and the time of deglaciation between 15 000 and 12 000 years ago. The staple and currency of paleoecology has been ecological change that can inform us about the range of ecological responses to past and future climates. Lamentably, we have been less interested in ecological stasis.

The persistence of essentially modern ecosystems in the face of relatively well-documented late-Quaternary climate change is important to know in forecasting ecological responses to future climate. Yet fossil records of ecological persistence or stasis seldom are published in high-impact journals; in fact, they often are deemed uninteresting and go unpublished. Surprisingly, to date, ecological persistence has been mostly a topic for pre-Quaternary paleoecologists and evolutionary biologists (DiMichele et al. 2004). A notable exception is Clark and McLachlan (2003), who used temporal and spatial information from pollen data from forests in eastern North America to show that forest diversity stabilized rapidly over broad geographic scales after the last deglaciation. This runs counter to the conventional wisdom of Holocene forests as randomly assembled and ever-changing species associations (the neutral drift model), and suggests the presence of strong stabilizing forces.

Woodrat middens in the southwestern United States generally show dynamic changes from pinyon-juniper woodlands to desert scrub at what are now desert elevations from the Pleistocene into the Holocene (Betancourt et al. 1990). Surprising stability in pinyon-juniper woodlands marks the period between 50 000 to 13 000 years ago, despite evidence from the marine and ice-core record for millennia-scale variability in Earth's climate system during this period (Holmgren et al. 2011). In the southwestern United States, this includes a period around 15 000 years ago when lakes and springs simultaneously went dry but large-scale vegetation changes did not happen until 2000 years later (Holmgren et al. 2006; Pigati et al. 2009). So how did the glacial-age plant communities persist during times of directional change in both surface-water and groundwater systems? Some biomes like the C_4 desert grassland of the Arizona/New Mexico borderlands with Mexico appear relatively stable in both flora and geography from glacial to interglacial (Holmgren et al. 2007). And at some sites, such as Joshua Tree National Park in southeastern California, except for a few notable species extirpations in the early Holocene, there is surprising stability in midden assemblages over the last 30 000 years (Holmgren et al. 2010).

So what exactly are the stabilizing forces in these forest, woodland, grassland, and desert-scrub ecosystems, and how might they operate on our humanized landscapes in the future? Quaternary ecologists should strive to identify times, places, and conditions under which ecosystems have been stable, even in times when we know the climate was changing. This may necessitate not only a thorough and purposeful search of the literature, but also an aggressive survey of living paleoecologists to unearth unpublished records that may prove useful for determining which plant communities are stable when the climate is not. Attention must be given to improving or developing precise measures of degree and rate of ecological change, persistence, resistance (to invasion), resilience, and variability in the late-Quaternary fossil record. It is equally important that we develop paleoclimatic reconstructions, regional master chronologies if you may, that are mostly independent from these measures of ecological stasis and change.

Questions about ecological stability versus instability in the context of significant climate change in the past and present renew debate about whether communities assemble through environmental filters and competitive interactions (Diamond 1975) or through neutral stochastic processes (Hubbell 2001) and the degree and ways in which ecological change can be predicted. Community assembly rules suggest that local extirpation and incumbency of old species and the assembly of new immigrants can take different directions, depending on phylogenetic structure, plant functional traits, the exact sequence of arrivals, species interactions, and regional source-sink dynamics. What traits (dispersal ability, reproductive potential, flowering season, pollination mode, seed mass, rooting volume, size in adulthood, etc.) led some species and functional types to invade new areas and expand their ranges quickly and others to delay? Following establishment, what traits allowed some species to persist until now while others subsequently disappeared? Did the presence or absence of certain species or functional types predict the presence or absence of others via facilitation or inhibition? Do community assembly patterns affect the invasibility of a locality? And, were past invasions typically more successful in species-poor compared to species-rich areas? The fossil record, provided that it is highly resolved taxonomically, temporally, and spatially, can help address many of these questions.

Finally, it seems to me that, in the face of a nonstationary world, history is alive and well, and historical ecology is more priceless than ever. There are still many sites to be studied, and many of our best historical sites and records should be restudied knowing what we know today. For example, in the western United States, there are several detailed migrational histories where we can specify the time of colonization at a given site with great resolution (e.g. Lyford et al. 2003). This allows us to examine in great detail the genetic consequences, local pathways and rates of expansion, and ecosystem-level responses associated with natural invasions (Gray et al. 2006). It also permits us to characterize the present structure and composition of local communities, adopt a chronosequence approach, and thereby determine the extent to which population, community, and ecosystem properties accumulate as a function of time since colonization or disturbance.

What is needed most of all are networks and clearinghouses for historical information useful for making ecological forecasts and managing ecosystems under continuously changing climate and land use. A decade ago, Alverson et al. (2001) proposed development of a formal Global Paleoclimate Observing System, and a paleoecological observatory network is under development for the northeastern United States (http://www.paleonproject.org/). Historical databases like the International Tree-Ring Data Bank (http://web.utk.edu/~grissino/itrdb.htm) have been invaluable, and new multiproxy databases are under rapid development (http://www.neotomadb.org/). We need to serve up historical information in formats useful for testing, parameterizing, and initializing ecosystem models driven by climatic projections. This could take the form of simple lookup tables for such things as degrees and domains of ecological synchrony, disturbance dynamics, rates of species migration and turnover, and even community assembly rules under specified climate conditions. This historical information also could be integrated directly in decision-support models linked to both near-term (season to decade ahead) and long-term (decades to century) climatic predictions, or less formal frameworks such as the USDA Forest Service's Template for Assessing Climate Change Impacts and Management Options (http://www.forestthreats.org/tools/taccimo). HRV and HS served us well in more stable times, but we now need to replace these paradigms with new ones that not only are still informed by historical information but are also customized for a changing world.

Finally, the point was made earlier in this chapter that hydrological nonstationary will require us now more than ever to maintain existing stream gauges, as we increasingly rely on historical baselines to adjust for the direction and magnitude of future hydrological change (Milly et al. 2008). The same should be said for sites rich in historical ecological data. Many of our long-term ecological research sites are being compromised by human impacts (Nijhuis 2008), and well-studied paleoecological sites are being destroyed by fire and land use (Gray et al. 2006). If history truly matters, and baselines are essential in a changing world, then we need to identify a reference network and step up efforts to protect the most important sites.

REFERENCES

Abatzoglou, J.T. & Kolden, C.A. (2011). Climate change in western U.S. deserts: potential for increased wildfire and invasive annual grasses. *Rangeland Ecology and Management*, **64**, 471–478. (special issue). doi: 10.2111/REM-D-09-00151.1.

Allen, C.D., Macalady, A.I., Chenchouni, H., et al. (2010). A global overview of drought and heat-induced tree mortality reveals emerging climate change risks for forests. *Forest Ecology and Management*, **259**, 660–684.

Alverson, K., Bradley, R.S., Briffa, K.K., et al. (2001). A global paleoclimate observing system. *Science*, **293**, 47–48.

Anderson, R.S. & Feiler, E.J. (2009). Holocene vegetation and climate change on the Colorado Great Plains, USA, and the invasion of Colorado piñon (*Pinus edulis*). *Journal of Biogeography*, **36**, 2279–2289.

Baker, V.R. (2008). Paleoflood hydrology: origin, progress, prospects. *Geomorphology*, **101**, 1–13.

Barnett, T.P., Pierce, D.W., Hidalgo, H., et al. (2008). Human-induced changes in the hydrology of the western United States. *Science*, **316**, 1080–1083.

Betancourt, J.L. & Van Devender, T.R. (1981). Holocene vegetation in Chaco Canyon, New Mexico. *Science*, **214**, 658–660.

Betancourt, J.L., Van Devender, T.R., & Martin, P.S. (1990). *Packrat Middens: The Last 40 000 Years of Biotic Change*. University of Arizona Press, Tucson, AZ, USA.

Betancourt, J.L., Schuster, W.S., Mitton, J.B., & Anderson, R.S. (1991). Fossil and genetic history of a pinyon pine (*Pinus edulis*) isolate. *Ecology*, **72**, 1685–1697.

Bradley, B.A., Blumenthal, D.M., Wilcove, D.S., & Ziska, L.H. (2010). Predicting plant invasions in an era of global change. *Trends in Ecology and Evolution*, **25**, 310–318.

Breshears, D.D., Cobb, N.S., Price, P.M., et al. (2005). Regional vegetation die-off in response to global change-type-drought. *Proceedings of the National Academy of Sciences of the United States of America*, **102**, 15144–15148.

Brown, P.M. & Wu, R. (2005). Climate and disturbance forcing of episodic tree recruitment in a southwestern ponderosa pine landscape. *Ecology*, **86**, 3030–3038.

Burkett, V.R., Wilcox, D.A., Stottlemeyer, R., et al. (2005). Nonlinear dynamics in ecosystem response to climatic change: case studies and policy implications. *Ecological Complexity*, **2**, 357–394.

Clark, J.S. & McLachlan, J.S. (2003). Stability of forest biodiversity. *Nature*, **423**, 635–638.

Cole, K.L., Ironside, K., Eischeid, J., Garfin, G., Duffy, P., & Toney, C. (2011). Past and ongoing shifts in Joshua tree distribution support future modeled range contraction. *Ecological Applications*, **21**, 137–149.

Cook, E.R., Woodhouse, C.A., Eakin, C.M., Meko, D.M., & Stahle, D.W. (2004). Long-term aridity changes in the western United States. *Science*, **306**, 1015–1018.

Craig, R.K. (2010). Stationarity is dead–long live transformation: five principles for climate change adaptation law. *Harvard Environmental Law Review*, **34**, 9–75.

Craine, J.M., Battersby, J., Elmore, A., & Jones, A. (2007). Building EDENs: the rise of environmentally distributed ecological networks. *BioScience*, **57**, 45–5H4.

D'Arrigo, R., Wilson, R., Liepert, B., & Cherubini, P. (2008). On the "divergence problem" in northern forests: a review of the tree-ring evidence and possible causes. *Global and Planetary Change*, **60**, 289–305.

Diamond, J.M. (1975). Assembly of species communities. In *Ecology and Evolution of Communities* (ed. M.L. Cody and J.M. Diamond), pp. 342–444. Harvard University Press, Cambridge, MA, USA.

DiMichele, W.A., Behrensmeyer, A.K., Olszewski, T.D., Labandeira, C.C., Pandolfi, J.M., Wing, S.L., & Bobe, R. (2004). Long-term stasis in ecological assemblages: evidence from the fossil record. *Annual Review of Ecology, Evolution and Systematics*, **35**, 285–322.

Feng, S., Hu, Q., & Oglesby, R.J. (2011). Influence of Atlantic sea surface temperatures on persistent drought in North America. *Climate Dynamics*, **37**, 569–586. doi: 10.1007/s00382-010-0835-x.

Frechette, J.D. & Meyer, G.A. (2009). Holocene fire-related alluvial-fan deposition and climate in ponderosa pine and mixed-conifer forests, Sacramento Mountains, New Mexico. *The Holocene*, **19**, 639–651.

Gangopadhyay, S., Harding, B., Rajagopalan, B., Lukas, J., & Fulp, T. (2009). A non-parametric approach for paleohydrologic reconstruction of annual streamflow ensembles. *Water Resources Research*, **45**, W06417. doi: 10.1029/2008WR007201.

Graham, N.E., Hughes, M.K., Ammann, C.M., et al. (2007). Tropical Pacific – mid-latitude teleconnections in medieval times. *Climatic Change*, **83**, 241–285.

Gray, S.T., Betancourt, J.L., Jackson, S.T., & Eddy, R. (2006). Role of multidecadal climate variability in a range extension of pinyon pine. *Ecology*, **87**, 1124–1130.

Holmgren, C.A., Betancourt, J.L., & Rylander, K.A. (2006). A 36 000-year history of the Peloncillo Mountains, southeastern Arizona, USA. *Palaeogeography, Palaeoclimatology, Palaeoecology*, **240**, 405–422.

Holmgren, C.A., Norris, J., & Betancourt, J.L. (2007). Inferences about winter temperatures and summer rains from the late Quaternary record of C_4 perennial grasses and C_3 desert shrubs in the northern Chihuahuan Desert. *Journal of Quaternary Science*, **22**, 141–161.

Holmgren, C.A., Betancourt, J.L., & Rylander, K.A. (2010). A long-term vegetation history of Joshua Tree National Park. *Journal of Quaternary Sciences*, **25**, 222–236.

Holmgren, C.A., Betancourt, J.L., & Rylander, K.A. (2011). Vegetation history along the eastern, desert escarpment of the Sierra San Pedro Mártir, Baja California, Mexico, *Journal of Quaternary Science*, **75**, 647–657.

Hubbell, S.P. (2001). *A Unified Neutral Theory of Biodiversity and Biogeography*. Princeton University Press, Princeton, NJ, USA.

Hunter, K.L., Betancourt, J.L., Riddle, B.R., Van Devender, T.R., Cole, K.L., & Spaulding, W.G. (2001). Ploidy race distributions since the Last Glacial Maximum in the North American desert shrub, *Larrea tridentata*. *Global Ecology and Biogeography*, **10**, 521–533.

Interagency Advisory Committee on Water Data. (1982). *Guidelines for Determining Flood-Flow Frequency: Bulletin 17B of the Hydrology Subcommittee, Office of Water Data Coordination, U.S. Geological Survey*, Reston, VA, USA.

Jackson, S.T., Betancourt, J.L., Lyford, M.E., Gray, S.T., & Rylander, K.A. (2005). A 40 000-year woodrat-midden record of vegetational and biogeographic dynamics in northeastern Utah. *Journal of Biogeography*, **32**, 1085–1106.

Jackson, S.T., Betancourt, J.L., Booth, R.K., & Gray, S.T. (2009). Ecology and the ratchet of events: climate variability, niche dimensions, and species distributions. *Proceedings of the National Academy of Sciences of the United States of America*, **106**, 19685–19692.

Keane, R.E., Hessburg, P.F., Landres, P.B., & Swanson, F.J. (2009). The use of historical range and variability (HRV) in landscape management. *Forest Ecology and Management*, **258**, 1025–1037.

Kitzberger, T., Brown, P.M., Heyerdahl, E.K., Swetnam, T.W., & Veblen, T.T. (2007). Contingent Pacific-Atlantic ocean influence on multi-century wildfire synchrony over western North America. *Proceedings of the National Academy of Sciences of the United States of America*, **104**, 543–548.

Landres, P.B., Morgan, P., & Swanson, F.J. (1999). Overview and use of natural variability concepts in managing ecological systems. *Ecological Applications*, **9**, 1179–1188.

Lanner, R.M. & Van Devender, T.R. (1998). The recent history of pines in the American Southwest. In *The Ecology and Biogeography of Pinus* (ed. D.M. Richardson), pp. 171–182. Cambridge University Press, Cambridge, MA, USA.

Lyford, M.E., Jackson, S.T., Betancourt, J.L., & Gray, S.T. (2003). Influence of landscape structure and climate variability in a late Holocene natural invasion. *Ecological Monographs*, **73**, 567–583.

Maass, A., Hufschmidt, M.M., Dorfman, R., Thomas, H.A., Jr., Marglin, S.A., & Maskew Fair, G. (1962). *Design of Water-resource Systems: New Techniques for Relating Economic Objectives, Engineering Analysis, and Government Planning*. Harvard University Press, Cambridge, MA, USA.

McLachlan, J.S., Hellmann, J.J., & Schwartz, M.W. (2007). A framework for debate of assisted migration in an era of climate change. *Conservation Biology*, **21**, 297–302.

Meko, D., Woodhouse, C.A., Baisan, C., Knight, T., Lukas, J.J., Hughes, M.K., & Salzer, M.W. (2007). Medieval drought in the upper Colorado River Basin. *Geophysical Research Letters*, **34**, L10705. doi:10.1029/2007GL029988.

Meyer, G.A., Wells, S.G., & Jull, A.J.T. (1995). Fire and alluvial chronology in Yellowstone National Park: climatic and intrinsic controls on Holocene geomorphic processes. *Geological Society of America Bulletin*, **107**, 1211–1230.

Millar, C.I., Stephenson, N.L., & Stephens, S.L. (2007). Climate change and forests of the future: managing in the face of uncertainty. *Ecological Applications*, **17**, 2145–2151.

Milly, P.C.D., Betancourt, J.L., Falkenmark, M., Hirsch, R.H., Kindzewicz, Z., Lettenmaier, D.P., & Stouffer, R.J. (2008). Stationary is dead: whither water management? *Science*, **319**, 573–574.

Morin, X., Augspurger, C., & Chuine, I. (2007). Process-based modeling of species' distributions: what limits temperate tree species' range boundaries? *Ecology*, **88**, 2280–2291.

Mueller, R.C., Scudder, C.M., Porter, M.E., Trotter, R., III, Gehring, C.A., & Whitham, T.G. (2005). Differential tree mortality in response to severe drought: evidence for long-term vegetation shifts. *Journal of Ecology*, **93**, 1085–1093.

National Research Council. (2007). *Colorado River Basin Water Management: Evaluating and Adjusting to Hydroclimatic Variability*. National Academies Press, Washington, DC, USA.

New, J. (2007). Holocene Charcoal-based Alluvial Fire Chronology and Geomorphic Implications in Caballero Canyon, Sacramento Mountains, New Mexico. Master's thesis, University of New Mexico, Albuquerque, NM, USA.

Nijhuis, M. (2008). Where research and tourism collide. *New York Times*, July 22, 2008.

Norris, J.T., Jackson, S.T., & Betancourt, J.L. (2006). Classification tree and minimum-volume ellipsoid analyses of the distribution of ponderosa pine in the western USA. *Journal of Biogeography*, **33**, 42–360.

Nowak, C.L., Nowak, R.S., Tausch, R.J., & Wigand, P.E. (1994). Tree and shrub dynamics in northwestern Great Basin woodland and shrub steppe during the Late-Pleistocene and Holocene. *American Journal of Botany*, **81**, 265–277.

Pederson, G.T., Gray, S.T., Woodhouse, C.A., et al. (2011). The unusual nature of recent snowpack declines in the North American Cordillera. *Science*, **333**, 332–335.

Pierce, J.L. & Meyer, G.A. (2008). Late Holocene records of fire in alluvial fan sediments: fire-climate relationships and implications for management of Rocky Mountain forests. *International Journal of Wildland Fire*, **17**, 84–95.

Pierce, J.L., Meyer, G.A., & Jull, A.J.T. (2004). Fire-induced erosion and millennial-scale climate change in northern ponderosa pine forests. *Nature*, **432**, 87–90.

Pigati, J.S., Bright, J., Shanahan, T.M., & Mahan, S.A. (2009). Late Pleistocene paleohydrology near the boundary of the Sonoran and Chihuahuan Deserts, southeastern Arizona, USA. *Quaternary Science Reviews*, **28**, 286–300.

Post, E. & Forchhammer, M. (2004). Spatial synchrony of local populations has increased in association with the recent Northern Hemisphere temperature trend. *Proceedings of the National Academy of Sciences of the United States of America*, **25**, 9286–9290.

Reuss, M. (2003). Is it time to resurrect the Harvard Water Program? *Journal of Water Resources Planning and Management*, **129**, 357–360.

Sherriff, R., Berg, E., & Miller, A. (2011). Climate variability and spruce beetle (*Dendroctonus rufipennis*) outbreaks in south-central and southwest Alaska. *Ecology*, **92**, 1459–1470.

Shuman, B., Henderson, A., Colman, S.M., et al. (2009). Holocene lake-level trends in the Rocky Mountains, USA. *Quaternary Science Reviews*, **28**, 1861–1879.

Stephenson, N.L., Millar, C.I., & Cole, D. (2010). Shifting environmental foundations: the unprecedented and unpredictable future. In *Beyond Naturalness: Rethinking Park and Wilderness Stewardship in an Era of Rapid Change* (ed. D.N. Cole and L. Yung), pp. 50–66. Island Press, Washington, DC, USA.

Swetnam, T.W. & Betancourt, J.L. (1998). Mesoscale disturbance and ecological response to decadal-scale climate variability in the American Southwest. *Journal of Climate*, **11**, 3128–3147.

Trouet, V., Taylor, A.H., Wahl, E.R., Skinner, C.N., & Stephens, S.L. (2010). Fire-climate interactions in the American West since 1400 CE. *Geophysical Research Letters*, **37**, L04702. doi:10.1029/2009GL041695.

USDA Forest Service. (2005). Forest Service Handbook 1909.12, Chapter 40, Science and sustainability, Section 43.13. Range of variation. National Headquarters, Washington Office, Washington, DC, USA. http://www.fs.fed.us/im/directives/fsh/1909.12/1909.12_40.doc.

Webb, R.H. & Betancourt, J.L. (1992). Climate variability and flood frequency of the Santa Cruz River, Pima County, Arizona. USGS Water-Supply Paper 2379.

Westerling, A.L., Hidalgo, H.G., Cayan, D.R., & Swetnam, T.W. (2006). Warming and earlier spring increases western US forest wildfire activity. *Science*, **313**, 940–943.

Williams, J.W. & Jackson, S.T. (2007). Novel climates, no-analog communities, and ecological surprises: past and future. *Frontiers in Ecology and the Environment*, **5**, 475–482.

Chapter 24

THE GROWING IMPORTANCE OF THE PAST IN MANAGING ECOSYSTEMS OF THE FUTURE

Hugh D. Safford,[1,2] *John A. Wiens,*[3,4] *and Gregory D. Hayward*[5,6]

[1]USDA Forest Service, Pacific Southwest Region, Vallejo, CA, USA
[2]University of California, Davis, CA, USA
[3]PRBO Conservation Science, Petaluma, CA, USA
[4]University of Western Australia, Crawley, WA, Australia
[5]USDA Forest Service, Alaska Region, Anchorage, AK, USA
[6]USDA Forest Service, Rocky Mountain Region, Lakewood, CO, USA

Historical Environmental Variation in Conservation and Natural Resource Management, First Edition. Edited by John A. Wiens, Gregory D. Hayward, Hugh D. Safford, and Catherine M. Giffen.
© 2012 John Wiley & Sons, Ltd. Published 2012 by John Wiley & Sons, Ltd.

24.1 INTRODUCTION

History is the discovery, collection, organization, and presentation of information about past events. History examines and analyzes the sequence of past events and attempts to investigate the patterns of cause and effect that determine events (Stearns et al. 2000). History has application to all facets of the human experience, including our interactions with the earth, its other inhabitants, and the ecological systems the earth supports. Historical ecology forms the basis for our understanding of ecosystem change through time, and the biological and physical components of this change. As the magnitude and rapidity of global change have become more and more apparent, a natural reaction for some has been to question the value of historical understanding in a fundamentally different world. But the world is never "fundamentally different." The sun drives ocean and atmospheric dynamics, climate drives geological and ecological processes, organisms compete, predate, and evolve, and humans struggle to comprehend it all.

Despite the fundamental consistency of solar-driven planetary processes, nature is dynamic, and unless one confines one's attention to a small place for a short time, this dynamism is a reality that must be acknowledged as we attempt to manage natural systems, whether for the persistence of native plants and animals (i.e. conservation) or the sustainability of ecosystem commodities and services (i.e. resource management). Although theory has sometimes led them to believe otherwise, practicing ecologists, conservationists, and natural resource managers have known about ecosystem dynamism for a long time. Over the past few decades, the importance of the natural variability of environments and the biological systems they contain has been more formally recognized, particularly in the concept of the historical range of variation (HRV) of ecological systems. This concept, which is the cornerstone for much of the work reported in this book, acknowledges that systems vary through time. By considering the history of these systems, one can determine a range of conditions that encompass this past variation. This range may then define an appropriate "comfort zone" for management. The presumptions are that (1) this historical variation is inherent to the natural dynamics of the system and its components; (2) consequently, the biota are evolutionarily adapted to environmental variation within this range; and (3) the variations occur around a long-term average condition that is relatively stable, at least over the period of history that is relevant to conservation and management. The concept has been widely applied, especially to the management of forest ecosystems in North America and hydrologic and oceanic systems worldwide (Morgan et al. 1994; Landres et al. 1999).

As with any broad concept, however, the devil is in the details. And in the case of HRV, the details are particularly bothersome. Some of these details are operational. For example, what are the appropriate scales in time (and space) for defining "history" and therefore bracketing the range of variation? Some of these details are almost philosophical. For example, what conditions in an HRV should serve as the targets for management or restoration – maintaining the system within the extremes, aiming for some "acceptable" range of conditions such as a standard deviation from a long-term mean, or shooting to be as close to the mean as possible?

As the reality of the magnitude and accelerating rate of environmental change brought about by the twin forces of climate change and land-use change has become more obvious and widely accepted, these details have been pushed into the background. Increasingly, managers, policymakers, and conservationists are asking whether history has any relevance to a future that is likely to be vastly different from the present or the past. Human impacts on the earth are so pervasive that scientists debate whether we have entered a new Epoch in the Quaternary Period, the "Anthropocene" (Steffen et al. 2007). In this coming New World (actually, it is already here), does history have anything important to tell us (Safford et al. 2008)?

The answer from the chapters in this book is a resounding, if qualified, "yes!" Historical ecology can tell us what environmental conditions ecosystems experienced in the past, how and by what means ecosystems responded to variation in those conditions, and how past events acted as "switch points" to redirect the temporal trajectory of a system. History contains information about the resilience of species and ecosystems, information that is essential as conservationists and managers realize that the status quo cannot be maintained or preserved. Priorities will need to be established for what to do and where and when to do it, and history can provide guidance about what is or what is not likely to produce results that managers and society can live with.

The preceding chapters explore the barriers to incorporating historical ecology into land management, describe approaches for more effectively examining the historical record, and provide examples illustrating a wide range of approaches to using historical ecology in land management. Although these chapters demonstrate that developing HRV assessments is difficult and the application of historical ecology requires rigorous attention to methodological details and underlying assumptions, none of the authors suggests that the tremendous changes occurring as we enter the Anthropocene render HRV or historical ecology impotent. Rather, the different chapters demonstrate the various roles that historical ecology can play in helping to navigate the current and future challenges facing resource managers and conservationists. Understanding spatial and temporal variation in ecosystems is critical to the development of effective risk-management strategies in an uncertain future. Historical ecology and well-crafted HRV assessments are essential components in exposing that variation and understanding its consequences.

In this closing chapter, we highlight several themes that have emerged consistently, albeit in various guises, in the preceding chapters. These include the difficulties of managing ecosystems in a dynamic world; issues with the quality and quantity of historical data; the role of humans as contributors to past and ongoing environmental change; key issues confronting the use of historical ecology in the Anthropocene; and, finally, some suggestions about what managers and conservationists can do to pursue historically informed approaches to dealing with future challenges.

24.2 INTRINSIC DIFFICULTIES IN MANAGING DYNAMISM IN ECOLOGICAL SYSTEMS

Although we have learned much about ecological dynamism from historical ecology, that knowledge has not translated to more effective management of variation. Several challenges frustrate the translation from historical understanding to management execution. These range from the scales of observation and management, which are rarely optimal and are often mismatched to the local resource issue; to obstacles stemming from a lack of social or political will to execute management plans with short-term costs that cloud recognition of long-term benefits; to economic

barriers; to intrinsic operational difficulties related to managing for disturbance, especially in ecosystems characterized by high disturbance frequencies or intensities. Indeed, the more we learn about variation in ecosystem processes and patterns, the more difficult it becomes to deal with such variation. The idea of stationarity, which has been central to hydrologic management (Milly et al. 2008) and calculations of oceanic and atmospheric chemistry (e.g. Codispoti et al. 2001), also provided the foundation for many applications of HRV, at least over the short term. Unfortunately, stationarity ignores long-term patterns, which are often directional or subject to various thresholds. Historical ecology cannot solve the inherent – and enduring – political, operational, and interpretational problems that plague management of ecological variation, but it can help provide a basis for understanding the nature and importance of ecological dynamism in a way that can inform conversations among a wide variety of stakeholders (see Chapter 12, this book). Recent developments integrating concepts of past, present, and future ecological variation with social acceptability (e.g. the "social range of variability") provide some hope that an explicit examination of the ways in which human societies constrain, drive, and benefit from ecological systems can lead to resource-management and conservation decisions that are implementable and sustainable through time (Duncan et al. 2010).

24.3 ISSUES WITH THE QUALITY AND QUANTITY OF HISTORICAL DATA

The assembly of comprehensive and complete descriptions of historical ecosystems is impossible. Whether one is seeking broad coverage of historical records across space or developing a complete time series for a given landscape, limited availability of spatial and temporal records is the norm. This problem becomes more acute as we look farther back in time (the "fading record" problem). In addition, more recent data usually have higher resolution than older data, but the older data may better represent variation relevant to the anticipated future. Recently, several authors (including some in this book) have underlined the importance of exploring "deep history," as longer time series may be necessary to represent climate and ecosystem processes important to managers planning for conditions at the end of this century (Chapter 7, this book; Flessa

& Jackson 2005; Dawson et al. 2011). The refinement of techniques to better mine paleoecological data for insights germane to future conditions is one of the major recent developments in historical ecology (Dietl & Flessa 2011). Data from "deep" ecological time, tens of thousands to even millions of years in the past, are providing insights into climate warming and its likely impacts on, among other things, fire regimes (Marlon et al. 2009), insect herbivory (Currano et al. 2008), and biological invasions (Dietl & Flessa 2011).

Another issue has to do with the selectivity of the historical data record. Only certain types of historical data are available, and these vary with the spatiotemporal scale of observation, ecosystem of interest, and the geographic location of interest. For example, most concrete applications of HRV have occurred in temperate, tree-dominated ecosystems in western North America (hence the unavoidable geographic bias of much of this book) because large, long-lived woody structures in semiarid, temperate climates with a relative minimum of human disturbance are among the most likely places to find intact vestiges of the ecological past. Because rigorous historical ecological data are difficult to sample and analyze, intensive local investigations represent the empirical foundation for most HRV assessments. Consequently, considerable ingenuity is often required to develop general products useful for managers. Careful syntheses of very different data sources and types, integrating broad ecological concepts with local empirical data, become necessary (Bigler et al. 2005; Dawson et al. 2011). This book presents numerous applications of variegated and integrated historical data sources to management planning in a wide variety of ecological settings, from East African savanna to California riparian forests, western Canadian rainforests to Fennoscandian taiga, and North American rangelands to Australian eucalypt woodlands. These examples demonstrate the need for close communication between practitioners and scientists to confront difficult resource-management problems. Through dialogue, scientists developing HRV assessments learn to focus their syntheses on the most relevant historical processes and patterns, and to communicate the knowledge gained through such syntheses using language accessible by the public. Similarly, managers learn to explore science more carefully to ask questions that more effectively inform their decisions.

Fortunately, issues with the quality and quantity of historical ecological data are becoming progressively less challenging, as historical data accumulate and practitioners develop better ways of looking into, interpreting, and applying information from the past. Cooperation among researchers has led to the development of historical ecological data networks that are beginning to shed light on long-term historical trends across broad regions. These include the United States Historical Climate Network (USHCN; Easterling et al. 1999), the International Tree Ring Data Bank (Grissino-Mayer 1997), and the European Phenology Network (EPN; Van vliet et al. 2003). The great value of networks of stations that measure standardized time-series data has become obvious as global change accelerates, stimulating national and international initiatives to develop new networks for use in the future. These include the USHCN and EPN (above), the United States National Ecological Observatory Network (NEON; Keller et al. 2008), and the United Kingdom Environmental Change Network (ECN; Lane 1997). There are many other examples (see, e.g. Hart & Martinez 2006).

Major advances in modeling have also allowed managers to examine temporal and spatial gaps in the coverage of primary historical data. Modeling can simulate historical ecological dynamics and provide estimates of trends and ranges of variation for most types of data. In North America, such models have been widely used in forest management, from single management units to nationwide efforts such as LANDFIRE (Rollins & Frame 2006). A major focus in terrestrial HRV modeling has been the quantification of vegetation patch dynamics and other measures of landscape structure (see Chapters 8 and 9, this book). Modeling using historical data has also been profitably used in many other contexts, including understanding patterns in fire-return intervals (Grissino-Mayer 1999), using "long memory" models to predict drought hazard (Pelletier & Turcotte 1997), and deciphering human roles in past, current, and future global nitrogen budgets (Galloway et al. 2004).

24.4 THE ROLE OF HUMANS IN ECOSYSTEM DYNAMICS

Over the past millennia, humans have altered ecosystem dynamics and the face of the landscape over much of the Earth. Historical ecology has helped us recognize the profound influences of humans on aquatic and terrestrial ecosystems, and the ways in which these influences have waxed (and sometimes waned) as human societies have evolved. This book is full of examples, from all over the world, of how intensive historical study was necessary to uncover the roots of modern ecosystem status and trends. In many cases, human actions (or sequences thereof) were identified that played fundamental roles in, for example, creating modern landscapes, molding the modern biota, or driving nutrient or hydrological budgets (see Chapters 4, 6, and 7, this book).

In the twenty-first century, management strategies that focus exclusively on "natural" ecosystem dynamics (i.e. excluding the actions and interactions of humans) seem increasingly outdated and misguided, especially in light of rapidly growing human populations, land-use changes, resource extraction, and climate impacts. Of the habitable continents, only in the Americas does human history extend back only to the beginning of the Holocene. Not coincidentally, the concept of "wilderness" developed in the Americas, as did the preservationist value system that led to the world's first National Park systems. We have gained much from the protection and study of these landscapes, but the rapidity and pervasiveness of anthropogenic global change require that we admit that in some (or many) of these places, human intervention may be necessary to promote the persistence of specific biological entities and important ecosystem processes (e.g. Hobbs et al. 2006; Cole & Yung 2010). Such interventions must be carefully considered, and they may range from delineation of new reserves or corridors between reserves (which is not an option in many highly settled parts of the world), to managed relocation of certain species (controversial and costly), to direct human intervention in ecological processes and patterns like fire, succession, hydrology, and habitat condition. In North America, it may be hard for us to accept the concept of "beneficial" human intervention in ecosystems, but in much of the rest of the world, such intervention is ubiquitous and unremarkable.

24.5 THE RELEVANCE OF HISTORICAL ECOLOGY IN THE ANTHROPOCENE

This book focuses squarely on the issue of the relevance of history to an unknown but rapidly changing future. If history teaches us anything, it is that the basic mechanisms of human and ecological interactions and response remain the same, even if the specific circum-

stances of the moment change. Because global change processes proceed at scales of time that are much longer than human lifetimes (at least until recently), our only chance at properly comprehending the workings of the world is to compare observations over time and to use those observations to build mental or mathematical models of those workings. Soldiers prepare for war by examining multiple scenarios that are based on the actions of adversaries in the past; students prepare for their final exams by poring over previous tests (if they can get them!); in sports, reviewing videos of opponents' past games or matches is an indispensable part of a team's preparation. Resource management in a rapidly changing world is no different. The missing ingredient – an ecology of temporal scaling developed from historical ecology – represents the game tape, the previous test, the historical battle needed by managers to develop a game plan for the future.

In our view, the principal issue is not the fundamental value of historical ecology, but rather the ways in which historical information is used. Historical ecology and its methods, including HRV assessments, constitute a body of knowledge that holds no explicit prescription for how that knowledge should be applied. The wise coach, insightful student, and careful general all know that history does not provide the answers, but it does stimulate the insight necessary to craft a responsible plan. That said, many past management applications of historical ecology made assumptions about ecosystem stability or ecological stationarity that are clearly untenable in a world in which fundamental parameters such as climate, land use, species pools, and nutrient inputs are changing at a rapid pace. There is no prescription for the application of historical ecology, but under rapid global change, there are approaches that are likely to succeed, and others that seem doomed to fail.

What are the key uses of historical ecology in a rapidly changing world? This question is either directly or indirectly posed in most of the chapters in this book. We see three principal areas of use, and we outline them briefly here (see Chapters 1 through 7 for more detail). The first fundamental use of historical ecology in a rapidly changing world is to allow us to understand patterns and mechanisms of temporal dynamics in ecosystems and their component organisms. Since the biological and ecological mechanisms by which organisms and ecosystems respond to environmental variation are basically the same over time, learning about past changes is essential to understanding future response. Historical (including paleo-) ecology is essential to study these past responses, as it can pinpoint the absolute or relative vulnerabilities of species and ecosystems to global change, and it can identify potential ecosystem states and trends and species niche spaces that are not currently extant or occupied and to which we would otherwise be blind.

Another important focus area for historical ecology in a rapidly changing world is the detection of human impacts on past and current ecosystem dynamics and the incorporation of this knowledge into more sustainable conservation and resource management. This is in contrast to the past focus of historical ecology, which was firmly centered on recognizing the characteristics of "natural" ecosystems that were outside the realm of human impact. We do not mean to suggest that study of undegraded historical ecosystems is not worth pursuing; indeed, information from this domain of historical study provides the baseline by which we now recognize the profound and growing influence of humans on ecosystems worldwide. However, current trends of human population growth, climate influence, resource acquisition, and so on suggest that restoration of many (if not most) ecosystems to some past "pristine" state is unlikely or impossible. The current and future status of most of the world's ecosystems depends on the outcomes of human interactions with those ecosystems. Historical ecology allows us to recognize when and how human-mediated degradation occurred in a target ecosystem, it can plot trends in ecosystem condition over time, it can identify how ecosystems and species have responded to human interactions or inputs, and it can help us discern what are realistic and attainable reference conditions for short-, medium-, and long-term sustainability.

A third focus area for historical ecology in a rapidly changing world is helping us to properly balance our desire to preserve certain species, species assemblages, or historic landscapes with the imperative to maintain the delivery of critical ecosystem functions and services. The point has been made that strictly preservationist policies, which dominate conservation and resource management in some regions today, are among those most likely to fail under rapid global change (Hunter et al. 1988; Hobbs et al. 2006; Craig 2010; Stephenson et al. 2010). A school of thought, based in historical ecology, is developing that is more focused on ecological function, resilience, or integrity of an ecosystem, rather than the identities of its component biota. Historical ecology can help us to identify

ecosystems that have persisted through previous environmental changes and may have the capacity – whether with or without human intervention – to preserve landscapes or ecological communities of relatively high historical fidelity. Historical ecology can also identify highly mutable ecosystems, where maintenance of certain ecosystem functions may be impossible in the future. Where species of interest or their habitat are unlikely to endure (information that will most probably come either directly or indirectly through historical ecology), historical ecology is necessary to define the parameters by which we will recognize "properly functioning" ecosystems, determine expected levels of ecosystem services, or know when we have made an ecosystem more resilient to climate change.

24.6 WHAT SHOULD RESOURCE MANAGERS DO?

The observation has been made in this book (and elsewhere) that the challenges of climate change are not really novel. Rapid, directional climate change has happened in the past and will certainly happen in the future. Species responded to climate change in the past, and they will respond in the future (although not perhaps in ways that we will like). Furthermore, the mechanisms of their response will be the same as they have been: migrate, modify behavior, adapt, or die. In contrast to previous punctuated global changes, however, the process this time is being driven by the consequences of human behavior, and the rapidity of change is on a scale only rarely seen in the earth's history (and then never when humans were around to watch, or perhaps even manage, the show). Over the next decades, standard approaches to conservation and land management (e.g. conventional reserve design, strict preservationism) are unlikely to meet the goals of biological conservation or retaining ecosystem services.

Faced with the intractable nature of the problem, many managers seem instinctively to recognize that learning through trial and error will be required. Since most human learning is experiential and driven by the unexpected rather than the expected outcomes (Kolb 1984; Zaghloul et al. 2009), experimental forays into novel realms of management will be called for. This will require enhanced efforts at outreach on the part of resource-management agencies and strong, multipartner collaborative efforts to share costs, knowledge, skills, and even blame if management actions go awry (which some unavoidably will). Global change may be a catastrophe in many ways, but it also presents a massive, global opportunity to learn.

Summarized from the chapters of our book, we offer 10 history-based recommendations to resource managers, restorationists, and conservationists in a world in which the environmental baseline is undergoing rapid change:

1. Do not ignore history. Indeed, learn everything you can about it (focusing on a tractable list of key variables and understanding both ecological process and mechanisms). To understand where an ecosystem is going, you must understand where it has been, in what direction it is currently moving, and what factors led to its past and current conditions. All notions of ecological integrity, resilience, and sustainability are unavoidably based on historical ecology. Climate change vulnerability analyses (Füssel & Klein 2006), which have recently come into management vogue, center on exercises that trace historical responses to past environmental variation (or model these responses, also based on historical data) and then translate these responses (and, ideally, the mechanisms that drive them) into future hypotheses of change.

2. Do not uncritically set historical conditions as your ultimate management objective, and avoid aiming for a single, static target. Develop your notions of desired future conditions based on ranges of ecological processes and structures and incorporate likely future trends in your projected range of variation. In some cases, historical conditions may be a useful short- or medium-term "waypoint" for management, but they will rarely suffice to prepare an ecosystem for an altered future.

3. Simple temporal extrapolations of historical trends are (highly) unlikely to provide a reliable forecast of the future on their own. Mine history for more than simple snapshots of ecosystem status at different times in the past. The investment in historical ecology provides the biggest payoff when it provides us insight into the way things work.

4. Use historical ecology to identify past anthropogenic disturbances that have affected ecosystem condition or trends. Even where management for direct climate-change adaptation is not possible or practical, use appropriate management practices to enhance ecosystem resilience and sustainability by removing or ameliorating other, nonclimate stressors.

5. It will likely be necessary to go beyond the last few centuries in a search for mechanistic relationships that aid in developing possible scenarios for a given planning area. Deep ecological history is proving more and more valuable to our understanding of the likely impacts of climate change on ecosystems and species. Scientists and practitioners should focus attention on historical ecology at a range of scales to develop an empirical foundation for predicting change in the immediate and long-term future.

6. Wherever possible, use information from contemporary reference systems in conjunction with historical systems. Ensure that you understand the history and current status of the contemporary reference system as well as your own.

7. Numerous comparisons of current ecosystem conditions with historical data sets have shown that environmental conditions common on certain parts of the landscape in the past have shifted to other parts of the landscape today; they will continue to shift in topographic and geographic location in the future as climates continue to change. Use information on these geoclimatic shifts to translocate "tried-and-true" restoration techniques from less appropriate to more appropriate (based on current and hypothesized future climate) locations on the management landscape.

8. Historical ecology, combined with ecological theory and mechanistic models, can light the way forward for collaborative efforts between resource managers and stakeholders. Managing for future change is by definition uncertain, and this uncertainty is exacerbated by economic, political, or logistical issues. Given this, collaborative project development, including balancing social acceptability with ecological variability, will be essential to success.

9. Use collaborative processes to reach agreement with management partners and stakeholders on more creative, even experimental (within reason!) management strategies. Use different, credible hypotheses of future change to develop different management procedures that can be carried out on different parts of the landscape. With your collaborators, monitor and evaluate ecosystem responses to these management experiments and use the results to modify your thinking and your actions. Share both credit for successes and blame for failures with your partners, and move on.

10. Plan for the future, not for the past, but do not forget that the past provides our only empirical glimpse into the likely course of the future.

The earth has undergone many pronounced climate cycles in the past, and historical ecology shows that only a few of these have truly been catastrophic events in the biological sense (Dawson et al. 2011). This suggests that biological organisms and ecological systems have substantial resilience. The current and projected pace and scale of environmental change are unprecedented, however, and they involve far more than only climate change. It seems unavoidable that the earth will look very different in 100 years than it does today. In the end, the causes and solutions of the current crisis rest in our hands. When confronted with novel circumstances, unknown lands, or unsailed seas, primitive people often turned to tradition, sages, or holy texts for guidance. These were the repositories of the culture's accumulated knowledge, and provided a foundation (however insecure) for action in the face of uncertainty. Today we may be more likely to rely on the more secure foundation of science, but our reliance on history is undiminished.

REFERENCES

Bigler, C., Kulakowski, D., & Veblen, T.T. (2005). Multiple disturbance interactions and drought influence fire severity in Rocky Mountain subalpine forests. *Ecology*, **86**, 3018–3029.

Codispoti, L.A., Brandes, J.A., Christensen, J.P., Devol, A.H., Naqvi, S.W.Q., Paerl, H.W., & Yoshinari, T. (2001). The oceanic nitrogen budget and nitrous oxide budgets: moving targets as we enter the anthropocene? *Scientia Marina*, **65**(Suppl. 2), 85–105.

Cole, D.N. & Yung, L. (eds). (2010). *Beyond Naturalness: Rethinking Park and Wilderness Stewardship in an Era of Rapid Change.* Island Press, Washington, DC, USA.

Craig, R.K. (2010). Stationarity is dead – long live transformation: five principles for climate change adaptation law. *Harvard Environmental Law Review*, **34**, 9–75.

Currano, E.D., Wilf, P., Wing, S.L., Labandeira, C.C., Lovelock, E.C., & Royer, D.L. (2008). Sharply increased insect herbivory during the Paleocene-Eocene Thermal Maximum. *Proceedings of the National Academy of Sciences of the United States of America*, **105**, 1781–1782.

Dawson, T.P., Jackson, S.T., House, J.I., Prentice, I.C., & Mace, G.M. (2011). Beyond predictions: biodiversity conservation in a changing climate. *Science*, **332**, 53–58.

Dietl, G.P. & Flessa, K.W. (2011). Conservation paleobiology: putting the dead to work. *Trends in Ecology and Evolution*, **26**, 30–37.

Duncan, S.L., McComb, B.C., & Johnson, K.N. (2010). Integrating ecological and social ranges of variability in conservation of biodiversity: past, present, and future. *Ecology and Society*, **15**(1), 5. http://www.ecologyandsociety.org/vol15/iss1/art5/.

Easterling, D.R., Karl, T.R., Lawrimore, J.H., & Del Greco, S.A. (1999). *United States Historical Climatology Network Daily Temperature, Precipitation, and Snow Data for 1871–1997.* ORNL/CDIAC-118, NDP-070. Carbon Dioxide Information Analysis Center, Oak Ridge National Laboratory, US Department of Energy, Oak Ridge, TN, USA.

Flessa, K.W. & Jackson, S.T. (2005). Forging a common agenda for ecology and paleoecology. *Bioscience*, **55**, 1030–1031.

Füssel, H.-M. & Klein, R.J.T. (2006). Climate change vulnerability assessments: an evolution of conceptual thinking. *Climatic Change*, **3**, 301–329.

Galloway, J.N., Dentener, F.J., Capone, D.G., et al. (2004). Nitrogen cycles: past, present, and future. *Biogeochemistry*, **70**, 153–226.

Grissino-Mayer, H.D. (1997). The International Tree-Ring Data Bank: an enhanced global database serving the global scientific community. *The Holocene*, **7**, 235–238.

Grissino-Mayer, H.D. (1999). Modeling fire return interval data from the American southwest with the Weibull distribution. *International Journal of Wildland Fire*, **9**, 37–50.

Hart, J.K. & Martinez, K. (2006). Environmental sensor networks: a revolution in the earth system science? *Earth-Science Reviews*, **78**, 177–191.

Hobbs, R.J., Arico, S., Aronson, J., et al. (2006). Novel ecosystems: theoretical and management aspects of the new ecological world order. *Global Ecology and Biogeography*, **15**, 1–7.

Hunter, M.L., Jr., Jacobson, G.L., Jr., & Webb, T., III (1988). Paleoecology and the coarse-filter approach to maintaining biological diversity. *Conservation Biology*, **2**, 375–385.

Keller, M., Schimel, D.S., Hargrove, W.W., & Hoffman, F.M. (2008). A continental strategy for the National Ecological Observatory Network. *Frontiers in Ecology and the Environment*, **6**, 282–284.

Kolb, D.A. (1984). *Experiential Learning: Experience as the Source of Learning and Development.* Prentice Hall, Englewood Cliffs, NJ, USA.

Landres, P.B., Morgan, P., & Swanson, F.J. (1999). Overview of the use of natural variability concepts in managing ecological systems. *Ecological Applications*, **9**, 1179–1188.

Lane, A.M.J. (1997). The UK Environmental Change Network database: an integrated information resource for long-term monitoring and research. *Journal of Environmental Management*, **51**, 87–105.

Marlon, J.R., Bartlein, P.J., Walsh, M.K., et al. (2009). Wildfire responses to abrupt climate change in North America. *Proceedings of the National Academy of Sciences of the United States of America*, **106**, 2519–2524.

Milly, P.C.D., Betancourt, J., Falkenmark, M., Hirsch, R.M., Kundzewicz, Z.W., Lettenmaier, D.P., & Stouffer, R.J. (2008). Stationarity is dead: whither water management? *Science*, **319**, 573–574.

Morgan, P., Aplet, G.H., Haufler, J.B., Humphries, H.C., Moore, M.M., & Wilson, W.D. (1994). Historical range of variability: a useful tool for evaluating ecosystem change. *Journal of Sustainable Forestry*, **2**, 87–111.

Pelletier, J.D. & Turcotte, D.L. (1997). Long-range persistence in climatological and hydrological time series: analysis, modeling and application to drought hazard assessment. *Journal of Hydrology*, **203**, 198–208.

Rollins, M.G. & Frame, C.K. (2006). *The LANDFIRE Prototype Project: Nationally Consistent and Locally Relevant Geospatial Data for Wildland Fire Management*. General Technical Report RMRS-GTR-175, USDA Forest Service, Rocky Mountain Research Station, Fort Collins, CO, USA.

Safford, H.D., Betancourt, J.L., Hayward, G.D., Wiens, J.A., & Regan, C.M. (2008). Land management in the Anthropocene: is history still relevant? *Eos*, **89**, 343.

Stearns, P.N., Seixas, P., & Wineburg, S. (eds.). (2000). *Knowing Teaching and Learning History. National and International Perspectives*. New York University Press, New York, USA.

Steffen, W., Crutzen, P.J., & McNeill, J.R. (2007). The Anthropocene: are humans now overwhelming the great forces of nature? *AMBIO*, **36**(8), 614–621.

Stephenson, N.L., Millar, C.I., & Cole, D.N. (2010). Shifting environmental foundations: the unprecedented and unpredictable future. In *Beyond Naturalness: Rethinking Park and Wilderness Stewardship in an Era of Rapid Change* (ed. D.N. Cole and L. Yung), pp. 50–66. Island Press, Washington, DC, USA.

Van Vliet, A.J.H., de Groot, R.S., Bellens, Y., et al. (2003). The European phenology network. *International Journal of Biometeorology*, **47**, 202–212.

Zaghloul, K.A., Blanco, J.A., Weidemann, C.T., McGill, K., Jaggi, J.L., Baltuch, G.H., & Kahana, M.J. (2009). Human substantia nigra neurons encode unexpected financial rewards. *Science*, **323**, 1496–1499.

INDEX

Page numbers in bold denote tables, those in italic denote figures.

Abies balsamea (balsam-fir), 97, *100*, 177
Abies concolor (white fir), 130, 139, 154, 254–5
Abies lasiocarpa (subalpine fir), 130, 139, 154–5, 254
Abies spp. (firs), 80, *100*
Accipiter gentilis (northern goshawk), 170, 187
Acer spp. (maples), 80–3, 177, *183*, 285
Africa, 56, 64, 102, 266–9, 271, 291, 295
agriculture, 7, 39, 55, 77, 79–81, 83–85, 87, 93–4, 102, 182, 209, 212, 219, 221, 271, 282, 290, 299
Alabama, 22, 83
Alachua Savanna, 84
alternative regimes, 293, *294*
American Indians *see* Native Americans
Anthropocene Epoch, 320–1, 323
archaeology, 257–8, 284
Artemisia spp. (sagebrush), **82**, **134**, 142, 226
Asia, 56, 77, 102, 266
Asimina triloba (pawpaw), 78, *98*
aspen (*Populus tremuloides*), 130–1, **134**, 154, 156, 177–8, 182, *183*, *185*, **187**, **189**, *251*, **252**, 254–7, 259
autocorrelation, 49, 123
  spatial, 120, 311
  temporal, 116, 120

Baltic Sea, 299, 302
banks, 222, 224, 234, 239, 253, 285, *292*
Bartram, William, 83–4
baseline(s), 53, 57, 86, 93–5, 101–3, 105, 213, 234, 277, 286, 294, 313, 316, 324–5
bear *see Ursus*
beaver *see Castor canadensis*
beech *see Fagus*
beetle outbreaks, 38–9, 42, 66, 67–8, 96, 142, **152**, 155–6, 162–3, 254, 257, 311
beetle, 34, 121, 132, 136, 154, 247, 283
Bering Landbridge, 77
Beringia, 77, 99
best available science, 21, 24–6, 169, 191
Betula alleghaniensis (yellow birch), 96
Betula spp. (birch), 99, 177, **187**, **189**, 285
Bialowieza Forest, 282
biodiversity, 95, 100, 129, 150, *168*, 169–70, 173, 187, 195, 202, 228, 233, 237, 239, 257, 274–9, 286, 290, 308, 314
biological legacies, 274–8
biomass, *103*, 179, 186, 192, 226, **250**, 255, 269, 274, 292, 303, 310
birch *see Betula* spp.
birds, 209, 222, 225–6, 228, **250**, 277–8, 285, 303
  post-fire recovery of, 277–8

Bison bison (bison), 83–4, 95, 102, 212, 253, 283
black grama *see Bouteloua eriopoda*
blowdown *see* windthrow
boreal-hardwood transition zone, 177
Bos spp. (cattle, cows, aurochs), 102, 283
boundary mismatches, 238
Bouteloua eriopoda (black grama), 293
British Columbia, 167, *168*, 169–74, 211,
Bromus tectorum (cheatgrass), 69, 249, 309
bulk commodities, 300
burning, 7, **11**, 12, 66, *103*, 122, **151–2**, 157, 162, 202, **252**, 254–8, 266–7, 276–9
  by Native Americans, 77, 80–1, **82**, 83–7, 94
butterflies, 285

California, *20*, 69, 85–6, 94, 207, 211, 219, *220*, 227–8
  Southern, 8, 12, 195, 315
Canada thistle *see Cirsium arvense*
canebrakes, 84
Canis lupus (wolf), 167, 251, 283
carbon, 51, 97, 104, 114, 129, 182, 269, 274, 313
Carpinus spp. (hop hornbeam), 99
carrying capacity, 80, 290, 301
Carya spp. (hickory), 81, 84

Historical Environmental Variation in Conservation and Natural Resource Management, First Edition. Edited by John A. Wiens, Gregory D. Hayward, Hugh D. Safford, and Catherine M. Giffen.
© 2012 John Wiley & Sons, Ltd. Published 2012 by John Wiley & Sons, Ltd.

Cascade Mountains (Cascade Range), 85, 200–1, *237*
Castanea spp. (chestnut), 82
Castor canadensis (North American beaver), 95, 234
cattle *see Bos*
Central Highlands of Victoria, 274
Central Valley (California), 219, *220*, 221, 228–9
Cervus canadensis (elk), 84, 142
Chaco Canyon, 94, 310
chaos theory, 65, 68
cheatgrass *see Bromus tectorum*
Chihuahuan Desert, 290, 293, *294*
chinook salmon (*Oncorhynchus tshawytscha*), 226, **227**
Chippewa National Forest, 177, 182, *184*, *185*, 189, **190**
Choristoneura fumiferana (spruce budworm), 72, 136, 142
Choristoneura spp. (budworms), 154, 162
Cirsium arvense (Canada thistle), 253, 255
clearance, 79, 93–4, 102, 283, 300
clear-cutting, 21–2, 201, 239, 274–7, *276*
climate
   and fire, 34, 55
   refugia, 5, 36, 59, 239
   variability, 34, 37–8, 93, 95–7
   climate change, 8, 10–11, 13, 26, 34, 36, 40–1, 47–59, 54–5, 64, 66, 69, 72, 78, 86, 94–5, 97
   adaptation, 58–9
   alternative regime models, 293–4
Clovis culture, 77, 102,
coarse-filter strategy, 173
Coast Information Team, 167, 169, 171
Coast Ranges (US Pacific coast), *15*, 85–6, 94, 219
Coastal Plain (US Atlantic coast), 84, 94
collaboration, 21, 24–5, 129, 148, 180–1, 188, 191–2, 202–3
Collaborative Forest Landscape Restoration Program, 153
Colorado, 96, 129, *130*, 154–6, *251*, **252**, 254–5, 310, 312

Front Range, 150–4, *151*, *154*, 156, *158*, 162–3
Colorado Plateau, 97, 258–9
Columbia River Basin, 117–8
composition
   landscape, 6, 114–5, 120, 124, 132–3, **134–5**, 138, 140, 142–3, *144*, 145, 179, 202, species, 16, 50, 84, 97, 104, 132, 195, 197, 199, 206, 212, 226–7, 247, 253–5, 275, 278, 290
condition classes, 133, 142, 195, *196*, 197
conifer-hardwood forest, 85,
connectivity, 4, 23, 188, 208, 212–3, 219, 224, 227, 234, 237, *238*, 239, 313
conservation
   optimization, 303
   paradigm, 270–1
   participatory, 186, 302
contingency, 42, 271
Contopus borealis (olive-sided flycatcher), 142
coppice, 282, 284–7
Cosumnes River Preserve, 219, *220*, 221, *222–3*, 226–7
Crataegus spp. (hawthorns), 78
crown fires, 155–6, 199, 248, 250
Cucurbita spp. (squash), 79
cultural landscape, 79, 94
current departure, *52*, 138, 140

dams, 24, 26, 209, 219, 234–5, *237*, 241–2
   removal, *237*
   re-operation, 237, *237*
data sources, 37, 131, 154, 182, 322
deer, 84, 102, 283–4, 286–7
dendrochronology, 16, 56, 310
Dendroctonus ponderosae (mountain pine beetle), 39, 121, 136, 156
Dendroctonus pseudotsugae (Douglas-fir beetle), 136
Dendroctonus rufipennis (spruce beetle), 132, 311
Dendroctonus spp. (pine beetles), 66

departure, 24, 37, *47*, *52*, 58, 64, 115, 118, 120–1, 129, **134–5**, 138, 140–3, 145, 155, 159, 195–9, *196*, 249–50, 255
desired conditions, 4, 23, 27, 33, 36, 41, 47, 54, 64, 85, 150, 153, 179–80, 186, 188–9, 241, 291–2
Diospyros virginiana (persimmon), 78
disease, 37, 131, 139, 150, 177, 182, 188, 192, 208, 253, 257, 285–7
   human, 83, 94–5, 132, 291
   Rinderpest, 266–7
disturbance
   ecological, 33
   herbivory, 78, 102, 199, 322
   natural *see* natural disturbance
disturbance regime, 5, 7, 12, 24, 38, 58, 69, 71, 78, *82*, 93, 95, 97, 103–4, 117, 123, 131–3, 136–8, 143–4, 163, 167, 169–74, 180–2, 195, 200, 206, 221, **250**, **252**, 254, 256–8, 274–5, 277, 279, 291, 308
   altered, 47, 84, 87
   fire, 177, 182, 258, *185*
   hydrogeomorphic, 172
   wind, 112, 122–3, 136, 167, 177, 182–3, *185*
Domesday Book, 283–4
dormouse, 285
Douglas-fir *see Pseudotsuga menziesii*
drought, 39, 41, 55, 59, 65, 72, 85, 93–6, 156, 159, 162, 200, 242, 247, 254–5, 257, 267, 269, 292
dynamic equilibrium, 8, 67, *67*, 132, 140, 144
dynamics
   temporal, 30, 36, 41, 54–5, 70, 73, 148, 323
   spatial, 38–9, 68, 120, 208

ecological
   dynamics, 54, 69, 103, 148, 269, 322
   function, 22, 57, 101, 227–8, 236, 324
   integrity, 5, 8, 11–12, 24, 50, 57, 170–1, 173, 234, 247, **248**,

249–50, **252**, 253, 256–9, 278, 325
  novelty, 40–2, 53–4, 58, 72, 93, 97, 99–100, *200*, 233–4, 237, *239*, 240–2
  potential, 290–1, 293, 300–1
  processes *see* ecosystem processes
  site, 290–1, 293, 295, 316
  resilience, 11, 50, 59
  thresholds, 50, 56, *67*, 68–9
ecological range of variability (ERV), 7, 13, *15*
Ecological Restoration Institute, 12
ecoregions, 77, 182,
ecosocial perspective, 298
ecosystem
  aquatic, 13, 50, 56, 139, 153, 182, 184, 224, 226, 291, 312
  terrestrial, 13, 93, 95, 97, 322
  disturbance, 95, 199
  management, 5, 22–4, 34, 47, 56, 150, 242, *251*, **252**, 312–13
  processes, 26, 51, 54, 57–8, 116, 118, 122, 150, 219, 258, 308, 321, 323
  riverine, 233, 235, 242
  services, 34, 36, 51, 57, 129, 150, 162, 173, 233, **250**, **252**, 253, 256, 298–9, 303, 324
  turnover, 97–101
edge effects, 122–3, *124*, 131
El Niño Southern Oscillation (ENSO), 65, 72, 159, 310, 313
elephant, 102
  culling, 266
  populations, 267, *268*, 269
environmental regime, 103, *234*, 235, 237, *238–40*, 242
  flow, 24, 219, 228, 233, *234*, 235, 236, *237*, 238–9, 240–2
  sediment, 234, 239
  temperature, 234
equilibrium, 30, 33, *35*, 38–9, 52, 58, 64, 71, 121, 140, 221, 228
  dynamics, 8, *67*, 132, 140, 144
Equus spp. (horses), 82–84, 95, 102, 132, 267, 283
Eragrostis lehmanniana (Lehmann's lovegrass), 294
European settlement, 5, 7, *15*, 23, 26, 35, 53, **82**, 84–5, 94–5, 131, 177, 186, 195, 199, 210, *211*, 212

eutrophication, 56
Everglades, 208–9
extinction, 72, 84, 94, 97, 101–2, *240*, 267, 300–1
  faunal, 31, 93
  megafaunal, 77–9
  species, 287, 301
extreme events, 38, 68, 137, 236,

fading record, 9, 321
Fagus spp. (beech), 80–4, 285
farming, 221, 282, 286, 301
faunal extinctions, 31, 93
fire
  adaptation, 78, 81, 258
  anthropogenic, 31, 82, 177, 195, 201
  area, 55, 117
  effects, 37, 117–18, 157
  frequency, 9, *52*, 68, **82**, 84–5, 116, **120**, *124*, 148, **151**, 153, 157, *158*, 159, 195–6, *200*, 201–2, 253, 258, 278
  history, 115, *158*, 159, 182, 195, 201–2, 277–9
  management, 23–4, 93, 150, 250, 276, 278
  recurrent, 276
  severity, 155, **157**, 275, 278
  stand-replacing, 11, 85, 131, 142, 155–6, 183, 199, *200*, 257, 274, 276, 310
  suppression, 12, 34, 37, 55, 70, 83–6, 132, **151–2**, 155–6, *161*, 195, 199, 202, 278–9, *294*, 313
  surface, 12, *122*, **151**, 155, **157**, 198–9, 250, 255–6
  wildland, 55, 150, 197, 314
  wildland fire use, 24, 202
Fire Learning Network, 198
fire regime, 8, 11–13, 23–4, 34, 55, 71, 78, 81, 83, 85, 94, 97, 105, 116, 119, **120**, *122*, 123, *124*, 147, **151**, 153, 155–6, **157**, *158*, 159, 163, 195, *196*, 197–9, 201–2, 209, 248–5, **252**, 253–4, 256, 259, 275–6, 279, 308, 322
  condition class (FRCC), 24, 155, 195, *196*

departure, 196
  gradients, 195
  fire scars, 9–10, 37–8, 56, 104, 119, 159, 201, 274
First Nations, 167, 169, 171
fish, 24, 56, 86, 179, 182, 188, 207–9, 213, 222, 226, **227**, 228–9, 233–5, 237, 239
fisher *see Martes pennanti*
fishing, 22, 79–80, 282, 299
  overfishing, 56
flood, 21, 24, *25*, 42, 49, 53, 95, 131, 148, 173, 177, 182, 209
Flood Pulse Concept, 227
floodplain, 49, 167, 212
  activation, *224*, 225
  restoration, 219–29
Florida, 83–4, 208–9
forest
  conifer-hardwood, 85
  exploitation, 299–301
  forest-based economy, 302
  industries, 167, 300
  mixed-conifer, 195, 198, 310
  montane ash, 274–5, *276*, *277*, 279
  oak-hickory, 81
  restoration, 150, **151**, 197–8, *222*, 228, 250, 312
  spruce-fir, *116*, 130–1, *154*, 155, 182, **187**, **189**
  subalpine, 34, 97, *100*, *116*, *154*, 155–6, 162–3
  wet ash, 274–7
Forest Land and Resource Management Plans (Forest Plans), 177, 179–82, 188–91
Forest planning, *183–4*, 191–2, 312
forest protection, 302
Forest Reserve Act of 1891, 21
Forest Service *see* US Forest Service
forest stand
  scale, 38, 93, 153, 157
  structure, 5, 9, 12, 37, 53, 104, 153, 198–9, 254–5, 259, 275, 312
  age class, 37, 118, 258, 275–6
forestry, 86, 106, **152**, 170, 172, 181, 285–6, 298, 300–2
fossils, 5, 10, 37, 94, 97, *99–101*, 104, 154, 206, 210, 269, *270*, 283, 309, 314–5

Four Corners region (US), 94, 96, *130*, 310
fragmentation, 4, 22–3, 47, 58, 70, 150, 201, 211, 213, 284, 308
FRAGSTATS, 137
framework, 8, 21, 23–4, 26, 50, 56, 148, 156–7, 159, 177, *178*, 180–2, 190, 195, 206, 228, 233, 235, 242, 308, 311, 314, 316
   application of HRV for management, 247–8, *249*, 250, *251*, **252**, 253–9
   collaborative planning, 186
Fraxinus nigra (black ash), 99,
FRCC *see* fire regime (condition class)
Fremont cottonwood *see Populus fremontii*
Frog, yellow-legged *see Rana muscosa*
future range of variability (FRV), 7, 11, 13

geomorphology, 222–5, 227, 291
giant sequoia *see Sequoiadendron giganeum*
Gleditsia triacanthos (honey locust), 78
Government Land Office Public Land Survey, 182
government-to-government, 169, 186
grants, 256, 285
grassland, 5, 11, 21–2, 68, 70, 82–6, 129–30, *154*, 159, *160*, 195, 199, *200*, 219, 249, **250**, 254, 269–70, 283, 285–6, 290, 293, *294*, 312, 315
grazing, 9, 12, 21–2, 33–4, 37, 55, 56, 68, 78, 83–4, 86, 93–5, 101–2, 132, 147, 195, 197, 199, 212, 221, 239, 267, 269, 282, 284–6, *292*, 293, *294*, 313
Great Basin, 69
Great Bear Rainforest, 167
Great Lakes region, 94, 210
Great Plains (North America), 84, 95
Greater Yellowstone Ecosystem, 250–1
greenhouse gases (GHG), 30, 47, 56
growth stage, 182–3, 187, 189, 276

Gymnobelidues leadbeateri (Leadbeater's possum), 274, 277
Gymnocladus dioicus (Kentucky coffee tree), 78

habitat
   fragmentation, 4, 213, 308
   supply threshold, 170–2
   template, 235
Healthy Forests Restoration Act, 150, 196, 314
heathland, 277, 283–4, 287
hedges, 282, 284
herbivory, 78, 102, 199, 322
heterogeneity, 34, 104, 142–3, 145, 202, 225, 242, 278
hierarchy, 104, *178*, *238*
hierarchy theory, 38–9
historical ecology
   Anthropocene, relevance in, 323–4
   case studies, Australia, 274–9
   case studies, Fennoscandia, 298–303
   case study, UK, 282–7
   definition, 48
   development of concepts, 20–7
   global perspectives, 286–7
   in forest plans, 189
   issues of application, 51–60
   land management application, 200–2
   regional application, 150–63
   wildlife management, 206–13
historical ecosystem function, 182
historical fire frequency, 153, 159
historical fire regimes, 12–13, 24, 85, 123, 153, 155, **157**, *158*, 248, 253
   application case studies, 195–203
historical range of variability, 180
   definition, 6–7
   modeling, 129–45
historical range of variation
   application case studies, 250–7
   arguments for abandoning, 40–1
   assessment, 33–42
   case study, 177–92
   challenges, 34–41
   climate change, and, 47–60
   definition, 6

   ecosystem-based management, utilization in, 167–74
   extended HRV (eHRV), 102–5
   framework for management applications, 247–50
   Front Range assessment, 153–63
   management, fire-prone ecosystems, 195–203
   Native American influence on baselines, 93–102
   premises of, 33–4
   rangeland management, 290–1
   scale, issues of, 64–73
   simulation modeling issues, 114–26
   target, 52, 171–2
   wildlife conservation, informing, 206–13
historical records, 11, 56, 72, 123, 180, 219, 235, 271, 313
historical reference conditions, 30, 35, 49, 51, 53–4, 59, 137, 197, 227–8
Holocene Epoch, 5, 31, 77–9, 85–7, 97, 99, 102, 199, 206, 210, 253–4, 309, 314–5, 323
horse *see Equus* spp.
HRV *see* historical range of variation
human history, 30–1, 53, 56, 77, 323
hunting/foraging economy, 94
hydrograph, 219, *220*, 236–7, 241–2
hydrologic stationarity (HS), 308, 311–13, 316
hydroriparian guidelines, 170
hyporheic zone, 239

ice ages, 300, 314
   Little Ice Age, 8, *47*, 59, 71, 132, 293, 298
immigration, 70, 97
Indiana, 102
Indians *see* Native Americans
indigenous peoples, 9, 39, 56, 82–5, 87, 102, 247
   see also Native Americans, First Nations
industrialization, 299–300
invasive species, 8, 10–11, 34, 36, 41, 50, 56, 58, 100, 189–90, 233, 239, 241, 247, 258, 294
Ips spp. (engraver beetles), 66

Iroquois, 81
ivory trade, 267–9

Jervis Bay Territory, 274, 277
Juglans spp. (walnut), 81, 84
Juniperus osteosperma (Utah juniper), 98–9, 130, 257, 309–10

Kentucky, 83, 210

Laetoli, 269, *270*
land use change, 4, 58, 64, 72, 87, 233–4, 287, 313, 320, 323
LANDFIRE, 123, 155, 196–7, 322
landscape
    composition, 6, 114–15, 120, 124, 132–3, **134–5**, 138, 140, 142–3, *144*, 145, 179, 202
    configuration, *103*, 132, **134–5**, 138, 140, 142, *144*, 145, 256
    departure, 142, 145
    disturbance-succession model, 133
    dynamics, 34, 114, 118, 123, 131, 133, 136–7, 143, 202
    ecology, 22–3, 41, 73, 202, 227
    ecosystems, 182, *183–5*, *187–8*, **189–90**,
    metrics, 132–3, **134–5**, 138, 141–2, 145
    perspective, 93, 301
    simulation, 10, 112, *114*, 115, 133
    structure, 12–13, 34, 37, 114, 129, **131**, 132–3, **134–5**, 137–43, 145, 178, **252**, 255, 259, 308, 322
Larix spp. (larch), 99, 177, 285, 300
Laurentian Mixed Forest Province, 177, *178*
Leadbeater's possum *see Gymnobelideus leadbeateri*
legacy, 96, 142, 190, 206, 221, 234, **250**
    effects, 95, 228
Lehmann lovegrass *see Eragrostis lehmanniana*
Leptographium wagneri (black stain root rot), 136
levees, breaching of, 221, *222–3*, 224, 227–8
Lewis and Clark, 206

lime, 284–6
Lithocarpus spp. (tanbark oak, tan oak), 85
Little Ice Age, 8, *47*, 59, 71, 132, 293, 298
livestock, 21–2, 86, 93, 212, 221, 253, 255, 290, 294, 313
lodgepole pine *see Pinus contorta*
logging *see* timber harvest
longleaf pine *see Pinus palustris*

Maclura pomifera (Osage orange), 78
Magnolia spp. (magnolia), 80, **82**, 84
Maize *see Zea mays*
management
    adaptive, 21, 24, 26, 162, 169, 173–4, 190, 200–1, 228, *240*, 308, 311
    direction, 143, 150, 179, 181, 188, 192,
    disturbance-based, 24
    ecosystem, 5, 22–4, 34, *47*, 51, 150, 242, *251*, **252**, 312–13
    panarchical approach, 86
    post-fire, 276, 278
    practices, 16, 21, 23, 54, 56, 72, 93, 102, 142, 150, 178, 192, 201–2, 235, 274, 279, 325
    reference states, 47
    target, 8, 30, 35–6, 247
    tool, 49
maps, 10, 37, 47, *101*, 111–12, 114–17, *122*, 123, 142, *158*, 196, 202, 206, 221, 282–4, 287, 290
Martes americana (American pine marten), 142
Martes pennanti (fisher), 210, *211*, 212–13
meadows, 83–4, 102, 130–1, *184*, 283, 286–7, *292*
Medieval Warm Period, *47*, 59, 71
megafaunal extinction, 77–9
megaherbivores, 78, 87, 102, 269
Mesa Verde, *251*, 256–8, 310
metapopulations, 70, 208, 212–13, 239, 311
Michigan, 81, 96–7, 177, *178*, *183*, 212, *236*
middens
    packrat, 5, 9–10, 47, 99–*100*
    woodrat, 309, 315

Minnesota Drift and Lake Plains, *178*, 182, *184–5*,
Minnesota Forests Resources Council, 180–1
Mississippi River, 24, *25*, 83–4, 94
Mississippi River Valley, 94
Mississippian Culture, 94
mitigation
    climate, 51, 242
    environmental, 58, 150, *238*, *249*
    fire, 12, 151, **152**, 153, 155, 159, 163
mixed-conifer forest, 195, 198, 310
mixed-crop agriculture, 94
models/modeling
    CRBSUM, 117
    DISPATCH, 118
    ecological, 39, 313
    equilibrium, 121
    FETM, 118
    FIRE-BGC, 117
    FIRESCAPE, 114
    FVS, 117, 119
    LANDFIRE, 123, 155, 196–7, 322
    LANDIS, 117, 139
    LANDSUM, 114, 117–18, 121, *122*, 125
    Markov chain, 136
    parameterization, 40, 112, 114, 119, 121, 133, 136, 139, 312
    PHENOFIT, 308
    RMLands, 129, 131, 133, 136–40, 142
    scenario planning, 41–2
    SIMPPLLE, 118
    simulation, 10, 39–40, 114–26, 129–45
    state and transition, 67, 118, 121, 182, 196–7, 202, 290–4
    VDDT, 118, 196–7, 199
    Zelig, 118
monitoring, 56, 126, 148, 153, 173–4, 191, 201, 219, 225, 228, 234, *239*, 240–2, 311–13
montane ash forests, 274–5, *276*, 277, 279
Muir, John, *20*, 21–3, 25, 298
multiple use, 23, 162, 179, 181
Multiple-Use Sustained-Yield Act, 21
Mycteria americana (wood stork), 208–9, *210*, 213

National Environmental Policy Act
(NEPA), *15*, 21–2, 186–7
National Forest
Arapaho-Roosevelt, *158*
Chequamegon-Nicolet, 23
Chippewa, 177, 182, *184*, *185*,
189, **190**
Conecuh, 22
Mark Twain, 23
Monongahela, 22
Pike and San Isabel, 153
San Juan, 129, *130*, **131**, 132–3,
134, 138, *141*, *144*
Superior, 177, *178*, 182, *183*,
188, 190
Willamette, 200, 202
National Forest Management Act of
1976, 22, 64, 186
National Hierarchical Framework of
Ecological Units, *178*, 182
National Park
Booderee, 274, 277–8
Joshua Tree, 308, 315
Kenya, 266
Mesa Verde, *251*, 256
Yellowstone, 9, 42, 250, *251*,
252, 253–4, 258–9, 310
National Park Service *see* US
National Park Service
Native Americans, 8–9, 31, 39, 56,
64, 177, 179, 181–2, 186,
212, 293, *294*,
and ecosystem development,
77–87
influence on HRV baselines, 93–5
Natura 2000, 302
natural disturbance, 23, 33–5, 37,
71, 131, 133, 136, 139, 143,
167, 169–72, 177–8, 189,
200–1, 206, 213, 274–7, 291
type (NDT), 169–70
natural landscape, 21, 35, 282
natural range of variability (NRV),
22, 24, 291
definition, 6–7
Neolithic, 102, 282–3, 285–6
neotropical migratory birds, 225–6
network, 119, 213, 241–2, 286,
29–302, 311–12, 315–6, 322
habitat, 242
river, 233–7, *238–40*
Nevada, 207

New England, 7, 94, 97, 102, 210,
212
New Mexico, 129, *130*, 293, 310,
315
New York, 51, 81, 97, *100*, 102,
212,
niche, 53, 55, 206, 225, 235, *239*,
204–1, 279, 308, 324
nitrogen deposition, 56
Nixon, Richard, 21
no analogue conditions, 53, 56, 72,
101, 102, 308
nonequilibrium dynamics, 67
non-governmental organizations
(NGOs), 13, 21, 23–5
non-native plants, 221, **252**, 253,
255, 257–8
nonstationarity, 41, 95, 104–5,
312–13
North Carolina, 26, 83, 209–10
northern goshawk *see Accipiter
gentilis*
Northern Superior Uplands, *178*,
182
Northwest Forest Plan (NWFP), *15*,
23, 200–2
northwoods transition zone, 177
novel climates, 99–100
novel ecosystems, 31, 53, 93,
99–100

oak *see Quercus*
oak-hickory forests, 81
oak-pine woodlands, 79, **82**
Odocoileus spp. (mule deer and
white-tailed deer), 84, 187
Ohio River Valley, 94
old growth/old forest, 11–13, *15*,
22, 71, 133, *157*, 167, 169,
171–2, 188–9, 275–7, 282,
298–9
Olduvai Gorge, 269–70
Oncorhynchus spp. (salmon), 8, 167,
234
Oncorhynchus tshawytscha (Chinook
salmon), 226, *227*
Ontario, 81, 94, 212
Oostvaardersplassen Reserve,
285
Oregon, *15*, 71, 85, 94, 195, *196*,
197–8, 200–2
Organic Act of 1897, 21–2

origin of ecosystems, 93, 96–7, 99,
103–4
Ozark-Ouachita Highlands, 87

Pacific Northwest, 12, *15*, 22, 85,
118
paleoclimates, 104, 309, 315
paleoecology, 41, 48, 64, 72, 78, 86,
93–5, 97, 99–*100*, 102, 104–5,
201, 206, 314–16, 322
of the Serengeti Plains, 269–70
Palmer Drought Severity Index, 132,
136
parkland, 105, 169, 283
patch dynamics, 40, 112, 199, 322
pathogens, 95, 99–100, 131–3,
142, 199
peatlands, *100*, 104, 177, 300
Pennsylvania, 81, 83, *99*, 212
pests, 100, 287
phase space representation, 301,
303
Phaseolus spp. (beans), 79,
photography
aerial, 16, 112, 159, 282
historical, 10, 37, 47, 56, 154,
159, *161*, 206
repeat, 37, 85, 159
Picea engelmannii (Engelmann
spruce), *130*, 139, 154–5,
254
Picea glauca (white spruce), 177
Picea mariana (black spruce), 177,
183
Picea pungens (blue spruce), 99, 254
Picea sitchensis (Sitka spruce), 167
Picea spp. (spruce), 94, 99–100, *178*,
183, 300–1
Piedmont (US), 94
Pinchot, Gifford, *20*, 21–3, 25
Pinus banksiana (jack pine), 97, 177
Pinus contorta (lodgepole pine), 9, 38,
97, 154, 195, 250
Pinus edulis (two needle pinyon), 5,
55, *98*, 130, 257, 309
Pinus flexilis (limber pine), *98*, 139,
154
Pinus palustris (longleaf pine), 22
Pinus ponderosa (ponderosa pine), 11,
34, 56, 71, 97, *98*, 130, 147,
153, 195, *200*, 250, 309
Pinus remota (papershell pinyon), 55

Pinus spp. (pines), 79, 81, 94, 99–100, 177, 284, 309
Pinus strobiformis (southwestern white pine), 139, 254
Pinus strobis (eastern white pine), 177
pinyon-juniper woodland, *251*, **252**, 257–9
Plains cottonwood see *Populus deltoides*
plantations, 83, 201–2, 282, 287
Platanus spp. (sycamore), 285
Pleistocene Epoch, 5, 59, 102, 206, 293, 298, 300, 315
    megafaunal extinctions, 77–8
Pogonichthys macrolepidotus (Sacramento splittail), 226, **227**
pollard, 284–5
pollen, 5, 10, 37, 47, 85, 94, 99, *101*, 104, 182, 199, 253, 270, 283–4, 314
    record, 78, 94, 270, 284
pollution, 47, 58, 301
ponderosa pine see *Pinus ponderosa*
population
    persistence, 170, 208
    viability, 209, 309
Populus deltoids (Plains cottonwood), 241
Populus fremontii (Fremont cottonwood), 221, 225–6
Populus spp. (poplar, cottonwood, aspen), 99
Populus tremuloides (aspen), 130–1, **134**, 154, 156, 177–8, 182, *183*, *185*, **187**, **189**, *251*, **252**, 254–7, 259, 300
Portulaca spp. (purslane), 81
postfire salvage logging, 274–5, *276*, 277
Potawatomi (tribe), 81
Potomac River, 82
Prairie Peninsula, 83
pre-Columbian times (Americas), 8, 105, 251, 253, 257
prescribed burning, 11, 33, 66, **151**, 189, 202, **252**, 254, 258, 277–8,
preservation, 5, 20–2, 31, 36, 47, 57, 150, 162, 226, *240*, 303, 313, 323–4
primary succession, 97
Prosopis spp. (mesquite), 78

Pseudotsuga menziesii (Douglas-fir), 12, **82**, 98–9, 139, 114, *116*, 120, *154*, **157**, 159, *161*, 254–5, 310

Quercus gambelii (Gambel oak), 130, 140, 257
Quercus lobata (valley oak), 219
Quercus spp. (oak), 65, 79, 81, 85, 99–100, 282

rabbits, 285–6
Rana muscosa (yellow-legged frog), 170, 207–8, 213
range of natural variation, 7, 23, 47, 58
    see also historical range of variation
rangeland, 21, 314, 322
    health, 290–3
    management, relevancy of HRV for, 290–5
reference conditions, 30, 35–6, 48–54, 137, 162, 172, 183, 191, 197–99, 202, 227–8, 235, 290, 313, 324
    definition, 6–7
    reference condition-based management, 58–9
reference ecosystems
    contemporary, 47, 57
    role of humans in, 8–9
reference period, 8–9, 5, 12, 47, 49, 53, 59, *103*, 132, 136–40, *141*, 143, *144*, 145, 154–5, 162, 182, 186, 195, 247, **248**, 250–1, **252**, 253, 255–7
refugia, 5, 36, 59, 239
regional planning, 150–63
resilience see ecological resilience
resource use, 21, 300
    extractive, 301
response lag, 95
restoration
    ecological, 16, 36, 49, 56, 150–1, **152**, 153, 155–6, 159, 163, 199, 227
    experiments, 219–29
    targets, 59, 101
rewilding, 101–2, 286
rinderpest, 266–7

riparian
    forest, 51, 131, 219, 221, 225–6, 234, 322
    vegetation, 139, 233, 237
    zone, 237
risk
    fire, 12, **151–2**, 156, 179, 197
    management, 311–12, 321
river, 24, 50, 81–4, 94, 131, 167, 212, 219, 224, 228, 233–42, 269, 282, 299–300
Robinia pseudoacacia (black locust), 81
Rocky Mountains, 36, 38–9, 41–2, 84, 97, 129, 132–3, 139, 150, 153, 156, 162–3, 254, 259, 312
Roman period (UK), 283–84
Roosevelt, Theodore, *20*, 21

Sacramento splittail see *Pogonichthys macrolepidotus*
Sacramento Valley, 86
sagebrush see *Artemisia*
Salix spp. (willow), 99, 221, 225
Salle, René-Robert Cavalier de la, 83
salmon see *Oncorhynchus* spp.
salvage logging, post-fire, 274–5, *276*, 277
San Francisco Bay Area, 86, 219
San Joaquin Valley, 86
San Juan Mountains, *251*, 254–5
satellite imagery, 112, 198
savannas, *80*, 81, 84–5, 87, 97, *100*, 129, *154*, 219, *294*
    African, 266, 268–9, *270*
scale
    domains, 65–6, 68–72
    extent, 65, 68, 70
    geographic, 314
    grain, 65, 68, 73
    spatial, 25, 38–9, 41, 64, 65, 66–8, 70–3, 93–4, 129–32, 153, 170, 173, 182, 238–9, 256, 310
    temporal, 23, 30, 37, 39–41, 58, 64–8, 70–1, 73, 87, 93, 95, 99, 102, 129, 131–3, 139, 148, 172–3, 195, 206, 212, 233, 247–8, 298, 311, 322
scale-specificity, 301
Scolytus spp. see bark beetles

seasonal flooding, 219
seasonal wetland, 219, 228
Second World War, 21, 284
sediment
    deposition, 221–2, 224, 228
    transport, 219
Sequoiadendron giganteum (giant
    sequoia), 71, 308
seral stages, 133, **134–5**, 136, 138,
    140, *141*, 142, 162, 167, *168*,
    169–71, 173, 181–2, 195–6
Serengeti, 266–9, *270*, 271
Shenandoah Valley, 83
shifting baseline syndrome, 286
shifting mosaic, *35*, 70, 140, 144,
    291
shrubland, 5, 11, *82*, 87, 142, *154*,
    189, 199, 249, **250**, 257, 290,
    294, 312
Sierra Nevada, 23, 69, 94, 118, 207,
    211, 219, 222, 298
silviculture, 33, 300
simulated range of variability,
    134
slash-and-burn cultivation, 299–302
smallpox, 94, 266
snapshot, 35, 57, 64, 138, 153,
    228, 325
social range of variability, 13, 24,
    248, 321
    definition, 7
Sorenson's Index, 115, 196
Soto, Hernando de, 83
source-sink dynamics, 235, 315
space, 5–6, 9, 23–4, 33, 35, 48, 52,
    65, *66*, *69*, 70, 72, 86–7, 94,
    120, 139, 145, 148, 195, 208,
    213, 239, 247, 302, 320–1,
    324
    climate, 104, 308–9
    phase, 301, 303
spatial heterogeneity, 34, 104,
    145
spruce *see Picea*
Spruce budworm *see Choristoneura*
    spp.
spruce-fir forest, *116*, 130–1, *154*,
    155, 182, **187**, **189**
squirrels, 285
St. Lawrence region, 94
stakeholders, 42, 153, 162, 169,
    186, 192, 198, 202, 233, 238,
    240, 241–2, 247, 321, 325

stationarity, 30–1, 40–1, 49, 52–3,
    59, 67, 235, 308, 311–13,
    321, 323
stochasticity, 118, 121, 123, *125*,
    133, 136–7, 315
subalpine forest, 34, 97, *100*, *116*,
    154, 155–6, 162–3
succession
    class, 196, 198
    models, 114, 117, 121, 133,
    198–200
    primary, 97
    secondary, 97, 285
sustainability, 5, 34, 36, 56, 57, 60,
    162, 179–80, 191–2, 198,
    282, 291–2, 298, 312, 320,
    324–5
sustainable use, 20
swidden *see* slash-and-burn
    cultivation
sycamore *see Platanus*

taiga, 298–303, 322
Taylor Grazing Act (1934), 21
technology transfer, 197
temporal dynamics, 30, 36, 41,
    54–5, 70, 73, 148, 323
temperate rainforest, 59, 85,
    167–74, 277, 322
Tennessee River Valley, 99
The Nature Conservancy (TNC), 23,
    50, 197–8, 219
thresholds, 23, 40, 49, 50, 56, *65*,
    67, 68–9, 72, 104–5, 139, 143,
    170, 201, 206, 213, *224*, 228,
    241, 291, 293, 321
    dynamics, 31, 65, 67
    risk, 171–3
Tilia spp. (basswood), 80, *82*,
    284–5
timber, 21, 170, 172, 179, 195,
    197, **250**, 274, 277, 285,
    301–2
    production, 12, 37, 255, 299–300
timber harvest, 11, 13, 21–2, 37,
    132, *155*, *168*, 178, 182,
    188, 201–2, 211–12, 239,
    256
tree
    establishment, 155, 157, 201,
    255
    mortality, 55, **152**, 155, **252**,
    257

rings, 37–9, 95, *96*, *100*, 104,
    154–6, 159, 197, 310, 313,
    316, 322
recruitment, 55, 96, 102, 104,
    221, 225–6, 228, 237, 253,
    279, 314
trees with hollows, 275–6, 279
trout, 207–8
Tsuga spp. (hemlock), 80, *82*, 99,
    167, 172, 195, 200

Ulmus spp. (elm), 83, 99, 284,
    286
uncertainty, *4*, *14*, 26, 58, 115, 121,
    133, 172–4, 190, 192, 203,
    221, 235, 242, 247, 313,
    325–6
Uncompahgre Partnership, **252**,
    256
Uncompahgre Plateau, 254, 256
United States
    eastern, 16, 21, 79, 94, 210,
    212
    midwest, 70, 94
    northeastern, 70, 316
    northwestern, 99
    southeastern, 94–5
    southern, 31
    southwestern, 55–6, 59, 94, *130*,
    313
    western, 7, 9, 50, *52*, 70, 94, 104,
    150, 156, 162, 195, 202, 212,
    241, **252**, 291, 308, 310–13,
    315
Ursus spp. (bears), 283
Ursus americanus (black bear),
    167
Ursus arctos ssp. *horribilis* (grizzly
    bear), 167, 251
US Fish and Wildlife Service, 64
US Forest Service (USFS), 20, 22–4,
    33, 36, 50, 65, 131, 181, 187,
    290
US Geological Survey (USGS), 24,
    64
US National Park Service, 50
Utah, 96, *99*, 120, *130*, 211, 310

Valley oak *see Quercus lobata*
variability
    spatial, 6–7, 70, 155
    temporal, 6–7, 33, 36, 129, 136,
    143, 155

Variable Retention Harvest System, 277

vegetation
  communities, 41, 171, 173, 177, 189
  composition, 37, 59, 97, 101, 182, 187, **189**
  structure, 93, *100*, 104, 145, 182, 198, 228, 278

Vera hypothesis, 283

Vermivora bachmanii (Bachman's warbler), 84

vertebrates, 57, 170, 206–7, 213, **250**
  cavity-dependent, 275
  post-fire recovery, 277–8

Virginia, 82–3

Washington (state), 198, *236–7*

watershed, 21, 50, 59, 123, **131**, 143, **152**, 162, 169, 171, 201–2, 208, 229, 233–5 237, *238*, 239, *240*, 241–2, 312

Weeks Act, 21

West Virginia, 22, 212

western hemlock (*Tsuga heterophylla*), **82**, 167, 195, 200

western red cedar (*Thuja plicata*), 167, 171

wet ash forests, 274–7

white fir *(Abies concolor)*, 130, 139, 154, 254–5

wildebeest, 266–7

wilderness, 9, 22, 79, 84, 87, 94, 105, 207, 250, 282, 286, 298, 323

wildlife, 4, 21, 38, 47, 77, 83, 85, 116, 139, 142, 162, 177, 179, 182, 186, 188, 192, 199, 206–9, 211, 213, 222, 238, 257, 266–7, 274–5, 285, 308

windthrow, 37–8, 95, 117, 122, 131, 173, 199

Wisconsin, 23, 81, 177, *178*, *183–4*, 212

Wolf *see Canis lupus*

wood, 211, 233, *234*, 235, 247, **250**, 277, 283, 285
  demand, 299
  products, 181, 197, 202, 287, 299–300
  shortage, 301

wood stork *see Mycteria americana*

woodland, 34, 79, **82**, 83–4, 87, 95–7, 99, 130–1, **134**, 142, *154*, 155, *160*, 197, 219, *251*, **252**, 257–9, 266–9, *270*, 277, 282–3, 310–11, 313–15
  ancient, 258, 284–7
  conservation, 286
  cover, 282–3, 286
  regrowth, 277, 284

wood-pasture, 283

Wyoming, 150, 155–6, 211, *236*, 250, 309–10

Wytham Woods (UK), 71, 282–7

Yellowstone National Park, 9, 250, *251*, **252**, 253–4, 258–9, 310
  1988 fires, 42

Yosemite Valley, 20, 94

Younger Dryas event, 79

Zea mays (corn, maize), 79, 81, 94

Zizania spp. (wild rice), 80

PRINTED IN U.S.A.